MENTALS
REEZING

FUNDAMENTALS OF FOOD FREEZING

Norman W. Desrosier
Editor-in-Chief

and

Donald K. Tressler
President

AVI PUBLISHING COMPANY, INC.
Westport, Connecticut

Frontispiece courtesy of USDA

Library of Congress Cataloging in Publication Data

Main entry under title:

Fundamentals of food freezing.

    Includes index.
    1. Food, Frozen.   I.   Desrosier, Norman W.
Tressler, Donald Kiteley, 1894–
TP372.3.F86   641.4'53   77-22059
ISBN 0-87055-233-3

Printed in the United States of America

06/10/85 $26.50

Contributors

WENDELL S. ARBUCKLE, Professor, Department of Dairy Science, University of Maryland, College Park, Maryland

A. BANKS, formerly Deputy Director, Torry Research Station, Ministry of Technology, Aberdeen, Scotland

FRANK P. BOYLE, formerly Head, Processing Technology Investigations, Fruit Laboratory, Western Utilization Research and Development Division, Agricultural Research Service, U.S. Department of Agriculture, Albany, California

L. J. BRATZLER, Professor of Food Science, Animal Husbandry, Departments of Food Science and Animal Husbandry, Michigan State University, East Lansing, Michigan

MICHAEL J. COPLEY, formerly Director, Western Regional Research Laboratory, U.S. Department of Agriculture, Albany, California

JOHN A. DASSOW, Deputy Division Director, Technological Laboratory, Bureau of Commercial Fisheries, Fish and Wildlife Service, U.S. Department of the Interior, Seattle, Washington

DONALD deFREMERY, Research Chemist, formerly of the Poultry Laboratory, Western Utilization Research and Development Division, Agricultural Research Service, U.S. Department of Agriculture, Albany, California

WILLIAM C. DIETRICH, Research Chemist, Vegetable Laboratory, Western Utilization Research and Development Division, Agricultural Research Service, U.S. Department of Agriculture, Albany, California

ROBERT V. ENOCHIAN, Agricultural Economist, Marketing Economics Division, Economic Research Service, U.S. Department of Agriculture, Albany, California

BERNARD FEINBERG, Assistant to the Director, Western Utilization

Research and Development Division, Agricultural Research Service, U.S. Department of Agriculture, Albany, California

LEONARD S. FENN, Director of Technical Services, National Association of Frozen Food Packers, Washington, D.C.

ERNEST A. FIEGER, Emeritus Professor, Louisiana State University, Baton Rouge, Louisiana (deceased)

A. M. GADDIS, Meat Laboratory, Eastern Utilization Research and Development Division, Agricultural Research Service, U.S. Department of Agriculture, Beltsville, Maryland

JAMES M. GORMAN, Technical Director, Seymour Foods Co., Division of Norris Grain Co., Topeka, Kansas

M. F. GUNDERSON, formerly Director of Microbiological Research, Campbell Institute for Food Research, Camden, New Jersey

ROBERT P. HARTZELL, formerly Packaging Coordinator, Packaging Program, Food Protection and Toxicology Center, University of California, Davis, California

ORME J. KAHLENBERG, Research Director, Seymour Foods Co., Division of Norris Grain Co., Topeka, Kansas (deceased)

ALVIN A. KLOSE, Head, Poultry Meat Investigations, Poultry Laboratory, Western Utilization Research and Development Division, Agricultural Research Service, U.S. Department of Agriculture, Albany, California

AMIHUD KRAMER, Professor, Department of Horticulture, University of Maryland, College Park, Maryland

SAMUEL MARTIN, Vice President, E. W. Williams Publications; Editor, Quick Frozen Foods, New York, New York

ARTHUR F. NOVAK, Head, Department of Food Science and Technology, Louisiana State University, Baton Rouge, Louisiana

ROBERT L. OLSON, Assistant Deputy Administrator, Western Utilization Research and Development Division, Agricultural Research Service, U.S. Department of Agriculture, Albany, California

JOHN A. PETERS, Deputy Laboratory Director, U.S. Department of Commerce, Northeast Fisheries Center, Gloucester Laboratory, National Marine Fisheries Service, Gloucester, Massachusetts

ARTHUR C. PETERSON, Director, Inspection Services, Campbell Institute for Food Research, Camden, New Jersey

JAMES D. PONTING, formerly Research Chemist, Fruit Laboratory, Western Utilization Research and Development Division, Agricultural Research Service, U.S. Department of Agriculture, Albany, California

WILLIAM D. POWRIE, Professor and Chairman, Department of Food Science, The University of British Columbia, Vancouver, Canada

T. L. ROTH, Quality Control Supervisor, Safeway Stores, Inc., Stockton, California

ROBERT N. SAYRE, Research Chemist in Food Technology, U.S. Department of Agriculture, Agricultural Research Service, Western Regional Research Laboratory, Berkeley, California

THOMAS J. SCHOCH, formerly Professor of Food and Nutrition, New York State College of Home Economics, Cornell University, Ithaca, New York (deceased)

JOSEPH W. SLAVIN, Assistant Director for Industrial Research, Bureau of Commercial Fisheries, Fish and Wildlife Service, U.S. Department of the Interior, Washington, D.C.

WILLIAM L. SULZBACHER, Chief, Meat Laboratory, Eastern Utilization Research and Development Division, Agricultural Research Service, U.S. Department of Agriculture, Beltsville, Maryland

BERNARD A. TWIGG, Professor of Horticulture, Horticultural Department of Horticulture, University of Maryland, College Park, Maryland

WALLACE B. VAN ARSDEL, formerly Assistant Director Emeritus, Western Regional Research Laboratory, U.S. Department of Agriculture, Albany, California (deceased)

J. J. WATERMAN, Ministry of Agriculture, Fisheries and Food, Torry Research Station, Ministry of Technology, Aberdeen, Scotland

BERNICE K. WATT, formerly Nutritionist, Consumer and Food Economics Research Division, Agricultural Research Service, U.S. Department of Agriculture, Hyattsville, Maryland

BYRON H. WEBB, Chief, Dairy Products Laboratory, Eastern Utilization Research and Development Division, Agricultural Research Service, U.S. Department of Agriculture, Washington, D.C.

FRANK H. WINTER, Extension Specialist, Department of Food Science and Technology, University of California, Davis, California

EVERETT R. WOLFORD, Research Microbiologist, Fruit Laboratory, Western Utilization Research and Development Division, Agricultural Research Service, U.S. Department of Agriculture, Western Washington Research and Extension Center, Puyallup, Washington

WILLIS R. WOOLRICH is Dean Emeritus and Emeritus Professor of Mechanical Engineering, University of Texas, Austin, Texas (deceased)

Preface

The remarkable growth of food technology in industry has been matched by an equal development of related educational programs in food science in colleges and universities in many countries.

A vast and growing body of reference books is now available to professionals in the field. They have at their fingertips the current state of the art and knowledge in the various areas of specialization embraced by the food industry. For example, excellent reference books are available in the general area of food freezing. *The Freezing Preservation of Foods* by Tressler *et al.* is a four volume reference work which covers the subject in detail.

Fundamentals of Food Freezing is a book written as a textbook. It represents the accumulated art and knowledge in the field of food freezing and draws upon the four volumes of *The Freezing Preservation of Foods* and the current literature in reference.

This new textbook is designed as a unit of instruction in food freezing. As such, it is presented in 16 chapters. The total effect we have attempted to develop is a rounded overall presentation for the student.

It is a pleasure to acknowledge the contributions of our many collaborators in preparing this text. These collaborators are identified in the list of contributors; to each, we are most deeply obliged. However, the undersigned are responsible for errors of omission or commission.

We had a difficult decision to make concerning the use of the metric system in refrigeration engineering nomenclature, since the matter has yet to be resolved adequately. Therefore, we retained the use of the British system in reference to refrigeration units in this edition. We expect to complete the conversion in the next edition.

We have included conversion tables for metric to British and British to metric in the Appendix.

NORMAN W. DESROSIER
DONALD K. TRESSLER

July 7, 1977

Contents

The Rise
of Frozen Foods

Robert V. Enochian
and Willis R. Woolrich

T he pleasure and benefits of iced foods and drink are as old as they are new. The storage and use of ice extends back into history to the use of mountain snows, pond and lake ice, chemical mixture cooling to form freezing baths, and the manufacture of ice by evaporative and radiation cooling of water on clear nights.

Some verse lines from the "She King" in Chinese poetry, 1100 B.C., reflect the community interest in the ice harvest in the second millennium.

> In the days of the second month, they hew out the ice with harmonious blows;
> And in the third month, they convey it to the icehouses
> Which they open in those of the fourth, early in the morning,
> Having offered in sacrifice a lamb with scallions.

In both Egypt and India, the evaporative cooling process supported by radiation to clear skies at night furnished ice for the Royal tables as early as 500 B.C. Protagoras of Greece wrote of the early Egyptians: "They expose the evaporative earthen ewers on the highest part of the house and two slaves are kept sprinkling their porous pitchers with water the whole night. By morning the water has become so cold it does not require snow to cool it to ice temperature."

THE BEGINNING OF MAN-MADE REFRIGERATION

Complementing the utilization of mountain snow, and pond and lake ice in warm climates was the production of cold by producing a bath solution of "frigoric" mixtures such as saltpeter mixed with water.

By the 18th century, some 10 or 15 similar mixtures were known to lower the temperature. Some mixtures, such as calcium chloride and snow, which made possible a temperature down to $-27°$ F. $(-32.8°$ C.) were introduced for commercial use. Chemical mixture freezing machines were introduced in Great Britain for the production of low temperatures, but by the time these inventions were available for exploitation mechanical ice making processes made the chemical mixture methods for freezing noncompetitive except for some batch processes like ice cream-making which used a mixture of common salt and ice.

In 1748, William Cullen of the University of Glasgow in Scotland made the earliest demonstration of the man-made production of cold when he evaporated ether in a partial vacuum. In 1805, Oliver Evans of Philadelphia, Pennsylvania, proposed a closed cycle of compression refrigeration, and in 1844, John Gorrie of Apalachicola, Florida, described in the Apalachicola Commercial Advertiser his new machine for the ice and air-cooling needs of his hospital. In 1851, he was granted U.S. Patent No. 8080. This was the first commercial machine in the world built and used for refrigeration and air conditioning. Gorrie received international recognition. His machine was built in New Orleans, Louisiana, and then used in his hospital of tropical fevers at Apalachicola.

These early developments resulted in the invention and operation of refrigeration machines in Texas and Louisiana. Since there was no southern technical press, their accomplishments were reported only in the daily newspapers.

The economic problems of the Australians in transporting surplus meats and other perishable foodstuffs were not unlike those of the stockmen of Texas and all of the other semi-tropical Gulf states: both Australia and Texas and the Gulf Southwest had to solve the problem of transporting surplus foodstuffs to Great Britain and Western Europe where there were insufficient supplies of perishable products, especially fresh meat. To move by water highly perishable meats from the 30th parallel south of the equator (the location of Australia) to London, Paris or Amsterdam, however, involved much more difficult conservation problems than carrying a shipload of beeves or mutton from Galveston to those same consumer centers, near the northern 50th parallel of Europe.

The accomplishments of the Australians in solving their transport problems at so early a date were of greater scientific significance than those of the Americans, since the Australians succeeded with a much more difficult task and accomplished their end only some months later. In their shipping trials, the Australians put first-grade meats on the tables of Europe.

The third area to promote the production of low temperatures, the United Kingdom and Western Europe, followed immediately with the invention and manufacture of commercial refrigerating machines and systems. Great Britain and Germany, with their wealth and manufacturing capacity, soon began to manufacture refrigeration equipment for use on both land and sea.

ICE MANUFACTURE

Ice by Aqua Ammonia Absorption System

The earliest machine method of producing ice in the southern United States was by the Ferdinand Carre aqua-ammonia absorption process as improved by Daniel L. Holden of San Antonio, Frances V. DeCoppet of New Orleans, Harrison D. Stratton of Philadelphia and Thomas Rankin of Dallas.

The aqua-ammonia absorption plant installations reached their zenith in North America by 1900 when more than 400 installations of over 25 tons were in operation for ice making. Fifty years later the existing ice plants still operating with aqua-ammonia refrigeration were mostly over 25 years of age, except those being installed in oil refineries and petrochemical plants. Electric drive, high speed vapor-compression units proved more acceptable to both owner and operator, since they cost less to purchase and maintain and were more reliable.

Ice by Ammonia Compression Machine

Many attempts to use concentrated ammonia in vapor compression units were not successful until 1873. David Boyle, a Scotsman, after spending all of his earnings in developing an ammonia compressor, opened his initial steam-driven ammonia-type ice plant in the shed of a lumber mill at Jefferson, Texas. He sold ice all summer long in northeast Texas at a profit, but the burning of the lumber mill destroyed his entire investment. With financial help in Chicago from R. T. Crane, he established the Boyle Ice Machine Company in that city, and by 1876-1877 established a plant at the Philadelphia Exposition, a second at the King Ranch in Texas and a third at Austin, Texas. Henceforth, ammonia refrigerant machines in the United States were destined to carry on much of the commercial refrigeration load for ice making, cold and freezer storage and brewery operations.

During the same period that progress was being made with refrigeration in Texas and Louisiana in the United States, W. James Harrison, born in Renton, Dunbartonshire, Scotland, and who migrated to Sydney, Australia, became convinced that the greatest need of the Commonwealth was refrigeration of the beef and mutton of the domestic ranges. He built his first compressor in 1851 on the Barwan River at Geelong, Australia, and used ethyl ether as the refrigerant. He actually launched his first machines by building two units in the shops of P. N. Russell and Co., of Sydney in 1859 and two others in the plant of Seibe and Gorman of London in 1861. Thomas Mort and his technically trained French engineer Eugene Domique Nicolle came into the picture to help solve Australia's problem. Mort is given credit for erecting the world's first cold storage plant in Darling Harbor, Australia, in 1861.

In Europe, a German, Carl von Linde, built his first ammonia compressor in 1874 and was granted British Patent No. 1458 in 1876. This was followed by his U.S. patent in 1880.

Linde did more to analyze the thermodynamics of the vapor compression process than any designer and builder of his period. His early machines were more scientifically correct and proportioned than any then available in Great Britain, Australia or North America.

Within the United States, the early brewing industry advanced with much success, especially in regions of new settlements of Europeans from Germany and Switzerland. Many of the imported brewmasters emigrated to the United States from the brewery institutes of Germany to manage the beer making of the new wort plants.

The most successful early manufacturers of ammonia refrigeration machinery for breweries subsequent to David Boyle were Fred W. Wolf, Ernest Vilter, George Frick, Edgar Penny, Charles Ball, John C. De LaVergne and Thomas and William S. Shipley. However, several American companies manufacturing ammonia compressors operated under Carl von Linde patents, and German brewmasters in America would point with Teutonic pride if they could obtain a Linde ammonia compressor to furnish the refrigeration for cooling the malt of the brewery under their control.

U.S. FROZEN FOOD INDUSTRY

In about 1880, ammonia refrigeration machines began to be used for freezing fish in the United States. By the end of the century, fish freezing was an important industry. Annual production in the United States had reached about 3,500 tons originating in the Great Lakes area, the Pacific Coast, and New England. By that time an important export trade in frozen salmon from the United States had developed, and fish also were being frozen in Europe, but in smaller quantities.

The freezing of shellfish originated much later than the freezing of fish but now is also an important industry. Shrimp, oysters, clams, scallops, lobsters, crabs, and frog legs are frozen in various coastal locations in the United States. Large quantities of shellfish, especially shrimp and lobster, are imported from other countries.

Poultry was first frozen commercially at about the same time as fish, also by use of a mixture of salt and ice for refrigeration. In 1870, six carloads of chickens were frozen in Wisconsin and shipped to New York for sale. Freezing of poultry gradually gained in commercial importance, but for young chickens (broilers) it is less important today than for turkeys and other kinds of poultry.

Egg freezing is of somewhat more recent origin. In the late 1890's, H. J. Keith conceived the idea of removing the egg meats from the shell and freezing them. This was done primarily as a method of marketing eggs that were unsuitable for sale because they had cracked or soiled shells. When bakers began using frozen eggs and learned that they performed as satisfactorily as freshly broken eggs, acceptance of frozen eggs was rapid. Today, frozen eggs are made from high-quality shell eggs and save the baker and other users the cost and mess involved in breaking eggs themselves.

The freezing of red meats in areas of cold winter climates was practiced by trappers, early settlers, and later by farmers. Farmers utilized snow and pond ice to preserve their killings for family and community use. Commercial freezing of meat probably began in New Zealand for keeping mutton in good condition during transport to England. In 1891, New Zealand exported over two million frozen mutton carcasses. During this early period, Australia and Brazil began shipping frozen beef to England and France.

The first successful shipment of frozen beef in the United States was made in 1867 by Dr. Peyton Howard. The beef was transported from Indianola, Texas, to New Orleans were it was served in hospitals, hotels, and restaurants. Commercial freezing of meats in this country since that time has grown steadily but represents only a small part of the total production.

The commercial freezing of small fruits and berries began in the eastern part of the United States in about 1905. The berry freezing industry of the Pacific Northwest started in about 1910. At first it grew slowly, but in 1922 began to expand rapidly, and by 1926 the pack had reached 41 million pounds. These commodities, frozen during the peak of the growing season, were used primarily for later processing into jams, jellies, ice cream, pies, and other bakery foods.

The commercial freezing of vegetables is of much more recent origin. The first recorded experimental work on the freezing of vegetables was in 1917. By the late 1920's many private firms were conducting trials. The results of these early attempts were generally poor, for enzymatic action was not checked sufficiently by low temperatures to prevent deterioration and development of off-flavors.

In 1929, Joslyn and Cruess came simultaneously and independently to the conclusion that it was necessary to scald vegetables briefly before freezing them. This treatment, called blanching, inactivates the enzymes that might cause deterioration during frozen storage. In 1930, H. C.

Diehl and C. A. Magoon of the Bureau of Plant Industry, United States Department of Agriculture (USDA) suggested to packers in the Northwest that they blanch peas before freezing them. A year earlier the Birdseye organization had installed at Hillsboro, Ore., a double belt freezer which had been developed by Clarence Birdseye. This system of quick-freezing, in which packaged food is frozen between two metal belts as it moves through a freezing tunnel, was more suitable for freezing fruits and vegetables than the earlier methods that had been successful for freezing fish and poultry.

Simultaneously with these developments, the Birdseye organization introduced the quick-freezing of consumer-size packages of food products. All of these developments would seem to have assured a rapid acceptance of frozen vegetables. It was not until 1937, however, that the freezing of vegetables became of commercial importance.

Of the commodities that are classified as frozen fresh foods, concentrated citrus juices are the most recent in origin. These did not achieve commercial importance until after World War II, during which a satisfactory method of concentration was developed. Since then the concentration and freezing of citrus juices, especially orange juice, have grown very rapidly.

Today, the only important fresh products not being frozen commercially are bananas, pears, tomatoes, lettuce, and other salad greens. Commercial trials are now being made on sliced tomatoes. Food scientists are continually making discoveries and improvements that may make it possible to successfully freeze all of these products in the future. In addition to frozen fresh foods, many frozen cooked and prepared foods have been introduced and have achieved varying degrees of commercial success over the years.

Growth of the Industry

Growth of the frozen fresh food industry was slow in its early years. Economic factors, such as the Great Depression of the 1930's, no doubt contributed to this slow takeoff. The time lag required to build up storage and distribution facilities, and the lack of venture capital for investing in an industry whose future was still uncertain, also had a delaying influence on industry growth.

Consumer prejudices and long established habits had to be overcome, or had to await the fresh outlook of a new generation of Americans before the industry could achieve a rapid rate of growth.

During the past 40 years frozen foods have made tremendous gains due to the pioneering and continuing research of frozen food processors, container manufacturers, the state colleges, and state and federal government agencies. This research has been concerned with improvements in the freezing process itself, and improvements in the product through the breeding of better strains, and better selection, preparation, and handling of raw product. Equal attention is being given to preservation of product quality during marketing. An important part of the research has focused on developing information on the effects of time and temperature on frozen food quality during marketing. An important part of the research has focused on developing information on the effects of time and temperature on frozen food quality during transportation, storage, and merchandising. As this information was translated into government and industry standards for maintaining good quality throughout the entire marketing process, as advertising and promotion attracted consumers, and as investments were made in facilities for handling frozen foods, the industry grew rapidly.

Growth of Individual Product Groups

Frozen food includes practically everything we eat: vegetables, fruits, juices, poultry, meats, seafoods, and a number of prepared specialties. These product categories, however, are not of equal importance in the foods we eat in all forms, nor are they in frozen form. Furthermore, the importance in dollar value does not necessarily correspond to poundage. Of all categories, prepared specialties had the greatest dollar value in 1970 but were third in poundage. Vegetables, including potatoes, but excluding other prepared vegetables, were first in poundage but fifth in dollar value.

The major reasons for this difference in ranking between dollar values and production are the amount of labor, raw materials and packaging costs in the different product categories. Prepared items such as dinners, bakery products, nationality foods, etc., generally include a costly tray or pouch, require a great deal more labor in their preparation, and contain higher priced raw materials than do most of the items in the vegetable category.

Growth trends can be characterized in many different ways. Dollar value of sales may be best for certain purposes and production or consumption figures for others. Because production and consumption figures are more readily available, and have greater significance in terms of requirements for storage, distribution, and display facilities, most trends in the remainder of this chapter are based on production and consumption figures.

Trends in production of each major product category are not uniform. Some categories achieved important commercial status early and then leveled off. Others grew slowly, but steadily. Still others have grown at remarkable rates. The growth trends of ten separate product categories from 1960 to 1975 are shown in Table 1.1. These product categories are discussed separately in later chapters. If frozen food follows previous growth patterns, 1980 sales in supermarkets will be approximately $9.08 billion. This reflects a 120.8% increase from 1970.

FOREIGN TRADE IN FROZEN FOODS

One of the identifying traits of the degree of economic development in a society today is the relative proportion of the fresh food supply that is preserved by freezing. There are two reasons for this. Only the highly perishable foods must be systematically preserved by man. These are generally the highest in price and, therefore, the least in demand in the less developed areas of the world. Another reason is that frozen foods require accompanying developments and facilities for transporting, storing, and marketing from the processing plant to the user's kitchen. In the poorer countries there are not, nor is there any prospect of there being soon available, large amounts of capital for investment in these types of facilities.

These factors, plus the self-sufficiency policies of many countries which encourage local production through high tariffs and quotas on imports of frozen foods, result in small shipments of U.S. frozen foods to foreign markets. There are a few notable exceptions to this general situation, however.

Imports

As indicated earlier, U.S. imports of frozen seafoods already exceed the domestic pack and are growing at a much more rapid rate. Comparative advantage of the fishing industries of other countries, due primarily to

TABLE 1.1

GROWTH TRENDS OF FROZEN FOODS

Top Ten Frozen Categories	Increases from 1959 to 1974 (%)
Cakes and pastries	198.9
Pies	137.7
Breaded shrimp	68.3
Toppings	67.5
Seafood	66.3
Grapefruit juice	64.3
Meat	60.4
Fruit drinks	52.9
Pre-cooked foods	52.7
Ice cream	51.5

Source: Williams (1976A).

lower wage rates, is the major reason for this trend. For similar reasons more than a third of our frozen strawberry consumption is imported—mostly from Mexico. U.S. imports of frozen strawberries from Mexico rose from an average of 23 million pounds in 1959–1961 to about 70 million pounds in 1966 and are continuing to grow. Total U.S. imports of frozen strawberries amount to over 86 million pounds, with most of the difference coming from Canada, which also ships over 6 million pounds of frozen blueberries to the United States a year.

For imports of other fruits and vegetables it is a different story. Frozen vegetable imports are practically nonexistent. At present, imports are confined to small quantities of specialties such as baby pod peas from Taiwan. Frozen fruit imports, other than strawberries, amount to about one per cent of the total domestic pack. This situation could change, especially for vegetables. Faced with high and continually increasing costs, U.S. companies have been investing in vegetable production and processing operations in Mexico where costs are lower than in the United States. Eventually large quantities of frozen and otherwise processed vegetables could come into the United States from Mexico just as has been the case with strawberries.

The only other frozen product category imported by the United States in significant quantities is meat. Meat import statistics are not reported separately for fresh, chilled, and frozen shipments, but since imports are from considerable distances—predominantly Australia and New Zealand—it can be assumed that they are mostly in frozen form. Over a billion pounds of frozen, fresh, and chilled meat and poultry of all kinds are imported by the United States. Of this total quantity nearly 70% was beef and veal, 13% poultry, 7% mutton and lamb, 4% each of pork and horsemeat, and 2% for all other kinds of meat. Most of these meats, with the possible exception of poultry meat and lamb, are used in further processing. Horsemeat is used for pet foods. The Meat Import Act restricts the amount of meat that can be imported by the United States, so that there is not an opportunity for much growth in imports of frozen meats.

Frozen eggs and frozen dairy products, especially cheeses of various kinds and cream, are imported in quantities sufficient to cause some concern to the U.S. dairy industry. The United States recently has taken a protectionist position on dairy products, and it is doubtful that there will be large gains in their importation in the near future.

Exports

Exports of all U.S. farm products have reached an all time high and a continued growth is estimated until at least 1980. It is unlikely, however, that frozen foods will share proportionately in this growth.

In the first place, food aid exports have been, and are likely to continue to be, nearly a quarter of total U.S. exports of farm products. These exports are mostly food and feed grains and oilseeds. Secondly, most exports for dollar sales have been in these same products and not in products requiring freezing for preservation.

The reason that most U.S. exports are food and feed grains and oilseeds, even to countries in which per capita incomes are relatively high, and in which there is an increasing demand for livestock products, is that the governments of these countries have adopted policies that encourage the internal production of livestock, especially poultry, for which feed grains and other feedstuffs are needed. The impact of such policies is exemplified by the data which show that the U.S. exports of most poultry products peaked in the early 1960's, but have since been falling off. For frozen and canned poultry, the European Economic Community, especially Western Europe, and Japan and Hong Kong are our most important markets even though the quantities imported by Western Europe have shown declines in recent years. Europe continues to be a most important export market for dried eggs, with most U.S. exports of shell and frozen eggs going to markets in North America, especially Canada.

Broiler production in Europe, especially in the countries of the Common Market, has shown a steady rise in recent years. This rise is encouraged by means of high entry prices, plus levies against broilers coming from countries outside the Common Market.

Similar situations are being repeated in other countries where, in spite of significant increases in per capita consumption of poultry products, the U.S. share is declining. However, increasing per capita incomes plus a desire on the part of large segments of the population in our major foreign markets to upgrade diets with more animal proteins and convenience foods will continue to provide some opportunities for export of poultry products, mostly in frozen form.

A third reason that frozen foods will not share proportionately in the growing export demand for U.S. farm products is because of competition from other product forms and from other production areas. U.S. exports of frozen fruits amount to about 2% of the total frozen fruit pack and only about ½% of all fruit exports. The story for frozen vegetables is the same. U.S. exports are less than 1% of the frozen pack.

Much of our frozen fruit and vegetable exports go to Europe. Both the processing and distribution of frozen foods in Europe have shown much progress in recent years. Facilities for handling frozen foods are being built up. The forecast by trade sources is that more homes will acquire refrigerators. There is a shortage of labor and many married women work outside the home. All of these conditions point to increased use of frozen food in Europe.

TABLE 1.2

ESTIMATES[1] OF FROZEN FOOD PER CAPITA CONSUMPTION
(1 KILO EQUALS 2.2 LB.)

Country	Kilos per Capita
U.S.A.	37–38
Sweden	20.8
Denmark	17.8
Great Britain	13
Switzerland	12.3
Australia	12
Norway	10.4
New Zealand	9
France	7.[2]
Finland	7.
Netherlands	5.5
W. Germany	5.3
Belgium	5.
Hungary	4.
Austria	4.0
Italy	1.44
Spain	1.20

Source: Williams (1976B).
[1]Most estimates include poultry but not ice cream.
[2]Includes both surgelé and congelé.

TABLE 1.3

U.S. FROZEN FOOD PACK (1974 AND 1975)

Commodity	1974	1975
Frozen vegetables		
(other than potatoes)	2,488,260	2,211,983
Frozen potato products	2,984,881	3,000,983
Frozen fruits and berries	603,972	566,825
Total—fruits and vegetables	6,077,113	5,779,791
Frozen concentrates, juices and purees		
(thousands of gallons)	186,911	193,482
Frozen seafood	517,411	212,181
Frozen poultry	2,222,294	2,069,690

Source: Williams (1976A).

But there are barriers to the United States sharing in this growing market. There are long established traditions and eating habits to be overcome. These traditions and habits are barriers to European processors as well as U.S. processors. Competition from European producers, who are closer to these markets, must be met. Frozen fruits offer to U.S. producers less opportunity for export than vegetables, because production costs for fruits are lower in Europe. Even for vegetables the main opportunities will be during periods of short supply resulting from unfavorable weather conditions. Air shipments of fresh fruits and vegetables—especially strawberries and asparagus—are beginning and will become more important as the growing use of larger aircraft results in lower air shipment rates. Frozen foods must compete with these developments.

It would appear, therefore, that although frozen food exports from the United States will continue to show some small gains, the opportunities for large growth are limited by protectionist policies of other countries, competition from other areas and other forms of food, and by the relatively low per capita incomes in most other countries of the world.

THE IMPACT OF FROZEN FOODS

The coordinated effort of many industries has been required to produce, process, package, store, transport, distribute, display, and sell the growing supply of frozen foods. New varieties were developed that were suitable for freezing. Equipment and methods for quick-freezing, packaging materials for meeting specific product and marketing requirements, and below-zero storage and transportation facilities for handling frozen foods all the way from the packer to the consumer were developed and are constantly being improved and expanded. Accompanying these developments have been important changes in the location of production of some commodities and changes in the marketing structure between the farmer and the consumer. These changes have been accompanied by, and in many instances are a direct result of, the growing importance of the large corporate retail chains.

Location of Production

Changes in the location of production of some fruits and vegetables are especially noteworthy. The trend has been for fruit and vegetable production to move West. In the 1945–1949 period an average of 50% of the frozen deciduous fruits and berries, and 54% of the frozen vegetables were produced in the West, mostly in California, Washington, Oregon, and Idaho. These averages have increased to 60 and 70%, respectively. Much of the increase in the deciduous fruit and berry category is accounted for by strawberries, which increased from 60% in the West in 1945–1949 to 90% at present. Frozen potato production, which is centered mainly in Idaho and Oregon, grew from 55% in the West in 1955–1957 to 70% today.

There are a couple of exceptions to the westward trend. Sour cherries and blueberries are and always have been produced entirely east of the Mississippi River. As processed orange juice replaced the consumption of oranges in fresh form there was a shift in the center of orange production from California to Florida.

There are a number of reasons why production of fruit and vegetables for processing has become centered in specialized growing areas. These reasons all add up to a comparative economic advantage of one region over another. Nearness to markets is sometimes a factor, but in the case of frozen fruits and vegetables over 60% of the western pack is estimated to move to markets east of the Mississippi River. Thus there are overriding reasons that make the West a more profitable area of production for many fruits and vegetables. The major reasons are higher yields of the varieties that are most suitable for processing, and longer operating seasons due either to ease of storage of raw product, or to the availability of a wide range of products that mature at different times and that can be processed in the same plant.

Frozen food processing plants are located near the source of supply because of savings in transportation of the finished product rather than the raw material and because of the greater possibility of freezing products at the peak of their freshness. Thus, most fruits and vegetables, except citrus fruits, cranberries, sour cherries, and blueberries, are frozen in the states of California, Oregon, Washington, and Idaho where they are produced. Likewise, all frozen ocean fish and shellfish are packed in the coastal and Gulf States and move to consumption centers throughout the country. Frozen eggs are packed mainly in the surplus production areas of the Midwest and move primarily to eastern markets. The relatively small quantities of frozen meats now being produced are packed in various market centers throughout the United States. Should frozen meats become important, adjustments would be required throughout the entire meat marketing system. Slaughter, breaking of carcasses, cutting, packaging, and freezing operations would, in all likelihood, tend to shift nearer to production centers than at present, particularly as old facilities were replaced with new ones.

As frozen food sales have expanded there has been an increase in demand for storage facilities at processing points. Thus, storage of frozen foods, especially fruits and vegetables, is also moving west. The reasons for this are economic. By purchasing directly from the packer at the production-point warehouse, the frozen food buyer can reduce storage and handling charges at an in-transit warehouse. Also, storage at point-of-production makes possible more orderly distribution of seasonally produced frozen foods in accordance with market demand.

Trade Associations

With the rapid growth of frozen foods, many trade associations wholly or partly devoted to various aspects of the industry have been organized. The regular meetings and conventions of these associations have become important parts of the machinery which moves the industry's products in the channels of trade. The officers of these associations and their standing committees often exercise great influence in developing standards and regulations and in affecting legislation pertaining to the industry. A comprehensive discussion of public and private regulations and handling codes is contained in Chapter 15.

Requirements for Facilities

Mass distribution of frozen foods has brought about a change in the type of storage facilities required at retail distribution points. The emphasis is now on rapid movement of stocks, not on long-term storage. The first commercial cold storage warehouse in the United States opened in 1865 with a minimum of facilities for mechanized handling. Today, refrigerated warehouses offer their customers electronic data processing, bulk handling of commodities, new redistribution concepts, and pallet exchange. All of these services result in reduced loading and unloading time and thus in better quality products and lower costs.

Refrigerated warehouse capacity at 0° F. (−18° C.) or below increased from about 135 million cubic feet in 1945 to nearly 640 million cubic feet in 1965. This represents an average increase of 25 million cubic feet each year for the past 20 years.

Public refrigerated warehouse space in the United States increased by 47,108,000 gross cu. ft. or 5.8%, from October 1973 to October 1975, according to USDA estimates. Gross cooler space increased from 171,676,000 cu. ft. to 172,817,000 cu. ft., while gross freezer space increased from 627,162,000 cu. ft. to 673,129,000 cu. ft. Total public refrigerated space of 845,946,000 cu. ft. in 1975 represented 52% of the nation's total. Total private and semi-private space (not including apple space) increased from 398,217,000 gross cu. ft. to 419,828,000 gross cu. ft. or 5.5%.

Assuming present growth rates and a two per cent replacement of present capacity each year, an annual investment of $50 million is required just for warehouse space to handle our frozen food supply. Should freezing of meats become important, a large part of the additional requirements for 0° F. (−18° C.) or below warehouse capacity will probably be provided by conversion of present meat refrigeration facilities, but additional investment will be required.

Similar developments have occurred in the processing plants themselves and in the industries that service these plants and provide them with packaging materials, lithographing services, refrigerants, etc. There has been an accompanying growth in mechanically refrigerated rail cars, truck-trailers, 0° F. (−18° C.) or below backroom storage in retail stores, retail display cabinets, and home freezers.

Total annual sales of all types of refrigerated display cases, for example, have remained relatively stable at about 100,000 cases per year for the past 10 years, but sales of 0° F. (−18° C.) display cases have increased from 25% to over 40% of the total.

In the home, zero storage capacity by the end of 1975 had reached well over the 5 billion pound mark. This did not include a large but unmeasured zero capacity in two-compartment refrigerators. According to surveys conducted by newspapers in major markets, however, well over ⅔ of U.S. families still do not own home freezers, and thus represent an area for large potential growth.

Public eating establishments have become large users of frozen foods. There are more than 400,000 public eating establishments and institutions with food service in the United States. Public eating establishments account for 93% of the total. The cubic feet of freezer space per establishment ranged from 19.2 for community programs, and 30.0 for sorority and fraternity houses, to 507.7 for dairy product stores, and 803.1 for colleges and universities, averaging 92.1 cu. ft. for all establishments.

It is hardly necessary to labor the point of this section further. The growth in frozen food sales has brought far reaching changes in our economy requiring large investments in many allied industries, and the end is not yet in sight.

A FROZEN FOOD HALL OF FAME

Some individuals who were instrumental in the development of the frozen food industry are mentioned in Table 1.4, but certain of those individuals, according to Fennema (1976), belong in a "Frozen Food Hall of Fame" because of their extraordinary contributions to the U.S. frozen food industry during its formative years:

Clarence Birdseye—Designer of freezing equipment, organizer of companies pioneering in the processing and marketing of quick-frozen foods, "father" of the frozen food industry in the U.S.

Donald K. Tressler—Professor, researcher on frozen foods, author and publisher of world-renowned books on food freezing.

TABLE 1.4

IMPORTANT EVENTS IN THE DEVELOPMENT OF THE U.S. FROZEN FOOD INDUSTRY

Year	Event
500 B.C.– 1800 A.D.	Cooling achieved by use of snow, natural ice, air in cold climates (weather freezing), evaporative cooling of water, and radiative cooling. The use of chemicals to lower the freezing point of water was known at least as early as 1550.
1820	By this time, natural ice had come into general use as an article of commerce and was used on a large scale for food preservation.
1842	H. Benjamin was granted British patent No. 9240 for a rapid freezing technique involving immersion of the article in a freezant.
1851	Jacob Fussell of Baltimore, Md., first sold ice cream on a serious commercial scale in the U.S.
1853–54	Captain H.O. Smith sailed to Newfoundland and conducted the first commercial "weather freezing" of fish.
1861	Thomas S. Mort established what is believed to be the world's first cold-storage plant in Darling Harbor, Australia. This facility was used primarily for freezing meat.
	Enoch Piper of Camden, Maine, was granted U.S. patent No. 31,736 (March 19), for the first practical method for artificially freezing fish and storing them in the frozen state. Freezing was accomplished in an insulated room by positioning pans of ice and salt over the fish. Glazing of the fish was also described.
About 1863	Artificial freezing of fish on a commercial basis began in the U.S., particularly in the Great Lakes region (salt and ice method).
1864	Ferdinand Carré patented an ammonia compression machine in France.
1865	A.&E. Robbins Co. of New York City, using ice and salt, was the first company to freeze poultry in the U.S.
1867	Dr. Peyton Howard made the first successful shipment of frozen beef in the U.S.
1868	William Davis developed a freezing system for fish (U.S. patent No. 85,913, Jan. 19, 1869) that involved packing fish tightly in metal pans, attaching a metal lid, and surrounding the pans with an ice-salt eutectic mixture. This system was later used extensively on a commercial basis.
1869	Pigeons were first frozen commercially in the U.S.
1870	Ammonia compression machines were brought to a level of practicality almost simultaneously by Dr. Carl Linde in Germany and David Boyle in the U.S. (Linde, British patent No. 1458, 1876). Until this development, frozen foods were relatively unimportant.
1876	Frozen meat was first shipped from the U.S. to England (ice-salt).
About 1880	Ammonia compression machines and insulated rooms began to be used in the U.S.
	Frozen meat was shipped from Australia to England using mechanical refrigeration. This is believed to be the first major use of mechanical refrig-

TABLE 1.4 (*Continued*)

Year	Event
	eration for preserving perishable food. About 30 tons of meat were transported on the Scottish steamer "Strathlaven," and the shipment arrived in good condition. Part of this meat was subsequently eaten by the royal family.
	Commercial freezing of fish, poultry, and meat became common in the U.S.
1889	Liquid eggs were first frozen commercially in the U.S.
1891	By this time, 2,153,000 frozen mutton carcasses had been exported from New Zealand.
1892	In Sandusky, Ohio, an ammonia compressor was first used in the U.S. for freezing fish.
1900	By this time, fish freezing had become an important industry in the U.S.
1905	Fruit was first frozen commercially in the U.S.
1907	By this time, the importance of proper packaging of frozen fruits was recognized.
	A histological technique was developed that enabled detection of the comparative physical effects of slow and quick freezing.
About 1910	Fruit was frozen commercially in the northwestern U.S. Some problems were encountered with fermentation of berries during slow freezing in large barrels.
About 1916	Scientific work on the methodology of freezing foods began in earnest.
1917	Experimental work was being conducted on freezing of vegetables.
1921	M.T. Zarotschenzeff was granted British patent No. 339,172 (Feb. 16, 1921) for a freezing device that involved spraying the food, either packaged or unpackaged, with cold brine. In the U.S., this method was used mostly for commercial freezing of poultry but also for freezing of fish and meat.
1923	The "Quick-Freezing" industry, as we know it today, had its origin with the founding of a freezing company by Clarence Birdseye.
1925	Some persons, notably C. Birdseye, were aware that the final quality of frozen food depended on proper selection, handling, preparation, freezing, storage, and distribution.
1928	About this time, efforts were made to use mechanical refrigeration for low-temperature display and storage boxes in retail stores.
1929	M.A. Joslyn and W.V. Cruess reported on the need to blanch vegetables prior to freezing.
	Blanched vegetables were first frozen commercially in the U.S. by the Birdseye organization in Hillsboro, Ore.
	A double-belt freezer was used by the Birdseye organization in Hillsboro, Ore., to freeze packages of fruits and vegetables.

TABLE 1.4 (Continued)

Year	Event
	The Birdseye organization was bought by Postum Co. (now General Foods Corp.), marking the real beginning of marketing of frozen foods through retail stores in the U.S.
1930	C. Birdseye was granted patents covering several aspects of food freezing processes, including a package and a double-plate freezer (U.S. patents No. 1,773,079; 1,773,080; and 1,773,081, Aug. 12, 1930). The plate freezer in modifid form is still used today for freezing foods in retail cartons. The package alluded to was a waxed cardboard carton, lined with waxed vegetable parchment paper or moisture-proof cellophane and overwrapped with waxed, heat-sealed glassine paper. This is not unlike the package being used today.
	On March 8th, the Birdseye Frosted Food Co. sold packaged meat, fish, oysters, vegetables, and fruit from low temperature display cases in selected stores in the eastern U.S.
	By this time, mechanical refrigeration had assumed an almost indispensable part of the food distribution system in the U.S.
1938	Quick Frozen Foods, an important trade journal, was first published by E.W. Williams.
1945–50	Many frozen prepared foods such as meat pies, baked goods, frozen dinners, breaded shrimp, fish sticks, potatoes, and orange juice concentrate had their commercial origin during this period.
1949	Mechanically-refrigerated railroad cars came into use about this time.
Early 1960s	Fluidized-bed freezers and individually-quick-frozen (IQF) foods began to assume a position of importance in the U.S.
1962	Liquid-nitrogen food freezers were first used commercially.
1968	Freezant-12 freezers were first used commercially.

Source: Fennema (1976).

William Boyle—Instrumental in the development and use of ammonia compressors in the U.S.

Maynard A. Joslyn—Professor, University of California; food chemist, researcher on frozen foods, consultant to the food industry, involved in early studies indicating that vegetables must be blanched prior to frozen storage.

Mary E. Penninngton—Chief, Food Research Laboratory, Bureau of Chemistry, USDA; promoter of frozen poultry products, consultant to the food industry.

Harden F. Taylor—Chief Technologist, U.S. Bureau of Fisheries, Vice-President for Scientific Research, The Atlantic Coast Fisheries Co., designer of freezing equipment, advocate of frozen foods.

E.W. Williams—Founder and publisher of Quick Frozen Foods, the first and most prominent trade journal serving the frozen food industry.

Jasper G. Woodroof—Professor, University of Georgia; conducted early studies on processing procedures for frozen peaches and other foods and on histological characteristics of frozen fruits and vegetables.

Contributions of Clarence Birdseye

Among this group of worthy individuals, Clarence Birdseye is, without question, the superstar. Some notion of his greatness can be gained from a statement he made in 1932:

> Products must be properly selected as to variety and quality, properly pre-treated and packaged before freezing, correctly frozen, suitably cold-stored, transported without being allowed to thaw, retailed efficiently at temperatures of about 5° F., and finally properly cooked. Any consideration of perishable food freezing which fails to take into account all of these factors is ill-advised.

Few of us today could make statements of similar importance and have them stand the test of time as well as this one has.

Accomplishments of Clarence Birdseye and the Birdseye Laboratories.—Although fish and meat had been frozen commercially for many years, frozen foods were considered to be inferior products prior to 1929 when Clarence Birdseye made certain improvements in the freezing and storage process which seemed to be trivial and of minor importance, yet turned out to revolutionize the entire frozen food industry. After the Great Depression, the industry became of great importance and has steadily increased in volume.

Clarence Birdseye had spent six years in the Arctic with Sir Wilfred Grenfell and had lived on frozen foods many long winters. He knew that frozen foods were delicious and was confident that he could establish an important industry. In fact, he was so certain of the possibilities of frozen foods that he devoted his entire energies to the establishment of the industry.

Accomplishments of the Birdseye Laboratories.—Clarence Birdseye and the Birdseye Laboratories gave a *big boost* to the frozen food industry in the following ways:

(1) Prior to Birdseye's days, frozen foods were considered very inferior to fresh foods and were usually thawed before sale to the public. Clarence Birdseye coined the word "quick-frozen" and publicized quick-frozen foods, emphasizing their high quality. The laboratory demonstrated that quick-freezing did not cause a loss of color, flavor, or nutritive value, provided the quick-frozen foods were held at 0° F. Freezing actually improved the color of many foods, e.g., peas, lima beans, broccoli, and other green vegetables.

(2) Further, the Birdseye Company began freezing foods in retail size

packages and selling them without thawing in Springfield, Mass., and Geneva, New York. The company proved that the quick-frozen products could be sold in grocery and meat markets when packed in attractive packages.

(3) Clarence Birdseye invented and perfected equipment designed for the "quick-freezing" of all kinds of foods in retail size packages. The first of these was a double-belt freezer. This consisted of two long, movable refrigerated stainless steel belts superimposed one over the other with just sufficient space between the two to permit the passage of the retail size moisture vapor-proof cartons of fish fillets or other foods.

The other freezer invented by Clarence Birdseye was a "multiplate" freezer which was portable and could be moved from plant to plant. This was an ideal freezer for freezing relatively small packs of various vegetables or other foods.

(4) The Birdseye Laboratory personnel under the direction of Dr. D. K. Tressler began studying the modification of commercial canning processes to perfect methods which would produce "quick-frozen" food of the highest quality. The Birdseye Laboratory personnel would move "plate frosters" into canning plants and freeze enough of a vegetable to demonstrate in Springfield, Mass. and some other markets that the public would by "quick-frozen" packaged vegetables if they were in the proper size packages and were of good quality. A large pack of lima beans was quick-frozen at Seabrook Farms in Bridgeton, New Jersey. Similarly, a commercial-sized pack of quick-frozen peas was made in Albion, New York. The commercial processes used for preparing peas and lima beans for freezing were not greatly different from those employed in preparing peas and lima beans for canning. The important steps were (a) mowing, (b) using hay loaders to load the vines onto "hay wagons," (c) hauling the vines to a viner station, (d) "vining" the peas or lima beans in "viners," (e) washing the vined vegetable, (f) blanching the peas or lima beans with steam or hot water to inactivate enzymes, (g) washing and cooling with water, (h) packing the peas or lima beans in retail size packages, (i) quick-freezing the packages of vegetables in "multiplate frosters," (j) packing the small cartons in shipping containers, (k) moving the products into storages maintained uniformly at 0° F.

In the case of the preparation of mushrooms for freezing, considerable study was required in order to obtain a frozen product which retained its color and flavor during quick-freezing and storage. The laboratory procedure was worked out in the Gloucester Birdseye Laboratories, but the world famous Dr. G. Raymond Rettew helped in perfecting the commercial process used in preparing mushrooms for freezing which was carried out in the plant of the Premier Mushroom Company near Philadelphia, Pennsylvania.

(5) The Birdseye Laboratories showed that nearly all foods could be packaged and frozen without great loss of vitamins or quality. Quick-frozen vegetables were shown to be superior in quality to canned foods, and, in many instances, the cooked quick-frozen foods were superior in color, flavor, and vitamin content to the cooked fresh product. The reason for this great improvement in quality was the fact that freezing enabled the product to be cooked in less time.

(6) The Birdseye Laboratories also showed the necessity of maintaining the quick-frozen product at a uniformly low temperature below 10° F. if the quality of the product was to be maintained during storage.

(7) The Birdseye Laboratories perfected the processes of preparing and packaging all kinds of precooked foods.

The notable innovations in preparing and marketing frozen foods completely revolutionized the industry. Today, 40 odd years later, frozen foods are on the whole considered superior to canned and dehydrated foods. It was Clarence Birdseye's vision and energy and determination that launched the industry we know today.

Readers desiring further information concerning the history of the frozen food industry will find the publications shown in the Additional Readings list especially helpful.

THE FUTURE OF FROZEN FOODS

The future growth of frozen foods will be influenced by a number of economic and technological factors. Among these are growth in population, changes in its composition and location, growth in personal incomes, relative costs of frozen versus other forms of foods, changes in food tastes and preferences, technological advances in methods of freezing as well as in other methods of food preservation, and consumer acceptance of new products and substitute or synthetic foods.

A large potential demand for frozen foods exists. Continued expansion of the industry will depend largely on the ability of processors, handlers, and distributors to maintain high standards of appearance, flavor, and nutritive value of frozen foods, and to continue to secure a satisfactory return on their investment. Improvements in merchandising and promotion and development of new convenience forms of frozen foods, as well as more information on relative costs and nutritive values of frozen foods, will contribute toward continued growth of the industry.

ADDITIONAL READING

FENNEMA, O. 1976. The U.S. Frozen Food Industry: 1776–1976. J. Food Technol. *30*, No. 6, 56–61, 68.

JOHNSON, A. H., and PETERSON, M.S. 1974. Encyclopedia of Food Technology. AVI Publishing Co., Westport, Conn.

PENTZER, W. T. 1973. Progress in Refrigeration Science and Technology, Vol. 1–4. AVI Publishing Co., Westport, Conn.

PETERSON, M. S., and JOHNSON, A. H. 1977. Encyclopedia of Food Science. AVI Publishing Co., Westport, Conn.

TRESSLER, D. K., VAN ARSDEL, W. B., and COPLEY, M. J. 1968. The Freezing Preservation of Foods, 4th Edition. Vol. 1. Refrigeration and Refrigeration Equipment. Vol. 2. Factors Affecting Quality in Frozen Foods. Vol. 3. Commercial Freezing Operations—Fresh Foods. Vol. 4. Freezing of Precooked and Prepared Foods. AVI Publishing Co., Westport, Conn.

UMLAUF, L. D. 1973. The frozen food industry in the United States—Its origin, development and future. American Frozen Food Institute, Washington, D.C.

WILLIAMS, E. W. 1963. Frozen Foods: Biography Of An Industry. Cahners Publishing Co., Boston, Mass.

WILLIAMS, E. W. 1976A. Frozen foods in America. Quick Frozen Foods *17*, No. 5, 16–40.

WILLIAMS, E. W. 1976B. European frozen food growth is now resumed—more new products launched. Quick Frozen Foods *17*, No. 5, 73–105.

2

Refrigeration Technology

Willis R. Woolrich
and Arthur F. Novak

There are a number of ways to accomplish quick-freezing, or, in more scientific terms, a rapid extraction of heat. The two most common methods are the vapor compression and the absorption refrigeration cycles. The former is compression by mechanical energy, the latter by the transfer of heat.

PRINCIPLES OF REFRIGERATION

The first law of thermodynamics states that different forms of energy are mutually interconvertible and that a definite numerical ratio exists for each conversion.

The second law stipulates that heat energy may be converted into work only when permitted to pass from one temperature to a region at a lower thermal value and, conversely, heat may be moved from a region of low temperature to one of high temperature only when work is done.

By definition, heat is a form of energy known by its effects. The effects are indicated through touch and feeling as well as by the expansion, fusion, combustion, or evaporation of the matter on which it acts. When heat is added to a substance its temperature is raised unless there is a change of physical state such as vaporization or melting. Likewise, when heat is removed from a substance there is a lowering of the temperatures except at the condensation and freezing. Heat is that form of energy which transfers from one system to another by virtue of the temperature difference which exists between them when they are brought into communication.

The molecular theory of heat assumes that molecules or particles of a substance are in continuous and irregular motion and that heat is the result of this motion. One of the first interests in the study of refrigeration is the transfer of heat. Heat may be transferred in any one of three different ways, or, more generally stated, it may be distributed in all three ways at the same time. The three modes for the distribution of heat are conduction, convection, and radiation.

Some engineers prefer to think of evaporation and condensation as a fourth method of heat transfer. Vaporization of either solids or liquids may be considered a separate phenomena. The vaporization of a solid without the intermediate formation of a liquid is defined as sublimation.

Definitions

Evaporation.—Evaporation of a liquid begins and may continue until all is entirely in the form of a vapor. During this period the temperature of evaporation at saturation remains constant at the saturation pressure. The heat added during this change of state is the change of enthalpy during evaporation, made up of the increase in internal energy and the mechanical work done in expanding the liquid to a vapor against the constant pressure. The mechanical work may amount to 5 to 10% of the total enthalpy of vaporization, depending upon the existing pressure. The sum of the enthalpies of the saturated liquid and the change of enthalpy due to vaporization is the enthalpy of the saturated vapor.

Condensation.—Condensation is the reverse of evaporation. It is the change of state of a substance from a gas to a liquid. In physics terminology, the velocity and distance between molecules of a gas are decreased by heat withdrawal, so that a substance condenses forming droplets of liquid such as dew and surface condensation. This process may proceed by direct contact or by dry surface condensation.

To prevent condensation on the outside of insulated equipment requires sufficient thickness of insulation to insure that the temperature drop from ambient air to insulated surface is less than the dew point depression. To stop condensation at 100% R. H. requires an infinite thickness of insulation, thus zero heat flow.

Conduction.—Conduction means the flow of heat through an unequally heated body or system of bodies from points of higher to points of lower temperature. It is exemplified in the heating of a metal rod by placing one end in a flame. The part in the flame soon becomes hot, the molecules of the adjacent parts have their motion quickened through the impact of those in the hotter part, and a transfer of heat takes place to points of lower temperature. In this way, a steady flow of heat is set up through a rod. Such a rod is said to be a conductor of heat, and the relative rate at which such a transfer is made is termed the thermal conductivity of the metal.

Convection.—By convection is meant the transfer of heat by the bodily movement of heated particles of matter. The heating of homes and buildings by steam or hot water are excellent examples of convection of heat.

Radiation.—By radiation is meant the transfer of energy from point to point in space by means of waves set up in the ether. The earth is heated by radiation from the sun. If one holds his hands over a heated object, the hands are heated both by convection through the air and by radiation, but if the hands are placed under the heated object, the heating of the hands is by radiation only.

Temperature.—Temperature may be defined as the thermal condition of a body. Temperature indicates how hot or cold a substance is; that is, it is a measure of sensible heat. Temperature, therefore, gives only the relative intensity of heat, and not the amount. Temperatures are measured by thermometers or pyrometers in the Fahrenheit, Centigrade, Rankine or Kelvin scales. The following methods of calculation permit the transfer from Fahrenheit to Centigrade scale or vice versa:

Centigrade degrees = $\frac{5}{9}$ × (Fahrenheit degrees − 32 degrees)

Fahrenheit degrees = ($\frac{9}{5}$ × Centigrade degrees) + 32 degrees

Absolute temperatures are based on absolute zero at which all molecular thermal energy is absent. Numerically it is 459.69° F. below the zero Fahrenheit and 273.16° C. below the zero Centigrade temperature. In equation relationship these become: $T = t_F + 459.69°$ R. for the Fahrenheit scale and $T = t + 273.16°$ K. for the Centigrade scale.

An ideal way to measure heat is to note its effect in raising the temperature of a measured body of water. The present generally accepted heat unit called the British thermal unit (abbreviated B.t.u.) is defined as $\frac{1}{180}$ of the heat required to raise the temperature of one pound of water from 32° F. (0° C.) to 212° F. (100° C.) at normal atmospheric pressure; in other words, in practice 1 B.t.u. is the measure of that heat which will raise the temperature of one pound of water one degree Fahrenheit.

Specific Heat.—The specific heat of a substance is the ratio of the heat required to raise the temperature of unit mass of the substance one degree to the heat required to raise the temperature of unit mass of water one degree.

Engineers consider the pound as the unit of mass, and a degree Fahrenheit as the unit of temperature. Since the specific heat of water is taken as the standard and is one, it may be said that the specific heat of a substance is the amount of heat required to raise or lower the temperature of one pound of the substance one degree Fahrenheit.

Forms of Heat Energy

Sensible Heat.—Sensible heat is so common to our everyday life that there is danger of our thinking of temperature change only as a measure of heat supplied. Sensible heat may be defined as that heat which produces a rise of temperature, as when a pan of water placed over a flame becomes hotter and hotter to the touch. One must carefully differentiate that type of heat from another type known as latent heat.

Latent Heat.—Latent heat is the quantity of heat required to change the state or condition under which a substance exists, without changing its temperature; e.g., a definite quantity of heat must be transferred to ice at 32° F. (0° C.) to change it into water at the same temperature. This definite quantity of heat is known as the latent heat of fusion in going from the solid to the liquid state, or the latent heat of evaporation when going from the liquid to the vapor state, as when water boils and forms steam.

It will readily be seen that latent heat is of great importance in the study and application of refrigeration. Water, when cooled, loses about 1 B.t.u. per pound for each degree decrease in temperature from whatever temperature it is at, until it reaches 32° F. (0° C.). Then 144 B.t.u. per pound (which is the latent heat of fusion for water) are extracted while the water is freezing, yet there is no change in the temperature of the water during this period. When the water is all frozen, the resulting ice then requires only approximately one-half B.t.u. per pound for each degree decrease in temperature below 32° F. (0° C.). The process is, of course, reversible, and therefore to change one pound of ice to water it is necessary to absorb 144 B.t.u. Naturally then, it is the latent heat of fusion of ice that makes it valuable for refrigeration.

The latent heat of evaporation is even more important in the study and application of refrigeration, for without this phenomenon it would be impossible to have mechanical refrigeration by the compression system. It is the latent heat of certain substances known as refrigerants that forms the basis of producing refrigeration by mechanical means.

Although thus far no real definition has been given for refrigeration, some points have been listed that do define it. In the literature numerous definitions are given, and a combination of a number of these would indicate that it is a process of removing heat from a confined space and material for the purpose of reducing and maintaining the temperature below that of the surrounding condition. Since refrigeration is a process whereby heat is removed, the quantity of heat removed is measured in British thermal units.

The Compression Cycle of Refrigeration

Figure 2.1 is a diagrammatic illustration of a compressor-type refrigeration unit with liquefiable gases and vapors in the compression cycle. The gaseous refrigerant is compressed, then passed through pipes to the condenser where it is cooled and condensed to a liquid, minus much of its original heat. Usually the liquid refrigerant is stored in a high-pressure cylindrical receiver with inlet and outlet valves in the continuous piping system.

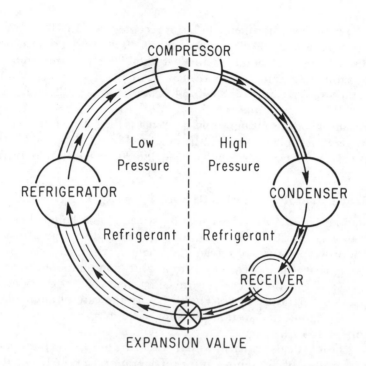

FIG. 2.1. REFRIGERATION CYCLE OF THE MECHANICAL REFRIGERANT COMPRESSOR FOR VAPOR RECIPROCATING, ROTATING, AND CENTRIFUGAL SYSTEMS

Under flow rate control the liquid refrigerant at high-pressure is passed from the receiver through an expansion throttling pressure reducing valve. The resultant action is to change the fluid to an atomized vapor-liquid mixture at low pressure as it enters the evaporator or cooler.

Actually the evaporator or cooler is a boiler in which the vapor-liquid refrigerant is completely vaporized by heat obtained from the enclosed refrigerator product and produce, which in turn warms the fluid refrigerant in the cooler coils to a gaseous condition. When the cycle has been completed by the refrigerant in the closed refrigeration system, the gaseous vapor starts again over the same path, changing by compression, cooling and heating from a gas to a liquid and returning again as a gas to the compressor for this and subsequent cycles.

Refrigeration Units

The standard unit of refrigerating capacity is known as a ton of refrigeration. The ton of refrigeration is derived on the basis of the removal of the latent heat of fusion from 2,000 lb. of water at 32° F. (0° C.) in order to produce 2,000 lb. of ice at the same temperature in 24 hr. The latent heat of fusion of ice (by calorimeter 143.4) is accepted as being 144 B.t.u. per lb. Therefore, with 2,000 lb. of water at 32° F. (0° C.) and the extraction of 144 B.t.u. from each pound, a total of 288,000 B.t.u. are removed to change a ton of water to a ton of ice at 32° F. (0° C.). The standard "ton" of refrigeration is therefore: 288,000 B.t.u. per 24 hr., or 12,000 B.t.u. per hr., or 200 B.t.u. per min.

An example of this rating would be: What is the refrigerating capacity requirement of a plant in which the heat load is calculated at 2,500,000 B.t.u. every 24 hr? Answer: 2,500,000 ÷ 288,000 = 8.69 tons of refrigeration.

Standard Ton of Refrigeration (U.S.A.)

A standard rating of a refrigerating system using liquefiable gas or spond to a suction pressure of 20 lb. gage and a discharge pressure of 169.7 adopted pressures of refrigerants, namely: the inlet (suction) pressure being that which corresponds to a saturation temperature of 5° F. (−15° C.) and the discharge pressure being that which corresponds to a saturation temperature of 86° F. (30° C.). In the case of ammonia, this would correspond to a suction pressure of 20 lb. gage and a discharge pressure of 169.7 lb. gage.

Since the text deals primarily with the freezing of foods, it is proper that those specialties of low temperature refrigeration be noted. The definition of a standard ton of refrigeration has been explained. However, the refrigerating capacity as noted is in tons of refrigeration at standard conditions; namely, suction temperature of 5° F. (−15° C.) or 20 lb. gage pressure for ammonia. To accomplish the problem of freezing foods as quickly as possible, it is necessary to utilize the lowest temperature obtainable with the equipment at hand. It is generally accepted that temperatures for quick-freezing should range downward from −28° F. (−33.3° C.). In order to obtain a temperature of −28° F. (−33.3° C.) it is necessary to operate the refrigerating system at a suction pressure of 0 lb. gage. For lower temperatures than this, correspondingly lower operating pressures must be employed. As the operating suction pressure of a system is reduced, its refrigerating capacity is also reduced. For example, if a system is rated at 250 tons of refrigeration at standard conditions, its capacity will be reduced to approximately 100 tons of refrigeration when operating suction pressure is lowered from 20 lb. gage to 0 lb. gage. In general, the refrigerating capacity at standard conditions is reduced about two and one-half times for operation at 0 lb. gage.

The above is predicated on the use of ammonia as the refrigerant. Present-day practice employs ammonia as the refrigerant in the majority of food freezing installations and all further examples and discussion in this chapter will be based on using it as the refrigerant. The evaporation of 1 lb. of liquid ammonia at 86° F. (30° C.) to a gas at 5° F. (−15° C.) will produce a refrigerating effect of 474.4 B.t.u. per lb. at standard conditions. No other commercial refrigerant can approach ammonia in this respect. For example, the evaporation of 1 lb. of carbon dioxide at standard conditions will only produce a refrigerating effect of 55.5 B.t.u. per lb. of refrigerant circulated in the system. To accomplish the same amount of cooling, 474.4 ÷ 55.5 or 8.55 times as much carbon dioxide must be circulated. Condensing pressures are low for ammonia, whereas those for carbon dioxide may be 6 times greater.

Coefficient of Performance

The coefficient of performance of a refrigerating system in the closed cycle denotes a measure of the efficiency of operation in utilizing the energy input. This is the ratio of the energy utilized in the evaporator to the energy input.

Compression Ratio

Under normal conditions, the refrigerant with the lowest compression ratio is preferable, since differences between the suction and discharge temperature reflect the boiling point characteristic of the refrigerant adopted, and further, since the temperature of the evaporator and the condenser are prescribed by system design and the available condensing fluid heat level, then compression ratio is determined by the refrigerant selected.

The compression ratio of a refrigerant compressor is the quotient of the initial to the final volume within the cylinder when the piston is at its maximum discharge position. Because volume varies inversely as the pressure under constant temperature conditions, the compression ratio may be defined as the final pressure divided by the initial pressure.

In the first instance the equation becomes: Compression ratio = initial volume/final volume. Under the second condition the equation becomes: Compression ratio = final pressure/initial pressure.

The latter equation is readily applicable to the computing of the compression ratios of rotary and centrifugal machines. Both equations are readily applicable to all reciprocating compressors (Figs. 2.2 and 2.3).

Physical Characteristics of Refrigerants

The physical, chemical and thermodynamic properties of the several hundred available refrigerants determine their practical usefulness. The more important of the physical characteristics will be considered first.

Boiling and Condensing Temperatures and Pressures.—In dealing with refrigerant fluids, the evaporator and condensing temperatures determine the pressures. For most applications, it is desirable to select a refrigerant whose saturation pressure at minimum evaporating operating temperature is maintained at a pressure a few pounds above atmospheric. This gives a positive differential pressure between suction and atmospheric, and prevents leakage of air with inherent moisture into the low pressure side of the system, especially at the compressor shaft and piston rod seals.

The maximum condensing temperature is largely affected by climatic conditions. It is preferable to adopt a refrigerant with a condensing pressure within safety limitations and within system weight acceptability. Usually an air cooled condensing system will inherently require higher condensing temperatures, especially in hot climates. High condensing pressures are conducive to more system leakage and accidents.

Freezing, Critical and Discharge Temperatures.—The refrigerant fluid should have a low freezing temperature to avoid operational obstruction by the refrigerant itself. At the other end of the scale it is most desirable that the critical temperature be well above the maximum condensing temperature. Exceptions to refrigerants not having critical temperatures above the temperature of the condenser are air and carbon dioxide, the latter when the condensing medium is above 87.8° F. (31.08°

FIG. 2.2. ROTARY BOOSTER COMPRESSORS OF THE FES FULLER TYPE INSTALLED AT THE CINCINNATI TERMINAL WAREHOUSE, INC.

FIG. 2.3. A BOOSTER COMPRESSOR; THE ECLIPSE 3-CYLINDER DESIGN

C.). Under these temperatures machines operate as "dry gas" systems at a much lower effective efficiency.

High discharge temperatures from the compressor cause some refrigerant breakdowns as well as poor lubrication effectiveness and should be avoided whenever possible.

Latent Heat of Vaporization and Specific Heat of the Refrigerant.— Since most refrigerants, as used, pass through a liquid and a vapor cycle from receiver to compressor and the heat absorbed by them per pound is mostly the heat of vaporization, the higher the latent heat capacity of the refrigerant, the less gas must be compressed. Other factors that must be given consideration are the specific heat of the liquid and the densities of both the refrigerant liquid and the vapor. For small units the latent heat capacity of the refrigerant is less a favorable factor than for large compressor operation.

High specific heat of the refrigerant vapor is a desirable factor. The heat that is added to the saturated vapor in the evaporator and suction lines will boost its temperature less if the specific heat is low. A high specific heat of the refrigerant vapor is an asset in practical operation. On the other hand, a low specific heat of the refrigerant liquid is an asset since less latent heat of vaporization in passing through the expansion valve must be used in cooling the liquid. This leaves more cooling capacity in the evaporator.

Compression Ratio and the Refrigerant.—A low compression ratio is recommended to hold down first cost of the compressor and to reduce the operation energy required. The compression ratio, as previously defined, affects design, construction and operating characteristics of the entire system. Rotary and centrifugal refrigerant compressors operate better on fluids having low ratios of compression, especially to reduce vapor leakage by their rotors (Fig. 2.4).

Liquid Densities and Viscosities of Refrigerants.—Density must be considered with viscosities in refrigerant compressor operation. Most designers prefer low vapor density refrigerants in order to justify high gas velocities in the suction and discharge lines and valves. Liquid densities affect the operation of float valves and, in some cases, refrigerant oil mixtures. The trend is towards higher liquid densities in refrigerant use.

For long lines, low viscosity of the liquid refrigerant is desired to reduce pressure drop in the orifices and lines. Likewise, to reduce pressure drop and line size low vapor viscosities are usually preferable.

Chemical Characteristics of Refrigerants.—The chemical characteristics of refrigerants are most important not only for thermodynamic considerations, but for fire, explosion, safety and odor considerations. These are especially important in freezer storage on land and on marine craft.

FIG. 2.4. A MODERN INSTALLATION OF A CENTRIFUGAL REFRIGERANT COMPRESSOR

Toxicity of Refrigerants.—The toxicity of refrigerants is rated by Fire Underwriters Laboratories. This rating is based on the toxic effect on human beings over specified periods. Carbon dioxide, air, nitrogen, nitrous oxide and the fluorocarbon refrigerants have preferred ratings on account of very low toxicity at normal refrigerating temperatures. Carbon dioxide, and Refrigerants 12, 22 and 502 have most acceptable ratings for marine service on account of their preferred Underwriter's ratings as having nontoxic, nonirritating properties on shipboard.

Flammability and Explosion Hazard.—The flammability and explosive hazard of many potential refrigerants is cause for rejection of many fluids of excellent thermodynamic refrigerating properties. The particular hazard of consequence is the possibility of leaks occurring in the refrigeration system bringing about explosive concentrations of the flammable vapor with air.

Ammonia, methyl and ethyl chlorides will burn but are explosive only under unusual conditions.

Sulfur dioxide, carbon dioxide, nitrogen, nitrous oxide and the fluorocarbon refrigerants are nonflammable and nonexplosive.

The hydrocarbon group of refrigerants are highly flammable and explosive. They are useful in the petro-chemical industries where special safety provisions and trained management in the processing of hydrocarbons is a continuing enterprise.

Detection of the location of refrigerant leaks is a chemical reaction search. Fluorocarbon leaks can be traced with a halide torch. Methyl chloride leaks can be traced by adding 1 per cent acrolein to the refrigerant, then detection is by the escaping odor. Ammonia leaks are readily noted by their odor, and final location is improved by burning a sulfur candle and noting the ammonium sulfite smoke at the leak location. Sulfur dioxide leaks are most readily detected by sponges or cloths soaked in aqua ammonia (25%).

Refrigerant Odors.—Odors of refrigerants can be both an asset and a hazard. Odors of a refrigerant make it easy to detect leaks, but the same odors may contaminate foodstuffs in storage at all temperatures.

The fluorocarbons, carbon dioxide, nitrous oxide, air, and nitrogen rank very high as nearly odorless refrigerants. Sulfur dioxide and ammonia are compounds that are not only toxic to human beings, but must be carefully isolated because very small percentages give an odor to foods and stored products.

Compression Refrigeration System

The vapor compression refrigerating system is the most widely used method of obtaining the low temperatures necessary to carry out the process of the quick-freezing of foods. Several different refrigerants may be used for systems that vary in design features, but all fluids must adhere to the same basic principles (Fig. 2.5).

In the simple compression system, refrigerant liquid at condensing pressure is throttled through an expansion valve into the evaporator or "low side." The reduction in pressure takes place ideally without the addition or removal of heat. As the refrigerant at low pressure is evaporated, due to its latent heat of evaporation, the substance being cooled gives up its heat to the refrigerant and a reduction in temperature is obtained. The refrigerant vapor is compressed from the low pressure of the evaporator to a pressure with a corresponding saturation temperature that is a few degrees higher than the cooling water circulating in the condenser. Heat will flow from the high pressure refrigerant vapor to the cooling water and the effect is the condensation of the refrigerant vapor to its liquid state. The liquid refrigerant is again ready to repeat its cycle, starting at the high pressure side of the expansion valve. The compression is said to be dry if there is no liquid refrigerant in the vapor as it is drawn into the cylinder of the compressor and wet if the vapor taken in by the compressor does contain particles of liquid refrigerant in suspension.

In Fig. 2.6 a diagram of the compression refrigeration cycle is presented. Specifically, the refrigerant, as illustrated, leaves the condenser at its lowest energy or enthalpy level, then passes through the expansion

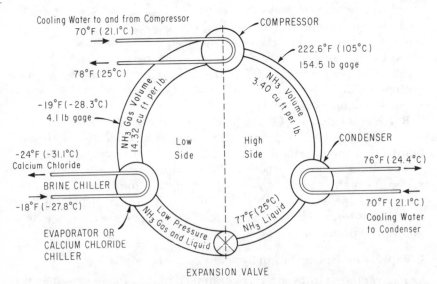

FIG. 2.5. THERMODYNAMIC CHARACTERISTICS OF A COMPLETE AMMONIA COMPRESSOR CYCLE FOR FREEZER WAREHOUSING

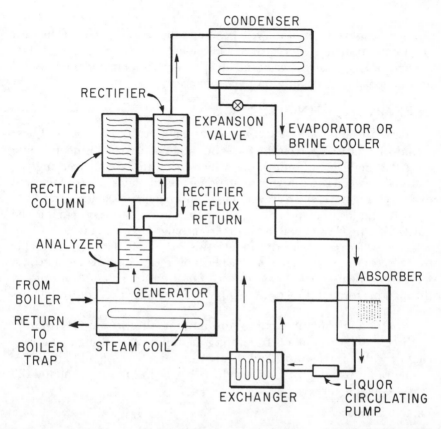

FIG. 2.6. THE AQUA-AMMONIA ABSORPTION THERMAL COMPRESSION REFRIGERATION SYSTEM WITH ITS ESSENTIAL PARTS

valve into the evaporator where it immediately begins to obtain heat from the foodstuff, water, air or other product until it is completely evaporated. In this evaporated form at or near the temperature of the evaporator, it enters the compressor as a low temperature and low pressure gas. The compressor raises both the pressure and the temperature to some superheated value of the refrigerant and delivers it to the condenser where it meets a low temperature produced by either available ambient air or condenser cooling water. This lowers the energy content of the refrigerant to the low level from which the cycle began.

Carnot and Reversed Carnot Cycles

The historical Carnot heat engine receives energy at a high level and converts some of it into work and rejects the remainder at a low level of temperature.

The Carnot refrigeration cycle performs the reverse effect of the heat engine and transfers energy from a low to a high temperature level. The reversed Carnot energy cycle is an ideal never to be obtained but is useful as a standard of refrigeration performance.

FREEZING TECHNOLOGY

There are a variety of techniques available to freeze foods. These include the following: freezing foods by immersion in cold brine; freezing on a flat metal plate underneath which brine flows; and freezing the material between two (moving or stationary) cold plates. The old method of air freezing has come back with many improvements, an important one being the rapid circulation of air, which then brought this system into the quick-freezing class. There are numerous types of air-blast freezers and, with the many recent improvements, this system of freezing is probably the most common one in use today. The most recent improvement in blast freezing is one in which the product is "fluidized" by a blast of air upward through perforated trays or a belt on which the unpackaged product has been placed.

All of the above methods may be grouped into three classes: (1) freezing by direct immersion in a refrigerating medium; (2) freezing by indirect contact with a refrigerant; and (3) freezing in a blast of cold air.

The rapid freezing of food products has been known for about 125 years, as noted earlier.

Early Systems

In 1861, Enoch Piper, of Camden, Maine, received a U.S. patent for the freezing and storage of fish. The fish were placed on racks in an insulated room; metal pans, containing an ice and salt mixture, were placed directly over the fish. However, at this early date quick-freezing and its value were not known, and Piper's patents did not indicate the intent of freezing by direct contact with a metal surface. It is interesting to note that this patent was held invalid by the Supreme Court of the United States, which took judicial notice of the well-known ice cream freezer as anticipating Piper's method.

In 1869, W. Davis and a few years later, in 1875, W. and S. H. Davis of Detroit, Michigan, were granted patents for freezing fish. The fish were packed in metal pans with tight fitting covers and in that way there was contact on all sides. The pans were placed in an insulated bin with alternate layers of salt and ice, the recommended proportions being the eutectic mixture. Hence, the freezing was done at about −6° F. (−21° C.), and the resulting frozen product was undoubtedly better than any produced up to that time.

As the interest in the freezing of fish and meat increased, other patents appeared both in this country and abroad. Notable among these were Howell's U.S. patent in 1870, Thew's British patent in 1882 which considered the packing of food products in cans and applying a refrigerant to the outside of the cans, and Hesketh's and Marcet's British patent in 1889 which proposed the freezing of meats in cold brine, by direct or indirect immersion. Most all of the patents issued during the early period required the use of salt and ice for a freezing medium, although some mentioned direct and indirect immersion in brine.

During all of the above period of time and for some years following, slow freezing, or freezing in uncirculated air, rapidly became an important industry. Until about 50 years ago, little scientific thought had been given to methods of freezing and the effects of freezing as a means of preservation.

Interest in quick-freezing of foods, and the invention of improved processes and equipment for rapidly freezing foods picked up after World War I, first with the introduction of direct immersion processes of freezing fish; and then about 1927 with the invention of methods of freezing packaged foods by indirect contact with the refrigerant (e.g., the Birdseye "Double Belt" Froster).

The latter method was given a great boost by the recognition of the importance of proper packaging as a means of retarding loss of quality during freezing and subsequent storage.

Sharp Freezing

Since its inception in 1861, sharp freezing has been the most widely used method of freezing. It has been of tremendous value to the world, and the industry in the United States has grown to huge proportions. In general, it consists of placing products to be frozen in a very cold room, maintained at temperatures in the range of +5° to −20° F. (−15° to −29° C.).

Although the air within the room will circulate by convection, usually little or no provision is made for forced circulation. The relatively still air is a poor conductor of heat and foods placed in even these low temperatures are frozen comparatively slowly, many hours or even days being required before the products are completely solidified.

The first products to be sharp frozen were meat and butter: they were seldom protected from evaporation and consequent desiccation during freezing and storage. The freezing of fish followed and these, being smaller items, were glazed immediately after freezing. A considerable amount of boxed poultry has been and still is frozen by sharp freezing methods, although in present-day practice the freezer rooms are main-

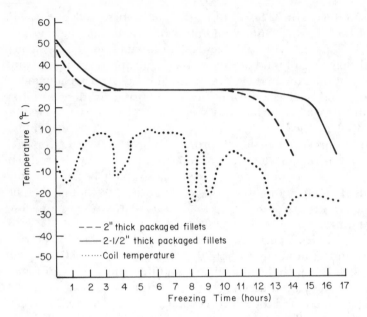

FIG. 2.7. RATE OF FREEZING PACKAGED FISH FILLETS IN A SHARP FREEZER

tained at $-10°$ F. to $-20°$ F. $(-23°$ to $-29°$ C.) or even lower, in contrast to the formerly used $0°$ F. $(-18°$ C.) temperature.

Sharp Freezers

Fundamentally, sharp freezers are cold storage rooms especially constructed to operate at and maintain low temperatures.

One type of sharp freezer room, which originated with fish freezing, has its cooling pipes so arranged as to make shelves on vertical centers of about ten inches. Galvanized iron sheets are often laid on the coils and the product is laid on these shelves. Ammonia is expanded through the coils to obtain the low coil temperature necessary for freezing.

In another type, the cooling pipes are arranged along the sidewalls and also suspended from the ceiling. In using this type of room, care must be taken in loading to be sure that sufficient space has been left between each container or group of containers to allow for air circulation.

More efficient air circulation may be obtained by the use of electric fans. However, this gentle air motion must not be confused with cold blast air freezing.

Sharp freezers in which the ammonia cooling pipes are arranged so as to form shelves (as described above) are often used for the freezing of round fish in pans or for fish fillets packaged in 5- and 10-pound boxes.

The time required for the freezing of packaged fillets is shown in Fig. 2.7 and 2.8.

Since it takes from 3 to 72 hr., depending on their bulk and particular methods and facilities employed, to freeze products in a sharp freezer, the term "slow freezing" has come to be applied to this method in comparison with the more advanced methods which are now termed "quick-freezing." The advantages of sharp freezing fish fillets are the following:

(1) The cost of freezing is lower than with the blast or plate freezer.

(2) The sharp freezer will give a fairly high output of frozen fish.

(3) A sharp freezer has a low maintenance cost.

The disadvantages of the sharp freezer are:

(1) It freezes products more slowly than blast, plate, or immersion freezers.

(2) Considerable handling of products is required.

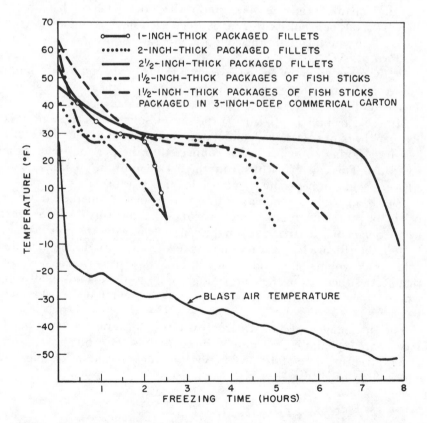

FIG. 2.8. RATE OF FREEZING PACKAGED FISH FILLETS AND FISH STICKS IN A TUNNEL-TYPE BLAST FREEZER IN AN AIR VELOCITY OF 500 TO 1,000 F.P.M.

(3) The coils must be defrosted at least once each six months.

(4) It requires that small packages of fish fillets have a weight on them to prevent them from bulging, owing to the expansion resulting from their being frozen.

(5) The loading and unloading of the products results in the coil frosting and increased freezing time.

The Continuous Conveyor-Type Sharp Freezer.—This type of freezer is similar to that previously described except for a continuous conveyor which moves the product through it. Such a freezer can be described as follows:

A freezer 110 ft. long, 29 ft. wide, and 10 ft. high has a capacity for 120,000 lb. of 10-lb. boxes of packaged fish fillets. The freezing is accomplished by 1¼ in. diameter steel coils, which are flooded with ammonia. The coils are arranged in 4 banks, 1 bank is placed against each wall, and 2 banks located in the center of the room. Each coil bank has 9 shelves (8 coils per shelf) with 7 in. between each shelf. Ammonia refrigeration is used to maintain the coils at temperatures of $-10°$ to $-20°$ F. $(-23°$ to $-29°$ C.).

A galvanized metal-mesh conveyor belt (12 in. wide and 335 ft. long) runs from the processing table through an opening in the freezer wall into the freezer between the bank of the wall and the center coils on one side of the freezer, around a 180° turn at the far end of the freezer, and back between the banks of wall coils and the center coils on the other side of the freezer. Another conveyor belt is connected to the previous one at the 180° turn. This conveyor runs from the back end of the freezer into an adjacent cold storage room to a truck loading platform.

The 10-lb. commercial boxes of fish are placed on the conveyor belt in the processing room. The conveyor carries the fish into the freezer, where one man loads the boxes on the coil shelves. After the products are frozen, the fish are loaded onto the conveyor, which carries them to the cold storage room. The products for shipment are placed on a conveyor that carries them from the cold storage room to the loading platform, where a portable conveyor is used to transport them into a refrigerated truck.

The advantages of this freezer over the conventional-type sharp freezer are (1) handling costs are reduced, (2) less defrosting is required, and (3) freezing time of the products is but little affected by loading and unloading of the freezer.

Air-Blast Freezing

Very cold air in rapid motion may also effect quick-freezing. To obtain very cold air, in very rapid motion, it is advisable to direct a blast of air through refrigerating coils, similar in principle to the unit type of heaters now being used for more rapid heating than is obtainable from a simple radiator. If still more rapid freezing is required, the cold air blast may be confined in an insulated tunnel.

In this method, the products to be frozen are placed on trays, either loose or in packages, and the trays are placed on freezing coils in a low-temperature room with cold air blowing over the product. In some installations of this system, the cold air that is in the low-temperature room is circulated by means of large fans, whereas in other installations the air is blown through refrigerated coils located either inside the room or in an adjoining blower room.

Tunnel freezing is possibly the most commonly used freezing system. Each packer of frozen foods using this system has his own ideas and little improvements, but in almost all cases a tunnel freezer is substantially a system in which a long, slow moving mesh belt passes through a tunnel or enclosure containing very cold air in motion. The speed of the belt is variable according to the time necessary to freeze the product. In some types of tunnel freezing, the products to be frozen are placed in wire mesh trays and loaded onto racks. The tray racks are passed into, and later pushed out of, the freezing tunnel. This method of handling is sometimes called the "tray and truck" system. Usually the cold air is introduced into the tunnel at the opposite end from the one where the product to be frozen enters, that is, the air flow is usually counter to the direction of the flow of the product. The temperature of the air is usually between 0° F. and −30° F. (−18° and −34° C.), although lower temperatures are sometimes used. The air is usually cooled by blowing it over bunker coils before it enters the tunnel, although some tunnels are equipped with coils throughout their length. The air velocity varies according to the ideas of the packer; however, if rapid freezing is to be had, it is necessary to recirculate a rather large volume of the air in order to obtain a relatively small rise in the temperature of the air as it touches and leaves the product. Air has a very low specific heat and for that reason a large volume must be carefully distributed through the system. Some packers blow cold air on both the top and bottom of the entire length of the freezing belt and in that way obtain very good distribution of the cold air.

Air velocities ranging all the way from 100 ft. per min. up to 3500 ft. per min. have been reported, and it is difficult to establish any speed as having more or less common usage. Possibly 2500 ft. per min. may be considered a practical and economical air velocity at −20° F. (−29° C.). A 15-lb. block of fish fillets (approximate dimensions 2¼ in. × 13 in. × 17 in.) requires slightly over 7½ hr. to reach a product temperature of 0° F. in −20° F. (−29° C.) air at a velocity of 500 ft. per min. The time is reduced to 5½ hr.

when the velocity is increased to 2100 ft. per min. At a velocity of 3200 ft. per min., the freezing time is slightly less than four hours.

Tunnel freezing of unpackaged foods has two major drawbacks, one being the problem of the dehydration of the product during freezing, and the other, caused principally by the first, being the constant need of defrosting the equipment, resulting in considerable lost time.

If the freezing is divided into two or more stages, loss of moisture in the product will be greatly reduced. If large volumes of air of high relative humidity are used in the first stage, the product may be frozen without excessive drying. In the second or later stage, the temperature differences and also the vapor pressure difference are not as great; hence, the freezing air at this point has considerably less desiccating effect, although the use of large volumes will result in faster freezing.

Finnegan designed an excellent air-blast freezer that is described as a multi-stage tubular freezer. This system reduces the moisture loss usually associated with ordinary air-blast and tunnel freezers by maintaining a relatively small temperature difference between the refrigerant and the air in contact with the product being frozen and also by maintaining a high relative humidity in the recirculated air used for freezing.

The Frick Spiro-flex Automatic Continuous Freezer.—The Frick Spiro-flex automatic continuous freezer (Fig. 2.9) has several advantages over the ordinary tunnel freezer. It consists of a tiered spiral belt conveyor which moves the product (Fig. 2.10) through chilled air of controlled velocity, volume, and temperature in an insulated enclosure of simple

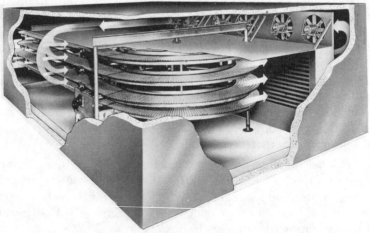

Courtesy of Frick Co.

FIG. 2.9. FRICK SPIRO-FLEX AUTOMATIC CONTINUOUS FREEZER

Courtesy of Frick Co.

FIG. 2.10. PACKAGED PRODUCTS BEING FROZEN IN FRICK SPIRO-FLEX AUTOMATIC CONTINUOUS FREEZER

panel construction. This enclosure can be erected inside a building or it can be weather-proofed for outside installation on a separate foundation. The refrigeration is provided by a two-stage system precisely balanced for operating efficiency and economy. The refrigeration coils can be arranged for liquid recirculation or for flooded operation with hot gas defrosting. The air is circulated by powerful fans. The Spiro-flex freezer is compact and requires much less floor space per ton of product frozen per hour than a long tunnel freezer. Because of its shape and design, there is relatively little heat leakage into it. Spiro-flex freezers are available in four standard belt lengths of 300, 600, 625 and 1125 ft.

The Spiro-flex freezer is well suited for quick-freezing many packaged products which are not commonly frozen in a multi-plate freezer. These include pies, and other bakery products, dinners, etc. Moreover, it is equally well suited for the continuous air-blast freezing of almost any packaged product.

The Gulfreeze Freezer.—The Gulf Machinery Co., of Safety Harbor, Florida, offers a type of air-blast freezer designed for the purpose of saving floor space. Although this freezer (Fig. 2.11) is only 13 ft. high and occupies a floor space of 9 × 27 ft., it has a conveyor belt one foot wide and 650 ft. long of which all but about 50 ft. is available for carrying the load of packaged products through a sub-zero air blast. The belt is run in a multi-level spiral fashion. This is made possible by the lateral flexibility of the "Omni-flex" belting used. This belt has the vertical flexibility normally associated with flat wire belts and in addition can turn right or left in a horizontal plane.

The Greer Multi-Tray Air-Blast Freezer.—A Greer "Multi-tray" air-

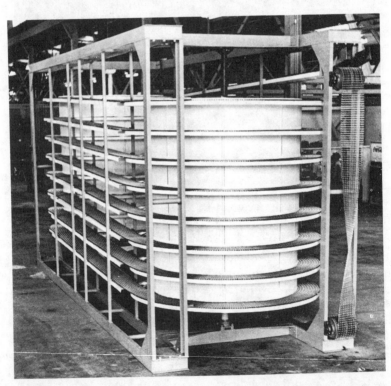

Courtesy of Gulf Machinery Co.

FIG. 2.11. GULFREEZE CONVEYOR BEFORE INSTALLATION OF REFRIGERATION AND INSULATED ENCLOSURE

blast freezer is used in large plants where a variety of precooked foods is frozen (including beef pies, dinners, casseroles, and fruit pies). This freezer provides both flexibility and efficient processing of several kinds of precooked frozen foods simultaneously.

The operation of this freezer can be described as follows:

Pie and casserole items are produced simultaneously and conveyed to an overhead −40° F. (−40° C.) freezing tunnel (Fig. 2.12) built into an adajcent −10° F. (−23° C.) holding room. This is accomplished by a Spivey elevating conveyor system with 6–8° slope. Conveyor sections from each of the product lines converge into a double-tier conveyor leading to the freezer. A similar double-tier discharge section conveys frozen goods back to the preparation area for cartoning and casing. Cases then go to the −10° F. (−23° C.) holding room for storage.

Tunnel-Freezer Operation.—The continuous freezer provides a 90-min. hold time for pies, casseroles and dinners. Due to product weight variation, however, there is a different production rate and hold capacity for each item:

Item	Frozen/Min.	Hold
Beef pies	72	6475
Dinners	51	4600
Casseroles	85	7650
Fruit pies	55	4950

To accommodate these various freezing rates, the unit has a special product-grouping system which is housed in an anteroom on the double-infeed conveyor and are delivered to an upper or lower tier of the grouping or metering conveyor. It groups a specified number of product units, as determined by freezing rate, and feeds the group single-file to the freezer's loading conveyor.

From the freezer's loading conveyor, a ram transfers the product group to a tray unit. This operation is sequenced by limit switches. As each tray is loaded in succession, it is progressively indexed upward, then pushed one tray space down the length of the freezer. These actions simultaneously cause a tray unit at the other end of the freezer to descend to the lower level, then to advance one tray space in the return direction.

A horizontal push-bar unit, the Greer tunnel freezer handles about 4,500 lb. of product per hr. Mounting this unit overhead in the 12,000 sq. ft. holding room saved some 1,875 sq. ft. of valuable floor space. The tunnel freezes to a core temperature of 0° F. (−18° C.) which requires about 120 tons of refrigeration. It is supplied by a Freon-12 system with −37° F. (−38° C.) evaporating temperature.

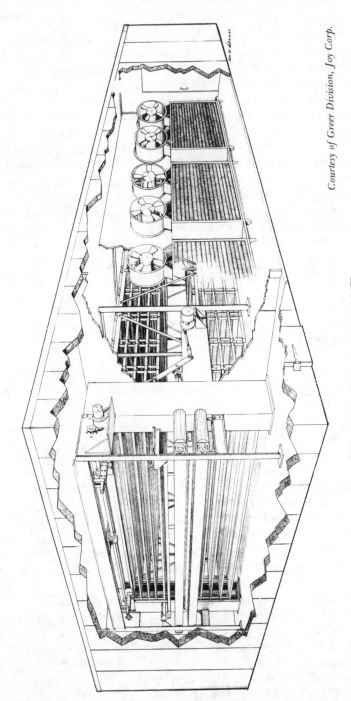

FIG. 2.12. GREER MULTI-TRAY AIR-BLAST FREEZER

Fluidizing Belt Freezers

Because of the great demand for IQF frozen vegetables, especially peas and lima beans, for bulk sale and for packaging for sale at retail (e.g., "boil-in-the-bag"), fluidizing belt freezers are extensively used. These operate on the same principle as fluidized bed driers in that a sufficiently powerful stream of air is forced up through a layer of the food to keep the product in suspension. In the case of the fluidized bed drier, the air is hot and dry, but in the fluidized bed freezer the air stream must be very cold.

Since each piece of food is kept loose and free flowing by air pressure, the freezing time is far less than that required for single plate, double plate and the usual air-blast freezing. For example, freezing time for peas and whole kernel corn is only about four minutes.

The Lewis Refrigeration Co. of Woodinville, Wash., and Frigoscandia, of Helsingborg, Sweden, have been leaders in the design and manufacture of equipment of fluidization freezing of foods. The Lewis fluidized loose freezing system was developed by that Company in 1961. Because of the relatively short time required to freeze small vegetables such as peas, lima beans, and whole grain corn, the mesh conveyor belt used is only about one-third as long as the belts used in conventional tunnels.

Products which fluidize easily and freeze in 3 to 5 min. (whole grain corn, peas, and lima beans) generally are loaded to a depth of 1¼ in. Green beans which are only partially fluidized are loaded to depths of 3 to 5 in. Larger products which are not fluidized at all may be loaded to a depth of 8 or even 10 in.

All sections include a combination air plenum and defrost water basin, which also acts as the main support for evaporator coils, conveyor belt tracks, and air guides. Two special high-pressure fans are mounted on the front of each section over access doors. High-pressure air blown into the base section goes up through the evaporator coils and through the conveyor belt where product fluidizing takes place.

The conveyor mesh belt is available in stainless or galvanized steel. The electric drive is a roller chain with variable speed push button control. This mesh belt is carried outside the main freezing chamber on each revolution. When outside the chamber it is subjected to automatic belt washing and drying. In some cases, continuous washing with special sanitizing agents may be used. This cleaning and drying is carried out without disturbing the continuous freezing operation.

Freez-Pak Fluidized Belt Freezer.—The Frick Company has developed an enclosed fluidized belt freezer called the *Freez-Pak*. It is a sectional freezer constructed in compact sections 10 ft. long by 10 ft. wide which permits flexibility in design, thus obtaining any desired capacity (Fig. 2.13). In operation, product is usually washed, dewatered, and

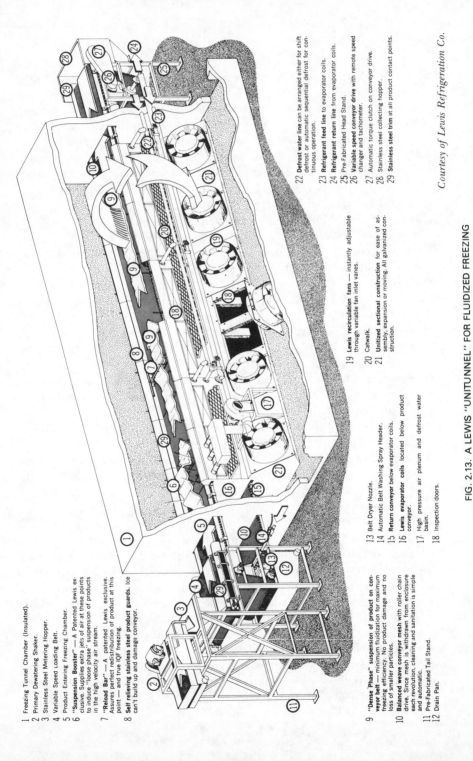

Courtesy of Lewis Refrigeration Co.

FIG. 2.13. A LEWIS "UNITUNNEL" FOR FLUIDIZED FREEZING

1 Freezing Tunnel Chamber (Insulated).
2 Primary Dewatering Shaker.
3 Stainless Steel Metering Hopper.
4 Variable Speed Loading Belt.
5 Product Entering Freezing Chamber.
6 "Suspension Booster" — A Patented Lewis exclusive. Supplies extra jets of air at these points to induce "loose phase" suspension of products in the high velocity air stream.
7 "Reload Bar" — A patented Lewis exclusive. Assures perfect redistribution of product at this point — and true IQF freezing.
8 Self relieving stainless steel product guards. Ice can't build up and damage conveyor.
9 "Dense Phase" suspension of product on conveyor belt — minimum fluidization for maximum freezing efficiency. No product damage and no loss of smaller particles.
10 Balanced weave conveyor mesh with roller chain drive. Since mesh is withdrawn from enclosure each revolution, cleaning and sanitation is simple and automatic.
11 Pre-Fabricated Tail Stand.
12 Drain Pan.

13 Belt Dryer Nozzle.
14 Automatic Belt Washing Spray Header.
15 Return conveyor below evaporator coils.
16 Lewis evaporator coils located below product conveyor.
17 High pressure air plenum and defrost water basin.
18 Inspection doors.

19 Lewis recirculation fans — instantly adjustable through variable fan inlet vanes.
20 Catwalk.
21 Unitized sectional construction for ease of assembly, expansion or moving. All galvanized construction.

22 Defrost water line can be arranged either for shift defrost or automatic sequential defrost for continuous operation.
23 Refrigerant feed line to evaporator coils.
24 Refrigerant return line from evaporator coils.
25 Pre-Fabricated Head Stand.
26 Variable speed conveyor drive with remote speed changer and tachometer.
27 Automatic torque clutch on conveyor drive.
28 Stainless steel collecting hopper.
29 Stainless steel trim at all product contact points.

distributed by a shaker-spreader from the hopper to a continuous belt; loose wet product is glazed in the precooler by an updraft of air at about +15° F. (−10° C.). Freezing is accomplished with low temperature air moving evenly upward through the belt, fluidizing the product and obtaining a product free from clusters.

For continuous operation, additional coils and fans are placed in the tunnel which is compartmented. All coils and fans can be run at maximum load, and one coil-fan unit can be defrosted while the others carry the design load.

The product, e.g., peas, is fed through dewatering and loading equipment on to the conveyor freezing belt. The conveyor carries it through the refrigeration chamber, where the air entering beneath the conveyor (through the meshes of the belt) fluidizes the product and thus freezes it in a loosely free-flowing form. At the far end of the tunnel when the product is frozen, it is carried outside the main freezing chamber to the unloading area where it falls onto a terminal conveyor, usually a pneumatic handling system moving it directly to a bulk storage.

Lewis Fluidized Bed Freezer.—Lewis Refrigeration states that the Lewis fluidized bed freezer has successfully frozen (IQF) the following list of fruits, vegetables and fishery products (See Table 2.1):

Fruits and Berries.—Strawberries, youngberries, Thompson seedless grapes, sliced peaches, diced peaches, diced pears, apple slices, pineapple cubes, paw paw balls, sweet melon balls (cantaloupe), paw paw cubes, sweet melon cubes (cantaloupe), guavas, apricot halves, grapefruit sections, goldenberries, blueberries and cherries.

Vegetables.—Cut green beans, sliced green beans, cut corn, Brussels sprouts, green peas, diced carrots, diced blanched potatoes, French-fried potatoes, potato-tots, potato patties, sweet potato slices and dice, lima

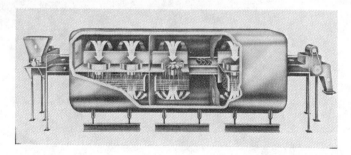

Courtesy of Frick Co.

FIG. 2.14. FREEZ-PAK AUTOMATIC BELT FREEZER. BY HIGH VELOCITY UP-DRAFT THROUGH THE LENGTH OF THE ENTIRE BELT, FLUIDIZATION IS ATTAINED, THUS ELIMINATING CLUSTERS, AND PROVIDING A LOOSE-FROZEN PRODUCT

CourtesyLewis Refrigeration Co.

FIG. 2.15. HIGH-PRESSURE FANS WHICH FLUIDIZE THE PRODUCT WITH AIR AT A LOW TEMPERATURE THUS EFFECTING VERY RAPID FREEZING

Courtesy of Lewis Refrigeration Co.

FIG. 2.16. FLUIDIZED PEAS ON LEWIS IQF FREEZER BELT

TABLE 2.1

TIME REQUIRED FOR FREEZING VARIOUS PRODUCTS IN LEWIS IQF FLUIDIZED FREEZER

	Min.
Peas	4
Whole grain corn	4
Lima beans	4–5
Cut green beans	5–12
French-fried potatoes	8–12
Diced carrots	6
Sliced and whole tiny carrots	8–10
Blueberries	4–5
Strawberries	9–13
Shrimp, small	6–8
Shrimp, large	12–15
Fish sticks	15
Fish fillets	30

Source: Lewis Refrigeration Co.

beans, cut okra, whole okra, southern peas, onion rings, and diced squash.

Fish.—Shrimp peeled and deveined, green headless shrimp, breaded shrimp, cooked fish sticks, raw fish sticks, cooked fish portions, raw fish portions, fish fillets, small fish (whole), small lobster tails, and fish cakes.

Poultry.—Cut-up chicken.

Because of the rapid rate of freezing, there is little loss of moisture. This has two great advantages: (1) a greater yield of product (on the order of 2%) is obtained; and (2) the frozen vegetable retains its quality better during storage because a film of ice still remains on the exterior of the product. The latter is particularly important in vegetables stored in bulk.

Other advantages of fluidized bed freezers are the following: (1) low initial and installation costs; (2) portability and ease of expansion; (3) ease of control and sanitation; (4) standard interchangeable parts; and (5) use of corrosion resistant materials that make painting unnecessary.

A Portable Fluidizing Freezing System.—In 1964, the Lewis Refrigeration Co. introduced a mobile IQF freezer mounted on two trailers each 40 ft. long, 8 ft. wide and 13 ft. 4 in. high. One trailer contains the freezing tunnel in which 5,000 lb. per hr. of vegetables can be frozen. The other trailer contains the entire refrigeration plant required to operate the freezing equipment. There is a two-stage ammonia system, expansion valves, and evaporator coils.

The motors on the freezer trailer are operated through connecting cables which run to the electrical control center on the equipment trailer. Flexible ammonia, water and drain connections also run between the two trailers; these connections are demountable for mobilization. The use of these mobile units, which are designed to operate at any food processing plant, can convert a vegetable cannery into a freezing plant in a relatively short time.

Flo Freeze Freezer.—Frigoscandia has perfected the *Flo Freeze* freezer which also operates on the fluidization principle. In this freezer, the end at which the product enters is slightly higher than the outlet; consequently, the food to be frozen flows slowly through the length of the freezing section. The fluidized product flows at a steady rate from the infeed end along the stationary perforated freezing tray to the discharge end. No conveyor belt is needed because of the slight incline. As the cold air passes through the fluidized bed, it removes the heat from the product. As the product reaches the discharge end of the freezer, the completely frozen product is discharged piece by piece without breakage or cluster formation.

The moisture which condenses on the refrigeration coils is washed off by a continuous spray of glycol. The glycol that drips off the evaporator coils is collected in a drip pan. A pump recirculates the glycol. Some water-diluted glycol is pumped to a concentrator where it is heated in order to evaporate the moisture that it has picked up. The concentrated glycol passes through a heat exchanger in which it is cooled, after which it is sprayed on the refrigeration coils. This continuous defrost system permits hour after hour operation at full capacity without shut-down for defrosting or fall-off in capacity because of frost build-up on the evaporator.

The velocity of the air entering the perforated bottom of the trough is sufficient to fluidize the product being frozen; however, the product is not blown out of the trough because the top of the trough is flared outwardly, thus lowering the velocity of the air in the upper part of the trough.

Freezing by Direct Immersion

Freezing by direct immersion in low temperature brine was the beginning of quick-freezing. Since liquids are good heat conductors, a product can be frozen rapidly by direct immersion in low temperature liquids such as sodium chloride brine and sugar solutions.

The principles of direct immersion were applied early in this century in freezers invented by Hasketh and Marcet, Rouart, Rappeleye, Henderson, Hirch, Kyle, Newton, Mann, Fyers and Watkins, Goér de Herve, Bull, Paulson, Dahl, and Pique.

The direct immersion principle has been used for the freezing of packaged fruit juices and concentrates.

The earlier direct immersion equipment such as the Ottesen brine freezer, the Harden F. Taylor brine spray system, the Zarotschenzeff "fog" freezer, the R. B. Taylor immersion freezer (sometimes called the "T.V.A." freezer), the University of Texas Polyphase Quick-Freezing

System, also known as "Bartlett's Freezer," and the Fisher "National Continuous Individual Berry Freezer," once of commercial importance, are not in use today.

The advantages claimed for immersion freezing are as follows: There is a perfect contact between the refrigerating medium and the product, wherefore the rate of heat transfer is very high; fruits are frozen with a coating of syrup which keeps the color and flavor of the products while they are in storage; and the resulting frozen product is not a solid block because each piece is a separate unit.

The disadvantages are as follows: Sodium chloride brine is a good refrigerating medium, but it cannot be used on fruits; it is difficult to make a syrup that will not be viscous at a low enough temperature; refrigerating temperatures must be carefully controlled, as at a high temperature the medium will enter the product by osmosis and at a low temperature the medium may freeze solid; it is very difficult to maintain the medium at a definite constant concentration, and also very difficult to keep it free from dirt and contamination.

A refrigerating medium, to be suitable for immersion freezing, must be edible and capable of remaining unfrozen at 0° F. and slightly below. Solutions of sugars, glycerol, and sodium chloride are the ones formerly used. Because of difficulties with aqueous media, their use for products other than canned concentrates has been discontinued. In some European freezing operations, some of these difficulties have been avoided by the use of a sugar-water-ethyl alcohol solution as the refrigerant in which the food is immersed. Recently, immersion in liquid nitrogen has been found to freeze faster than any other method. Although costly, this method has many advantages and is described later in this chapter.

Freezing by Indirect Contact with Refrigerant

Indirect freezing may be defined as freezing by engaging the product with a metal surface which is cooled by the freezing brine or other refrigerating media. In principle, this is an old method of freezing, since it is used in the can-method of making artificial ice as well as in the well-known ice cream freezer. Included in this classification are those for the packing of fish or meats in cans and immersing these in brine (Petersen, or Birdseye) or in cells with hollow walls, cold brine being pumped into the spaces in the walls (Hesketh and Marcet, Cooke). In other methods, canvas bags (Hesketh and Marcet), rubber sheaths (Davis and Skuce) or metal tubes (Haslacher) are used. Metal plates cooled from below, either moving or stationary (Cooke) are used, or floating pans (Kolbe) are employed in other systems. Metal shelves (product covered) sprayed with brine (Hendron, Mathews, and Bloom) and metal shelves made of brine coils with an air-blast blowing across the top (Murphy) and Zarotschen-zeff's "Flexible Froster" consisting of superimposed flat elastic bags are used in other methods. A later method of Birdseye provides refrigerated plane surfaces in the form of moving metal belts, or stationary metal plates, between which the food in packages is frozen. Stone's method provides for refrigerated plates radiating from and revolving around an axis.

Other methods of indirect freezing between refrigerated plates include the Amerio, FMC Continuous Sliding Contact Plate Freezer, Knowles Automatic Package Freezer and the Patterson Continuous Plate Freezer. Freezing products in tin cans where the container acts as a separating metal surface was the method used by Finnegan, Sorber, the Dole Freze-Cel, and the FMC Continuous Can Freezer.

Multiplate Freezers.—*Birdseye Freezers.*—The Birdseye multiplate freezer (Fig. 2.17) consists of a number of superimposed hollow metal plates actuated by means of hydraulic pressure in such a manner that they may be opened to receive products between them and then closed on the product with any desired pressure. The entire freezing apparatus is enclosed in an insulated cabinet. The smaller machines, six stations and less, are self-contained and have compressor, compressor motor, condenser, hydraulic lift cylinder, hydraulic oil tank, and pressure pump located beneath the insulated freezing chamber. The larger machines require separate refrigerating systems.

The plates are made of rolled aluminum alloy and, although they may be considered hollow, actually they are provided with sinuous passages. The ammonia, Freon, or brine is circulated through these passages. From one end of each plate a rubber hose connects to a header which feeds the plates with the refrigerant. The other end of the plate is also connected by a rubber hose to another header which carries the gas (in an ammonia type) to a surge drum located on top of the machine. Both headers are connected to the surge drum as it also acts as an accumulator for the liquid refrigerant.

In the older, as well as the smaller types of machines, the plates are actuated by means of pantagraphs or lazy tongs, the hydraulically operated cylinder, located under the bottom plate, being the means for imparting motion to, and pressure on, the plates. The present day machines are so constructed that as pressure is applied on the under side of the first plate, it lifts its load until it meets the second plate, which in turn is raised with its load, and so on up.

Before the product to be frozen (usually in packages) is placed in the machine, the plates are cooled to the desired low temperature. After loading (Fig. 2.17), the hydraulic cylinder is raised to squeeze the product

Courtesy of General Foods Corp.

FIG. 2.17. LOADING A BIRDSEYE "MULTIPLATE" FROSTER

between the plates, the cabinet doors are closed and the product left therein until its temperature reaches about 0° F. (−18° C.). The freezing time varies with the thickness of the package as well as the nature of the product to be frozen. In general, two-inch packages of fish and meats can be completely frozen in less than 90 min. Fruits and vegetables require about two hours.

Usually sticks of wood, as long as the depth of the plate and slightly less in height than the height of the packages that are to be frozen, are placed on each side of the machine between each plate to prevent any excess pressure on the packages. In this way, sufficient pressure to obtain the desired results is exerted, yet excess pressure which might break the cartons is eliminated.

All sizes of multiplate machines are portable in that they can be placed on a truck or rail car and moved at will. However, only the six-station and the smaller units are considered really portable, since only these sizes are self-contained units, and hence require only electric current and cold water connections for their operation.

The Birdseye multiplate freezer has been designed primarily for the purpose of freezing products in the packages in which they are to be marketed, although individual bulk products may be handled equally well.

The very rapid increase in the retail sales and distribution of frozen foods has increased the demand for flat, well-formed, quality frozen packages that stack compactly in the retail cabinet and take up a minimum of space. This, in turn, has expanded the use and manufacture of controlled pressure-type multiplate freezers such as the Amerio contact plate freezer.

This versatile, portable freezer consists of a number of refrigerated, movable metal plates encased in a heavily insulated cabinet. The refrigerated plates are made of two parallel sheets of steel, seam welded together on all edges under vacuum. These plates are zinc coated on the outside to prevent rusting. Metal tubes, through which the refrigerant flows, are encased between the two sheets of steel. One end of the tube in each plate is connected by a hose (rubber in ammonia types, metal in Freon types) to a header which supplies the refrigerant; the other end of the tube is connected by a hose of the same material to the other header, which returns the refrigerant to the system.

A double-acting hydraulic cylinder mounted on top of the freezer and operated by an electrically driven hydraulic pump opens the plates, creating a space between them known as a freezing station. The product to be frozen is placed in metal trays and the trays are then loaded into the freezing stations. After the freezer is fully loaded, the hydraulic cylinder lowers the plates and pressure is applied to make proper contact with the product. Pressure on the packaged product is controlled by using spacer sticks the same height as the package being frozen; one spacer stick is used at each end of each plate. This assures a flat, uniform, frozen package at all times.

Amerio Freezers.—The Amerio contact plate freezers are built in sizes ranging from the Junior Model with six freezing stations 24 in. × 44½ in. to the Model C with 15 freezing stations 55 in. × 72½ in. (Fig. 2.18). These freezers are available in various types for use with Refrigerant-12 or -22, ammonia, or brine refrigerants.

The Junior Model with 6 freezing stations is self-contained. This model is equipped with either a Freon-12 or -22 condensing unit which is

Courtesy of Amerio Contact Plate Freezers, Inc.

FIG. 2.18. MANUALLY OPERATED AMERIO PLATE FREEZER

mounted on the frame beneath the freezing cabinet. The other models (B and C), being large production units, require a separate ammonia or Freon condensing unit. Both of these models are available with a varying number of freezing stations, ranging from 7 to 15. The number of freezing stations required is based upon (1) the desired output per day and (2) size of the package being frozen. The normal opening range is from 1 in. minimum to 3½ in. maximum. However, as the number of freezing stations increases, the opening range decreases; thus, the opening range for a 200-station freezer would be from 1 in. minimum to 2¼ in. maximum. The main difference between the Model B and C is the size of the freezing plates; the Model B freezing plate is 42 × 55 in., (accommodating 100 retail packages 5¼ × 4 in. per station), while the Model C freezing plate is 55 × 72½ in. (accommodating 132 retail packages 5¼ × 4 in. per station).

The ammonia-type freezers are available for use with ammonia fully flooded system or ammonia recirculating system. A surge drum (accumulator) and a float valve is mounted on top of each ammonia fully-flooded-type freezer. The surge drum holds the ammonia and the float valve controls the flow of the liquid ammonia refrigerant through the plates. In the ammonia recirculating system, the liquid ammonia is

Courtesy of Amerio Contact Plate Freezers, Inc.

FIG. 2.19. AMERIO AUTOMATIC CONTACT PLATE FREEZER

pumped through the plates and returned to an accumulator for recirculating through the system. The latter type of freezer can also be used for recirculating cold brine through the plates instead of ammonia. The ammonia or brine freezers can be loaded or unloaded from either front or rear of freezer; the Freon freezers can only be loaded and unloaded from the front.

The efficiency of the plate freezer is dependent upon the extent of contact between the plate, product being frozen, type and temperature of the product, and temperature of the refrigerant. For best results, packages should be well-filled, or slightly over-filled, with no slack space. Solid, compact products, such as meat or fish fillets, freeze more quickly than vegetables where the individual pieces are separated from each other by small air spaces. With a temperature of −28° F. (−33° C.) on the plates, a compact product in well-filled packages of 1 in. to 1½ in. thick will be frozen in 1 to 1½ hr.; 2 in. to 3 in. thick packages will require a proportionately longer time to freeze.

A recent development in frozen food processing is the portable, continuous, automatic double pressure plate freezer by Amerio (Fig. 2.19). The heart of this method is the Amerio automatic loading-freezing-unloading system with Amerio portable cabinet freezers with double contact plates for fast, quality freezing and uniform packages.

These freezers provide for continuous operation, are automatically loaded from a conveyor from the overwrap machines, frozen under pressure and automatically discharged for casing with elimination of manual labor. Each freezer is entirely portable and can be moved from one location to another, or incorporated in any floor plan that may be changed from time to time as there is no built-in construction. Maximum head room required is about 13 ft. Floor space occupied per freezer is approximately 7 × 10 ft.

Dole Freze-Cel.—The Dole *Freze-Cel* is a modern double contact freezer having a new type of plate with the Dole "thermo-film" feature which is said to increase freezing speed up 10%. Each of the "thermo-film" plates is made of two heavy gage sheet steel surfaces, affording perfect flatness on top and bottom and enclosing a coil of square steel tubing. The Dole *Freze-Cel* consists of two separate parts—the freezing mechanism itself is made up of the plates, supporting frame-work, liquid and suction headers with flexible connections to the plates, and hydraulic cylinder, all mounted in one assembly. The other part of the freezer is the insulated cabinet with doors on front and back to permit trays of products to be put in one side and removed at the other.

Advantages of Multiplate Freezers.—To obtain fast, efficient freezing, the following precautions should be taken: (1) Maintain the plates at a low temperature prior to loading the freezer. (2) Keep the freezer doors closed, except when loading or unloading, thereby preventing excessive frost from building up on the plates. (3) Use proper-sized spacers. (4) Package fish fillets properly, leaving as few voids as possible (see Fig. 2.20 and 2.21).

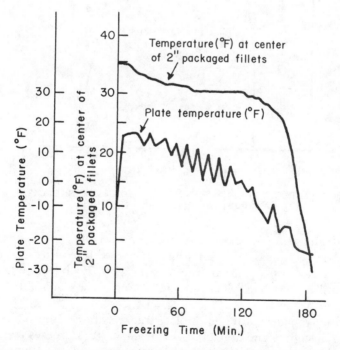

FIG. 2.20. FREEZING 2-INCH-THICK PACKAGED FISH FILLETS IN A MULTI-PLATE COMPRESSION FREEZER

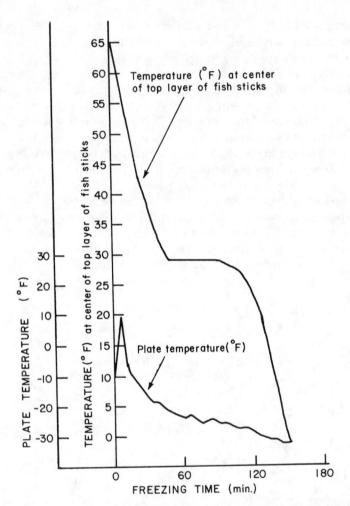

FIG. 2.21. FREEZING 1½ INCH-THICK (10 OZ.) PACKAGES OF FISH STICKS IN
A MULTIPLATE COMPRESSION FREEZER

The advantages of the multiplate freezer are: (1) It produces a uniform well-shaped package with a minimum of voids. (2) It requires a minimum amount of floor space. (3) It freezes packaged fish fillets quickly and economically. (4) It does not require defrosting of the plates if the freezer is operated properly.

The disadvantages of the multiplate freezer are: (1) It requires much handling in loading and unloading the products. (2) It freezes very slowly those products with a dead-air space in the package. (3) It requires large storage space for pans and spacers.

Single-Contact Freezers.—In addition to the several multiplate freezing machines described above, one must not neglect the many freezing systems using a single plate or single contact. In one of the very early fish freezers (Brooklyn Bridge Freezer), trays of fish were placed on refrigerated coils. This basic idea of a single plate freezer was later modified by Murphy by welding the brine coils and thereby obtaining a flat surface.

Another arrangement of "pipe shelves" is the Kold-Hold Serpentine Plates and Dole Vacuum Cold Plates. Many engineers consider these plates the most efficient cooling unit ever developed, and they are not only ideal as a quick-freezing surface but are also valuable cooling plates for refrigerated trucks and cold storage rooms. They are used in some locker plants. Dole plates are thin and compact and defrosting is merely a matter of brushing off the frost.

Cost Data for Plate Versus Bulk Freezing

The relative cost of freezing vegetables in a plate freezer, air-blast belt freezer, and fluidized bed freezer shown in Fig. 2.22 can be summarized as follows:

The lowest cost figures are generally for air-blast freezing of unpackaged materials, including turkeys which are usually tightly wrapped in plastic film. Typical costs are about a half-cent per pound where volume is large, season is long, and operation is efficient.

Figure 2.22 presents a summary of cost data supplied by one engineering firm. Costs are shown for freezing a vegetable like peas at a rate of 4000 to 5000 lb. an hour in packages in plate freezers and in bulk on belts and trays. In plate freezing, the higher costs related to fixed investments in the automatic system are more than offset by lower labor costs. The

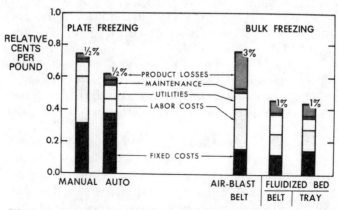

FIG. 2.22. SUMMARY OF RELATIVE COST DATA FOR FREEZING VEGETABLES IN DIFFERENT TYPES OF FREEZERS

automatic method is less costly than the manual. Both estimates allowed ½% product loss.

Costs can be reduced by freezing in bulk, on belts or trays, but at 3% evaporation loss on the ordinary belt freezer more than offsets the lower freezing cost. With only a 1% evaporation loss on the fluidized systems, total costs for both belt and trays are about 27% less than the automatic plate freezer.

While the savings can be stressed, actually these differences are very small, perhaps inconsequentially so, when other production and marketing factors are considered.

Similar conclusions concerning the cost of freezing turkey pieces and strawberries are shown later in this chapter.

Other Methods

Freezing Liquids.—*Ice Cream.*—Two common types of ice cream freezers are in common use today (1) batch freezers and (2) continuous freezers. The former are used principally as a counter freezer for soft-serve ice cream. Both types use direct expansion R. (Freon) 12 (or ammonia) as refrigerant.

The making and freezing of ice cream is a highly specialized field of food freezing adequately covered in books devoted exclusively to the subject (Arbuckle 1976) and so will not be discussed here.

Fruit Juices.—Large quantities of fruit juice concentrates are continuously slush frozen in a Votator. Slush-freezing and other methods of freezing juices are discussed in detail in Chapter 4.

Cryogenic Freezing

Cryogenic freezing may be defined as freezing at very low temperatures, e.g., below −75° F. (−60° C.). The refrigerants which can be used to produce these low temperatures were described earlier. The advantages of very rapid freezing of foods have long been recognized. Most foods give satisfactory products when "quick-frozen" by methods commonly used, viz. air-blast, double plate, and brine immersion, but a few products require ultra fast freezing in order to obtain a product of excellent quality. Some of these are mushrooms, sliced tomatoes, whole strawberries and raspberries.

Even if the high quality of the cryogenically frozen food is not considered, the ultra fast freezing processes have special advantages. One of these is that the equipment required is simple and usually not costly. Another is that most cryogenic freezers require less floor space than conventional food freezing equipment. For these two reasons, a cryogenic food freezing plant usually has a higher production for invested dollar than conventional quick-freezing plants.

Freezing with Nitrous Oxide.—The I. G. Farbenindustrie of Frankfurt, W. Germany, has been a leader in the study of immersion freezing in liquid nitrous oxide.

Shortly after the close of World War II, that company built a pilot plant with a capacity of one ton of food per hour in which the immersion freezing of many kinds of vegetables and fruits was studied. The nitrous oxide which boiled off during the freezing operation was reliquefied and used over and over again. At present prices, freezing using nitrous oxide is very costly unless provision is made for the collection and reliquefaction of the nitrous oxide vaporized when foods are frozen. This equipment is complicated and costly, and consequently this refrigerant has only been used for freezing foods on a pilot plant scale.

Freezing by immersion in liquid nitrous oxide has been found to require less than one-tenth as much time as that required for air-blast freezing. The quality of many foods frozen so rapidly was superior to those frozen by conventional methods.

There has been a report in the industry that a new, more economical means of recycling the nitrous oxide has been perfected. If this becomes practical, freezing with nitrous oxide may become of commercial importance because of the high latent heat of vaporization of this substance.

Freezing with Liquid Nitrogen.—Liquid nitrogen, a by-product of the manufacture of oxygen from air, is relatively cheap, and also is available in large quantities in many parts of the United States and other industrialized countries. The boiling point of liquid nitrogen is much lower than that of nitrous oxide (see Table 2.2), and consequently effects even faster freezing. It is nontoxic, but, of course, is suffocating. In most locations it is cheap enough that it does not pay to collect and reliquefy it.

Liquid nitrogen freezing systems are of three general types: (1) immersion; (2) a spray of liquid nitrogen; and (3) circulation of very cold

TABLE 2.2

PROPERTIES OF CERTAIN LIQUEFIED GASES USED FOR IMMERSION FREEZING OF FOODS

	Nitrous Oxide	Nitrogen
Boiling point (1 atm, °F.)	−127.237	−320.454
Latent heat of vaporization (at b.p., B.t.u./lb.)	161.78	85.67
Sensible heat (gas to 70°F.) B.t.u./lb.	40.00	96.99
Total heat to 70°F. (21°C.)	201.78	182.68
Liquid density at b.p. (lb./cu. ft.)	76.54	50.44
Gas density at 0° C., 1 atm (lb./cu. ft.)	0.1148	0.07
Specific volume, standard conditions (cu. ft./lb.)	8.711	13.80
Specific heat ratio (K) at 70°F., 1 atm., K = CP/CV	1.26	1.40
Specific heat, constant pressure at 70°F. (21°C.)	0.2095	0.2464
Specific heat, constant volume at 70°F. (21°C.)	0.1609	0.1774
Molecular weight	44.016	28.018
Color, odor	None	None

nitrogen vapors over the product to be frozen. At the present writing, more food is frozen by (2) a spray of liquid nitrogen than by other cryogenic methods.

Immersion in Liquid Nitrogen.—Much of the early work on cryogenic freezing employed immersion freezing. Whole strawberries were found to retain their shape after freezing by immersion in liquid nitrogen, storage at 0° F. (−18° C.) and thawing.

TABLE 2.3

THERMODYNAMIC CHARACTERISTICS OF NITROGEN

Temp.		Pressure-Enthalpy Table Volume			Enthalpy		
		Pressure,	Liquid,	Vapor,	Heat of Liquid	Latent Heat	Total Heat,
°F.	°C.	p.s.i.a.	cu. ft./lb.	cu. ft./lb.	B.t.u./lb.	B.t.u./lb.	B.t.u./lb.
−320.4	−195.78	14.7	0.01985	3.47	0.0	85.64	85.64
−315.0	−192.78	20.8	0.02018	2.51	2.57	84.02	86.59
−310.0	−190.00	27.5	0.02053	1.93	5.08	82.32	87.54
−305.0	−187.22	36.0	0.02090	1.50	7.61	80.51	88.13
−300.0	−184.44	46.5	0.02129	1.19	10.18	78.59	88.77
−295.0	−181.67	59.0	0.02171	0.94	12.78	76.54	89.32
−290.0	−178.89	73.9	0.02215	0.76	15.41	74.36	89.77
−285.0	−176.11	91.2	0.02263	0.62	18.07	72.02	90.09
−280.0	−173.33	111.3	0.02316	0.51	20.8	69.45	90.25
−275.0	−170.56	134.3	0.02326	0.42	23.63	65.88	90.20
−270.0	−167.78	160.7	0.02444	0.35	26.56	63.37	89.93
−265.0	−165.00	190.5	0.02520	0.29	29.59	59.85	89.44
−260.0	−162.22	224.1	0.02604	0.24	32.75	56.01	88.76
−255.0	−159.44	261.6	0.02700	0.20	36.14	51.64	87.82
−250.0	−156.67	304.0	0.02820	0.17	39.80	46.82	86.62
−245.0	−153.89	350.9	0.02997	0.135	43.80	40.96	84.77
−240.0	−151.11	402.8	0.03238	0.110	48.09	33.75	81.84
−235.0	−148.33	460.4	0.03623	0.083	53.87	23.78	77.65
−232.42	−146.89	492.9	0.05092	0.059	66.19	00.0	66.19

Processes Using Nitrogen Sprays.—Equipment and procedures for the freezing of foods with nitrogen sprays were perfected at the Illinois Institute of Technology. The so-called Cryotransfer process and equipment perfected at that institution is as follows:

The food to be frozen enters the chamber on a conveyor belt. The unit is insulated by a two-inch vacuum envelope to withstand −320° F. (−196° C.), the temperature of liquid nitrogen. The cylinder is open-ended; the supercold environment is maintained by high-velocity curtains of gaseous nitrogen at each end, as well as by the slight pressurization of the interior of the unit.

Liquid nitrogen is supplied to the Cryotransfer unit from an external storage tank to a sump adjacent to the freezing unit. Here, the liquid nitrogen (pressurized in the supply vessel) is brought to atmospheric pressure. A pump brings the nitrogen from the sump to a spray header,

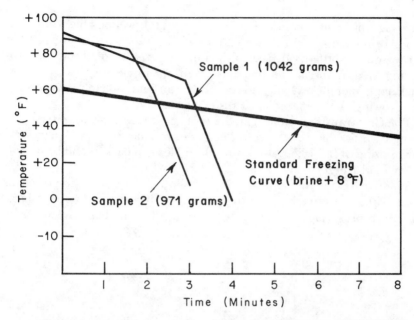

FIG. 2.23. RATE OF FREEZING OF CUT-UP CHICKEN BY IMMERSION IN LIQUID NITROGEN (SAMPLES NO. 1 AND NO. 2) VS. BRINE AT +8°F. (13°C.) (STANDARD FREEZING CURVE)

which is a series of flooding nozzles. These disperse droplets of nitrogen around the product as it passes on a conveyor, freezing the food and creating gaseous nitrogen. The gaseous nitrogen is collected and recirculated into the freezing area; some of it is also directed to both ends of the unit to maintain the gaseous curtains. A portion is also used to precool the product, when it first enters, to a temperature just above the freezing point or change in state of the product.

Over-all efficiency of the system is increased by utilizing the cold exhaust gas to perform secondary cooling functions, such as for maintaining a cold storage area, air conditioning a facility, or additional precooling of a food product.

The prototype unit at the Illinois Institute of Technology Research Institute freezes approximately 600 lb. of food per hr. Production models will have several times that capacity. For individual applications, custom units could easily be designed, but considerable flexibility would also exist for standard models. The operator would probably have a chart for the exposure time for different classes of foods. The operator would then set the Cryotransfer unit to a predetermined setting.

Possibly the greatest potential use of Cryotransfer lies in its mobility— its ability to be located *immediately at the food source*. Two trucks carrying the freezing unit, and a liquid nitrogen supply tank will be able to process foods being harvested in the field, obviating the necessity during the off-season for little-used freezing plants scattered throughout the producing area. Thus, by adding flash freezing to cleaning, sorting and packaging operations, fruit and vegetables can be completely processed at the growing site for rapidity of distribution.

The Cryotransfer system (Fig. 2.24) of spray freezing with nitrogen is being used for the freezing of baked goods by a large chainstore bakery, for freezing of fish fillets by a Boston Fish Pier fish packer, and by several other food freezers (Fig. 2.25, 2.26, 2.27, 2.28).

The Air Reduction Company has developed equipment for freezing foods either by immersion in or spraying with nitrogen by the "Magic Freeze" process (Fig. 2.27). Pilot Airco Spray Freezing Units are available

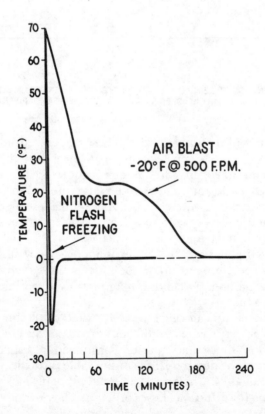

FIG. 2.24. COMPARISON OF RATE OF FREEZING OF 9 IN. APPLE PIES BY AIR BLAST AT −20° F. (29° C.) WITH NITROGEN FLASH FREEZING (CRYOTRANSFER PROCESS)

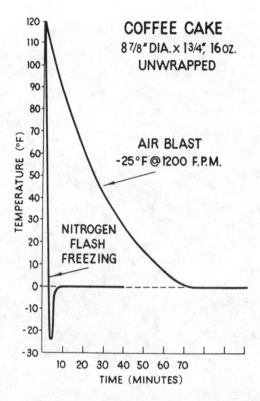

FIG. 2.25. COMPARISON OF RATE OF FREEZING OF COFFEE CAKE (8⅞ IN. DIAM. × 1¾ in., 16 OZ., UNWRAPPED BY AIR BLAST AT −25°F., AND NITROGEN FLASH FREEZING (CRYOTRANSFER PROCESS)

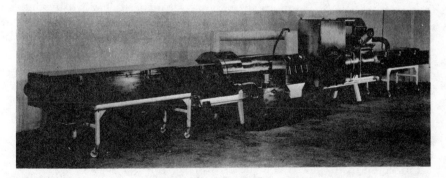

Courtesy of Liquid Carbonic Division of General Dynamics

FIG. 2.26. THE CRYOTRANSFER LIQUID NITROGEN FLASH FREEZER

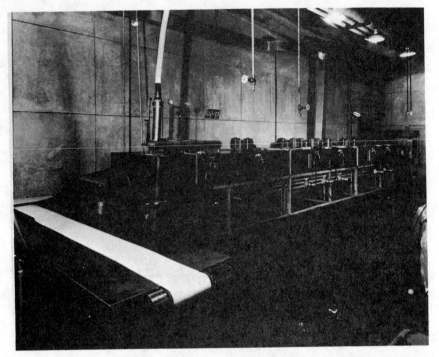

Courtesy of Airco Industrial Gases Division

FIG. 2.27. THE AIR REDUCTION MAGIC FREEZE NITROGEN FREEZER

in 2-ft. widths. The commercial equipment is made both in 4- and 6-ft. widths. Throughput can be controlled by the length of the tunnel and the speed of the belt. The units are of stainless steel construction with foamed-in-place polyurethane insulation four inches thick. Mushrooms, whole strawberries, sliced peaches, avocado halves, fish fillets and shrimp have been successfully frozen in "Magic-Freeze" equipment.

The Singleton Packing Corp. of Tampa, Fla., has perfected a continuous operation for spray freezing with nitrogen and glazing shrimp on a belt moving at a speed of about 3½ f.p.m. The entire unit is 7 ft. wide and 52 ft. long. The "tempering" or precooling is carried out in 5 modules each 4 ft. long. The freezing section has 8 such modules. Freezing is begun by circulating cold gaseous nitrogen in the "tempering" section and completed by a mist of liquid nitrogen from a series of spray nozzles.

Emerging from the freezer, the shrimp pass into a tank of cold water in which the shrimp pick up a coating of ice held to 2 oz. of ice per pound of shrimp by controlling the temperature of the entering shrimp and that of the water in the glazing tank and the residence time in the bath.

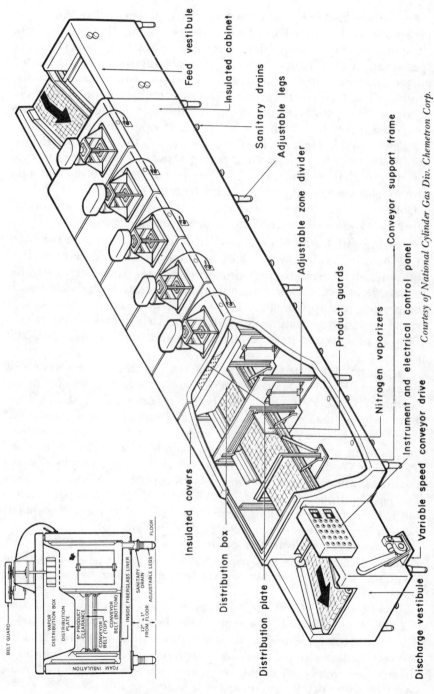

FIG. 2.28. THE NCG "ULTRA-FREEZE" NITROGEN FREEZER

Courtesy of National Cylinder Gas Div. Chemetron Corp.

Freezing with Nitrogen "Vapor".—In the Greer "Ultra-Freeze Jet Freezer" and the NCG Ultra-Freeze process, liquid nitrogen does not come in contact with the product. It is vaporized by the product heat in a special vaporizer. Circulation of cold nitrogen vapors, which are blasted on the product at high velocities in each zone, "wipes" heat from the product. The heat of the product controls the consumption of liquid nitrogen. The temperature and flow of the cold nitrogen vapors are controlled through each zone to achieve high operating efficiency.

As illustrated in Fig. 2.28, the freezer is a well-insulated tunnel divided into precooling, freezing and tempering zones. As the product is conveyed through these zones, it is quickly reduced from the input temperature to its freezing point, then quickly frozen.

Conveyor and blower speeds are variable, permitting flexibility of processing and a wide range of products with negligible changeover time.

Freezing with Carbon Dioxide.—Carbon dioxide is probably the safest of all the primary fluids which have been used, with the exception of air and water. Its thermodynamic properties are not particularly desirable for refrigeration except in cryogenic applications. Carbon dioxide gas is compressed to approximately 1,200 p.s.i.a. in two or more stages. A typical operation of compression to 1,200 p.s.i.a. would be a first stage to 100 p.s.i.a., a second stage to 300 p.s.i.a., and the final stage to 1,200 p.s.i.a.

Carbon dioxide is ued to freeze foods in two different ways. One of these sprays liquid CO_2 on the food. It consists essentially of a slowly rotating helical screw in a cylindrical housing, working upward at an angle from the horizontal. The CO_2 snow is formed as the carbon dioxide liquid under high pressure expands to dry ice snow at the spray nozzles. About one pound of snow is formed from two pounds of liquid carbon dioxide.

Freezing Foods with Dry Ice.—Since a given weight of solid carbon dioxide will absorb more than twice as many B.t.u. of heat in vaporizing as does liquid nitrogen (solid CO_2 246.3 B.t.u. vs. 85.67 B.t.u. per lb.), it is only natural that dry ice should be proposed for the cryogenic freezing of foods. The properties of dry ice are given in Table 2.4.

At atmospheric pressure carbon dioxide does not exist as a liquid. Dry ice may appear as a compact frozen solid or in the form of compressed CO_2 snow. The compressed snow is the form most generally known by the public.

Liquid nitrogen at 760 mm. pressure boils at a far lower temperature (−320.8° F. or −195.8° C.) than the sublimation point of dry ice (−109.3° F. or −78.5° C.). Nevertheless, if an ample amount of powdered dry ice is in direct contact with the food being frozen, very rapid freezing can be effected and unless nitrogen is very cheap, there will be a considerable saving by using dry ice to freeze foods instead of nitrogen.

TABLE 2.4
PROPERTIES OF SOLID CARBON DIOXIDE

Temp.		Pressure	Volume, cu. ft./lb.		Enthalpy, $-40°$ F.$(-40°C.)$, B.t.u./lb.		
			Solid or		Solid or Liquid,		
°F.	°C.	p.s.i.a.	Liquid	Vapor	°F.	°C.	Vapor
−145	−98.33	2.43	0.01005	32.40	−122.8	−86.01	128.5
−125	−87.22	6.98	0.01015	11.56	−117.4	−83.00	131.3
−109.4	−78.55	14.67	0.01025	5.69	−112.9	−80.51	133.4
−80	−62.22	50.70	0.01049	1.70	−102.3	−74.61	135.8
−69.9	−56.60	75.1	0.01059	1.16	−97.8	−72.01	136.0
−40	−40	145.87	0.01437	0.6113	0.0	−17.78	137.9
−30	−34.44	177.97	0.01465	0.5025	4.7	−15.25	138.3
−20	−28.89	215.02	0.01498	0.4165	9.2	−12.58	138.7
−10	−23.33	257.46	0.01533	0.3465	13.9	−10.16	138.9
0	−17.78	305.76	0.01571	0.2905	18.8	−7.43	138.9
10	−12.22	360.4	0.01614	0.2435	24.0	−4.44	138.8
20	−6.67	421.8	0.01662	0.2048	29.6	−1.33	138.5
30	−1.11	490.6	0.01719	0.1720	35.6	2.00	137.8
40	4.44	567.3	0.01786	0.1442	41.8	5.45	136.8
50	10.00	652.7	0.01867	0.1204	48.5	9.16	135.1
60	15.56	747.4	0.01970	0.0995	55.7	13.17	132.2
70	21.11	852.5	0.02109	0.0800	63.7	17.61	127.8
80	26.67	969.3	0.02370	0.0600	74.0	23.33	119.0
87.8	31.00	1072.1	0.03453	0.0345	97.1	36.17	97.1

Thermice Process.—The Thermice process (Figs. 2.29 and 2.30), used for freezing mushrooms and some other products, is very simple, involving merely the comminution of the cakes of dry ice and the mixing of the comminuted dry ice with the product being frozen in the interior of a slowly rotating, slightly inclined, insulated cylinder. The level of dry ice in the freezing cylinder is maintained by an automatic dry ice feeder which proportions it to the amount of product being frozen. By inclining the cylinder several degrees from the horizontal, gravity progressively moves the product forward. The action is gentle enough to permit the freezing of a fragile product such as whole mushrooms. Excess dry ice is separated from the product and returned for reuse.

The interior of a Thermice cylinder is hexagonal; as the freezing cylinder turns it effects good blending of the dry ice and the product being frozen. The intimate mixing of the fine dry ice at −110° F. (−79° C.) with the product brings about rapid freezing. The rotation of the barrel turning the product over and over in the dry ice also helps to break up any gas film on the product which might otherwise impede freezing. The freezing drums or cylinders are made in different sizes. A cylinder 12 in. in diameter and 10 ft. in length will have a capacity of approximately 1000 lb. per hr. A 24 in. cylinder of the same length will freeze 3 or 4 times as much.

Obviously, for a maximum heat exchange rate, a sufficiently high dry

Courtesy of Thermice Corp.

FIG. 2.29. THE THERMLFREEZE DRY ICE FREEZER SHOWING AUTOMATIC FEED GRINDING AND INPUT MECHANISMS. BLOCKS OF DRY ICE ARE PRELOADED ON A CONVEYOR WHICH IS PRESET TO DELIVER THE ICE INTO THE MACHINE TO BE GRANULATED AS REQUIRED TO MATCH THE FREEZING RATE

ice level is necessary to completely envelop the product as the barrel rotates. To control the temperature, or to accommodate production volume changes, the pitch of the cylinder (or the r.p.m.) may be changed; this changes the residence time of the product in the cylinder or "barrel" as it is sometimes called. The normal rotational speed of the freezing cylinder should be such that a gentle rolling action of the product in the dry ice takes place. If the rotation is too fast, centrifugal forces may carry the product up too high in the "barrel" with the possibility of breakage due to dropping. If the rotation is too slow, the capacity of the barrel is reduced.

When dry ice is used, it is important to make certain that all of it is separated from the frozen product before the latter is packaged. If pieces of dry ice remain in the product, the packages may explode because of the pressure of the gaseous CO_2 in the headspace.

Thermice equipment is very compact, requiring no more than about 30 sq. ft. of floor space to freeze up to 2 tons of product per hour. But space is also required for storage of an adequate supply of dry ice.

Courtesy of Thermice Corp.

FIG. 2.30. END VIEW OF THE THERMLFREEZE DRY ICE FREEZER. EXCESS DRY ICE IS SEPARATED FROM THE PRODUCT BY A SCREENING DEVICE AND RETURNED TO THE BARREL INLET THROUGH A POSITIVE FEED AUGER

The same equipment may be used to freeze food with snow from liquid carbon dioxide. When the snow is sprayed into the cylinder or barrel, it mixes with the product being frozen and freezes it rapidly.

Thermal Efficiency of Cryogenic Freezing of Foods.—Determination of thermal efficiency in a mechanical refrigeration system is relatively simple.

$$e = (T_1 - T_2)/T_1$$

expresses the ideal efficiency related to T_1, the low temperature maintained, and T_2, the high temperature at which heat is rejected. No such simple equation can be used to express the thermal efficiency of the cryogenic freezing process because in being vaporized from a liquid at $-320°$ F. $(-196°$ C.$)$ to a gas at $-320°$ F. $(-196°$ C.$)$, each pound of nitrogen absorbs 86 B.t.u. Each pound of gas at $-320°$ F. $(-196°$ C.$)$ absorbs another 80 B.t.u. in being heated to $0°$ F. $(-18°$ C.$)$, 90 B.t.u. in being heated to $40°$ F. $(4°$ C.$)$, according to the relationship:

Enthalpy (Ref. Liq. at NBP $= 85.9 + 0.249*$ (t $+ 320.5$) B.t.u./lb.

---

*Specific Heat at STP, Pressure at 14.7 p.s.i.a. and temperature $70°$ F. $(21°$ C.$)$.

Thus, 1 lb. of liquid nitrogen absorbs 176 B.t.u. in being heated to 40° F. (4° C.), enough to freeze 1 lb. to 2 lb. of food, depending upon the food itself. Thus it is essential to consider these factors in an overall expression of refrigeration system efficiency.

Liquid Nitrogen Consumption.—To estimate the liquid nitrogen consumption to freeze foods (pounds of liquid nitrogen per pound of product) consideration must be given to the following parameters: (1) initial and final temperature of the product; (2) specific heat, moisture and enrobing characteristics of the product; (3) cool-down requirements of the products; (4) length of time of production run; (5) rate of production; (6) steady state heat gain of the system; (7) temperature of the exhaust gas. As an example, calculations of nitrogen consumption per pound of shrimp follow.

Assumed Basis: 1,000 lb./hr. of shrimp (173,000 lb. of shrimp per month)

Initial shrimp temperature	40° F. (4° C.)
Final equilibrated temperature of glazed product	0° F. (−18° C.)
Glaze of 2 oz. water per lb. of shrimp meat, or	
0.125 lb. water per 1.0 lb. of shrimp meat.	
Glazing tank water temperature	33° F. (0.5° C.)
Shrimp moisture content	83%
Freezing point	28° F. (−2.2° C.)
Specific heat above freezing point	0.86
Specific heat below freezing point	0.45

Theoretical Refrigeration Available per Pound of Liquid Nitrogen

Temperature rise is from −320° F. to + 40° F. (−195.5° to + 4.4° C.)

Latent heat, $Q_1 = 86$ B.t.u./lb.

Sensible heat, $Q_s = 0.25 (40 + 320) = 90$ B.t.u./lb.

Total heat absorbed, $Q_t = 86 + 90 = 176$ B.t.u./lb.

Theoretical Refrigeration Required per Pound of Shrimp

Cool shrimp at 28° F.	$Q_1 = (1) (0.86) (40−28)$	=	10.3 B.t.u.
Freeze shrimp at 28° F.	$Q_2 = (1) (0.83) (144)$	=	119.6 B.t.u.
Cool shrimp to 0° F.	$Q_3 = (1) (0.45) (28−0)$	=	12.6 B.t.u.
Cool glaze to 32° F.	$Q_4 = (0.125) (1.0) (33−32)$	=	0.1 B.t.u.
Freeze glaze at 32° F.	$Q_5 = (0.125) (144)$	=	18.0 B.t.u.
Cool glaze to 0° F.	$Q_6 = (0.125) (0.48) (32−0)$	=	1.9 B.t.u.

Total required refrigeration $Q_t = 162.5$ B.t.u.

Theoretical Liquid Nitrogen Consumption

$C_t = 162.5/176 = 0.923$ lb. of liquid nitrogen/lb. shrimp meat

System Losses and Efficiency Factors

(a) Freezer efficiency is 80%; this includes items such as: (1) heat leak through the freezer walls; (2) exfiltration of cold nitrogen gas; (3) infiltration of warm room air (maintained at minimum) fan heat input losses.

(b) Storage tank filling losses are approximately 3% of the liquid nitrogen transferred (97% transfer efficiency).

(c) Storage tank heat loss is equivalent to 0.5% per day loss based on full tank capacity. For a 7,500 gal. tank the loss is 7,500 lb. of liquid nitrogen per month.

(d) Transfer losses from storage tank to freezer (25 ft.) are 375 B.t.u. per hour, which is equivalent to 2.13 lb. liquid nitrogen per hour of operation. For one shift, 5 days per week operation, this loss is equal to 370 lb. liquid nitrogen per month.

(e) Freezer cleaning requires that the equipment be warmed and then cooled to operating temperatures once a day. This cool down consumes approximately 4,200 lb. liquid nitrogen per month.

Calculated Liquid Nitrogen per Month

Theoretical liquid nitrogen consumption:

(173,000 lb. shrimp/month) (0.923 lb. liquid nitrogen/lb. shrimp)
= 160,000 lb. liquid nitrogen/month

Actual liquid nitrogen consumption:

(370 + 160,000)/0.80 + 7,550 + 4,200 = 212,213 lb. liquid nitrogen/month

Actual Liquid Nitrogen Consumption per lb. Shrimp:

$$C_a = 212,213/173,300 = 1.22 \text{ lb. of liquid nitrogen/lb. shrimp}$$

Summary of Liquid Nitrogen Consumption

Daily   = 10,100 lb. liquid nitrogen (140,000 cf. gas)
Monthly = 212,213 lb. liquid nitrogen (2,929,000 cf. gas)

There are available today several liquid nitrogen freezing tunnels of varying designs. These vary from (1) vacuum-jacketed high speed tunnels to (2) somewhat more complicated tunnels for obtaining (3) simpler, foam insulated tunnels. Each has claimed advantages over the other. Principal differences lie in investment cost and efficiency of utilization of liquid nitrogen.

The liquid nitrogen tunnel can also be used to advantage in conjunction with existing freezing equipment by producing, rapidly and relatively inexpensively, a crust-freeze on the food which works to reduce dehydration during the remainder of the freezing in conventional equipment. This can be accomplished at ⅓ to ½ the cost of complete freezing and further suffices to increase the productivity of the conventional freezer by as much as 25%.

Cost of Cryogenic Freezing.—*Liquid Nitrogen.*—There is little agreement among those who have calculated and compared the cost of freezing foods by conventional methods and by cryogenic methods (liquid nitrogen and dry ice). Many writers have indicated that freezing with liquid nitrogen costs several times as much as sharp freezing or quick-freezing by conventional methods. Others, who take into consideration the loss of weight (moisture) during freezing, find that there is little difference in the cost of quick-freezing and freezing in liquid nitrogen. This may be true if

TABLE 2.5

LIQUID NITROGEN FREEZING DATA ON LABORATORY QUANTITIES OF BAKERY PRODUCTS

Product	T_1, °F.	T_2, °F.	Freezing Time, min.	Unit Weight, lb.	Test Weight, lb.	Test Freezing Rate, lb./hr.	Liquid Nitrogen Consumption, lb. of Liquid Nitrogen/lb. of product
1. Bismark raspberry-filled, 4 in. diam. × 2 in.	65	0	1.0	0.125	72.71	485	0.767
2. Cinnamon streusel coffee-cake, 10 in. diam. × 1½ in.	50	0	1.0	0.945	115.56	358	0.963
3. Devil's food cream cake, 7 in. diam. × 3½ in.	56	0	3.4	1.62	79.38	198	1.02
4. Butter-filled strip coffee-cake, 12 in. × 6 in. × 1 in.	62	0	0.90	0.995	56.8	311	0.707
5. Cinnamon pull-apart coffee cake, 10 in. diam. × 1 in.	65	0	1.0	0.975	48.75	489	0.554
6. Whipped cream layer cake, 7 in. diam. × 3 in.	56	0	4.5	1.035	114.0	127	1.23
7. Apple strudel, 10½ in. × 4 in. × 1½ in.	63	0	1.15	1.025	60.5	303	0.795
8. Pecan coffee-cake, 9 in. diam. × 1½ in.	72	0	1.7	0.820	64	382	0.665

Note: All products except No. 1, 3, 4, and 6 were in uncovered aluminum pans.

TABLE 2.6

LIQUID NITROGEN FREEZING DATA ON PRODUCTION QUANTITIES OF BAKERY PRODUCTS

Product	T_1, °F.	T_2, °F.	Freezing Time, Min.	Unit Weight, lb.	Total Production Weight, lb.	Production Freezing Rate, lb./hr.	Liquid Nitrogen Consumption, lb. of Liquid Nitrogen/lb. of product
Chocolate cream square, 8 in. × 6 in. × 1 in.	89	0	4.75	0.75	4,550	2,640	1.09
Devil's food cake with icing, 8 in. × 7 in. × 1¼ in.	84	0	4.75	0.75	1,000	1,800	0.988
Chocolate brownie, 8 in. × 6 in. × 1 in.	86	0	5.7	0.812	500	2,150	1.01
Coffee cake, 10½ in. × 2 in.	107	0	5.1	0.75	500	2,160	0.892

Note: All products were in uncovered aluminum foil pans. Preliminary results.

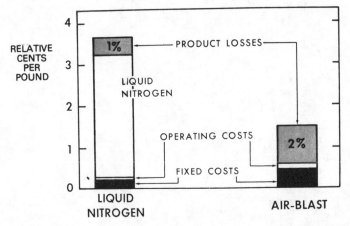

FIG. 2.31. COST COMPARISON FOR TURKEY PIECES; LIQUID NITROGEN VS.
AIR-BLAST

Product—cut-up breaded 4-oz. portions. Output—2,500 lb. an hr.
5,000,000 lb. a year. Dehydration losses assumed—2% in air-blast, 1%
in liquid nitrogen. Liquid nitrogen—1.12 lb. required per lb. product.

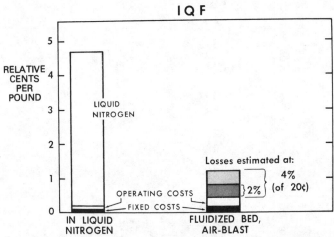

FIG. 2.32. SUMMARY OF COST DATA FOR FREEZING STRAWBERRIES LIQUID
NITROGEN VS. AIR BLAST IN FLUIDIZED BED FREEZER.

This comparison assumes liquid nitrogen at three cents per pound on a
use ratio of 1.5 lb. nitrogen per pound of strawberries. In spite of the
greater cost of operating the air-blast freezer, the only factors of significance in this comparison are the cost of liquid nitrogen and the evaporation losses in the air-blast operation. If freezing costs and evaporation
losses were the only two factors considered, the losses would have to
be over 20% with strawberries at 20 cents per lb. for the total costs to be
equal.

the foods are not packaged, but this would hardly be the case if the food were packaged in moisture-proof packages prior to freezing, and consequently no loss in weight during freezing occurred. Studies on the freezing of fish and shellfish with liquid nitrogen indicate that smaller ice crystals form and there is less protein concentration and consequently less loss of drip when the products are thawed. Concerning the amount of nitrogen required to freeze shrimp, if we assume the price of the nitrogen to be about 4 cents, the cost of nitrogen per pound of shrimp will be 5 cents.

Dry Ice.—Since dry ice may absorb approximately twice the amount of heat that liquid nitrogen does, the cost per pound of food frozen will often be much less when dry ice is used unless a very cheap source of nitrogen is available.

Additional data on the relative cost of freezing foods by cryogenic and conventional methods can be obtained from the Western Utilization Research and Development Division of the USDA, Albany, California. Summarized, their studies show that freezing with liquid nitrogen costs more than twice as much as freezing in an air blast and more than three times as much as freezing in a fluidized bed freezer.

ADDITIONAL READING

ASHRAE. 1974. Refrigeration Applications. American Society of Heating, Refrigeration and Air-conditioning Engineers, New York.

ASHRAE. 1975. Refrigeration Equipment. American Society of Heating, Refrigeration and Air-conditioning Engineers, New York.

ASHRAE. 1976. Refrigeration Systems. American Society of Heating, Refrigeration and Air-conditioning Engineers, New York.

ARBUCKLE, W. S. 1972. Ice Cream, 2nd Edition. AVI Publishing Co., Westport, Conn.

ARBUCKLE, W. S. 1976. Ice Cream Service Handbook. AVI Publishing Co., Westport, Conn.

HALL, C. W., FARRALL, A. W., and RIPPEN, A. L., 1971. Encyclopedia of Food Engineering. AVI Publishing Co., Westport, Conn.

HARPER, W. J., and HALL, C. W., 1976. Dairy Technology and Engineering. AVI Publishing Co., Westport, Conn.

HELDMAN, D.R. 1975. Food Process Engineering. AVI Publishing Co., Westport, Conn.

HENDERSON, S. M., and PERRY, R. L. 1976. Agricultural Process Engineering, 3rd Edition. AVI Publishing Co., Westport, Conn.

JOHNSON, A. H., and PETERSON, M. S. 1974. Encyclopedia of Food Technology. AVI Publishing Co., Westport, Conn.

PENTZER, W. T. 1973. Progress in Refrigeration Science and Technology, Vol. 1–4. AVI Publishing Co., Westport, Conn.

PETERSON, M. S., and JOHNSON, A. H. 1977. Encyclopedia of Food Science. AVI Publishing Co., Westport, Conn.

TRESSLER, D. K., VAN ARSDEL, W. B., and COPLEY, M. J. 1968. The Freezing Preservation of Foods, 4th Edition. Vol. 1. Refrigeration and Refrigeration Equipment. Vol. 3. Commercial Freezing Operations—Fresh Foods. Vol. 4. Freezing of Precooked and Prepared Foods. AVI Publishing Co., Westport, Conn.

UMLAUF, L. D. 1973. The frozen food industry in the United States—Its origin, development and future. American Frozen Food Institute, Washington, D.C.

WILLIAMS, E. W. 1976. Frozen foods in America. Quick Frozen Foods *17*, No. 5, 16–40.
WOOLRICH, W. R. 1965. Handbook of Refrigerating Engineering, 4th Edition. Vol. 1.
  Fundamentals. Vol. 2. Applications. AVI Publishing Co., Westport, Conn.
WOOLRICH, W. R., and HALLOWELL, E R. 1970. Cold and Freezer Storage Manual.
  AVI Publishing Co., Westport, Conn.

3

Freezing Vegetables

William C. Deitrich, Bernard Feinberg,
Robert L. Olson, T. L. Roth and Frank H. Winter

M any vegetables can be frozen to closely resemble garden-fresh products on the consumer's table. Retention of fresh quality is a principal objective of freezing. The fresh vegetable is the standard of comparison even though new generations are growing up who are only slightly familiar with fresh vegetables.

Technical knowledge exists to freeze many vegetables with a premium quality that is not always achieved in commercial operations. The compromise of some quality to cost considerations is realistically faced by industry. Complicated and costly hand operations are simplified and mechanized. Raw material specifications are flexible to avoid waste of good food. Maximum volume throughput in factories prevents careful attention to each piece as it passes over sorting and trimming belts in the production lines. However, in the *best* commercial packing practice, a small increment of further quality improvement is possible but may be obtained only at a higher cost. The marginal improvement that might be achieved would not always reach the consumer undiminished, because of uncertain and variable handling and cooking that vegetables may receive after they leave commercial hands.

In the home kitchen, a few extra minutes on the stove will cause losses of color and texture that overshadow the significance of added quality that processing improvements could achieve. Thus, correct handling and cooking instructions for consumers are at least as important as improved processes.

In this chapter we shall discuss many of the factors that influence the quality of frozen vegetables before they reach the consumer. The subject matter is diffuse and complicated because "vegetables" is a generic term including certain roots, bulbs, corms, stems, blossoms, buds, seeds, seed pods, fruit, etc. Yet, in spite of the fact that the types of plant tissues are so different and the number of frozen vegetables that are widely used so great, the material can be organized under broad terms.

Raw material is basic to the quality and character of frozen vegetables. Growing area, variety selection, cultural practices, and maturity at time of harvest each has specific effects on the retention of garden freshness. It is not within the scope of this chapter to cover these topics in detail, yet some general knowledge of the area is essential to an overall understanding of vegetable freezing. Suggested readings are listed at the end of this chapter.

In the relationship between the processor and the producer, it is the role of the processor's fieldmen to provide communication and make important decisions concerning raw material for processing. Therefore, something is also to be said here about fieldmen activities.

CROP PRODUCTION AND HARVEST

Growing area and variety selections for vegetables evolve on the basis of cost and quality competition. Year-to-year reliability of abundant harvest and freedom from insects and diseases determine the suitability of growing areas and selection of varieties. Plant breeders are consistently striving to develop and select better varieties that can extend the harvest season and the prime growing areas for production.

Vegetable varieties for freezing should be selected by judging processed products as well as yields per acre. Pea varieties that are best for canning are generally too tough when given the milder freezing process. Blue Lake snap beans, however, have been considered the standard for comparison for both canned and frozen products because texture can be controlled by blanching. Golden Cross Bantam corn is also considered near ideal for both canning and freezing as whole grain corn. However, other corn varieties can outproduce Golden Cross Bantam and there is a need for earlier and later varieties to lengthen corn harvest for processing and to give a good quality of corn-on-the-cob. Longer harvests spread the fixed manufacturing costs over a greater production and reduce unit costs.

Some prolongation of harvest is achieved by staggering the planting dates. Blossoming, seed set, and maturation are greatly dependent on day length and temperature, so it may become difficult to harvest each field at its best time when maturation is telescoped inward by irregular climate. An early cold spell or a late hot spell or both may foreshorten the differences achieved by staggered planting dates as separate fields approach harvest.

Varieties that are ideal for one area may not be as satisfactory for another. Henderson Bush lima beans require high temperature for maximum yields. This condition ruins yield and quality of Fordhook limas. Even over short distances, serving a single processing plant, localized soil and climatic conditions may require different corn varieties. On the other hand, some vegetable varieties grow in satisfactory yield and quality over wide areas (e.g., Processor green beans, Thomas Laxton and Dark Skin Perfection peas, Russet Burbank potatoes, and many others).

All variety testing should be continued for several years so a broader statistical base for judgment than a single year may be used. A successful commercial operation is a continuing one, and a single good or bad year for growing a particular variety should not be the basis for determining variety—an important factor to the success of a freezing operation.

There is no single source of complete information on varietal selection for freezing. The patterns of production, harvesting practice, and growing areas continually change. One must use information sources that are alert to such changes. Large processors develop much of their own information on varieties. Seed companies, who supply a large proportion of the seed for commercial production of freezing varieties, have research and development staffs to keep abreast of industry needs. The USDA and State Agricultural Experiment Stations and Extension Services conduct research and demonstration trials to provide a continuing flow of information on new variety developments. Results of public research and development are readily available to any who are interested. A competitive advantage is obtained by companies who operate their own research programs on new varieties.

When the time for harvest approaches, the plans for processing schedules are based on predictions of maturity for most vegetables. Processing can be efficient only if the operation remains steady at a high level once it starts for the season. It is no easy task to keep the lines full, but not overburdened, when the raw material is extremely variable, ideal maturity is fleeting, and the processed product has maximum value only if it consistently makes the top quality grade. Inevitably, some fields must be harvested too soon or too late for optimum maturity and maximum yield if the flow of processing is to remain reasonably constant during the entire season.

A system of "heat unit" accumulation based on time and temperature is used for some vegetables to approximate harvest schedules. The first estimate, based on knowledge of the vegetable's growth and maturation habits and the climatic records for the area, serves as a basis for planting.

As the plants grow and the climate varies from normal, adjustments are made in the estimates until harvest is very near at hand. More precise predictions are then necessary.

Instruments and tests have been developed to predict maturity so processing schedules can be planned as accurately as possible. Some of the instruments that measure maturity are also used to judge quality at the time of harvest as a basis for paying the grower for his crop. The Tenderometer is an instrument that has been used extensively to judge pea maturity. The Lee-Kramer shear-press has also been used extensively, and has been tested on a large number of vegetables. Both instruments have been used to test the texture of vegetables prior to optimum maturity as an aid in predicting ideal harvest dates. Other machines with various modifications to measure texture by resistance to crushing and shearing are available.

Various chemical and physical measurements that are related to the loss of starch and accumulation of sugars that accompany maturation of some vegetables have been used to evaluate maturity and predict suitable harvest dates—for example, refractive index of plant juice, alcohol-insoluble solids, specific gravity, etc.

Much of the judgment of maturity for some products remains highly subjective. An experienced fieldman can frequently feel and look at a crop and make wise decisions on harvest scheduling.

The fieldman of the processor is generally the principal contact between grower and processor. He negotiates contracts with growers, advises on cultural practices, determines harvest time and procedures, and is responsible for the adequate but not too large flow of raw material from the field to the plant. The grower wants to maximize his yield by harvesting as late as possible while still meeting premium quality specifications. The processor's production manager wants harvest to be scheduled so as to: (1) minimize processing costs (e.g., reduce trimming and sorting labor); (2) obtain a maximum pack of "A" Grade (e.g., reduce incidence of overmature and imperfect raw material); and (3) above all, keep a steady flow of raw material delivered to the plant that will utilize the maximum production capacity uniformly throughout the harvest season. Basic conflicts of interest exist between the grower and processor and they must usually be resolved by the fieldman.

Preprocessing Handling of Raw Material

Freezing preserves garden-fresh quality for the consumer. Garden-fresh quality to the highest degree is obtained when the time between harvest and cooking pot is only the time required for washing, cutting, sorting, or other handling that makes the vegetable ready for serving. Only those who raise vegetables within easy reach of the kitchen achieve this degree of freshness. It is never achieved for many kinds of vegetables in the channels of trade. The time required for hauling from fields, filling into containers, shipping to markets, delivering to stores, displaying for customers, and carrying home requires hours and days. The delays impose a great chance for loss of garden freshness.

Freezing only approaches retention of garden freshness because the mere operations of freezing and thawing cause changes. Freezing can arrest certain quality losses and essentially stop many of the deteriorative changes that occur in the handling and distribution of fresh produce.

Vegetables vary in their susceptibility to loss of freshness. Squash can be held for many days at ambient temperatures without obvious quality change. Potatoes are stored for months at low temperatures without serious deterioration. Potatoes will slowly lose some of their ascorbic acid during storage, and their content of reducing sugars may increase so as to sweeten their flavor and cause them to turn dark when they are fried. Asparagus loses sweetness and natural flavor in a short time after harvest. The tips become flabby and the butts woody. Corn also loses its sweetness in a matter of hours after harvest. Green peas are reasonably stable if they are plucked with intact pods. If refrigerated, peas that are not shelled retain fresh quality for a reasonable period of time. However, when peas are vined, either by a mobile viner or at a viner station, they are bruised and separated from their pods and become very susceptible to development of off-flavors and loss of fresh flavor.

In the interest of efficient plant operations, much quality can be lost. Plants in central California process Fordhook lima beans grown on the coastal plain. Kentucky processors use green snap beans grown in Florida. A pea or bean viner station, operating two hours away from the processing plant, can add an extra hour or two for a truck load to accumulate and another hour at the receiving platform to provide a surge collection of raw material to keep the lines running at a steady pace; better to hold the raw material a little longer it is thought, than to force inefficient plant operations. These, along with many other factors, result in lost garden freshness. How important such losses are depends upon the vegetable used, the variety that has been planted, and the type of harvester, as well as the determination and the ability of the entire staff responsible for operations from the field to the processing lines.

All steps possible should be taken to reduce the delays between harvest and processing and to reduce the effects of delays. Careful scheduling of harvests to minimize build-up of raw materials should be the rule and so should the precise dispatching of hauling trucks. Cooling vegetables by cold water, air blast, or ice will often reduce the rate of post-harvest quality losses sufficiently to provide extra hours of high-quality retention for transporting raw material considerable distances from the fields to the processing plant.

Root vegetables, squash, and melons generally can be handled with considerably less dispatch than the vegetables we have been discussing. Carrots can be stored for several weeks in cold storage. Potatoes in temperature-controlled warehouses are processed as long as eight months after harvest. However, bruising, mold and other fungus infection, and insect infestation may be intensified by prolonged storage.

PREPARATION OPERATIONS

Cleaning

Vegetables are prepared for freezing by the same general operations that prepare them for the table. Cleaning and washing remove field dirt, debris, and surface residues, and provide the first step in control of microbial contamination. Inedible parts are removed by trimming and, where appropriate, vegetables are cut, sorted, size-graded, or whatever is usual for the particular item. Mechanical devices are used where possible to reduce costs. The mechanical methods of washing vegetables for processing use high-pressure jet sprays and flotation and do a more thorough job than is usually done in the home kitchen.

Grading

Most vegetables for freezing are bought today under a contract agreement between the grower and processor. In the Northeast and Pacific regions of the United States, 86 and 93% respectively of the raw product supplies for freezing of fruits and vegetables are obtained by contracting. Such an agreement usually includes specifications which set maximum or minimum limits for such factors as size and shape, color, texture, density, blemishes, rot, insect infestation, etc. If the lot delivered falls below the specification requirements, the processor may either reject the load or assess a penalty; on the other hand, a bonus may be given for lots of exceptional quality.

An initial grading is usually performed at the time a load is received to determine compliance with the specifications. Representative samples are taken. They may be graded either by personnel of the plant or by a neutral third party.

Peas are frequently graded in the plant for maturity. For example, high-quality peas will float in a brine of 1.04 specific gravity, whereas substandard peas will sink in 1.07 specific gravity brine.

Material Handling

One of the biggest operations in the freezing plant is the physical moving of large quantities of vegetables, both before and after processing, from one part of the plant to another. This is done in a variety of ways: belts, air conveyors, flumes, screw conveyors, tote bins, etc. All of these have special advantages and disadvantages of cost, convenience, space, and sanitation. Moving vegetables by water in a flume or by pumping has special attractions to the freezer because it simultaneously moves, cools, and washes. Unfortunately, immersion in water can also result in leaching of sugars and flavors, as well as water disposal problems. In recent years, air conveyor systems have been advantageously used for moving such items as peas, corn, diced carrots, string beans, and lima beans. These are used both to convey the raw product from receiving bins into the plant and to carry both frozen and unfrozen product within the plant from one operation to another.

Peeling

Such vegetables as turnips, yams, potatoes, and carrots require peeling. There are three principal means of peeling: abrasive, steam, and lye-peeling, all of which are currently used for peeling white potatoes. Lye or steam peeling is used for carrots; flame-peeling is frequently used for peppers and onions.

Inspecting

Inspection and hand sorting of vegetables is a continuous process, beginning with the moment raw material is received at the plant and continuing until the processed product is ready for the final packaging operation. Mechanical harvesting has in recent years greatly increased the necessity for careful inspection in the plant. Sticks, stones, twigs, leaves, miscellaneous trash, etc. are always mixed in with the load from the field. Although metal-detecting devices and various ingenious sorting, screening, and vibrating equipment are used, a vigilant inspection and hand sorting is still required for all vegetables.

Blanching

Blanching consists of heating vegetables for a sufficient time to stabilize them for prolonged subsequent storage at subfreezing temperature. Blanching inactivates a portion of the enzymes and affects color and texture. It reduces microbial populations.

Enzymes are natural constituents of living material. They control chemical changes in metabolism of live tissue and in decay of dead tissue. Enzymes in raw frozen vegetables are responsible for undesired colors and flavors that developed during storage. Blanching has long been used to stabilize flavor and color of dehydrated vegetables.

There are many different enzymes, and their specific reactions that are responsible for flavor and color changes are not positively known. Most enzymes are inactivated rapidly as temperature rises to 180° F. (82° C.). Some enzymes persist at higher temperatures and are presumed to cause some of the chemical changes in vegetables during storage in a freezer. Vegetables are stabilized by heating and being held for a time ranging from a few minutes at 190° F. (88° C.) to a half-minute or so at 212° F. (100° C.). Catalase and peroxidase are two enzymes that resist heat inactivation and lose their reactivity in the range of importance for stabilizing frozen vegetables. These two enzymes have been widely used to tell whether or not blanching has been adequate. Specific chemical tests have been developed for each of these enzymes. When heat treatment is sufficient, the vegetables no longer give a positive test.

Neither catalase nor peroxidase has been specifically indicted as an initial causative agent of frozen vegetable deterioration. Therefore, their use in testing for adequacy of blanching is presumptive. Correlation of enzyme inactivation with quality stability under anticipated storage conditions had to be determined before the tests could be relied upon.

Peroxidase inactivation is very commonly used in determining the adequacy of a blanching operation.

Other enzyme tests, in addition to the chemical measurement of peroxidase and catalase, can be used in some cases. Brussels sprouts, which are difficult to blanch properly because individual sprouts differ in size and density, develop pink centers and off-flavor during storage if they have been underblanched. Pink centers in underblanched sprouts can be induced at the time of processing and thus used as a guide to adjustment of blanching to an adequate degree. The sprouts to be tested are cut in half and a dilute (1 to 3%) solution of hydrogen peroxide is applied to the cut surface. The appearance of pink discoloration in 1 to 5 min. indicates inadequate blanching.

A rapid, qualitative test for peroxidase is available in the form of a dry test paper containing an enzyme substrate and a dye indicator. The vegetable to be tested is placed in contact with the test paper so as to wet it with tissue fluids. Too much peroxidase is indicated by the development of a blue color in 1 to 15 sec. There is also a rapid method for peroxidase based on development of a color spot on a paper disk.

Underblanching of vegetables can cause changes that are even less desirable than the changes in freezer storage that occur with no blanching at all. A more moderate heat treatment than required to completely inactivate enzymes can disrupt vegetable tissues and cause enzymes and natural substrates to react. Instead of destroying most enzymes, the moderate heat may destroy some and increase activity of others, thus causing an imbalance and accelerating deterioration.

Large pieces of vegetables (e.g., Brussels sprouts, cauliflower florets, and broccoli spears) may require prolonged heating to inactivate enzymes clear to the center of each piece. Consequently, the outer surfaces may be overblanched while the center still contains active enzymes. To counter this problem in Brussels sprouts, a preliminary treatment in warm water is used to bring large sprouts to a temperature of about 125° F. (52° C.). Thus warmed, sprouts can be adequately blanched at high temperature without the excessive exposure to heat at the surface. Total exposure to the high temperature may be reduced by as much as 20% by this method.

Underblanching green snap beans can induce a toughness that cannot be overcome easily by subsequent cooking. Overblanching, on the other hand, can cause quality losses as well as impose extra, unneeded processing costs. The green pigment, chlorophyll, is slowly converted to the olive-green to brown pigment, pheophytin, when green vegetables are not living and respiring. The rate of conversion is very high at blanching temperatures, so some of the desired bright green color may be lost during blanching. Overblanching leads to unnecessary conversion of chlorophyll. Overblanching may also cause undue softening of the surface tissues. Surface tissues from the pods of green snap beans may slough if beans are overblanched.

Vegetables should be cooled immediately following blanching to control heat effects and to minimize changes in soluble and heat-labile nutrients.

Processing control to avoid under- and overblanching requires testing for adequacy of blanch and attention of the processor's quality control staff and production manager. Failure to maintain precise control may result in loss of grade and, consequently, loss of sales value of products.

There are some vegetables that can be frozen and stored without blanching. Onions, rhubarb, and green peppers do not seem to develop undesirable flavors and colors during storage even when they are frozen without blanching.

Blanching times recommended for various vegetables are included in this chapter; however, these guidelines must be used with caution. The blanching operation is necessarily a compromise between destruction of enzymes and undesirable changes in texture. A factor which has considerable influence on the time and temperature used for blanching is the final use to be made of the frozen product, which will affect the buyer's requirement. The "boil-in-the-bag" vegetable is a case in point. Whereas most frozen vegetables are placed in a saucepan with a little added water and boiled for periods ranging from 4 to 15 min. depending on the vegetable, boil-in-the-bag vegetables are heated only to a serving temperature, and little or no additional cooking takes place during the heating process. Therefore, most vegetables which go into the boil-in-the-bag package are essentially fully cooked before freezing. Frozen vegetables which are to be used for freeze-drying may require special blanching treatment and freezing techniques. Packers who assemble TV dinners will also have specific requirements for texture which will affect the blanching operation.

Methods for testing adequacy of blanch are described in reference. Although these tests are simple in themselves, interpretation of the results is as much an art as a science. Too often the complete inactivation of peroxidase has become accepted by some as the sole test for adequacy of blanch. For some vegetables, the blanch necessary to obtain a completely negative peroxidase test results in a loss of flavor, and such vegetables as snap beans and asparagus could show a positive peroxidase test and still be adequately blanched. Blanching remains an empirical procedure today. This is due in part to the common fault of overblanching. There is, however, considerable literature on blanching, and a more complete discussion of blanching can be found later in this chapter.

Blanching can continue after the vegetable has emerged from the blancher unless it is promptly and adequately cooled. However, although cooling in cold water flumes is an easy and convenient technique, excessive contact of the blanched vegetable with water could result in leaching losses of nutrients and flavor. Some processors halt the blanching operation by blowing cold air over the product or by spraying with cold water.

Manufacturers of food processing equipment offer a great variety of types and sizes of blanchers. However, freezers frequently design and build blanchers for their own individual requirements. Small freezing plants may be content with a batch-type blancher consisting of a vat of hot water into which vegetables are placed in a mesh basket and immersed for an appropriate interval. The continuous blanchers used in large plants come in several types. One common type consists of a wire mesh belt which carries the vegetable material either through a bath of hot water or through a steam chest or through a combination of both. Another blancher commonly used in the freezing industry is the direct-injection screw heater; the principal component of such a blancher is a helix or screw rotating in a trough (Fig. 3.1). The screw may be hollow or solid, the trough may be jacketed or plain. Steam is injected into the product compartment containing the vegetables, which are suspended in water. Still another kind of blancher consists of a long tube through which hot water flows, carrying the vegetables. This tube is usually folded back upon itself several times and provides considerable blanching capacity in a relatively small amount of space. The efficiency of steam use in blanchers varies with the amount of steam leakage and vent losses, as well as heat loss through blancher walls, and can range from 33% in units that are poorly designed and operated, to as high as 90% in well-insulated, well-designed, and properly operated units.

Vegetables are blanched either in steam or hot water. Where vegetables

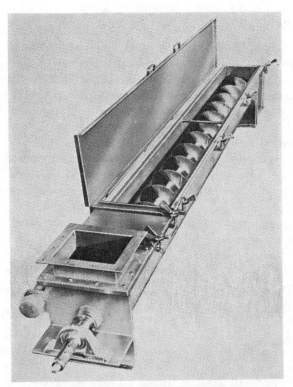

FIG. 3.1. ROTARY SCREW-TYPE BLANCHER

have a considerable quantity of cut surface, such as sliced rhubarb, and water blanching may result in leaching of soluble solids and flavors, steam is the preferred heating agent. On large vegetables, such as artichokes, where it is difficult to get complete penetration of steam, hot water blanching may be preferred. The term "steam" as used hereafter in this chapter will refer to steam at atmospheric pressure. The term "hot water" will mean a water temperature of 200° to 205° F. (93° to 96° C.)

So far, only heat has been used to inactivate enzymes for stabilizing frozen vegetables. Attempts to use chemical, enzyme-blocking reagents have not been successful. Studies have revealed the nature of enzymes and indicated lines of research that would elucidate their intricate reactions. Both hot water and steam blanchers are used. Water blanching seems to inactivate enzymes faster at a given temperature than steam blanching, although no theoretical basis has been developed to explain the experimental evidence. For large pieces of vegetables which require long blanching times that may cause product deterioration, heating by microwave energy transfer may offer some quality benefits to sustain extra processing costs. Microwave blanching is still in the experimental stage.

FREEZING METHODS

Retention of fresh quality is partly determined by the rate at which the vegetable is frozen. Some deteriorative changes take place rapidly in unfrozen vegetables, and the texture, in particular, may suffer pronounced damage if the freezing step itself is not completed very rapidly. Some institutional buyers of frozen vegetables therefore specify a maximum allowable time for the product temperature to fall to 0° F. (−18° C.) during the freezing.

Little opportunity should be allowed for the quality deterioration that can occur in the temperature range above freezing. Commercial freezing of vegetables is quite frequently done by blowing blasts of cold, −30° to −40° F. (−34° to −40° C.), air over unpackaged or compactly packaged products. Packaged products are also frozen in contact plate freezers in which heat is transferred from the package through cooling plates held tightly against the package surface so that heat transfer will be rapid. It is possible for larger packages of vegetables to be somewhat damaged in the course of freezing because the center of the large mass of product does not attain the desired low temperature quickly enough.

Results of study on freezing rates over a broad range, from freezing in still air at 0° F. (−18° C.) to freezing with dry ice, reveal that only asparagus was markedly improved in texture by very fast freezing rates. In 1938 Woodroof described the effects of freezing rate on ice crystal size and textural quality of certain frozen products. He pointed out that with very rapid freezing, vegetables suffer very little damage to tissue structures, but that slow freezing caused separation of cell walls and considerable tissue damage. Long since Woodroof's investigations, very rapid freezing (e.g., by use of liquid nitrogen, solid carbon dioxide, very low temperature air blast, and fluidized bed techniques) has come into the range of commercial feasibility. It was necessary to re-evaluate these pioneer investigations with a higher degree of precision and a more rigorous evaluation of quality effects as judged subjectively.

Recent research indicates that green snap beans can be frozen to give a product which, when cooked, has a texture closely resembling that of fresh beans if freezing to the center point is completed in seven minutes or less. Slower freezing resulted in tissue rupture and loss of the tender crispness that is found in fresh green beans. The experiments were done with liquid nitrogen to bring about rapid freezing. Liquid nitrogen freezing also resulted in better textured asparagus than could be obtained with conventional freezing methods.

It is not possible to freeze packaged materials as fast as unpackaged products. On the other hand, there are some disadvantages to freezing unpackaged products. Some vegetables may lose significant weight and value by surface drying in an air-blast freezer. Some are not well adapted to handling and packaging after they have been frozen. Broccoli and asparagus spears may be broken and cannot be packed compactly after they are frozen. The same is generally true for cauliflower and Brussels sprouts. Freezing spinach or squash before packaging does not appear to be commercially feasible. They could be frozen in forms adjusted to package size or large frozen blocks could be cut to fit a particular package size. Vegetables that must be packaged before freezing do not lend themselves well to liquid nitrogen freezing, although a liquid nitrogen spray over the packages would provide more rapid freezing than conventional air-blast freezing.

In general, the more rapid the freezing rate, the more costly the method; however, equipment costs for liquid nitrogen and solid carbon dioxide freezing are less than those for conventional mechanical refrigeration. Slightly over a pound of liquid nitrogen is required to freeze a pound of vegetables in the most efficient freezers yet developed, so costs must be relatively high.

There are some exceptional products that do not lose their natural crispness when they are frozen and thawed. Chinese water chestnuts, bamboo sprouts, and lily root (also used in Chinese cookery) retain a high degree of their original textural quality after cooking and freezing.

The processor's choice of which of the various freezing methods to use depends upon many factors: the kind of vegetable to be frozen, capital limitations, quality desired, whether material is to be bulk stored or not, maintenance, space, etc. In general, most large processors of items such as diced carrots, shelled corn, green peas, green lima beans, and cut green beans use either a belt freezer or fluidized bed. Large fragile vegetables such as asparagus spears, cauliflower, and broccoli, while still in a soft, limp condition, are hand packed (wet pack) in cartons and frozen in plate freezers. Contact plate freezing may be used with practically all types of frozen vegetable products. It is unfortunate that in some installations case freezing in the same cold room in which the vegetables are stored is still practiced. Since this kind of freezing takes 1 to 3 days it almost always results in poor texture quality. The relative merits of cryogenic freezing vs. cold-air freezing are still being debated and at the time of this writing it appears that such freezing methods as liquid nitrogen freezing will be limited to special high-cost items.

PACKAGING, SIZES, AND METHODS

Frozen vegetables have been packed in waxed or plastic-coated fiberboard cartons, either with or without a waxed or plastic-coated overwrap. In the early years of the industry a 10-oz. net weight was standard;

however, the consumer can now find frozen vegetables packed in net weights of 5½, 6, 7, 8, 9, 10, or 12 oz. Cartons are usually made of bleached sulfate or sulfite Fourdrinier board. With the development of a carton coated on both sides with extruded polyethylene and capable of being rapidly hot-air sealed by machine, the use of an overwrap to protect against moisture loss may be eliminated. In recent years bulk-packed frozen vegetables packed in heat-sealed 2.5 ml. (0.0025 in. thickness) polyethylene bags holding 1¼, 1½, 2 or 3 lb. have become popular (Table 3.1). Some of the newest packages for vegetables include a "heat 'N serve tray" in which cabbage wedges, broccoli spears, and cut leaf kale are packed in aluminum foil trays inserted in opaque polyethylene pouches. The tray is converted into a double boiler by placing in a deep skillet with ¾ in. of boiling water and steaming for several minutes, and is brought to the serving table in the same tray. A more complete review of packaging materials and packaging equipment for frozen foods may be found in Chapter 12.

A substantial percentage of frozen vegetables is packed in portable bulk bins holding from 1,000 to 2,000 lb. to be repackaged during the off-season. The bins are usually large polyethylene-lined fiberboard cartons, sometimes supported by a rectangular metal frame (Fig. 3.2). Polyethylene liners used for tote bins are 1½ ml. in thickness. Even larger quantities are stored in "silos" (wire mesh partitions lined with polyethylene). Bulk storage is advantageous since it gives the freezer an opportunity to keep his packaging lines running for a good part of the year. Only free flowing items such as peas, cut green beans, lima beans,

TABLE 3.1

TYPICAL FROZEN VEGETABLE PACK, BY SIZE OF CONTAINER

Container Size	Volume (Lb.)
10 oz.	504,106,466
12 oz.	43,113,704
Polyethylene bags	155,759,200
Boil-in-bag	47,624,023
Other retail sizes (1 lb. and under)	329,989,959
2 and 2½ lb.	387,805,163
4 and 5 lb.	603,012,230
Other small sizes (10 lb. and under)	74,055,658
30 lb.	6,186,171
50 lb.	101,776,823
Other large sizes (over 10 lb.)	276,545,584
Bulk	488,555,206
Total	3,018,530,187

Source: Compiled by National Association of Frozen Food Packers.

Courtesy of Marshburn Farms

FIG. 3.2. BULK STORAGE OF FROZEN VEGETABLES
Note cut-outs near bottom, for removal of samples.

diced carrots, and cut corn are bulk packed in this manner for packaging after freezing. Such loose silo-type storage of IQF vegetables has several advantages: more efficient use of storage space due to elimination of containers and roadways for forklifts; lower refrigeration requirements; and minimum dehydration in storage because air is not circulated in the bins. There is no air circulation in the room for this type of storage, and freezing temperatures are maintained by blowing cold air through air ducts surrounding the bin wall. Vegetables such as broccoli or asparagus are almost always frozen after packaging since they must be hand placed in the carton in a soft, limp condition.

Inadequately protected frozen vegetables will lose quality through surface desiccation, which can be unsightly and may advance to a stage where textural quality changes are noticeable even after the product has been cooked. An old practice of packaging vegetables with a lightweight, unmarked waxed paper for subsequent overwrapping with a heavier labeled waxed paper is giving way to greater use of bulk storage of frozen vegetables. Products taken from bulk storage may be packaged and labeled throughout the year to fill specific orders. Plastic film-lined tote bins that hold up to a ton are in common use. Some packers are using huge metal bins in their freezer storages. Such bins may hold 20 tons or more of frozen peas.

Bulk storage for frozen vegetables become necessary with their expanded use for remanufacture. Rather than guess at harvest time the inventory distribution required for a large number of different combination products, the components are stored in bulk. Specific orders are filled throughout the year with much less risk of misjudging sales predictions for each item. Inventory costs are less, too, because they do not include labor and materials costs of packaging.

During storage, frozen foods may lose surface moisture. Refrigeration coils are colder than the surrounding air, so moisture condenses on them. This lowers the relative humidity and causes moisture to evaporate from the food surfaces. The temperature differential among refrigeration coils, packages, and food products and the amount of airflow in the storage space affect the rate of moisture loss from the products. The resistance of containers to vapor passage is a protection against product moisture loss, although ice may accumulate on the inner surface of the package, signifying some loss from the frozen product.

Some frozen foods (especially fish and fried chicken) tend to oxidize during storage and have relatively short shelf-life at temperatures usually encountered in commercial practice. There is not much tendency for adequately blanched vegetables to lose quality by oxidation. Ascorbic acid oxidation is an exception to this statement, but its oxidation is of little significance unless temperatures are allowed to rise to unadvisable levels during storage. Early in the development of vegetable freezing, products were sometimes frozen in brine, which reduced oxidation. However, the added cost of freezing the brine overbalanced the slight improvement in resistance to deterioration that could be obtained. The current practice of evacuating boil-in-pouch vegetables reduces oxygen tension, and sauces added to these products offer a medium for adding antioxidants.

VARIETIES AND PREPARATION FOR FREEZING

Artichokes

Practically all of the artichokes frozen in the United States are grown in California in the coastal counties of San Mateo, Santa Cruz, Monterey, San Luis Obispo, and Santa Barbara. Monterey County alone accounts for more than half of the total artichoke acreage in California. The only variety grown commercially is the Globe. Artichokes are harvested by hand. The pickers wear heavy leather gauntlets to protect their hands from the stiff, spiny leaves and must use clippers to cut the artichoke buds from the plant. The "chokes," as they are usually called in the plant, are brought to the packing house in large trailers. Here they are sized by rope graders into large and small sizes. This size classification is necessary not only to obtain a uniform pack but to enable better trimming, since the trimming machines are set to handle only one size. Various sizes are collected in overhead bins and fed through a chute to a line of Hydrout coring machines. Here they are hand-positioned by women who place the stem end in the machine; this simultaneously removes the stem, and separates the outer leaves from the heart.

After trimming, the hearts are washed and elevated to a shaker screen which separates the loose leaves. This operation is followed by a visual inspection where damaged artichokes and foreign matter are removed. The artichoke hearts are then carefully trimmed to the desired size and conveyed to a blancher. Artichokes are very susceptible to a discoloration which is accelerated by contamination with iron; therefore stainless steel equipment, including the blancher and Hydrout knives, is essential. Water blanching is used almost exclusively. Citric acid is periodically added to the blanch water to maintain a concentration of approximately 0.5%. Blanching time will vary according to size and maturity but usually ranges from 5 to 9 min. The artichokes are quickly cooled in cold water, dewatered, packaged in suitable cartons, and frozen.

Asparagus

About half of the asparagus produced in the United States is grown in the peat and sandy loam soils of Central California. The crop there is harvested over a few months' period in early spring. The varieties of asparagus preferred for freezing in eastern states, such as New Jersey and Michigan, are Palmetto and Martha Washington; Mary Washington and University of California strains 500W, 711, 66, and 72, lead in the West. Although considerable effort has been devoted to designing a mechanical harvester for asparagus, practically all asparagus is still cut by hand. Asparagus is usually purchased on a 7- or 9-in. basis. The asparagus spears, customarily called "grass," are hauled from the field trimming-sorting stations in standard 50-lb. cannery lugs which are double cleated to prevent damage to the tender asparagus heads.

Since asparagus deteriorates rapidly after cutting, it should be processed promptly. If it must be held at the processing plant for any length of time it should be kept in a cold room maintained at about 34° F. (1° C.) and 90% R. H. When asparagus fields are located at some distance from the freezer, it is common practice to hydro-cool the cut asparagus spears immediately after cutting. Ice water containing about 5 p.p.m. chlorine is circulated over the asparagus and the temperature of the spears is rapidly reduced from field temperatures, which may be as high as 80° F. (27° C.), to 34° F. (1° C.).

Because asparagus is rather brittle and the fragile heads are subject to damage, considerable hand labor is necessary for cleaning, sorting, and packing. The first processing step consists of cutting the asparagus to uniform length. The spears are placed crosswise on a conveyor belt so that the tips are flush against a guard rail guide at the edge of the belt; care is taken to prevent overlapping and misalignment. The spears then pass under a rotating knife blade or past a band saw which cuts them into six-inch lengths. The butt ends are discarded and the spears are realigned. A second cut is made so that the spears are now in five-inch lengths. The one-inch end cut, which is rather fibrous, is frequently discarded but there is a small market for these as "center cuts." The spears are thoroughly washed to remove all silt and sand by first soaking in detergent and water, and then rinsing under powerful water sprays.

After washing, the spears are elevated on a drain belt to the sizing operation. Here they are graded into four sizes, either by hand sorting or by a mechanical size grader using spaced rollers. Commercial sizes are: pencil (less than ⅜ in. diam.), medium (⅜ to ⅝ in.), jumbo (⅝ to ⅞ in.), and colossal (⅞ in. and larger). Each size grade is diverted to its own specific flume of fresh water to be conveyed to a blancher. Pencil and medium sizes are usually blanched in one blancher; jumbo and colossal in another. Both steam blanching and water blanching have been successfully used, with blanch times ranging from 2 to 5 min., depending upon size and texture. As the asparagus spears emerge from the blancher they are quickly cooled and conveyed to sorting belts where broken, mishappen, and discolored spears not meeting specifications are removed.

The spears are hand placed into cartons; a special colossal pack is sometimes packed with half of the heads facing in one direction and half in the other to compensate for the tapered shape. Sort-outs are sliced and packed as "cuts." Some IQF jumbo size spears are used in a special pack for high-quality restaurants. A rapid freezing rate, such as is obtained by plate freezing of individual cartons, is recommended because asparagus is susceptible to texture deterioration sometimes known as "freeze rupture," when frozen at a slow rate (Fig. 3.3).

Beans, Green

The commercial freezing of green beans is divided among bush beans which are raised primarily in New York and the Midwest, and pole beans which are raised in Oregon and Washington. Although several varieties of bush beans are used, the only important pole beans are the high-quality Blue Lake variety or "Blue Lake type" varieties which are considered highly desirable for processing. Although pole beans have a yield of 9 to 10 tons per acre compared to the 1 to 4 tons usually obtained with bush beans, the labor cost of growing and harvesting pole beans is very high. In recent years because of the development of mechanical harvesters for bush beans (Fig. 3.4) and the lack of a similar harvester for pole beans, many acres formerly in pole beans in California and Oregon have switched to bush types. The principal bush type green bean varieties

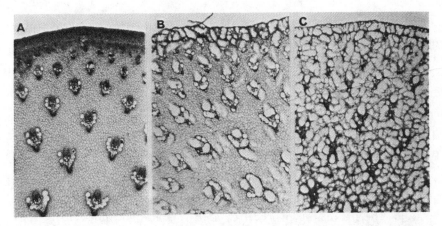

Sections and photos prepared by Dr. Milford Brown, Western Regional Research Laboratory, USDA, Albany, Calif.

FIG. 3.3. PHOTOMICROGRAPHS OF ASPARAGUS SECTIONS SHOWING DAMAGE TO NORMAL TISSUE STRUCTURE AT VARIOUS FREEZING RATES. MAGNIFICATION: × 180

FIG. 3.4. MECHANICAL HARVESTING OF BUSH BEANS

Bean pods are pulled from vines by rubber rollers or fingers

grown in the West are Tendercrop, Cascade, Gallatin Valley 50, Asgrow 274, and Cornelli 14. In some areas of California it has recently been demonstrated that it is possible to grow two crops of bush beans on the same field in one year with a total annual tonnage of up to nine tons per acre.

The Italian green bean, or Romano variety, has become increasingly important in recent years. These are grown primarily in California and the Pacific Northwest. Like Blue Lake green beans, Romano's are usually pole grown, but new bush varieties have been developed that can be machine harvested. Romano pods should be smooth, succulent, and well-filled without pronounced swelling of the seed cavity at the time of harvest. The bean width should be from ¾ to 1 in.

Unlike the harvesting procedures used for peas and lima beans, the green bean vine is not cut; instead the bean pods are pulled off the vine by a series of rubber rollers or fingers. This inevitably results in a certain amount of trash, stems, leaves, etc., mixed in with the bean pods and necessitates elaborate cleaning operations at the plant. Beans are mechanically transferred from the harvester into either large bins holding approximately 1,000 lb., small dump trucks, or trailers, and hauled to the plant. Here they are dumped onto a belt and fed into a shaker screen and a blower to remove field dirt and extraneous material (Fig. 3.5).

The processing of green beans prior to freezing is almost identical to the procedures used in the canning industry. Some beans as they come from the harvester are in bunches with several bean pods attached to one stem. The pods are separated in a cluster-cutter, a machine consisting of a reel and wire fingers which break up the clusters. After separation the beans are conveyed to a reel washer and then to size graders. Size grading is an elaborate operation requiring much machinery and space. Beans have been classified into six commercial sizes by the USDA; the dimension given is thickness in 64ths of an inch: size 1, less than 14½; size 2, 14½ but not including 18½; size 3, 18½ but not including 21; size 4, 21 but not including 24; size 5, 24 but not including 27; size 6, 27 or more. Sizes 1, 2, and 3 are frequently packed as whole beans; sizes 3, 4, and 5 are cross-cut; size 5 and larger are utilized as French style, with the beans sliced lengthwise into slivers.

After size grading, beans are fed into snippers which remove the stems and most blossom ends from the pods. (Some processors prefer to snip beans before size-grading). The pods then pass through an "unsnipped end remover" which separates those beans which have not been snipped and returns them to be rerun through the snipper. This eliminates hand sorting of beans or cuts with stems attached. Those beans to be used for cuts pass to mechanical cutters where they are cross-cut to lengths of 1 to

Courtesy of Commercial Manufacturing and Supply Co.

FIG. 3.5. SHAKER SCREEN AND BLOWER

Used for cleaning beans and peas. Heavy material, such as stones and dirt, falls through the shaker screen, while light trash is blown upward by strong air blast.

1½ in. Beans to be used as a boil-in-the-bag item are frequently cut diagonally at a 45° angle. The cut beans pass over a series of vibrating screens which remove nubbins and small pieces; these are either discarded or used for lower grades, or as an ingredient in soups. After the cutting operation the whole beans or cross-cuts are blanched in steam or water from 2 to 3 min. The large size No. 5 and No. 6 beans used for French cut are first blanched and then cut. Although smaller size beans can also be cut after blanching, a technique which reduces the flavor loss that results when cut beans are blanched, this practice frequently leads to sanitation problems. In either case the product is quickly cooked after blanching, then is sorted and packed. Cross-cut beans are frequently IQF frozen on belts and then bulk stored.

Beans, Lima

Lima beans are usually packed as: baby limas which are usually of the Emerald Fordhook, Henderson Bush, S-1, or Clark variety; large limas, which are predominantly Fordhook varieties; and speckled butter beans, such as Jackson Wonder. About 85% of the Fordhook lima beans packed in the United States are grown in Southern California. Delaware and Wisconsin are important producers of other lima bean varieties.

The methods used in harvesting and processing limas are almost identical to those used for peas. The same types of harvesters and processing lines are usually used, with minor changes, for both items. While the term lima beans will be used throughout the remainder of this section, most of the procedures described will apply equally to peas.

Lima beans may be harvested in several different ways: the vines may be mowed, loaded on trucks and hauled to viner stations, either in the field or at the freezer, where they are threshed in stationary viners; they may be threshed directly on the field with mobile viner-sheller combine harvesters which leave all refuse on the field and transfer the field run beans to wating dump trucks or large bins which go to the freezer; or the pods only may be picked with a mobile pod-picker and hauled to a shelling station or directly to the freezer for threshing. Stationary viners at field vining stations are rapidly disappearing in favor of mobile viners; the pod picker is still in development. At one time bean vines were dried and baled for cattle fodder or fed green to cattle. Current regulations on pesticide residues, especially as they apply to feeding of dairy cattle, have almost eliminated this practice in California, although it is still common in some other areas.

Where mobile viners are used, it is usually possible to schedule harvest and delivery so as to effect a continuous flow. Where field vining stations at some distance from the processing plant are used, there is frequently some delay between the threshing operation and delivery at the plant. Sometimes truck loads of beans have to wait their turn at the plant. Since lima beans and peas develop off-flavors if held at warm temperatures or for long periods after vining, it is advisable in such cases to hydrocool or to add ice at the vining station. On arrival at the receiving platform, beans which have to wait more than four hours should be unloaded and held in a cooling room at about 35° F. (2° C.).

After unloading, the initial processing step is usually a pneumatic air cleaning and screening to remove loose stems, leaves, trash, etc. The beans are spread on a wide stainless-steel mesh belt, pass under cold water sprays for an initial rinse, and are conveyed to specific gravity separators which remove, by flotation, any remaining small bits of pod or vine material. The beans then go through a second cleaning operation, a froth flotation washing, which not only washes the product but takes out night-shade seeds and pieces of pod and stem as well as deformed beans not removed in the previous operation. This is followed by still another cleaning operation, the pneumatic separator, where once more bits of broken beans, pods, and stems are removed, this time by means of an air blast. On completion of these mechanical cleaning and sorting operations, the beans go to sorting belts for visual inspection and cleaning. Here a crew of trained women inspects and removes overmature beans or any that have been damaged during the harvesting operation and have not been removed by mechanical cleaning and sorting.

Lima beans suspended in water are well adapted to pumping, and it is common practice to transport the beans from the sorting operation to the blanchers by pumping. Direct steam-injection screw heaters are frequently used for blanching beans. Baby limas require approximately a 2-min. blanch in hot water while the larger Fordhook lima requires approximately 3 min. As they emerge from the blancher, the blanched beans drop into a flume of cold running water which conveys them to specific gravity separators consisting of specially-designed tanks of salt brine maintained at a specific gravity of 60°–70° salometer (1.1162–1.1362 specific gravity). This brine concentration is much higher than that used for peas. The tender, less mature beans float across the top of the salt water while the overmature starchy beans sink to the bottom of the tank, from which they are continuously removed by a pump. These more mature beans may be canned, or they may be packed as frozen butter beans. Some freezers use a preliminary 70°–80° salometer (1.1362–1.1582 specific gravity) separation to eliminate shriveled beans. After the specific gravity separation, the beans pass over sizing screens which sort them by size; alternatively, beans may be sized after freezing. Frequently they are not sized at all. They then pass through a final series of pneumatic separators which remove by air blast any split or broken beans created during the blanching operation, after which they go over a series of conveyor belts for a final visual inspection.

Lima beans and peas may be either frozen in the package or on a continuous belt freezer. When IQF frozen, the beans are frequently bulk-stored to be packaged later, or to be included in such items as mixed vegetables or succotash, or for packaging as "boil-in-the-bag."

Broccoli

Of the 122 million pounds of frozen broccoli packed in the United States in 1965, more than 100 million pounds were grown in the San Joaquin Valley and central coastal area of California. The green sprouting variety of broccoli was introduced into the United States from Italy in the latter part of the 19th century, but it made little impact on the American market until after World War II. Since then it has grown to be one of our leading frozen vegetables.

Several varieties of broccoli have proved to be satisfactory for freezing; these include Atlantic, Coastal, Topper 43, and Medium Length. In England, the most popular variety is the purple Italian, which turns green upon cooking.

Broccoli is harvested by hand and transported to the freezer in bins holding approximately 1,000 lb. Since this product deteriorates rather rapidly, hydrocooling or icing of bins in the field is common. Broccoli, like cauliflower, is essentialy a flower bud and when held at warm temperatures for even a short time will begin to blossom and show an undesirable yellow color. Samples of broccoli are taken when received at the plant and growers may be docked for the following defects: (1) over length—most broccoli is bought on the basis of a 7 in. cut and anything over this may be docked; (2) excessive leaves; (3) damaged heads; (4) insect infestation; (5) overmaturity; (6) off-color. The broccoli is dumped from the bin onto a conveyor belt and taken to a sorting line. Here women remove damaged, blossomed, insect-infested heads, and large oversize leaves. The side shoots are removed and used for chopped broccoli. The main stalks are placed crosswise on a belt and passed under a knife which trims the stalk to the desired length, usually five inch. The stalk is split lengthwise once or twice, depending on size, to form uniform units. These are washed in a tank of highly agitated water, followed by a final spray rinse. Commonly used blanch times are either 2 to 3 min. in water or 3 to 4 min. in steam.

Chopped broccoli and cut broccoli are made from raw material that, because of misshapen or broken pieces, will not make acceptable spears. For chopped broccoli, the material is diced on Urschel or other cutters with a knife setting of ¼ × ¼ in. or ½ × ½ in. Cut broccoli is chopped with a knife setting of 1 × 1 in. USDA Standards for Frozen Broccoli allow not more than 25% leaf material or less than 25% head material (by weight) in cut or chopped broccoli. There is also a high-quality floret pack made using a three-inch knife setting and eliminating some of the stalk.

Brussels Sprouts

Brussels sprouts are grown primarily in the cool coastal areas of California, although there is some acreage on the eastern seaboard and in the Pacific Northwest; an increasing amount is produced in lower California and Mexico. The principal varieties now grown are Jade Cross, Sanda, and Long Island; Jade Cross is stronger in flavor and has a tougher core than other varieties. Brussels sprouts are harvested weekly and this results in a high labor cost. Consequently, much effort has been expended to find a variety of Brussels sprouts in which all the buds on a single stalk will mature at the same time and so be adaptable to mechanical harvesting.

Jade Cross, a hybrid from Japan, permits mechanical harvesting (Fig. 3.6). Although Unilever in England has a model which shows considerable promise, the truly successful mechanical harvester has not yet been developed. One of the problems in mechanical harvesting is the coupling of a debudding device with a deleafer and stalk cutter in the field. Brussels sprouts will yield approximately 7 to 8 tons per acre when hand harvested and 4 to 5 tons when machine harvested (Fig. 3.7). This is because machine harvesting is usually a one-time-through harvest, whereas hand harvesting permits several pickings.

In California, the production of Brussels sprouts is covered by a state marketing order which sets certain minimum quality requirements and allocates the quantity to be picked and processed. Of the total quantity of Brussels sprouts grown in California approximately 25% are used fresh

FIG. 3.6. JADE VARIETY BRUSSELS SPROUTS

Note the large percentage of mature sprouts. This habit makes this variety useful for mechanical harvesting.

FIG. 3.7. PARTIALLY MECHANIZED HARVESTING OF BRUSSELS SPROUTS

Stalks, with attached sprouts, are cut by hand, elevated by belt to special hand-fed rotating knives which strip sprouts from stem. Sprouts are deposited into bins to be hauled to freezer.

and 75% are frozen. Sprouts are designated by size as No. 1—¾ to 1 in.; No. 2—1 to 1¼ in.; No. 3—1¼ to 1½ in.; No. 4—1½ in. plus. Marketing order size limitations set a minimum of 14 and a maximum of 26 sprouts per pound as received at the processing plant.

The product is received at the freezer in tote bins and the sprouts are run through a pregrader where extraneous material and small buds are removed. The sprouts are trimmed by Hydrouts which remove the butt end and outer leaves (Fig. 3.8). The latter are removed by a blower. The trimmed sprouts are conveyed to a size grader and separated into small, medium, and large sizes. The size grading is important not only from the standpoint of a uniform pack, but to enable an optimum blanch for each size sprout. Instead of maintaining separate blanching lines, it is common practice to let the predominant grade go directly to the blancher, while the other two grades are held in tote bins to be blanched later. Careful control of blanching time is especially important in Brussels sprouts because an overblanch results in a poorly colored soft product while an underblanch will result in a pink center. A 4-min. water blanch or 5 min. in steam is usually appropriate for average size sprouts. Additional information on blanching of Brussels sprouts can be found later in this chapter.

Courtesy of Magnuson Engineering, Inc.

FIG. 3.8. BRUSSELS SPROUTS BEING TRIMMED BY WATER-POWERED
ROTATING KNIFE

After further use of blowers and suction devices to remove loose leaves, the sprouts are mechanically filled into cartons or are IQF frozen and bulk stored for later packaging or use as a boil-in-the-bag product.

Carrots

More than 100 million pounds of frozen carrots are packed in the United States, of which about 90% are packed in the western states of Oregon and California. Carrot freezers in California are fortunate in that, by staggering plantings, it is possible to contract with growers to harvest carrots the year around. The Imperator variety of carrots is a favorite for freezers in that state.

Carrots are mechanically topped and dug in the field and loaded into large bulk trailers holding about 50,000 lb. On arrival at the plant the carrots are washed in a presoak tank followed by a tumble washer. The washed carrots are mechanically separated into three basic sizes: *small*, to be used for whole carrots; *medium*, to be used for sliced; and *large*, to be used for diced. The carrots are lined up on belts, pass between rotating knives which cut off the tip and butt ends, and are flumed by water to storage hoppers. After inspection, trimming, and peeling (usually by steam), the larger carrots are mechanically diced or sliced. The slices or dice pass first to an air cleaner, where chips and small pieces are removed by air blast, and then to a blancher. While water blanching is sometimes used, carrots are usually blanched by steam; blanch times vary from 2 to 8 min. depending on size, maturity, and texture (Fig. 3.9).

Diced and sliced carrots lend themselves well to fluidized freezing or other IQF techniques. Carrots are unique in that the prepared product is frequently graded for size after freezing. This grading operation takes place in a cold room held at about 30° F. (−1° C.). Frozen sliced or crinkle-cut carrots pass over grading screens which divide them into three diameter sizes: small, medium, and large. The frozen diced carrots pass over screens which remove chips and small pieces which are sold separately for use in soups or stews. Carrots are marketed in containers of various sizes, including 10-oz. carton, 2-lb. poly bags, 20-lb. bags, and 50-lb. bags.

Small whole carrots, which have become quite popular in recent years, are processed in much the same way as large carrots. Smaller carrots are not younger than large carrots, since both the large and small ones are harvested at the same time. Where plantings are thicker, more small carrots are obtained. Whole carrots are frequently sold by count: medium are 80 and over per pound, small are 200 per pound, and tiny are 300 per pound. Some packers cut off the tip ends of the smaller size carrots and run these through an abrasive peeler to round off the cut end. The finished product looks like a tiny carrot and is sold as such. Special tiny carrots from Belgium and Holland may have counts ranging from 375 to 450 per pound. These are the "Paris" variety, whose seed does not seem to be obtainable in this country. Because of the growing market for this item, U.S. seed companies are developing varieties of "baby carrots."

Cauliflower

An unusual feature of cauliflower culture is the fact that the developing "curd" or flower bud must frequently be protected from light by tying the outer leaves to form a cover approximately ten days before cutting.

Snowball, an important processing variety, requires 50 to 60 days to

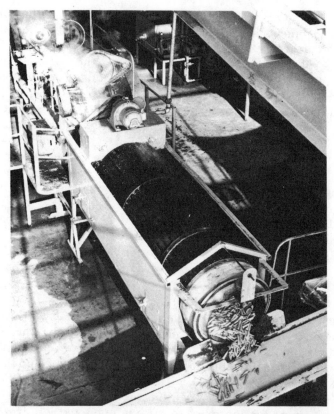

Courtesy of Marshburn Farms

FIG. 3.9. STEAM PEELING OF CARROTS

mature and is very desirable in areas having a short growing season. By contrast, the winter cauliflower varieties grown in the coastal area of California require up to 150 days to mature. By judicious choice of varieties it is possible for California processors to grow cauliflower during a nine months' season. The cauliflower is cut by hand and loaded into bulk trailers and then hauled to the processor. Rapid handling from fields to freezer is essential because discoloration resulting from the inevitable bruising during loading will show up after a fairly short holding period. If it is necessary to hold cauliflower for more than a few hours, it should be kept under refrigeration at 32° to 34° F. (0° to 1° C.) and a relative humidity of 85%.

The trailer full of cauliflower is unloaded onto a receiving belt and the heads are fed to "ballers," men with heavy knives who "ball" the cauliflower by cutting off the base. The trimmed heads go to a line of large Hydrout machines. This is a mechanical coring device which removes the core and detaches the bud cluster in one operation. Alternatively, some processors split the heads in half and scoop out the core with a special ringed knife. The curd is broken into individual florets by jamming the whole curd onto a metal ball. The broken cluster passes down an inspection belt where a sorter cuts any florets larger than 2¼ in. to smaller sizes. The florets pass through a cylindrical rod cleaner which eliminates small pieces of cauliflower, loose leaves, and foreign material. A grader-shaker removes all pieces smaller than ¾ in. A flume carries the florets to washers, where they are thoroughly cleaned by violent agitation and high-pressure cold water sprays. The washed material is blanched for about four minutes in steam or water and is cooled in water. The cooled, blanched florets are dewatered and either IQF frozen or mechanically packed into cartons to be frozen in a plate freezer or blast freezer. When packed as "boil-in-the-bag" the florets are usually machine sliced.

Celery

Most celery is frozen to be used as an ingredient for such items as peas and celery mix or as a component in frozen stew vegetables. Much of the trimming of celery takes place in the field. It is there that the butts are chopped off and the leaves removed. The hearts are taken out to be sold on the fresh market and only the large celery stalks come into the plant. Celery is washed, inspected, and run through an Urschel or similar slicer to be cut into appropriate sizes according to the final use. The cut celery is blanched in water for approximately two minutes and IQF frozen. It is ordinarily stored in large bins to be used later in the season for incorporation into the various mixes described above.

Corn

Frozen corn-on-the-cob was first produced commercially by Birdseye Frosted Foods in 1931. It has shown a steady growth since then and more than 40 million pounds are produced currently. It is not as popular, however, as cut corn, which reaches a pack in excess of 220 million pounds.

Yellow corn varieties such as Golden Cross Bantam are usually preferred for freezing. Other varieties which have been recommended are listed in Table 3.2. Corn should be harvested while still young and tender and the kernels full of "milk." The ears are mechanically harvested, promptly hauled to the processing plant, and automatically dehusked and desilked. Five pounds of unhusked corn are required for production of one pound of kernels. Probably more than any other vegetable, sweet

TABLE 3.2

YELLOW SWEET CORN VARIETIES SUGGESTED FOR FREEZING

Small to very small early kinds, 65–74 days[1] to harvest	Small to very small early kinds, 75–80 days[1] to harvest
Earliking	Carmelcross
Golden Beauty	FM Cross
Golden Rocket	Golden Bantam
Marcross	Golden Freezer
North Star	Gold Rush
Seneca Dawn	Northern Cross
Seneca Golden	Sugar King
Seneca 60	Tendergold
Spancross	
Spring Gold	
Medium to large midseason kinds, 81–89 days[1] to harvest	**Large late kinds, 90 days[1] or over to harvest**
Aristogold Bantam Evergreen	
Asgrow Golden 50	Golden Security
Calumet	Iochief
Golden Bounty	Prospector
Golden Cross Bantam	Prosperity
Golden Crown	Tendermost
Golden Nugget	Victory Golden
Golden Security	
Ioana	
Iobelle	**Varieties developed for mechanical harvesting**
NK 199	Experimental 5843
Merit	(Northrup-King)
Seneca Chief	NK 51036 (Northrup-King)
Seneca Wampum	Sugar Daddy (Ferry-Morse)
Tender Freezer	

[1] Days to harvest are the approximate number of days from planting to harvest when planted about the frost-free date in a region or season having a monthly mean temperature of 70° to 75°F. (21° to 24°C.) during most of the growing season. Mean growing season temperatures as low as 65°F. (18°C.) will increase the time to harvest by about 15 to 20 days for most varieties.

corn loses its quality rapidly after harvest and should be frozen within a few hours after it has been picked.

Corn-on-the-cob is a particularly difficult vegetable to freeze. Because of its large size, an ear of corn is difficult both to blanch and to cool. After the ear of corn has been dehusked and desilked, it is thoroughly washed and then blanched, usually in steam, for 6 to 11 min., and promptly cooled. Even an 11-min. blanch in steam does not completely inactivate the enzymes in corn-on-the-cob, particularly in the cob portion. Practically all commercial frozen corn-on-the-cob samples show positive peroxidase test in the cob area. It is believed that off-flavors frequently found in frozen corn-on-the-cob develop either from the remaining enzyme activity or from off-flavors developed in the cob which migrate out to the kernels. However, at least one variety (Barbecue), when cooked on the cob, retains high edible quality for 18 months at 0° F. (−18° C.).

Whole-kernel corn is produced in a number of ways: the corn may be completely blanched on the cob before cutting; it may be partially blanched on the cob "to set the milk," then cut and blanched again; or it may be cut and subsequently blanched. This last practice results in loss of flavor, lower yield, and microbiological problems, but it is still often used. The split blanch is recommended to keep bacterial counts at a minimum. After the corn is cut, the kernels must be cleaned to remove bits of small particles, husk, silk, light fiber, and dry immature kernels. Both froth washing and the brine flotation graders used for peas are excellent for this purpose. The corn kernels are removed at the sink or discharge end, while the trash floats off the top discharge or is picked up by a rotating reel.

Whole kernel corn is usually IQF frozen and either packaged directly into cartons or bulk stored.

Mushrooms

The only mushroom used for processing in the United States is the cultivated mushroom commonly known as *Agaricus campestris.* Freezers may find the USDA Standards for Grades of Mushrooms For Processing useful aid in the purchase of raw material. Mushrooms which are destined to be sold as whole select mushrooms, packed with or without a sauce, are necessarily of higher quality than mushrooms which are going to be sold as diced or random cut. The grower should perform at least a preliminary grading of the raw stock before shipping to the processor.

Mushrooms are commonly supplied by the growers in 9-lb. baskets, but other containers of various sizes are also used, with some tote boxes holding up to 30 lb. The harvested mushrooms should be precooled by the grower as soon as possible to 35° to 40° F. (2° to 4° C.). One plant has successfully cooled mushrooms by vacuum. On receipt at the freezer, diseased, decayed, and other cull mushrooms are sorted out and the remainder carried by belt to the washing step. Where mushrooms are to be packed as whole buttons, the base of the stem is mechanically cut off, leaving a small stub of about $^3/_{16}$ in.

Since many frozen mushrooms are sold to reprocessors, such as freeze-driers, microbiological counts have become increasingly important in recent years. To obtain low counts a preliminary soaking followed by a gentle washing, using either a rod washer or tumbling in a tank washer, is essential. High levels of chlorine, up to 50 p.p.m. in the wash-water, have been found to be useful in reducing the microbiological load. After washing, the mushrooms are graded for size by passing through a stainless steel rotating cylinder, with various size openings, which operates under cold water (Fig. 3.10). This technique results in a minimum of bruising during the grading operation.

Courtesy of A.K. Robins and Co.

FIG. 3.10. UNDERWATER GRADER FOR SIZING MUSHROOMS

The mushrooms are carried by a current of water and float up through holes of different sizes. This technique eliminates bruising.

Mushrooms contain a highly potent polyphenolperoxidase. It is this enzyme which is the cause of the undesirable discoloration that appears in bruised or cut mushrooms. During the washing operations soluble, oxidized phenolic compounds resulting from enzymatic action are washed away and in some cases the wash water may be a distinct red. Prevention of this enzymatic discoloration is one of the most important problems faced by the would-be mushroom freezer. A steam or water blanch adequate to destroy the enzyme system unfortunately results in an appreciable shrinkage, as high as 30% by weight, and may also produce an undesirable gray color. If a processor decides to use additives rather than blanching to halt discoloration, the mushrooms are still usually given a short blanch in hot water, primarily to aid in reducing microbial contamination rather than for enzyme inactivation. It is advisable to follow such a blanch immediately with a cold water quench.

The maintenance of a desirable white color in frozen mushrooms is still somewhat of an art. Probably the most important step in the processing is speed, from the time the mushroom is received at the freezer until it is fully frozen. A rapid freezing rate is essential. One processor aims for a center temperature of $-10°$ F. $(-23°$ C.) to be reached in 3 to 5 min. in diced, or in 20 min. in whole mushrooms. Because such rates are difficult to obtain in the freezing of packaged mushrooms, IQF freezing is recommended to obtain a high-quality frozen mushroom. Mushrooms have been successfully frozen on a semifluidized belt, approximately 1 layer deep for caps and 4 to 5 layers deep for the diced. Both liquid nitrogen and powdered dry ice have also been successfully used to freeze mushrooms rapidly.

Mushrooms packed in pouches for boil-in-the-bag may be first IQF frozen and then filled into the bag, or they may be filled in the unfrozen state. The latter technique is considered less desirable. In either case a sauce may be added, the bag sealed, placed in a carton, and the freezing completed in a plate freezer.

Two additives have proved useful in retarding development of undesirable brown pigments in frozen mushrooms. One is a dip in an ascorbic-citric acid solution (approximately 0.5% of each acid); unfortunately, the protective effect of this dip soon disappears. An alternative is the use of an acidified sodium bisulfite dip, with a concentration of approximately ½ to 1%. While this has a more permanent effect, the sulfite has a flavor which is detectable and objectionable to some people.

Vacuum packaging of mushrooms is desirable in order to eliminate discoloration from oxidation. A sauce coating may serve the same purpose. Mushrooms are particularly sensitive to time-temperature relationship in storage. One processor insists on storage at temperatures below $-10°$ F. $(-23°$ C.).

Okra

The pack of frozen okra has increased rapidly in recent years, rising from approximately 18 million pounds in 1960 to more than 30 million pounds today. Virtually all production is in the East and South.

The principal variety of okra frozen is Clemson Spineless. The harvested product is sorted at the freezing plant to remove culls and old fibrous or woody fruits. The pods are sized either by hand or mechanically, with both very small and overly large pieces going to a cutting machine to be packed as cut okra. The whole pods are trimmed either by hand or on Hydrouts to remove stems. After a thorough washing, the pods are blanched either in steam or in hot water for 2 to 3 min. depending on the size of the pod. Perforation of pods with a needle board prior to blanching will reportedly reduce the incidence of rupture during blanching. The blanched pods should be quickly cooled by spraying or immersion in cold water and then be drained in a dewatering reel or vibrating screen before packaging and freezing.

Onions

Both whole onions and chopped onions have become important frozen vegetables in recent years. Since the characteristic pungency and flavor of onions are developed by an enzyme reaction which occurs when the onion is crushed or eaten, and since the enzyme involved would be destroyed by blanching, it is fortunate that unblanched frozen onions are relatively stable.

White varieties of onions are preferred for freezing. These would include such varieties as Southport White Globe, White Creole, and White Sweet Spanish.

Onion sizing is important; for the diced and chopped styles large onions are preferred because of the economy of peeling; for boiling onions a diameter of ¾ to 1¼ in. is preferred; and for stew vegetables a diameter of 1½ in. is customary. Roots and tops may be removed before or after peeling. One method of doing this consists of hand-feeding individual onions to a rotating turret which conveys the onions between two rotating circular saws. The saws cut off the top and root ends of the onions. The onion bulbs then pass through revolving washers with high-pressure sprays which remove most of the outer skin. In another method, the onions are first conveyed through a flame-peeler which burns off the outer "paper-shell" and hair roots. A flame-peeler consists of an endless conveyor which simultaneously carries and rotates the onions through a rectangular refractory-shell furnace fired to a temperature of more than 2000° F. (1093° C.). The charred skin is removed by high-pressure washing; the tops, crown roots, and cores are then removed by hand-positioning each end of the onion against a water-driven rotating knife or by semi-automatic equipment which carries the onion between two parallel revolving circular blades.

For diced onions a Model RA Urschel dicer set to cut ⅛ or ¼ in. cubes may be used. Diced onions may be either individually quick-frozen or broken up after freezing but prior to packaging to make them free flowing. Plastic bags are usually used for packaging.

Pearl onions are a special variety of onions which develops small white bulbs desirable for pickles and for specialty items. In recent years they have been incorporated into several of the increasingly popular gourmet-type frozen vegetables such as "green peas and pearl onions." Most of these small onions are imported from Holland, Italy, and Germany. In those countries it is a common practice for bushels of these small onions to be distributed to homes, where the trimming of the rootlets and neck is a family project. Several mechanical trimmers are on the market. One resembles in action an electric razor. The onions pass over a slotted metal plate while rapidly rotating blades shave off the rootlets and neck portion which protrude through the slots. The thin paper skin can usually be removed by a vigorous tumbling action so that flame or lye peeling is not usually required. Whole onions are frozen in belt or plate freezers. A small acreage of pearl-type onions is reportedly being grown in various parts of the United States.

The literature on small onions is somewhat confused. Some are grown by heavy seeding of "short day" onions grown in latitudes where the days are long during the growing period or "long day" onions where the growing conditions are reversed. The aerial bulblets of certain species or varieties of onions are used for *Perlzweibeln* (pearl onions) in Europe.

Peas

Frozen peas have virtually replaced fresh peas in the American diet (Fig. 3.11), and they are second only to processed potato products in their importance as a frozen vegetable. More than 448 million pounds are produced currently. The principal growing areas for peas are

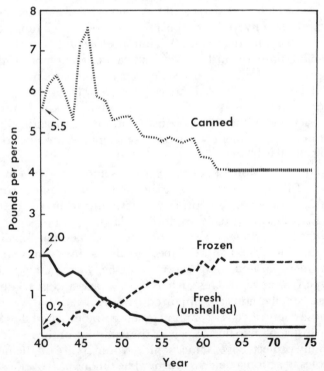

FIG. 3.11. CHANGES IN U.S. PER CAPITA CONSUMPTION OF FRESH, CANNED, AND FROZEN PEAS. 1940–1975

Washington and Oregon in the West, and Wisconsin, Minnesota, and Illinois in the Midwest.

Two varieties of peas favored by freezers are Dark Skin Perfection and Thomas Laxton.

Peas deteriorate in flavor soon after they have been vined (removed from the pod). During the vining operation juice or sap is inevitably expressed from the leaves and stems, some of which is picked up by the peas. This vine juice has been accused of being responsible for off-flavors in peas, although there has been little evidence to prove this. Peas should be processed as soon as possible after vining; when this is not possible, hydrocooling or icing is advisable. To obtain a high-quality product it is essential that peas be harvested before they become overmature and develop a starchy texture and tough skin. Growers naturally like to wait until the maximum yield is obtainable from a field, but freezers' specifications usually name such criteria as tenderness, measured by a tenderometer, to help decide the optimum harvest time which will combine maximum yield with maximum quality.

The harvesting and processing of green peas are almost identical with the same operations for green lima beans, which have been discussed earlier. Except for changes in sieve sizes and specific gravity of brines used in separators, little modification is needed to switch from peas to lima beans.

Processors use various blanching times for peas; 1 to 2 min. in hot water is a common treatment. The blanched peas are cooled rapidly in a cold water flume and from this are transferred to a specific gravity quality grader. This allows starchy, overmature peas to sink while the Grade A peas float and are separated. The brine concentration varies according to the speed at which the peas are moving, the temperature of the water, the final quality desired, etc.

Peas are not always sized. When they are, the sizing operation usually comes before blanching because peas become soft during the blanching process. However, special vibratory sorters which do not bruise or damage blanched peas can be used. This allows the unsized peas to go through the blancher together and then be sized following the blanching operation.

Peas may be packed and frozen in the carton or they may be IQF frozen for packaging or bulk storage. Freezing of peas in a fluidized bed has been very successful; freezing rates so rapid that the peas are brought down to an internal temperature of 0° F. (−18° C.) in 4 min. are easily obtained.

Peppers, Bell

Large peppers with thick walls and small placentas are desired as raw material for freezing bell peppers. One strain of Yolo Wonder, a variety used by processors in California, can average 3½ to 5 in. in diameter and weigh 8 to 10 oz. per pepper. Frozen peppers are sold both red (mature) and green (immature). It is usually desirable to pack either all-red or all-green peppers. Pods which are not fully colored or are part green, part red are considered to be of lower quality.

Bell peppers are harvested by hand and brought into the plant in thousand-pound bins or bulk trailers. They are washed, inspected, and graded for color and size. The peppers are fed by hand into an automatic coring machine which removes and ejects the stem, pithy placenta, and seeds; at the same time the peppers are cut into halves (Fig. 3.12). The halves are washed under strong water sprays to remove any remaining seeds or pithy material. Some operators slice the top off of the stem end and mechanically squeeze out the placenta and seeds. Other operators pass the whole pepper through rollers which crush the pepper, and then remove the placenta and seeds by hand or by screening.

Bell peppers are processed either as whole green, in which case they are usually not blanched, cored and halved (unblanched or blanched), or diced (unblanched or blanched). Both green and red bell peppers will freeze well unblanched but tend to develop objectionable flavors after about six months' storage. If a freezer is required to supply peppers with low bacterial and mold counts, it is frequently desirable to blanch peppers

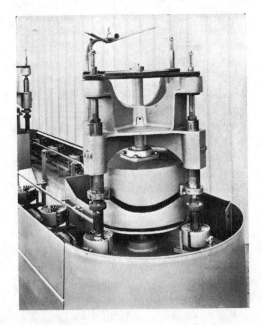

FIG. 3.12. CORING MACHINE
FOR BELL PEPPERS

in water or steam for approximately 2 min. Rapid freezing methods, such as belt or fluidized bed freezers, are recommended for diced peppers because slow freezing for this item results in bleeding and adverse texture changes. Diced bell peppers, either red or green, are usually packed in 3-lb. cartons, while whole or halved peppers are usually packed in polyethylene bags.

Pimientos

The pimiento (or pimento) is a thick-walled, tough-skinned, heart-shaped, brilliant red sweet pepper. Unlike bell peppers, pimientos are never harvested while green.

Because of their thick skin, pimientos are usually peeled by flame, hot oil, or lye. They are cored and seeded either before or after the peeling operation. Pimientos are processed either whole, as shoestring strips, or diced. Their bright red color and characteristic flavor make them a useful ingredient for remanufacture in such foods as potato salad, stew, luncheon meats, cheese, etc.

Perfection is the pimiento variety usually grown in the United States. There is a very large placenta or seed core in pimientos and much breeding research has been devoted to obtaining special strains with a small placenta to minimize coring loss. Because of this large core and the relatively large proportion of peel to flesh, preparation losses from coring and peeling may amount to over 60% of the weight of the harvested peppers. Pimientos are harvested when fully mature in a manner similar to that used for bell peppers. They are washed, sorted, peeled, cored, cut into strips or dice, and frozen.

Potatoes, Sweet

Frozen sweet potatoes are usually marketed in a precooked state. They are therefore covered in detail later.

Potatoes, White

The most important of all frozen vegetables, French-fried potatoes and similar products are covered in Chapter 11. This section will therefore discuss only frozen peeled whole potatoes and diced or shredded hash-brown style potatoes. Small tubers (under 1½ in. in diameter) are preferred for frozen whole peeled potatoes. The potatoes are peeled by steam, lye, or abrasive peeling. Since the peeling operation is an important factor in determining the yield of finished product, the operator must carefully consider the various options available to him. Frozen whole potatoes offer a desirable outlet for small tubers which are uneconomical to peel and slice for French fries. Because high-solids potatoes (above 20%

solids), such as the Burbank Russet, are preferred for frozen French fries, the small tubers which come in with the field-run will also be the same variety. If a freezer desires to pack only whole peeled potatoes he must seek some economical supply of small tubers; these are usually obtained from the packing and grading sheds where potatoes are packaged for fresh market. Either high-solids potatoes such as the Burbank Russet, or low-solids potatoes such as the White Rose variety have been successfully used for freezing; low-solids potatoes are the less subject to sloughing. Potatoes discolor rapidly from enzymatic browning after peeling. If a peeled tuber has to be held for any length of time prior to blanching, it should be kept under water or a dilute sulfite or sodium chloride brine (less than 0.5%) to retard this discoloration. A dip in a one per cent sodium acid pyrophosphate solution has been found useful in preventing after-cooking darkening.

Small, whole peeled potatoes are blanched to a depth of about ⅜ in. and then water cooled. Diced potatoes are made from peeled, trimmed, large potatoes diced to about ⅜ in. The cut dice must be thoroughly washed to remove free surface starch. The dice are then blanched in steam or hot water for about two minutes. Shredded potatoes are prepared by steaming or boiling small whole potatoes, allowing them to cool, and cutting or grating into strips of ⅛ in. cross section. In one successful operation the peeled potatoes are placed in nylon mesh bags, hung on a rack which is wheeled into a retort, and pressure cooked for 15 min. at 7 lb. steam pressure. After cooking, the rack is removed and the potatoes are allowed to cool. Some water—10 to 15% of the original weight—evaporates during the cooling operation and the starch is conditioned so that the cooked potato is cut rather than mashed during the shredding operation. All three forms of potatoes—whole, diced, or shredded—may be either blast- or belt-frozen.

Rhubarb

Rhubarb is brought in from the fields either in lug boxes or large tote bins. This vegetable is relatively easy to pack, since it requires no peeling, husking, threshing, or sizing. Most of the waste material, such as the large leaves, is removed in the field and only the stalks are brought in for processing. The stalks are washed by passing under high-pressure sprays and fed to a transverse slicer for cross-cutting into one inch slices. Slicing machines handling 4,000 to 6,000 lb. of rhubarb per hour are available.

The cut slices are conveyed into a tank of agitated water for preliminary wash and are spray-rinsed. The washed pieces are run over sorting belts where defective units such as rot, cuts, and broken pieces are removed.

Unblanched rhubarb is slow to develop off-flavors and this item has sometimes been packed. However, blanched rhubarb not only has a longer storage life but is also easier to pack because of its softer texture. Because of the large surface area in cut rhubarb, steam blanching (1 to 2 min.) is preferable to water blanching.

Before packing and freezing, the cut rhubarb is mixed with sugar in the proportion of 6 parts of rhubarb to 1 part of sugar. The sugar-rhubarb mixture is filled into cartons and is usually plate frozen.

Southern Greens

Collard greens, turnip greens, mustard greens, and kale are leafy green vegetables popular in the Southern states. It is only in recent years that commercial freezing of these vegetables has become common. Popular commercial varieties of turnip greens are Purple Top, Seven-Top, Sho-goin, and Just-Rite. Vates is a popular variety of collards, while Siberian and Vates are the usual commercial varieties of kale. All of the vegetables named above are harvested and processed in the same manner as described later for spinach. Kale is usually packed in whole leaf style, but collard greens and turnip greens are usually packed as chopped.

Southern Peas

Black-eyed peas and related types (greens, purple hulls, brown-eyes, and brown-crowders) are not only a popular and important food in the south, but increasingly so in the north. In many southern homes and restaurants it is traditional to serve black-eyed peas on New Year's Day to insure good luck and success during the coming year. The pack of frozen black-eyed has risen from 573,000 lb. in 1948 to more than 26 million pounds currently.

Most southern peas are harvested by hand, on single-family or share-cropper types of farms. Varieties of peas adaptable to mechanical harvesting, such as those used for green peas and lima beans, are now being developed. Such varieties require uniform maturity at harvest time, easy release from the plant, and an erect plant with pods set high. The yield of shelled peas per ton of mechanically harvested pods should equal the 50% yield now obtained with hand-harvested varieties. After harvesting, the pods of southern-type peas are handled and processed in almost the same way as green lima beans and green peas.

Spinach

Spinach harvest is usually divided into two periods of the year, spring and fall, with California producing most of the spring harvest. Two important varieties of spinach used for freezing in the United States are Savoy, grown in the East, and Viroflay, grown in California. Yields of spinach are considerably higher in California, with an average of 9 tons per acre, as compared with 4 tons frequently reported in other states.

Spinach should be cut before the seed stalks have appeared and before stems become overly long. Buyer specifications may call for maximum tolerances, some as low as 15%, for stems in the processed product. The harvesting operation is frequently under the control of the processor, who will own the mechanized equipment required for harvesting. A mechanical cutter cuts the spinach a few inches from the ground, carries it up on a conveyor belt and loads it into a trailer. Since a large mass of tightly-packed spinach will quickly heat up in the truck, it is common practice to cut spinach at night while the air is cool and haul it to the processing plant for processing the next day. At the plant the spinach is raked from the trailer onto a receiving belt. From there it is elevated into a circular mesh reel where the spinach is agitated by means of vanes to remove trash. A powerful fan may be used to blow a strong stream of air through the reel to rid the spinach of insects, bits of dirt, sand, grit, broken leaves, etc. After this "dry-cleaning" operation, the spinach is dumped onto an inspection belt, so arranged that the leaves are tumbled onto progressive belts, thus permitting the product to be frequently turned over for inspection and sorting. Sorters remove damaged and yellow leaves as well as extraneous material such as weeds.

The spinach proceeds by belt into a wash tank for a vigorously agitated wash followed by high-pressure water sprays. This wash involves several changes of water and each time is followed by passage over a mesh belt to drain dirty water from the previous wash. Washing of spinach must be particularly thorough so that all traces of sand and silt are removed.

The washed spinach is then wilted and blanched. Spinach may be either wilted and blanched in a single operation, or wilted in a wilting tank using water maintained at about 130° F. (54° C.), then blanched in boiling water or a steam tunnel for about 2 min. From the blancher the spinach passes into a cold water quench. Yellow leaves will frequently show up much better after blanching, so there is a final sorting of the blanched spinach in order to remove such discolored material. Some operations may float the leaves in a water trough to enable sorters to select out discolored leaves. The cooled, blanched spinach is elevated by a mesh belt to facilitate draining. Sometimes a roller is used on the conveyor belt to help press out free water; however, such an operation, if not carefully controlled, will press out natural juices. A common defect in frozen spinach is the presence of free liquid in the package and buyers therefore sometimes specify a maximum percentage of free liquid.

Spinach is packed in three principal styles: whole leaf; cut leaf (cut on a special Urschel cutter with knives set 2 × 2 in.); and chopped (cut ¼ × ¼ in.). Chopped spinach may be pumped into a filler machine and mechanically filled into paperboard cartons, an operation that is impossible with the whole-leaf style. Sometimes spinach, in various styles, may be bulk frozen in 50- or 100-lb. blocks for later use as a boil-in-the-bag item.

Squash

Both summer crook-neck and zucchini squash are commercially frozen. Summer squash must be harvested before it is overmature with resulting large seeds and hard rind. Squash is washed thoroughly in a flood-type washer, and sorted for size, and defective units are removed. The stem ends are cut off and discarded. In the case of zucchini squash the product is placed on a cup belt which runs past a cutter that removes the blossom end. The squash, trimmed at both stem and blossom ends, is then sliced into ½-in. slices or diced into ½-in. dice. The product is blanched in hot water for about 3 min., spray cooled, dewatered, and sorted. Sliced squash is usually packed in 10-oz. retail packages and is plate- or blast-frozen. Diced squash is usually IQF frozen and packed in 30-lb. cases.

A small quantity of frozen cooked pulped squash is packed. Boston Marrow, Golden Hubbard, Golden Delicious, or similar variety squash is washed, cut, cleaned of seeds, and cooked in a continuous screw cooker. The cooked pieces are run through a pulper with a 0.060 in. screen and the resulting pulp is pumped through a heat exchanger where it is cooled. It is then mechanically filled into containers and frozen.

Tomatoes

If a frozen tomato with the characteristic color, texture, and flavor of field-ripened tomatoes were commercially available, it would be a strong competitor for the box-ripened, poorly colored, low-flavored tomatoes that are usually found in the markets during the winter months. One large American processor first test-marketed such a frozen sliced tomato in the 1960's. A variety of tomato specially adapted for freezing was developed. The tomatoes as received at the plant were carefully selected—reportedly, only a small percentage of the tomatoes received were suitable for freezing. The tomatoes were carefully peeled (probably with hot lye), cored, and cut into ½-in. slices. Only the two center slices from each tomato were used. These were first pre-cooled in cold nitrogen gas and then completely immersed in liquid nitrogen. The length of time the slices were in the liquid nitrogen was found to be critical.

Eight frozen slices were packed in a Mylar-polyethylene pouch, heat sealed, and inserted into a self-locking wax carton. To defrost, the pouch was placed in cold water for 15 min., then turned over for another 15 min.

Consumers were advised that the product was best served while still icy cold. This product was in a test market stage in 1966, then withdrawn.

Other processors have tried to pack a similar product but the sliced tomatoes, which are not blanched, soon developed off-flavors and broke down in texture rapidly after thawing. It appears that a suitable tomato variety and technique for freezing tomatoes has yet to be developed.

Vegetables-in-Sauce

Vegetables-in-sauce are packed in three different ways: *boil-in-the-bag,* wherein vegetables and sauce are combined in a plastic bag which is immersed in boiling water to be heated for serving; *packed in a carton* with added cubes of prepared sauce, which requires the addition of water or milk and a slight cooking in a sauce pan; or *a separate pouch* of sauce packed in the carton with the frozen vegetables. The most popular of these items is the boil-in-the-bag. Because there is no pan to clean after cooking, because of the convenience, and because of the extra flavor contributed by the sauce, the boil-in-the-bag frozen vegetable has become a very popular item.

The introduction of the boil-in-the-bag concept has been credited to Mr. Ken Singer, who in 1945 packed meat on trays in sealed parchment bags at his plant in New York. This venture proved unsuccessful because the difficulties of obtaining a tight seal which would remain sealed during the heating process had not yet been solved. The first commercially successful plastic boil-in-the-bag frozen vegetable was reportedly packed by Seabrook Farms in 1955. The first large-scale marketing of frozen vegetables in a plastic bag, however, came about with the introduction by the Green Giant Company of Le Sueur, Minnesota, of their prepared vegetables in butter sauce, frozen in boilable pouches. In 1962, the first year of marketing of this item, the Green Giant Co. sold the equivalent of 35 million boilable pouch units of prepared vegetables. By 1963, this had reached 50 to 60 million units. The remarkable success of this item encouraged many other companies to follow with similar products. In 1964, approximately 225 million units of boil-in-the bag frozen vegetables were packed for retail use and 20 million for institutional use: Currently, billions of "heat-in-bag" frozen foods are packed annually.

Films used for packaging boil-in-the-bag vegetables must have special properties. These include the ability to withstand both the low temperatures encountered in freezing and the high temperature of the boiling water. Commercial films now available can withstand a temperature range of $-70°$ F. $(-57°$ C.) to $+240°$ F. $(116°$ C.). The package must also be able to maintain its seal strength in boiling water. Boil-in-the-bag frozen vegetables are, above all, a convenience food, and, like most convenience foods, must have no more than a short cooking time to please the consumer. Since the plastic material may act as a thermal barrier, it is advisable to keep the film as thin as possible while maintaining all the qualities necessary for satisfactory service. Present film now in use is 2 ml. (0.002 in.) thick.

The consumer will heat, rather than cook, vegetables which are packed as boil-in-the-bag items. Therefore most of the "cooking" must be done before freezing. One rule of thumb sometimes used is to "triple-blanch" such vegetables as peas and beans, since this will frequently give the desired texture. Practically all boil-in-the-bag vegetables are IQF-frozen, filled into the bag with specialized equipment, sauce added, the vacuum drawn, and the package sealed. A common proportion of sauce to vegetable is 1½ oz. of sauce to 8½ oz. of vegetables for a 10-oz. package. The sauce when used should be cold enough (55° to 70° F.) (13° to 21° C.) so that when added it does not thaw the vegetable with which it is packed. It is obviously necessary then to freeze the bag as quickly as possible in order to freeze the sauce. If sauces are brought to a temperature lower than 55° F. (13° C.), there is a danger that the butter or other components may separate from the sauce while it is still in the filling hopper, so that when the sauce is added to vegetables it will not be uniform in proportion of the ingredients. At temperatures higher than 70° F. (21° C.) the sauce may thaw the vegetables.

Sauces may be either smooth in texture (obtained by passing the mixture through a homogenizer), or coarse-textured with noticeable particles of butter scattered through the mixture. Sauces are usually made by cooking a mixture of butter and starch to form an emulsion with water and adding appropriate flavorings.

A basic formula used by some processors is:

	Pounds
Water	60 to 70
Butter	10 to 25
Starch	0.5 to 2
Sugar	0.5 to 3
Salt	1.5 to 2.5

Miscellaneous flavorings such as dehydrated onion, MSG, hydrolyzed vegetable protein, etc., to taste. The average commercial butter sauce contains about 12% butter.

The sauce formula given above is relatively salty but this is because the salt required to season the vegetables is contained in the sauce. Lima beans and peas which have been graded for density in brine solutions may pick up some salt; in this case the sauce formula must be adjusted accordingly.

The starch used for making sauces must be freeze-thaw stable. A processor formulating his own butter sauce should consult with his starch supplier, since starches that are excellent for this purpose are now available.

Although butter sauce is most popular in present-day frozen boil-in-the-bag vegetables, other types of sauce are constantly being developed. These include cream sauce for spinach, cheese sauce for broccoli, Brussels sprouts, and cauliflower, Hollandaise sauce for asparagus, etc. Cheese sauce formulas will include either fresh or dehydrated cheddar cheese. High viscosities in sauces other than butter are desirable so that the sauce will "cling" to the product.

Some processors pack vegetables in sauce in conventional waxed paper cartons. Vegetables are filled into the carton and small cubes of solidified sauce mixture are added. The cubes are obtained by preparing a relatively concentrated sauce, freezing in a slab, and cutting the slab into desired size cubes. Alternatively, cubes may be made by placing a thick mixture in a specially designed machine which extrudes the sauce in a manner similar to a sausage stuffer and cuts the extruded sauce into cubes which then fall directly into the filled carton of vegetables.

Vegetables, Mixed

Frozen mixed vegetables are now available in an ever increasing variety of mixtures. In addition to the long-time favorites, carrots and peas, succotash, and mixed vegetables, the consumer can now purchase peas and onions, peas and sautéed mushrooms, peas and diced white potatoes, French green beans and sautéed mushrooms, French green beans and almonds, corn and peas with tomatoes, and many others. Only the first three items will be discussed herein. Virtually all the mixtures described are made from vegetables which have been frozen and packed in bulk during the season. In the off-season the free-flowing vegetables are mixed and packaged while still in the frozen state.

Succotash.—This is a mixture of whole kernel corn with either lima beans or green beans. When green beans are used the product is designated as "frozen green bean succotash." The USDA Standards for Frozen Succotash recommends that the proportion of ingredients be between 50 and 75% for the corn and between 25 and 50% for the beans.

Peas and Carrots.—The ingredient proportion for this item is 50 to 75% peas and 25 to 50% diced or sliced carrots.

Mixed Vegetables.—The USDA Standards for frozen Mixed Vegetables describes this item as a mixture containing three or more of the following basic vegetables:

Beans, green or wax	cut-style ½-in. to 1½-in. cuts
Beans, lima	either large whole green limas or baby limas
Carrots	⅜-to-½-in. cubes

| Corn, sweet | whole-kernel yellow corn |
| Peas | sieve sizes 3 through 5 |

When 3 vegetables are used, no one vegetable should be more than 40% by weight of the total; when 4 vegetables are used, no single vegetable should be more than 35 or less than 8% by weight of the total; when 5 vegetables are used, no single vegetable should be more than 30 or less than 8% by weight. A good quality mix used by one processor is 22% green beans, 22% corn, 22% peas, 22% carrots, and 12% limas. Lima beans almost always are the smallest proportion. Inasmuch as frozen mixed vegetables are combinations of ingredients of different sizes and composition, care must be taken to see that all of the vegetables are so blanched that when the mixture is cooked all pieces will be of uniform "doneness." Lima beans for mixed vegetables should be "double-blanched" and diced carrots should similarly be blanched longer than usual. Frozen mixed vegetables are almost always packed by packers of peas and/or green beans. They may have to order those ingredients which they do not pack themselves from other packers. It is usually not economical, however, for a carrot packer, for example, to buy other components from other freezers.

Vegetables, Stew

This mixture of vegetables has become very popular in recent years because it considerably simplifies and eliminates much of the labor necessary for the consumer to make a meat and vegetable stew. Although there are no standards for this item, practically all mixes now on the market consist of large pieces of potatoes, carrots, onions, and celery. These vegetables are all IQF frozen separately and bulk-stored to be mixed and packaged in the off-season. The potato component is either composed of whole potatoes 1 to 1½ in. in diameter, or large pieces of irregular size and cut about 1 oz. in weight. The carrots are transverse slices of ⅜ to ½ in. in length, or they may be irregular cut pieces. Onions are almost always whole onions 1 to 1½ in. in diameter. The celery component is in the form of transverse ½ in. cuts. One popular mix now being packed consists of:

Potatoes	55% large pieces
Carrots	28% large pieces
Onions	12% whole
Celery	5% transverse cuts

A 1½-lb. polyethylene bag is a common package for this item.

PRODUCT STABILITY

Quality Changes as Affected by Storage Temperatures

For nearly 40 years, the problem of frozen vegetable stability as a function of storage temperature has been under investigation. Studies of frozen vegetables have led many to the opinion that 0° F. (−18° C.) is satisfactory for storage, based on compiled data on approximate storage life of 14 vegetables at 3 temperatures. At zero, storage life ranged from a minimum of 8 months for asparagus, snap beans, Brussels sprouts, corn-on-the-cob, and mushrooms to 24 months for cut corn, carrots, pumpkin, and squash. Lima beans, broccoli, cauliflower, peas, and spinach were reported to have a storage life of 14 to 16 months.

Extensive studies have been conducted by the Western Regional Research Laboratory.

The limited storage life of frozen vegetables at 0° F. (−18° C.) indicates that changes occur. We can measure changes in chemical and physical attributes and observe them subjectively in vegetables held at zero and lower temperatures.

Chemical changes in vegetable constituents coincide with color and flavor losses. Green vegetables lose their bright, fresh appearance as chlorophyll is converted to pheophytin, an olive-green to brownish compound. Ascorbic acid is oxidized by steps to dehydroascorbic acid and then to diketogulonic acid—chemical changes not directly correlated to aesthetic quality. Cauliflower darkens and tans but pigments formed are not yet clearly defined. Reflectance color measurements of vegetables and slurries made from them, and color absorbance of extracts also reveal the nature and amount of frozen vegetable instability. Flavor changes have not been directly associated with specific chemical reactions except that enzymatic reactions for underblanched vegetables are responsible for flavor changes.

Flavor and color deterioration can be evaluated by comparing samples from the same lot held under different conditions. Samples that have been stored at various temperatures are compared with a sample from the same lot that has been protected by storage at or below −20° F. (−29° C.). The storage duration required to produce a perceptible difference in flavor or color can be evaluated reproducibly for given lots for a given temperature, thus providing data for comparing different storage conditions and for making correlations with chemical and physical changes in the products. Such comparisons are extremely useful in research, but in commercial operations there is usually no sample that has been held at a protective low temperature available to serve as the base for comparison. Without the direct comparison, flavor and color deterioration must become substantial before subjective judgments can really tell that an adverse change has taken place. The time required to reach a perceptible difference may not have great commercial significance because the quality change involved is usually small. In most cases this slight deterioration in quality would cause no consumer complaint nor would it be observed

TABLE 3.3

MONTHS TO REACH A PERCEPTIBLE FLAVOR OR COLOR DIFFERENCE

	Temperature		
Product	0°F. (−18°C.	10°F. −12°C.	20°F. −7°C.)
	Flavor		
Green beans	10	3	1
Peas	10	3	1
Spinach	6	2	0.7
Cauliflower	10	2	0.5
	Color		
Green beans	3	1	0.2
Peas	7	1.5	0.3
Spinach[1]	—	—	—
Cauliflower	2	0.5	0.2

[1] Color deterioration of spinach varied. Two lots that were judged by a highly trained panel differed; one of the lots was as unstable as cauliflower, the other was more than twice as unstable as green beans.

even by a trained judge if he did not have the protected sample from the same lot for comparison.

Time required at various temperatures to reach a perceptible change of flavor or color was determined by trained panels judging stored samples against protected controls for 4 vegetables (green beans, peas, spinach, and cauliflower) in the Time-Temperature Tolerance (T-TT) of Frozen Foods project at the Western Regional Research Laboratory of the USDA. Frozen vegetables used in the project were commercially packed and represented several harvest years, growing areas, and varieties. They were tested over a range of temperatures, both steady, and with controlled fluctuations, from −30° F. (−34° C.) to 30° F. (−1° C.). Regression values (log time for a perceptible difference in color or flavor on temperature) for the vegetables are given in Table 3.3 over a temperature range of 0° to 20° F. (−18° to −7° C.). Tressler and Evers estimates of storage life for these vegetables are longer than the time shown here for perceptible changes, particularly color changes. At 0° F. (−18° C.), their estimates of minimum storage life were 8, 14, 14, and 8 months; and at 10° F. (−12° C.) 4, 6, 6, and 6 months for green beans, peas, spinach, and cauliflower, respectively.

For selected lots of frozen vegetables in the T-TT project, judgments were obtained from industry people whose function is concerned with frozen food quality (buyers, quality control staff, etc.). Without reference to a protected control sample the judgments were based on the appearance of coded samples that had been stored at various temperatures for various times. The panels judged samples as being acceptable or having deteriorated to the degree that they would give rise to consumer com-

TABLE 3.4

MONTHS[1] OF STORAGE REQUIRED FOR A 10% DECREASE IN CHLOROPHYLL

Product	Temperature		
	0°F. (−18°C.	10°F. −12°C.	20°F. −7°C.)
Green beans	10	3	0.7
Spinach, leaf	30	6	1.6
Spinach, chopped	14	3	0.7
Peas	43	12	2.5

[1] Regression values of 2 lots each for green beans and chopped spinach and 4 lots each for leaf spinach and peas.

plaints. In summary, this evaluation indicated that for vegetables at least 2 or 3 times the "first perceptible difference" would be required to reach a degree of deterioration that might lead to "consumer complaints." The toleration of consumers for quality loss caused by storage of frozen vegetables remains unknown, as is also the loss of subsequent sales because of consumer encounters with mediocre quality at some stage in the development of buying habits,. In spite of the inexactitude of the definitions and differences in the way judgments were acquired, it is of interest to note that 2 or 3 times the "first perceptible difference" for these 4 vegetables is in the range of the Tressler *et al.* "storage life" estimates. In general, the T-TT investigations would predict a somewhat shorter "storage life" for cauliflower and green beans.

The T-TT findings indicate that fluctuating storage temperatures do not, *per se,* accelerate changes in frozen vegetables. Deterioration observed under fluctuating temperatures was what would be expected from an integration based on the temperature experience and the known rates of deterioration for a series of constant temperatures in the range of the fluctuation.

A tendency exists for surface desiccation and interior package ice formation in storage, and one might expect these defects to be accelerated by fluctuations of temperature. However, over a wide range of gradual fluctuations in large rooms, frozen vegetables did not develop a serious condition of "cavity" ice or frost formation over several weeks or months in the T-TT study. Ice crystal and frost formation are often observed on the inner surfaces in packages of frozen vegetables held for a few weeks in household refrigerators in the frozen food compartment. In prolonged storage at temperatures above 0° F. (−18° C.), surface desiccation effects become noticeable.

Chemical Measurement of Changes as Affected by Storage Temperature

Chemical measurements were used in the T-TT study to evaluate the rate of product deterioration as affected by temperature. The chemical

TABLE 3.5

MONTHS[1] FOR A 50% LOSS OF ASCORBIC ACID

	Temperature		
Product	0°F. (−18°C.	10°F. −12°C.	20°F. −7°C.)
Green beans	16	4	1.0
Peas	48	10	1.8
Spinach	33	12	4.2
Cauliflower	25	6	1.7

[1] Average regression values for several lots each on storage time of ascorbic acid concentration (mg./100 gm.) for green beans, percentage retention of ascorbic acid for peas, and log concentration of ascorbic acid for cauliflower and spinach.

changes measured were oxidation of ascorbic acid, conversion of chlorophyll to pheophytin for green vegetables and development of brown pigments for cauliflower (Table 3.3). Correlations were established between chemical changes and subjectively judged changes in frozen vegetable quality.

The measurement of conversion of chlorophyll to pheophytin is rather tedious and not suitable for routine plant operations using present analytical methods. However, it is admirably suited for research on frozen vegetables because it reveals much about product deterioration and is so obviously related to the important color quality. It is of particular value in determining quality loss in processing and subsequent storage and handling, because essentially all the green pigment in vegetables at the time of harvest is chlorophyll. Thus, any measured amount of pheophytin represents an unmistakable change from original quality. There may sometimes be a slight conversion of chlorophyll in unprocessed vegetables if they are held for a long time or at a high temperature. The chlorophyll conversion that might take place under such conditions would probably be accompanied by other adverse changes that would downgrade the product. Raw material of this kind would not produce a high-quality frozen product.

Blanching vegetables causes considerable conversion of chlorophyll, and so do the abusively high temperatures that are occasionally encountered in storage, handling, and distribution of frozen foods. Different vegetables vary in the rate of chlorophyll conversion under a given set of conditions. Of the vegetables included in the T-TT study, green beans were least stable. In increasing order of stability were chopped spinach, leaf spinach, and peas. (See Table 3.4). The green beans of this illustration had been blanched so as to retain a high level (about 93%) of chlorophyll through the blanching treatment. Four other lots of green beans, blanched at lower temperatures for longer times, retain less chlorophyll, averaging about 73%. The latter lots were less stable in storage and converted chlorophyll at higher rates that were closely correlated with the retention during blanching.

The relation of the amount of chlorophyll converted in stored green vegetables to the occurrence of a "perceptible difference" in color from protected samples of the same lot was obtained indirectly. The average difference in percentage chlorophyll retained between samples judged to be different by test panel analysis was about 4% for green beans, 1.5% for peas, and 1 to 3.5% for different spinach lots. While these values represent a large sampling of commercially packed vegetables, there was a high degree of variation and they should be considered approximations for general use only.

At any point in the distribution chain a sample of a frozen green vegetable may be taken for chlorophyll analysis. A high conversion to pheophytin may indicate that the product was not blanched properly or had been allowed to deteriorate in storage or distribution (too long a holding time or too high a temperature). The packer should know his own blanching practice and by package code know the date of packing. Thus, it is possible for him to identify the existence of deterioration caused by storage and distribution, allowing for the amount of conversion that was caused by the blanching operation. A distributor or chain buyer, by knowing the characteristics of frozen vegetables when they enter his hands, can do much the same thing if he wants to know how much deterioration has taken place in various transfer operations and holding periods.

The ascorbic acid loss that is frequently measured in storage studies of frozen vegetables is regarded as more than loss of vitamin C activity. Ascorbic acid is easily measured, and its disappearance is assumed to parallel other oxidative changes of an adverse nature; that is, the loss of ascorbic acid is considered to be an indicator of general product deterioration. At 0° F. (−18° C.) ascorbic acid losses are moderate after 6 and 10 months for peas, lima beans, and asparagus. In the T-TT study, losses in peas and spinach were about 15% in a year at this temperature. Cauliflower and green beans were less stable. At higher temperatures, ascorbic acid loss occurred at a greater rate (See Table 3.5).

Discoloration of cauliflower is perceptible in about 2 months at 0° F. (−18° C.) The rate of color change has been measured by comparing the optical density of water-acetone extracts of cauliflower at different storage times. At 10° F. (−12° C.) and at 20° F. (−7° C.), rates of discoloration were 3.6 and 10.7 times faster than at zero.

Storage and Handling of Frozen Vegetables

As for all frozen foods, quality changes occur during storage and are greatly accelerated by increases in storage temperature. At the generally accepted 0° F. (−18° C.) storage, perceptible color changes occur in a few months and perceptible flavor changes occur in less than a year. However, these changes may have little commercial significance because they would not be detectable by consumers under usual conditions of use. Color changes in cauliflower and green beans, and flavor changes in spinach would be approaching or beyond "consumer complaint" levels after a year in storage at 0° F. (−18° C.). For the portion of the pack that is not to be consumed by the next harvest season, it may be desirable to store the more sensitive vegetables at somewhat lower temperatures. The accelerating effect of increased storage temperatures on product deterioration is such that all efforts should be used to avoid or limit experiences in which frozen vegetables are allowed to rise above 0° F. (−18° C.).

ADDITIONAL READING

AM. FROZEN FOOD INST. 1971A. Good commercial guidelines of sanitation for the frozen vegetable industry. Tech. Serv. Bull. *71*. American Frozen Food Institute, Washington, D.C.

AM. FROZEN FOOD INST. 1971B. Good commercial guidelines of sanitation for the potato product industry. Tech. Serv. Bull. *75*. American Frozen Food Institute, Washington, D.C.

DeFIGUEIREDO, M.P., and SPLITTSTOESSER, D. F. 1976. Food Microbiology: Public Health and Spoilage Aspects. AVI Publishing Co., Westport, Conn.

GOULD, W. A. 1977. Food Quality Assurance. AVI Publishing Co., Westport, Conn.

HAARD, N. F., and SALUNKHE, D. K. 1975. Postharvest Biology and Handling of Fruits and Vegetables. AVI Publishing Co., Westport, Conn.

HARRIS, R. S., and KARMAS, E. 1975. Nutritional Evaluation of Food Processing, 2nd Edition. AVI Publishing Co., Westport, Conn.

JOHNSON, A. H. and PETERSON, M.S. 1974. Encyclopedia of Food Technology. AVI Publishing Co., Westport, Conn.

KRAMER, A., and TWIGG, B. A. 1970, 1973. Quality Control for the Food Industry, 3rd Edition. Vol. 1. Fundamentals. Vol. 2. Applications. AVI Publishing Co., Westport, Conn.

LUH, B. S., and WOODROOF, J. G. 1975. Commercial Vegetable Processing. AVI Publishing Co., Westport, Conn.

PANTASTICO, E. B. 1975. Postharvest Physiology, Handling and Utilization of Tropical and Subtropic Fruits and Vegetables. AVI Publishing Co., Westport, Conn.

PENTZER, W. T. 1973. Progress in Refrigeration Science and Technology, Vol. 1–4. AVI Publishing Co., Westport, Conn.

PETERSON, M. S., and JOHNSON, A.H. 1977. Encyclopedia of Food Science. AVI Publishing Co., Westport, Conn.

RYALL, A. L., and LIPTON, W. J. 1972. Handling, Transportation and Storage of Fruits and Vegetables. Vol. 1. Vegetables and Melons. AVI Publishing Co., Westport, Conn.

SACHAROW, S. 1976. Handbook of Package Materials. AVI Publishing Co., Westport, Conn.

TRESSLER, D. K., VAN ARSDEL, W. B., and COPLEY, M. J. 1968. The Freezing Preservation of Foods, 4th Edition. Vol. 2. Factors Affecting Quality in Frozen Foods. Vol. 3. Commercial Freezing Operations—Fresh Foods. Vol. 4. Freezing of Precooked and Prepared Foods. AVI Publishing Co., Westport, Conn.

TRESSLER, D. K., and WOODROOF, J. G. 1976. Food Products Formulary. Vol. 3. Fruit, Vegetable and Nut Products. AVI Publishing Co., Westport, Conn.

UMLAUF, L. D. 1973. The frozen food industry in the United States—Its origin, development and future. American Frozen Food Institute, Washington, D.C.

WILLIAMS, E. W. 1976A. Frozen foods in America. Quick Frozen Foods *17*, No. 5, 16–40.

WILLIAMS, E. W. 1976B. European frozen food growth. Quick Frozen Foods *17*, No. 5, 73–105.

WOOLRICH, W. R., and HALLOWELL, E. R. Cold and Freezer Storage Manual. AVI Publishing Co., Westport, Conn.

4

Freezing Fruits

Frank P. Boyle, Bernard Feinberg,
James D. Ponting and Everett R. Wolford

C onventional methods of freezing tend to disrupt the structure and destroy the turgidity of the living cells of the fruit tissues. Of course, the same thing happens to vegetable tissues, but most vegetables have fibrous structures which tend to hold them together when they are thawed. In addition, a great many of the vegetable products are frozen at an immature stage of growth. Fruits, on the other hand, are harvested for freezing at the fully ripe state and are soft in texture even before freezing and thawing.

Before frozen fruits can match the spectacular growth in production of the frozen vegetable pack, the problems of poor texture and liquid losses must be solved. This will require a basic, long-term study of the effects of freezing and thawing on fruit tissues.

In this chapter we discuss ways in which desirable characteristics of frozen fruit, as well as fruit juices and purées, are affected before they reach the consumer. As with vegetables, the material is presented under the general topics of production and harvesting, handling, freezing, and, finally, the stability of these frozen products.

FRUITS

Fruits for freezing have certain characteristics in common with vegetables; for example, enzymes which cause deterioration if not controlled, but in other important characteristics they are vastly different. Fruits have delicate flavors which are easily damaged or changed by heat, so they are at their best when raw and become lower in quality with processing. Vegetables, on the other hand, are ordinarily eaten after cooking. Not only is their flavor improved by heating but their texture is softened from an over-firm to a desirable point. Fruit texture, like flavor, is best in raw fruit and tends to become too soft when the fruit is heated. This situation arises from the fact that fruit flavors are not developed fully until the fruit is soft-ripe, while immature vegetables are considered better in flavor than mature ones. Thus immature vegetables containing starch and other firming substances can be cooked with improvement in both flavor and texture, while fruits must be picked after they have lost their firming substances, and cooking degrades both flavor and texture.

Addition of materials to increase fruit firmness, such as pectin, algin, etc., has been proposed but has not been successful. Sugar has a mildly beneficial effect on texture although its main use is as a flavor enhancer. Since frozen vegetables are usually cooked and eaten hot, thawing is much simpler than with fruits. Plastic bags of frozen vegetables can be put in boiling water without opening, if desired, to thaw and cook in one operation. In contrast, frozen dessert fruits must be thawed slowly, usually at room temperature or below, and eaten just after thawing to retain their best quality. The difference in convenience of thawing, along with the factors mentioned above, may explain much of the difference in growth between the frozen fruit and the frozen vegetable industries.

An attractive color is important in frozen fruits. To maintain a bright color in fruits containing enzymes which cause oxidative browning or other color changes, chemical treatments or additives are often used in place of blanching by heat to inactivate enzymes. In some fruits a combination of chemical treatment plus a mild heat treatment can be used. In others, the enzymes are only partially inactivated, especially on the surface of the fruit, and reliance is placed on quick consumption of the fruit after thawing.

Changes affecting texture, flavor, and color of the fruit can occur at any of several stages of handling. Varietal selection, cultural practices, climate, method of harvesting (mechanical versus hand), transport and storage of raw material, processing, packaging, frozen storage, shipping, and thawing all must be considered in the overall operation of supplying the consumer with a high-quality frozen fruit product. A poor variety for freezing cannot be made into high-quality product; on the other hand, the best raw material can be quickly downgraded by poor processing or other handling practices.

Nutritive values of frozen fruits are important in some cases but not in others. For example, frozen concentrated orange juice is advertised and sold as a good source of vitamin C, as well as being a delicious beverage, but frozen apple slices, on the other hand, are sold primarily for flavor and texture in pies. Except for vitamin C, frozen fruits in general are not as high in vitamin (or mineral) content as frozen vegetables.

The acidity of frozen fruits limits their microbial flora largely to yeasts and molds. These are not a problem in fruits once they are frozen, but are very important in the preprocessing stage. A small amount of mold contamination in fresh strawberries can produce an off-flavor in the final product. Thus great care must be taken to select sound fruit and to wash and sort it carefully. If contamination is widespread it may become un-economical to process such fruit, and in fact whole growing areas have been abandoned because of this problem.

Production and Harvest

The characteristics of raw fruit for freezing are of primary importance in determining the quality of the frozen product. These characteristics are shaped by an array of factors including genetic makeup, climate of the growing area, type of fertilization, irrigation and other cultural practices and maturity at harvest.

Genetic Makeup.—New varieties of fruit are under constant development in experiment stations and by private growers. Plant breeders strive for improvement in characteristics that are of economic importance. These include yield, toughness or ability to withstand rough handling; resistance to virus diseases, molds and physiological defects; color; flavor; and uniformity of ripening. The latter, along with toughness, is of growing importance because of a trend toward higher cost of labor and consequent rapid development of mechanical harvesting. Since mechanical harvesting is done only once or twice in a season, fruit of a wide range of maturity is obtained. Unless the fruit ripens more or less uniformly, a great deal of green or over-ripe fruit must be sorted out, and the labor required as well as the loss of fruit may make the whole operation un-economical. Nevertheless, the trend seems to be toward more use of mechanical harvesting because where it is used at present it is as cheap as hand harvesting or cheaper. It should become relatively cheaper still in the future as intensive research increases efficiency of harvesters and trees or vines are trained to shapes more suited to mechanical harvesting. Harvesting of red sour cherries has become largely mechanized in Michigan, where nearly 70% of the U.S. supply is produced and most of this is processed. A particular stimulus toward mechanical harvesting was the low price for cherries, coupled with production and hand-harvesting costs of fruit per pound. Also, there was a shortage of pickers. Harvesting costs were reduced for mechanical harvesting, and the number of workers required was reduced to 5% of the number needed for hand picking.

Parallel to development of mechanical harvesting of fruits there has been some development of transportation systems which minimize deteriorative changes between harvesting and processing. Red tart cherries are now largely collected and transported in tanks of cool water, a practice which decreases respiration rate and scald from bruising in addition to firming the texture. Deep bins are widely used for peaches, cherries, and other fruits since these requie less transferring of fruit and may actually cause less bruising during trucking than smaller boxes. A logical extension of such procedures would be the moving of more prefreezing operations to the field so as to minimize changes between harvesting and processing. Specially designed equipment for this has been developed for some vegetables as well as fruits. Figure 4.1 shows a type of pallet tank now widely used for transporting cherries in water.

Blueberries, blackberries, plums, and prunes are well along the way to mechanical harvesting, but mechanical harvesting of other fruits is only beginning. These others include apples, peaches, apricots, pears, grapes, citrus, boysenberries, raspberries, and strawberries. In one study of mechanical harvesting of apples (Fig. 4.2) it was found that at a volume of 5,000 bushels or above, mechanical harvesting was cheaper than hand-picking. The mechanically harvested apples were considered satisfactory for processing, although about half were bruised and nine per cent were seriously damaged. Mechanically harvested apples were not considered satisfactory for fresh storage because of the bruising and too high a percentage of rotten fruit after storage. However, other experiments have shown that much of the bruising could be eliminated by padding equipment and using decelerator strips. Improvements must be made in

FIG. 4.1. PALLET TANKS FOR HANDLING CHERRIES IN WATER

Each tank holds about 1,000 lb. of fruit.

FIG. 4.2. MECHANICAL HARVESTING OF APPLES IN MASSACHUSETTS

the design of collecting units and in the manner of tree pruning. The latter is even more true for citrus fruits.

Experiments in Arkansas with mechanical harvesting of blackberries gave the somewhat surprising result that quality was actually better with mechanical harvesting than with handpicking. The shaking by the harvester evidently discriminated between ripe and green fruit better than did the human pickers.

Climate.—Climate, of course, has a large effect on most of the fruit characteristics mentioned above, and for this reason most fruits are grown in particular areas where they are suited to the climate. These areas can be altered to some extent by development of varieties more suited to a new location, but this is a slow process when all the desired characteristics are considered. Consequently, there has been considerable shifting of major growing areas from one section of the country to another and even to foreign countries. For example, the Southeastern states were once a major strawberry producing area, but the high prevailing humidity caused a serious mold problem. The main production areas then shifted to Washington and Oregon, but production there was limited to one crop by a relatively short growing season. California, with so-called everbearing varieties and a long growing season, then became the dominant producing area. Raspberries are grown mostly in the Pacific Northwest, not only because of a suitable climate but because harvesting has been done fairly cheaply and reliably by school children on vacation. There is some fear that this supply of labor will become more scarce and expensive, and that mechanical harvesting will be necessary to enable the industry to survive.

Cultural Factors.—Factors such as fertilization practices and irrigation have not only a quantitative but a qualitative effect on fruit crops. For example, both apricots and peaches have been shown to be lower in acid and astringency and firmer in texture when highly fertilized with nitrogen. In the case of apricots, low-nitrogen fruit had 1½% acid and a pH of 3.53, while high-nitrogen fruit had 1% acid and a pH of 3.86. This is an easily tasted difference and shows that flavor can perhaps be more or less tailored to a standard by control of fertilizer application. Perhaps also the high-nitrogen fruit could be harvested at an earlier date without having an excessively high acidity. Apples with high-nitrogen fertilization have been found less susceptible to core browning and softening in storage than those with low-nitrogen fertilization.

Maturity at Harvest.—It has always been desirable to know when a fruit has reached optimum maturity for harvest, but with the trend toward mechanical harvesting this knowledge is more important than ever. Unfortunately, however, in spite of a great deal of research to establish objective tests for maturity, a trained fieldman's judgment is still the best criterion. Single, simple tests such as the pressure test do not correlate very well with subjective maturity but can be used to exclude fruit below a certain maturity level. Thus, peaches testing more than 12 lb. pressure on pared cheeks would not ripen to give a high-quality product. Ground color, weight, and size did not provide a means for determining maturity at which peaches should be harvested. For want of a better criterion, they considered a combination of pressure test and ground color to be the best index for pickers.

Fluorescence of chlorophyll in peaches correlates fairly well with ripeness, but this kind of test is of little value to orchardists or pickers. Similarly, apple maturity was found to be indicated best by the elapsed time from full bloom. Changes in ground color, starch pattern, and pressure test were unreliable. With mechanical harvesting, especially if it is done only once or twice in a season, the judgment of proper average maturity of an orchard becomes more critical economically than with hand-picking. Perhaps some training or retraining will be required to estimate optimum maturity for mechanical harvesting, which may be different from that for hand harvesting.

Handling of Raw Material for Processing

Controlled Atmosphere Storage.—Since the finding that apple storage life could be greatly prolonged by an atmosphere relatively high in carbon dioxide and low in oxygen, the use of controlled atmosphere storage has become very widespread. Some varieties of apples can by this means be stored in good condition from one year to the next, other varieties have had their storage life approximately doubled. For example, Gravenstein apples can be held in good condition in air only about 2 or 3 months, but in controlled atmosphere (CA) storage they can be held about 6 months. After this length of time they still have a good color and firm texture but their flavor has begun to decrease noticeably.

While CA storage has been used mainly for fresh apples, it is also used for processing apples in two ways: first, packing house culls from CA storage of apples for the fresh market are often processed; and second, some processors of frozen apple slices have found it profitable to store their apples fresh under a controlled atmosphere rather than in the frozen state. In this case quality differences are not significant but the processing season is extended.

Controlled atmosphere storage of other fruits is in its infancy but shows promise for extending the fresh storage life of many fruits. Since the principle of using a controlled atmosphere high in carbon dioxide and low in oxygen content is to slow down the rate of respiration, it should extend the life of any respiring fruit. Such extension may be only for a few days or weeks with perishable fruits, but this may be economically significant. Pears, peaches, nectarines, apricots, sweet cherries, strawberries, citrus fruits, and grapes have been stored successfully under a controlled atmosphere. A modification of CA storage combined with rapid cooling is achieved by blowing cold nitrogen gas (from liquid nitrogen) over the fruit, resulting in both a low temperature and a low-oxygen environment. Although this system was developed for shipment of fresh lettuce, it should be applicable to fruits as well.

Ripening.—Most fruits for freezing are picked as near eating-ripe maturity as possible, since they do not ripen appreciably after picking. Exceptions are apples, which ripen slowly in storage, and Bartlett pears which are always picked at a hard-green but sweet stage. They are stored under refrigeration until a few days before use, when they are brought to a temperature of about 70° F. (21° C.) for ripening. Some tropical fruits are similar to pears in this respect, but very little of such fruit is frozen in the United States. There is no satisfactory method for "degreening" or ripening uniformly the common temperate-zone fruits which are frozen. If fruit is ripe on one side but green on the other, it sometimes can be used in a sliced product or jam after trimming, or it can be used for purée or juice after passing it through a finisher to remove the unripe portion along with skin, seeds, etc.

Mold Control.—Although mold occurs on all fruits, it is a particularly bad problem with strawberries. They provide a highly suitable medium for mold growth as well as a humid environment close to the source of soil molds. For this reason great care must be taken to prevent a buildup of mold in picking and processing equipment. Berry picking hallocks are steamed or treated with a fungicide such as orthophenylphenate after use.

Since ripe fruits in general are subject to infection with molds and other microorganisms, a constant program of control by the processor is required. This includes chlorination of wash water, protection of fruit from bruising during handling, and sorting out of damaged and moldy fruit, as well as frequent and thorough cleaning of equipment. Spraying or dipping the fruit to cover the surface with a mold inhibitor has been tried with varying success. One of the most effective control methods for preventing growth of microorganisms is rapid cooling nearly to the freezing point of the fruit. This procedure has been used extensively for fruit; often the only cooling is that obtained by exposing boxed fruit to the air overnight to dissipate field heat. The cooling system mentioned above which uses liquid nitrogen might be effective in preventing mold growth in fruits, since it not only cools but lowers oxygen concentration greatly.

Sorting.—Besides removal of moldy fruit, sorting is necessary to separate green or over-ripe fruit, under-or oversized fruit, and fruit with physical defects such as bird-pecks, hail damage, bruises, etc. The amount of such sort-out fruit can be crucial to the profitability of a freezing operation. Sound but off-sized, off-colored, or over-ripe fruit can sometimes be used in making purée or juice, but unless there is a steady supply, profitable marketing may be difficult. With the advent of mechanical harvesting, the sorting operation becomes more important than ever because the whole crop or a good part of it may be harvested at one time. Hand-sorting labor cost is constantly increasing, and unless sorting is mechanized along with harvesting, the extra sorting cost can nullify the saving in labor from mechanical harvesting. Therefore, there is a strong incentive to develop automatic sorting devices. Mechanical size graders, etc., have long been used, but several more sophisticated devices have been developed recently. These include photoelectric instruments to sort lemons, apples, cherries, and other fruits by color. Flotation baths to separate fruit by density difference and various specialized instruments for particular products, as for example a particle size classifier for powders based on their electrostatic properties are also used.

Processing Problems and Product Quality Related to Raw Material

Flavor and color are usually best in fruit that is frozen when soft-ripe, but processing of such fruit is difficult. The fruit is easily bruised or crushed and does not slice well or retain its shape after freezing and thawing. Thus a compromise is usually made between the characteristics which give the best flavor and color and those which make processing easiest. Selection and development of varieties for processing may be based on the intended product. For example, three varieties of peaches are used for three different frozen products by one packer: for sliced dessert peaches a variety is used which has a red pit cavity; for preserves a variety is used which lacks red color around the pit because the red turns brown when cooked; and for pie fruit a firm-textured variety is used which also has good resistance to oxidation.

Raspberries are subject to inferior fruit coherence, or "crumbliness," as well as "mushiness" when processed. Mushiness is due to impact during processing which abrades and crushes the fragile pulp tissues, but crumbliness involves genetic and pathological problems affecting normal fruit structure. Virus infection is apparently responsible for crumbliness, and susceptibility to virus infection is partly genetic in character. Thus, in breeding raspberries for processing this factor should be taken into account.

The group of polysaccharides known as *pectins,* or pectic substances, are important both in determining the texture of whole fruits and the quality of fruit juices in cold storage. Pectins are long chains of polygalacturonic acid molecules whose carboxyl groups are partially esterified with methyl alcohol. If all of the galacturonic acid groups have been esterified, it is theoretically possible to have a pectin with 16% methoxyl groups by weight. Natural pectins, however, only contain from 9 to 11% methoxyl groups. Methoxyl groups are easily removed by naturally occurring *pectinesterase.* When de-esterification has resulted in pectins with less than 8% methoxyl content (degree of methoxylation under 50% of saturation), these low-methoxyl pectins will form gels in the presence of calcium ions and other polyvalent cations. By contrast, high methoxyl pectins require a low pH and high sugar concentration to form the characteristic jelly. Gelation in frozen juices, therefore, results from two different special situations: (1) when juice has been concentrated to about 40% soluble solids at which time a true "jelly" forms with the high methoxyl pectin; and (2) the formation of low-methoxyl pectin gels in single-strength juices as low as 5% solids or in concentrates. Such low methoxyl pectins may result from activity of enzymes released from bruised or moldy fruit, since many molds are known to be high in pectinesterases. This sometimes happens with strawberry juice. In orange juice, pectinesterase occurs only in the solid particles suspended in the juice or if the vigorous extraction procedure used in modern processing results in the introduction of considerable enzyme.

In the early years of the orange juice concentrate industry there were two common defects, loss of "cloud" and "gelation." Citrus juices are naturally cloudy because of finely divided particles of cellulose and pectic materials. *Loss of cloud* is the descriptive term for the condition in a reconstituted juice where insoluble particles rapidly settle out to form an opaque layer at the bottom of the glass while the supernatant juice is clear. *Gelation* is the term for a concentrate which has formed a gel that does not reconstitute to a homogeneous mixture with added water, but forms clumps on stirring. Both of these defects are believed to be due to the action of the enzyme pectinesterase as previously described. This enzyme acts as a catalyst for the hydrolysis of the methyl ester bond of the pectin molecule, forming pectic acid and methanol. This is illustrated by the following formula:

$$
\underset{\text{Pectin}}{R-\overset{\overset{\displaystyle O}{\|}}{C}-O-CH_3} + H_2O \xrightarrow{\text{pectinesterase}} \underset{\substack{\text{Pectic}\\\text{acid}}}{R-\overset{\overset{\displaystyle O}{\|}}{C}-OH} + \underset{\text{Methanol}}{CH_3OH}
$$

where R represents the polygalacturonide chain of the pectic molecule. As a result of this action of pectinesterase on the pectic substances in orange juice, the resulting low-ester pectinic acids and pectic acids form insoluble compounds with calcium and magnesium in the juice. If the original pectin content is high, sufficient quantities of these compounds may be formed to make a stable gel. If the pectin content is relatively low, the pectinic acids will precipitate, carrying with them the suspended colloidal material which gives orange juice the characteristic desired turbid appearance. The pH of concentrated orange juice is favorable for this type of gel. The action of the enzyme is highly dependent upon temperature: orange concentrate stored at $0°$ F. $(-18°$ C.) will remain stable; in concentrates stored at $5°$ F. $(-15°$ C.) the cloud suspension of the reconstituted juice will break in about 3 months; at $15°$ F. $(-9.4°$ C.) in about 20 days; and at $+25°$ F. $(-4°$ C.) the suspension will break in 4 days.

It can be seen that frozen orange concentrate stored at the recommended temperatures of $0°$ F. $(-18°$ C.) or lower can avoid gelation or loss of cloud. Unfortunately, during distribution or in careless handling by the consumer higher storage temperatures are encountered.

As an insurance to provide additional stability for these periods of inadequate refrigeration, most processors apply heat to the evaporator feed juice to inactivate at least a portion of the pectinesterase. Temperatures of $145°$ to $160°$ F $(63°$ to $71°$ C.) for 5 to 30 sec. have been used commercially. Heat is applied by steam injection, as well as by tubular and plate-type heat exchangers.

Although there is a definite association between enzyme activity and cloud stability, there are other unknown factors which also play important roles in maintenance of cloud in frozen orange concentrate.

At the present state of our knowledge of textural and other changes in fruits occurring during preservation by freezing, reliance must be placed mainly on selection of raw material by experience. Varieties that are known to freeze best are used commercially when available, and improved varieties are compared with these in each of several characteristics mentioned earlier. Only if the overall comparison is favorable to the new variety is it likely to be adopted, and then only if production as well as processing factors are favorable.

Freezing

The preparation of fruits for freezing, involving the operations of peeling, cutting, treating to control enzymatic browning, syruping or sugaring, etc., are discussed elsewhere (see Additional Reading list). In this section, discussion will be limited to changes of the product taking place during and subsequent to freezing.

A great deal of research and observation has been devoted to the process of freezing and its effects on biological materials including fruits. The effect of rate of freezing on texture and other quality factors has been especially thoroughly studied. Strangely enough, there is still disagreement today on the desirability of fast *vs.* slow freezing for various fruits. This is at least partly explained by the wide variability in types of tissue encountered, with their different biochemical, physiological, and physical makeup.

Preservation of fruits by freezing depends on retardation of postharvest physiological changes, along with retardation of microbial action, by the action of low temperature. For prolonged storage it has been found that the temperature must be well below the freezing point of water (0° C., 32° F.), since many reactions, especially enzyme-catalyzed reactions, proceed at a measurable rate at 0° F. (−18° C.) or below. In a complex mixture of substances, such as the contents of plant cells, there is not one freezing point, but a number of eutectic temperatures where mixtures of a particular composition will freeze or solidify. The freezing of water is only the first of a series of these eutectic points, but of course it is an important one because a large fraction of the fruit is water. However, although freezing of water immobilizes it, the unfrozen components are free to interact, and in a concentrated form because of water removal.

Data (Fig. 4.3) showing deterioration of flavor and color of frozen strawberries, as well as of other frozen foods, is a straight-line logarithmic function of the temperature. The data presented cover temperatures between 30° and 0° F. (−1° and −18° C.) but there appears to be no reason

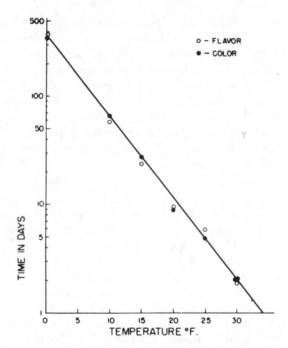

FIG. 4.3. EFFECT OF STORAGE TEMPERATURE ON SUBJECTIVE COLOR
AND FLAVOR CHANGES IN FROZEN STRAWBERRIES

it could not be extrapolated to lower temperatures. Thus, there is no
special temperature at which the product is preserved in a suspended
state of deterioration; but at temperatures near 0° F. (−18° C.) a small
decrease in temperature increases the storage life markedly. For example, the time for a detectable change in strawberry color and flavor is
about 160 days at +5° F. (−15° C.), 360 days at 0° F. (−18° C.), and 1,000
days at −5° F. (−20.5° C.).

Freezing of water out of the cell content mixture has a deleterious effect
on texture. Water removed by freezing from a biochemical system in
delicate equilibrium is not entirely replaceable on thawing: Colloidal
solutions become irreversibly dehydrated in cell membranes and this
causes a change in their permeability and elasticity. One result of this is a
loss of rigidity upon thawing, so that the fruit may be soft and perhaps
somewhat rubbery. At the same time there is excess fluid outside the fruit
from the irreversible dehydration, and less juiciness inside. The least
damage to fruit should occur when the least amount of water is frozen
out, at a low enough temperature to retard enzymic reactions sufficiently.
The least water is frozen out at a given temperature when the concentra-

tion of soluble substances is high, since dissolved solids cause a lowering of the freezing point. This is one reason that freezing certain fruits in syrup or sugar has been found to improve quality. However, the sugar should be dissolved in the cell fluid to be effective; packing with dry sugar and quick-freezing could produce a harmful osmotic dehydration instead of just a lowering of the freezing point.

Tests with viability of various cells as well as with stability of lipoproteins have confirmed that increasing the concentration of soluble solids within the cell stabilizes it to "freezing" and thawing. Viability of spermatozoa was found to increase greatly after frozen storage when glycerine was incorporated during freezing. There is always some unfrozen material in cells at normal frozen-storage temperatures. Not all the water in foods is frozen above −67° F. (−55° C.). In normally frozen foods there is a great deal of supercooling before there is a change in state from liquid to solid. It has been found that microorganisms survived better and proteins remained more soluble if they were supercooled than if they were frozen at the same temperature. Therefore ice crystal formation rather than temperature *per se* is important in causing damage.

The effects of ice crystal formation are both mechanical and chemical. Mechanical damage can arise from the increase in volume when water is frozen. Damage of this type is more severe in fruits with large intercellular air spaces and much water, such as apples, than in more solid foods such as beans. The cellular membranes become torn and slough off in pieces, causing a mushy appearance and texture. Chemical damage is caused by reactions of concentrated nonaqueous constituents as mentioned above, resulting in changes in pH and salting out of calcium and other minerals from proteins. This in turn can cause instability of colloids, oxidation of lipid material, etc. Fruit changes during freezing have not been studied extensively, but it has been shown in the case of peas that the pH decreases and increases during freezing and storage over a range of 0.6 unit, and acidity fluctuates likewise. This pronounced change in pH indicates that the salt-acid equilibria are undergoing large shifts, with probable large effects on colloid stability, etc.

In fruits it is of interest that orange and other citrus juices acquire a "cardboard," "tallowy," or "castor oil" flavor which may be worse at −30° F. (−34° C.) than at 0° F. (−18° C.), probably because of the greater concentration of lipid flavoring components and the enzymes which catalyze their oxidation.

Rate of Freezing.—As mentioned earlier, the effects of rate of freezing on quality characteristics of frozen fruit are controversial. So-called "quick-freezing" was developed more for freezing vegetables than fruits, because in general there is more benefit for vegetables. One reason for this is that the enzyme activity which causes objectionable off-flavors is more pronounced in vegetables and therefore anything that will minimize enzyme action, such as fast freezing, will aid in retaining quality. On the other hand, nowadays most vegetables are blanched to inactivate enzymes and thus the benefit of rapid freezing on flavor is decreased. In fruits, enzymes producing off-flavors are not especially active; darkening of color by oxidative enzymes such as polyphenol oxidase is more important. Where enzyme-catalyzed changes in color are not pronounced, speed of freezing is of lessened importance. Berries are in this category and they make up a large part of the frozen fruit pack.

In the early days of berry freezing, the fruit was put in barrels with or without sugar and stored in a cold room to freeze. It sometimes took days or weeks for the fruit in the center of the barrel to freeze by this method. Nevertheless, the product was usually of satisfactory quality for making ice cream, etc., and if not, the reason was more than likely the growth of microorganisms. Quick cooling of the berries to a temperature below 40° F. (4° C.) was considered more important to quality than quick-freezing. There are reported no marked differences in texture, microscopic appearance, or palatability of strawberries, raspberries, and peaches packed in syrup, whether they were frozen slowly or rapidly. Slow-freezing was in insulated boxes in still air at 0° F. (−18° C.), an intermediate rate was attained in open cartons in still air at 0° F. and fast-freezing was in liquid air. Data indicate no difference in the condition of apple and cherry pie fillings when they were frozen slowly (14 hr.) or "rapidly" (2 hr.).

Very rapid freezing in liquid nitrogen has been found beneficial to the texture of strawberries and sliced tomatoes. Very likely the decrease in amount of drip found after thawing was related to the decrease in mechanical damage sustained by keeping the ice crystals fine and evenly distributed, and also to the decreased movement and interaction of nonaqueous constituents, as described above. This beneficial effect occurs only in a few fruits, even at the rapid freezing rate attained with liquid nitrogen, which is another indication that there is little distinction in effect on quality between fruits "slow-frozen" or "quick-frozen" at the usual temperatures.

Nevertheless, great advances have been made in the techniques for freezing fruits rapidly. The present individually quick-frozen (IQF) and cryogenic frozen fruits are superior in quality and stand up better upon thawing than the fruits frozen slowly in packages, cartons, or bulk containers.

Methods used for various fruit products will be mentioned in the discussion under each particular fruit.

The Preparation of Fruits for Freezing.—Details of the specific preparation operations for different fruits will be discussed under the sections on each fruit. General unit operations, common to most fruits, are covered in the following paragraphs.

Cleaning.—Fruits are delivered to the freezing plant in a variety of lugs, baskets, bulk bins, or bulk truckloads. Debris picked up in the orchards, storage yards, and handling equipment must be removed before processing. Dumping of the containers into flumes of water has become more and more common in the industry. This not only helps to clean the fruit but permits the movement into the plant with less damage. The use of detergent washes and spray rinses is usually effective for removing dust, spray residues, and other surface contaminants.

Handling.—A continuous, smooth movement of raw fruit into the plant and of finished products out of the processing area is desired. From initial fluming operations, the whole fruit can be conveyed to subsequent operations by means of belts or screw conveyors, or may be carried in pans, boxes, or bins.

Grading.—Samples of raw products are taken at the receiving station as a basis for determining grade. The grade is used as a basis for payment of the grower, and can include specifications for size, maturity, freedom from rot and blemishes, and absence of insects and foreign matter. Reasonable allowances are made but the schedule of payments rewards the grower who delivers the highest quality to the processor.

Pitting and Peeling.—Most stone fruits are pitted by means of halving and pit removal (peaches, apricots) or by punching out the pits (cherries). Peaches may be peeled by scalding in hot water, steam, or hot lye solution. Details will be discussed under the individual fruits.

Sorting, Grading, Inspecting.—After initial grading when received, most fruits pass through additional inspection procedures during their preparation for freezing. Size grading may be performed mechanically and a great deal of foreign material may be removed by screening, shaking, blowing, and other devices. However, the final removal of blemished, damaged, rotten, overripe, immature, off-color, misshapen, and other defective fruit must be accomplished by constant and careful sorting and inspection. Objective methods for color sorting have been developed but have not been applied as yet to any of the fruits which are commonly frozen.

Commercial Procedures for Individual Varieties

Apples.—Nearly all of the apples which are frozen commercially are packed in large containers and are utilized by the baking industry, mainly as pie-stock. A small pack of frozen applesauce in both retail and institutional containers is put up each year. Frozen baked apples have been available off and on for the past 20 years or so.

For freezing it is essential that apples have a low tendency to brown, have good flavor, and a texture which will not disintegrate or become mealy.

Varieties.—For many years the leading variety for processing in the eastern states has been the Rhode Island Greening. Baldwin, Northern Spy, Wealthy, Ben Davis, McIntosh, Cortland, Monroe, and Webster varieties are used in the frozen pack also, but not all of them are ideal freezers. McIntosh and Cortland tend to disintegrate even when packed soon after harvest.

In the Pacific Northwest, the varieties best for freezing are Jonathan, Stayman Winesap, Yellow Delicious, and Yellow Newtown. In Colorado and Utah the Jonathan, Rome Beauty, Stayman Winesap, and Yellow Newtown are important. In California only the Yellow Newtown is suitable for pies, Gravenstein can be processed for frozen sauce.

The varieties for freezing in Michigan are much the same as those used in New York and the Northeastern states, with Rhode Island Greening and Jonathan topping the list as to quality. York Imperial has proved to be one of the most desirable freezers in the Mid-Atlantic states.

Processing.—After sorting and size grading, apples are prepared for freezing by peeling mechanically, coring, trimming by hand, and slicing. To prevent browning the peeled or sliced fruit is held in a salt solution, containing 1 to 3% NaCl. Sulfite or sulfurous acid treatments have been used extensively on slices for frozen pie stock. This has the disadvantage of giving an undesirable flavor and a poor texture to the slices.

A third method to prevent enzymatic action causing browning consists of blanching in steam or hot water. The disadvantage of blanching is that leaching causes loss of solids and flavor compounds when the fresh apple slices are subjected to moist heat. Finally, it is possible to prevent browning by a treatment with ascorbic acid.

Other treatments include the addition of a sugar syrup containing ascorbic acid and insuring penetration into the slices by putting them under vacuum for several minutes. Apple slices treated in this manner have a better flavor and texture than those treated by steam blanching or sulfite dipping.

Soft varieties of apples or over-mature fruit may be improved in texture by dipping in solutions of calcium salts. The lactate, malate, and phosphate salts have less salty flavor than calcium chloride, but are less soluble and can be used only in low concentrations.

A good scheme for the packing of frozen pie-stock apples is illustrated in Fig. 4.4.

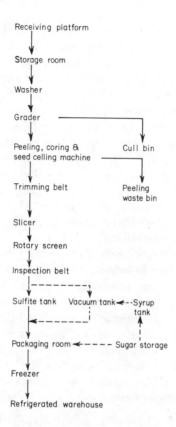

Receiving platform

Storage room

Washer

Grader ———————————————

Peeling, coring &   Cull bin
seed celling machine ————————

Trimming belt   Peeling
                         waste bin

Slicer

Rotary screen

Inspection belt

Sulfite tank   Vacuum tank ◄--Syrup
                               tank

Packaging room ◄ — — — Sugar storage

Freezer

Refrigerated warehouse

FIG. 4.4. FROZEN APPLE PROCESSING FLOW SHEET

Apricots.—Although the pack of frozen apricots has never been large, it has increased several fold in the last 10 years or so to something over 16 million pounds. Apricot halves are preferred to sliced fruit and can be packed with or without peeling. Practically all of the pack is frozen in 30-lb. cans for the baking, jam, and preserve trade.

Apricots for freezing should have low browning tendency, even ripening, good color, good flavor, a tender and smooth skin, and firm texture.

Varieties.—In California, where most of the commercial apricots are grown and processed, Tilton and Blenheim are still the most important varieties, although Hemskirk and Moorpark are frozen in small amounts. In the Northwest several varieties have been tested on a small scale for freezing, such as Blenheim, Tilton, Earliril, Blenril, and Perfection.

Processing.—Upon arrival at the processing plant, the apricots are first inspected on a conveyor belt and then passed through a halving and pitting machine. After separation of the halved fruit from the pits, the halves are inspected again. Then they are washed and finally treated to prevent browning before they are syruped, packaged, and frozen.

Three methods have been successful for the prevention of browning: the fruit may be blanched in hot water or steam, dipped in a bisulfite solution, or treated with ascorbic acid. Blanching in a single layer on a mesh belt for 3 to 4 min. in steam is satisfactory for firm fruit. It is probably best to treat softer fruit with SO_2 or ascorbic acid. The latter may be incorporated at a level of 0.05% or more into the syrup used for packing. The SO_2 treatments should be adjusted to leave a residue of 75 to 100 p.p.m. in the frozen apricots.

For the baking, jam, or preserve trade, the SO_2 treatment is satisfactory and results in better color. The packing medium can vary from 15° Brix syrup, to dry sugar at 3 parts of apricots to 1 part of sugar or even a higher proportion of fruit, depending on the specifications of the buyer. When dry sugar is used, it may be sprinkled on the fruit as it is filled into the container or can be distributed evenly by light mixing. Apricots which are to be reprocessed and filled into cans should not be treated with SO_2 since the SO_2 will be reduced by tin to hydrogen sulfide.

Retail packages of apricot slices or halves have been marketed from time to time, but have never been produced in any great volume. Unless removed, the fruit skins give a tough texture. Browning has been a serious problem, too, and can be controlled by a combination of blanching and packing in heavy syrup containing 0.1% by weight of ascorbic acid.

Avocados.—Freezing has been the only successful method for preserving avocados. Growers freeze halves and slices using liquid nitrogen, and pack them in polyethylene bags for sale. The Fuerte and Haas varieties still dominate the California production, accounting for about ¾ of the more than 60,000 tons grown; however, appreciable amounts of MacArthur and Nabal are produced. Florida produces some 15,000 tons of avocados annually, but almost all of it is sold as fresh fruit. The principal varieties grown in Florida are Lula, Booth 7 and 8, Collinson, and Waldin. An introduction to the literature on the Florida industry may be found in the bibliography. Hawaii and Texas each have a small production and have developed variations in the preparation of frozen guacamole salad.

The basic processing modifications introduced consist of acidifying the purée to about pH 4.5 by adding larger amounts of lemon or lime juice and extra salt. This treatment permits retention of natural flavor and light green color for at least a year in frozen storage.

Blackberries, Boysenberries, and Other Bramble Berries.—There are many types of berries of the bramble berry group. These berries vary in size, shape, flavor and color, drupelet size, and seed size and shape. Dewberry is a name referring to the types of blackberries having trailing vines. Many dewberry varieties, such as the Boysen, Logan and Young, are referred to as Boysenberries, Loganberries, and Youngberries as though they were a distinct kind of berry.

Characteristics of desirable varieties for freezing include even ripening and resistance to bruising during transportation and handling. Absence of flinty seeds or woody cores, and rich flavor are essential. Blackberries which tend to revert to red color when frozen are not desired, as they lack the customary black appearance and may lose grade by being judged as fruit of mixed maturity.

The commercial freezing of blackberries is largely a Pacific States operation, since upwards of 80% of the U.S. production in recent years has come from the Pacific Coast. The major blackberry producing area in the country is Western Washington and Western Oregon. Evergreen and Himalaya (Theodor Reimers) are the two most important varieties of this region, although a number of small-seeded, high-flavored varieties developed by George Waldo, of the USDA, cooperatively with the Oregon State Experiment Stations in Corvallis and Aurora, have earned some favor. These include the Chehalem, Cascade, Pacific, Olallie, Marion, and Aurora varieties.

In California the Olallie and Boysenberries are grown in greatest quantities with Boysen dominating the blackberry picture.

Blackberries are handled in much the same way as raspberries in pre-freezing treatment. The berries are picked in flats which are delivered to the processing plants. In plants so equipped, the berries are run through the air cleaner before the washing operation is performed. From the washer the fruit goes over the dewatering shaker and onto inspection belts. The fruit may be packed in plastic lined corrugated boxes in 30-lb. lots. Other containers widely used are 30-lb. slip cover enameled cans and plastic-lined steel drums. There is a very small production for the retail trade. The berries are packed straight or in a five plus one sugar pack. While the presence of the sugar is believed to "set the flavor" better, many industrial users of frozen blackberries, for reasons of economy, prefer to add all the sugar at their plants.

Blueberries.—The blueberries of commerce have their origin in the native blueberries of North America. Two native species have been frozen commercially: *Vaccinium lamarckii,* the Low-bush blueberry of New England, the Atlantic Provinces of Canada, and to a lesser extent Michigan and Wisconsin, and *Vaccinium ovatum,* the Evergreen blueberry of the Pacific Coast areas north of San Francisco Bay. Of these, the Low-bush blueberry is the more important. It makes up much of the Eastern blueberry pack. The Evergreen Blueberry was packed in good volume prior to World War II, but more recently has been packed only in decreasing tonnage. Insect infestation has been a major reason for the lessening importance of this berry.

Most of the varieties now being cultivated are descendants of the wild high-bush blueberry. Rubel was the only important commercial variety not the result of breeding and is the only one of the early selections from the wild to be important in today's blueberry market. The Rancocus and June varieties, developed through hybridization, are the only commercial blueberries not of pure high-bush parentage.

Varieties.—There are many varieties of blueberry. The varieties vary greatly in size, color, stem scars, firmness, and flavor. Desired varietal characteristics for freezing are: tender skin, large size, and sufficient acidity to give a sprightly flavor. The retention of the natural bloom on the berries is also a desirable characteristic. The leading varieties include:

Maine: Wild low-bush blueberries.

East: Jersey, Rubel, Bluecrop, Stanley.

Michigan: Jersey, Rubel, Bluecrop.

Pacific Northwest: Jersey, Bluecrop, Stanley, Rubel, Earliblue.

Characteristics of the high-bush varieties listed above are:

Bluecrop is a bush with average vigor; upright spreading; productive; fruit cluster loose; berry large, oblate, light blue; firm, resistant to cracking, slight aroma, above medium in dessert quality; small scar.

Earliblue is a vigorous, upright, productive bush; loose fruit cluster, large, light blue, firm berry; resistant to cracking, good dessert quality. Good scar.

Jersey is a vigorous, erect, productive bush; fruit cluster very long and very loose; berry medium, round-oblate, good blue color, firm, lacking aroma, medium in dessert quality; scar good; season late.

Rubel is selected from wild. Erect, vigorous, productive bush, very loose fruit cluster; small berry, oblate, medium blue, firm with slight aroma, fair dessert quality; scar good; late.

Stanley is an erect, vigorous, productive bush; medium, loose fruit cluster; medium size berry, oblate, of good blue color, very firm, very aromatic, of high dessert quality; scar above medium size; early mid-season.

Two varieties not on this descriptive list, Atlantic and Coville, are said also to be excellent for freezing.

In some areas, especially in the Pacific Northwest, many varieties are planted and blueberries are delivered to the processors with the varieties mixed. In some ways this is good, as the flavor effect of several varieties in a blend may be more pleasing than that of some single varieties.

In the future, it is possible that the prime factor to be considered in selecting blueberry varieties for planting will be how easily the varieties can be harvested. With harvest labor becoming more expensive and more difficult to obtain, the grower must turn to mechanical harvesting of high-bush blueberries in order to get his crop on the market. Factors such as a concentrated ripening of berries, ease in shaking the fruit from the bush, and resistance of the bush to damage from shaking must be considered in the mechanical harvesting of blueberries. USDA Selection 1613A is being widely planted in the Pacific Northwest because it is one of the easiest varieties to harvest by machine. Mechanical pickability is not the only attribute of this berry, but plays a big part in the increased planting of the selection. What may end in the future as the best available variety could very well be a variety whose chief attribute is that it is easily harvested by machine.

Freezing of Blueberries.—Blueberries are frozen in several ways. Much of the frozen blueberry pack is bulk for the preserve or bakery trade. The berries are washed, inspected, and filled into polyethylene-lined fiberboard containers holding 30 lb. of fruit, into 20-lb. lots in slip covered enameled cans, or in 60-lb. bags. For the retail trade, 10 oz. cartons are packed either as straight dry berries, as a 4:1 dry sucrose pack, or with a 40–50% syrup. For blueberries which are to be eaten as such, berries packed in 50 to 60% syrup or 4:1 dry sucrose have superior texture and appearance. Blueberries frozen in 60% sucrose syrup are too firm when baked into pies, but all others, whether dry berry water-pack, 20°, 30°, 40°, or 50° syrup, or 4:1 dry sucrose pack, have good texture.

Skin toughness can be a problem with frozen blueberries. Immersing of berries in boiling water for 30 to 60 sec. can eliminate skin toughness. The effect of scalding varies with time in storage. After a year in storage there is little preference for scalded over unscalded fruit.

Cherries, Tart.—Tart cherries are responsible for the second largest pack of frozen fruits, being surpassed only by strawberries. The trend in size of the pack has been generally upward in the past 20 years, interrupted by years of low production due to biennial bearing, spring freezes, and shortage of harvest labor. Mechanical harvesting has progressed rapidly in recent years, especially in Michigan, where about half of the crop was shaken from the trees in the last 2 or 3 years. Processors who were at first reluctant to accept mechanically harvested fruit now find that high grade cherries can be delivered to their plants at lower cost than hand-harvested fruit.

Tart cherries should be harvested when they are bright red in color. Immature fruit is too pale and lacks good flavor. Overmature fruit is too soft and too dark in color. The best operation which the author has seen was in Michigan where the fruit was shaken from the trees and then passed over a short sorting belt right in the field. At the end of the belt the cherries fell into a tank of cold water and were taken immediately to the processing plant. These cherries made 92% or better U.S. No. 1 in grade. One problem resulting from mechanical harvesting was the need to remove stems from the cherries at the plant. Several ingenious devices have been built to cut or knock the stems off the cherries without damaging the fruit. Figure 4.5 shows a harvester in action. It consists of 2 inertia shakers on 2 self-propelled units, and is capable of harvesting 30 to 40 trees per hour. The conveyor is at the bottom center, elevator along the right front. A 1,000-lb. capacity orchard tank (not shown) receives the fruit from the end of the conveyor. With this system 3 men can harvest as many cherries as 100 hand pickers could.

Varieties.—One variety dominates the red tart cherry (or red tart pitted, RTP) industry—Montmorency. Some Morellos and even fewer Richmonds are grown, but are of minor importance to freezers.

Processing.—Transportation and holding of cherries in ice cold water have improved the quality greatly. This technique not only prevents crushing and bruising, but results in firmer fruit and less juice loss during pitting.

From R. T. Whittenberger, Eastern Utilization Research and Development Division,
USDA, Philadelphia

FIG. 4.5. MECHANICAL TART CHERRY HARVESTING UNIT

Two inertia shakers, two self-propelled units. Thirty to forty trees per hour. Conveyor at bottom, elevator in right background. Missing (from right foreground) is 1,000-lb. capacity orchard tank of cold water, into which the elevator dumps the cherries. About 350 of these units brought in 47% of the 1967 crop in Michigan. Three men take the place of 100 pickers.

The cherries can be held in water at 32° F. (0° C.) in the processing plant for about a week, which helps to even out harvesting and production schedules. A small amount of calcium chloride in the cold water helps to firm up soft or over-ripe fruit.

The cherries are drawn from the holding tanks as required and are conveyed or flumed to the size grader, where fruit less than ⅝ in. in diameter drops through the belt. The larger fruit comes onto an inspection belt; the smaller cherries are graded for size again into those larger and smaller than ½ in. in diameter (Fig. 4.6). A recent innovation in color sorting has been the introduction of automatic electronic machines. Next the fruit goes to the pitting machines (Fig. 4.7). Studies have been made on the effects of various harvesting, handling, storage, and processing conditions on the firmness of tart cherries as related to pitter loss. The pitted cherries are packed with sugar in 30- or 50-lb. enamel-lined cans, according to the specifications of the buyer. Finally, the cans of cherries are frozen in a sharp freezer.

Cherries, Sweet.—Sweet cherries are produced and processed in the United States to a larger extent than tart cherries, but most of the sweets are canned or brined, not frozen. Sweet cherries tend to oxidize more easily than tart cherries and they benefit from being packed in sugar syrup containing ascorbic acid.

The dominant varieties of sweet cherries for freezing are Bing, Black Republican, Lambert, Napoleon (Royal Ann), and Windsor. In addition to the extensive studies on mechanical harvesting of tart cherries discussed earlier, a considerable amount of work has been done on the mechanical harvesting of sweet cherries. Although this work was concerned more with canning and brining varieties, the results obtained should be applicable to the freezing varieties. Processing and freezing procedures for sweet cherries are essentially the same as for tart cherries, both for the institutional and the retail markets.

Coconuts.—Many years ago it was reported that fresh coconuts could be preserved by freezing. Being such a flavorful and nutritious food, it seemed a waste that most coconut production went into the nonfood product, so workers at the Food Processing Laboratory at the University of Hawaii initiated a study of a number of food products made from coconut. They developed a frozen coconut milk and frozen shredded coconut of excellent quality.

Cranberries.—Cranberries freeze and maintain quality better than any other fruit, having a shelf-life of several years in 0° F. (−18° C.) or lower storage. Large quantities of whole fruit are frozen in bulk for later processing into juice, sauce, or jelly.

Varieties grown in Massachusetts and New Jersey include Early Black,

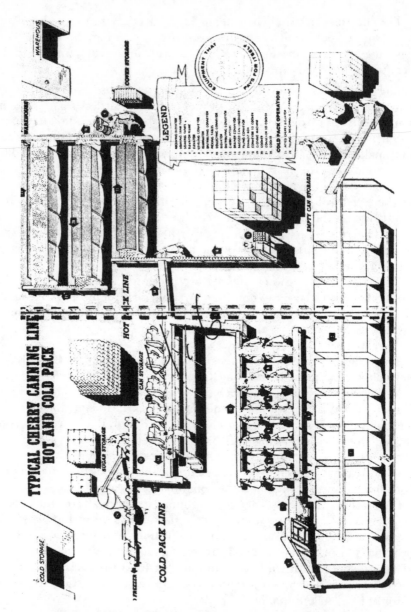

FIG. 4.6. TYPICAL CHERRY PROCESSING LINE, HOT AND COLD PACK

1. Receiving elevator. 2. Soaking tank flume. 3. Soaking tanks. 4. Elevator flume. 5. Elevator. 6. Leveling conveyor. 7. Eliminator. 8. Distributing conveyor. 9. Picking tables. 10. Collecting conveyor. 11. Elevator. 12. Distributing conveyor. 13. Cherry pitters. 14. Drainer conveyor. 15. Bagged can unloader. 16. Filling conveyor. 17. Exhaust box. 18. Conveyor to cooker. 19. Closing machine. 20. Cooker. 21. Conveyor to cooler. 22. Cooler (Cold Pack Operation) 23. Pitted cherry elevator. 24. Filing, weighing, and sugaring unit.

FIG. 4.7. CHERRY PITTER

The Dunkley cherry pitter will pit ¾ to 1 ton of cherries per hour. It works at a speed of 96 strokes per minute, pitting up to 53 cherries at each stroke.

Howes, and McFarlin; in Oregon, Washington, and Wisconsin the McFarlin is the principal variety grown.

The cranberries are harvested in much the same way as blueberries, by using a rake or a scooper. The main operation in preparing them for freezing consists of cleaning by blowing out leaves and chaff in a fanning mill. Soft or rotten fruit is eliminated by causing the sound berries to bounce over a barrier, then they are graded according to size. After washing in a tank of cold water, the berries are stemmed, drained, inspected, filled into containers, and frozen.

Currants, Black.—Products made from black currants are very popular in Europe, but are little known in the United States. In most states the cultivation of black currants is prohibited because this is the alternate host for white pine blister rust. Varieties recommended are Baldwin, Kenny, and Saunders for growing in Canada.

After harvesting, the berries are cleaned, washed, sorted, and stemmed. They may be packed whole, without sugar or syrup, and frozen at 0° F. (−18° C.).

Currants, Red.—Red currants are handled for freezing in the same manner as blacks. Most of this fruit is frozen for subsequent manufacture into jam and jelly, including the Red Lake and Wilder varieties for New

York and London Market, Perfection, Red Cross, and Wilder for Michigan. Perfection dominates the Northwest production, with Perfection, Red Lake, and Stephen's No. 9 as the best freezing varieties for Ontario.

Currants, White.—There is no reported manufacture of frozen white currant products, but two varieties have shown possibilities—White Imperial in New York and White Grape.

Dates.—The USDA Date Field Station in Indio, Calif., states that Barhee, as well as Khadrawy, Halawy, and even Deglet Noor varieties, make excellent frozen products. The Barhee variety may be frozen and held for 2 or 3 years at $-20°$ to $-30°$ F. ($-29°$ to $-34°$ C.).

Elderberries.—Several attempts have been made to commercialize frozen elderberries for the making of jelly and pie. One variety, Adams, was selected and cultivated about 40 years ago in New York. In general, the preparation and freezing steps for pie stock are similar to those given for blueberries, except for the stemming operation.

Figs.—Most of the varieties of figs grown in the United States are preserved by canning or drying. Calimyrna is the principal variety dried in California, but its thick skin does not permit a very good frozen product. Likewise, Kadota, a canning variety, is not very suitable for a frozen dessert fruit. Black Mission figs make an exceptionally good frozen product.

Good color of figs is maintained by using ascorbic acid and citric acid, or by dipping whole figs in a 2,000 p.p.m. solution of bisulfite for 2 to 3 min. before packing the fruit in 35° Brix sugar solution to preserve flavor and color.

Gooseberries.—These berries freeze very well for subsequent use in jams and pies. The major commercial packs are made from Downing, Poorman, and Oregon varieties in the Northwest.

Grapes.—A good-sized pack of "grapes and pulp" is frozen for remanufacture into jam and jelly. The principal variety is Concord. Muscadine-type grapes are frozen also for dessert products, consisting mainly of the Hunt variety and smaller quantities of Scuppernong and Thomas.

Since frozen pulp retains flavor and color better than whole grapes, most of the product packed in bulk for later conversion into jam and jelly is heated to 140° to 150° F. (60° to 67° C.), then the seeds are removed in a finisher and the pulp is frozen in large cans or barrels.

Guavas.—See guava purée in the section on fruit juices and purées later in this chapter.

Lychees (or Litchis).—This fruit is very popular in the Orient and finds a ready market among Chinese-Americans. Frozen in the hard shell, as a peeled fruit, or as a pitted fruit, lychees have excellent quality.

Brewster, Kwai Mi, Hak Ip, and Groff are well-known varieties grown in Hawaii; Brewster (Chen-tze) is the only variety of consequence in Florida, although it is sometimes sold as "Royal Chen."

Mangos.—The mango is one of the most delicious of all fruits and, fortunately, it maintains its excellent qualities when frozen as slices or chunks in syrup. Although it is a tropical fruit, there is some production in both Florida and Hawaii. Varieties such as Irwin, Keitt, Kent, Sensation, Smith, and Zill have been planted in Florida, but Haden remains the dominant variety. Pirie and Haden dominate the production in Hawaii, but a number of seedlings, hybrids, selections, and sports have been planted.

Melons.—Muskmelon and Honeydew melon flesh cut into 1⅛ in. balls and mixed in about 50:50 proportions makes a delicious frozen product; the mixture is now preferred to 100% Honeydews. After washing, the melons are cut into halves, deseeded, and cut into round balls with spoons. After inspection to remove imperfect pieces, the fruit is washed with water sprays and filled into either retail or institutional-sized containers, along with 28° Brix sirup. Racks of the packages are transported into an air-blast tunnel for freezing.

Nectarines.—This fruit is a smooth-skinned sport of the peach. A number of varieties, both white- and yellow-fleshed, are suitable for freezing, including Gower, Humboldt, Kim, New Boy, and Stanwick. As with peaches, it is important that nectarines for freezing have pronounced flavor, firm texture, and resiswnce to oxidative browning. The preparation of nectarines for freezing is similar to that for peaches.

Olives.—Although it is possible to freeze ripe olives and retain the flavor of freshly cured olives such as Mission, the skin texture becomes tough and the flesh texture may be too soft upon thawing. However, the time may be right to investigate again the possibility of marketing in the frozen form and to try newer methods of freezing which may result in superior quality.

Papayas.—In Hawaii the attempts to diversify agriculture have been most successful in enlarging the planting of papayas. Along with increased supplies for the fresh fruit market, more and more fruit has been processed in various forms. Frozen purée and frozen chunks or pieces have been produced in limited quantities.

Several lines of a variety named "Solo" are grown for the fresh fruit market, but are not well-suited for processing. It requires a thicker, firmer flesh. Another problem which must be solved before processing on a large scale would be feasible is the development of an improved machine for peeling the fruit.

Some years ago a delicious frozen fruit salad containing papaya chunks was marketed by the Dole Company, but it is no longer available. By careful selection and proper ripening of the fruit, the papaya chunks could be kept intact during the freezing and thawing process. If necessary the pieces could be dipped in a dilute calcium chloride solution to increase firmness.

To prevent the gelling of frozen papaya purée, flash pasturization is effective, and adding sucrose can inhibit the pectinesterase enzyme responsible for gelling.

Peaches.—In 1928, Culpepper, Caldwell, and Wright of the U.S. Bureau of Plant Industry recommended the freezing preservation of crushed peaches and peach pulp for use in ice cream manufacture. However, when the freezing of Georgia peaches was first undertaken on a large scale in 1930, sliced peaches were packed in consumer-size cartons. This merchandising of small cartons of frozen peaches was not financially successful, largely because of marketing difficulties. After the first year there was a shift to large containers to supply the wholesale trade.

In 1932, Birdseye Frosted Foods began, on a relatively small scale, to pack sliced peaches in heavy syrup in small cartons. Soon other companies began packing sliced peaches in syrup for sale at retail. The product was not altogether satisfactory because it was difficulty to thaw peaches without marked discoloration of the upper layer of fruit. This difficulty was eliminated by certain companies by the addition of ascorbic acid to the syrup.

From the 1940's, when 15 to 20 million pounds of peaches were frozen commercially, the pack increased to 40 to 50 million pounds annually in the 1950's. Over the last several years the frozen pack has grown to 60 to 70 million pounds per year and has remained in this range, with annual fluctuation due to freezes, short crops, etc.

Varieties.—Several dozens of freestone peach varieties have been found to be satisfactory for freezing, but none has all the characteristics considered desirable for the several types of frozen peach products. An ideal variety for freezing might have the shape, bright color and firmness of Rio Oso Gem, the flavor of Elberta, and the nonbrowning characteristic of Sunbeam.

The well-known varieties such as Elberta, Early Elberta, Fay Elberta, J. H. Hale, Halehaven, Redhaven, Kirkman Gem, and Rio Oso Gem still dominate the peach freezing industry in the major producing areas.

Freezers need three different types of peaches. For retail packages of slices, varieties with red around the pit cavity, such as Rio Oso Gem, are desirable. Peaches frozen in bulk for later manufacture into preserves should have good flavor and should not have red color around the pit cavity since it would detract from the appearance of the jam or marmalade. Fay Elbertas picked on the immature side to reduce the red center have proved to be satisfactory for preserves. For the institutional market, principally pies, a highly flavored, firm-textured variety with resistance to oxidation is required. Pie and dessert markets in particular need new varieties of the Gem type which will maintain good flavor through the entire freestone harvesting season.

To satisfy these requirements, Grant Merrill of Exeter, Calif., has been breeding peaches with characteristics suitable for freezing for many years. At the Western Regional Research Laboratory, a number of named varieties and numbered seedlings have been evaluated as frozen slices in syrup. Several of these newer selections scored as well or higher than standard varieties such as Kirkman Gem, Fay Elberta, and Rio Oso Gem.

Processing.—Peaches must be pitted, peeled, and sliced before freezing. Figure 4.8 illustrates one type of pitter currently used by the freezing industry. In the right foreground of the figure is shown the cup on which the peach is placed, and which swings up into the cutting position. The fruit is held here gently by spring-loaded arms while cutting blades from

Courtesy of Filper Corp., San Ramon, Calif.

FIG. 4.8. FREESTONE PEACH PITTER

four directions sever it into halves, push the pit out, and move the halves on to the next operation. After halving, the peaches may be peeled by passing them through a 10% lye bath at 140° F. (60° C.) for about 2 min. The loosened peel may then be removed by water sprays or by rubbing or brushing, and the remaining fragments are dislodged by hand. Following peeling, the peach halves are rinsed in clear, cool water and later by a 2% solution of citric acid to neutralize excess alkali and to retard browning. Slicing takes place next by means of rotary disks or fixed-blade knives which cut each half into 5 to 10 slices depending on the size of the fruit.

An alternate method of peeling consists of scalding the peaches in steam for 1 to 2 min., depending on the size, maturity, and variety of the peach. After scalding, the flesh is cooled by a water spray to loosen the skins and to prevent softening and darkening by the heat.

Sliced peaches for dessert use are packed in 40° syrup containing 0.1% by weight of ascorbic acid. A ratio of 2 to 4 parts fruit to 1 part syrup generally is acceptable. For the bakery and ice cream trade, slices are packed in syrup in 30-lb. enamel-lined slip-cover cans. Freezing takes place on racks or before air blasts in most freezing plants.

A great deal of literature not mentioned in the above paragraphs is available for consultation on various aspects of processing peaches for freezing, such as: maturity and ripening of peaches for freezing; the oxidation of phenolic compounds and the effect of tannin content on astringency of frozen peaches; the volatile components of the peach; a modification of the lye peeling method; and the varieties, harvesting, ripening, storage, and processing of freestone peaches.

Pears.—Pears are not frozen commercially in the United States, but there have been reports in the past of packers in Great Britain. The same methods used for apples may be used on pears, although the finished product is apt to be grainy due to the large number of stone cells in pears.

Persimmons.—The Japanese Hachiya variety of persimmons maintains good color, flavor, and texture when frozen as a purée. With sugar added this purée has been used as an ice cream flavor and for persimmon pudding.

Pineapple.—Frozen pineapple chunks have been an item of commerce for quite a few years. The Smooth Cayenne variety, grown in Hawaii and to a limited extent in Puerto Rico and other tropical regions, is the principal variety frozen. The Red Spanish variety of Cuba develops off-flavors when it is frozen. More than most other fruits, pineapple has a good texture when thawed due to its slightly fibrous structure.

In recent years other cuts of pineapple, such as crushed and tidbits, have been made available as frozen products for the institutional market, principally the bakery trade.

Pineapple for freezing is prepared in about the same way as for canning except that the cylinders usually go through an additional coring operation to remove the last vestiges of the fibrous core. Bins of fruit arriving from the plantation are unloaded onto conveyor belts or into flumes of water. Next they are washed thoroughly, graded into 3 or 4 sizes, and peeled and cored on the Ginaca machine. The cored cylinders are inspected and trimmed and then diverted to a second coring machine and a fixed blade chunk cutter. The chunks are filled directly into 211 × 414 or No. 10 cans, syruped, seamed, and conveyed directly to a tunnel blast freezer at −30° F. (−34° C.). The bulk-frozen packages are filled directly from the product line (crushed or tidbits), syruped and frozen in a low-temperature blast freezer.

Plums and Prunes.—In New York, the Damson, Redwing, and Yellow Egg plums, nd the Italian, German, Imperial Epineuse, and Stanley varieties of prunes are considered best for freezing. Michigan freezes some Green Gage plums and some Stanley prunes. In California the Santa Rosa plum and the French prune are frozen. In Oregon and Washington the only variety frozen is the Italian prune.

In the 1940's large quantities of prunes were frozen, principally in 30-lb. tins for remanufacture by fruit processors and bakers. The frozen pack has dwindled since World War II to a small percentage of what it was in the peak marketing years. Recently, the development of high-moisture and pitted prunes has led to an increasing demand for frozen prunes in transparent packaging material.

Preparation for freezing of plums and prunes is similar to that for peaches, except that the fruit need not be peeled.

Raspberries, Red.—Of the U.S. commercial pack of frozen red raspberries, about 95% are packed in the West, principally in Oregon and Washington. This puts the red raspberry second only to strawberries among the berries in amount commercially frozen.

Raspberries retain fresh fruit flavor to a greater degree than do most other fruits. Good, sound fruit of firm ripe maturity, properly harvested, and promptly prepared and frozen has a natural fresh raspberry appearance and flavor after thawing.

The desirable varietal characteristics are: deep red color, small seeds, rich flavor, and resistance to bruising, crumbling, and collapse of drupelets after thawing. Leading varieties of raspberries in the Pacific Northwest include Willamette, Sumner, Puyallup, Canby, Fairview, and Washington. During the past ten years, the Washington variety has dropped from the leading variety to a minor one among frozen raspberries because of increasing susceptibility to disease. The Meeker variety shows special promise for freezing.

In 1964, Willamette accounted for 58% of the acreage of red raspberries in Washington. Other varieties and the acreages planted at that time were: Sumner 19%, Puyallup 15%, Canby 3%, and Washington and Fairview 2% each. The varietal situation in Oregon was similar, but a larger percentage of the acreage was planted to Canby and Fairview than in Washington.

The *Willamette* is a very large berry and is rated very firm to firm in texture. It is a dark red berry which tends to become purplish upon standing. It ranges from mildly acid to acid. It stands handling well. It is a fairly good berry for freezing and would rate higher if its color were more uniform.

Sumner is a large, medium red, sweet, firm berry. It has very good quality for freezing, but when used in jam the seeds absorb color from the surrounding syrup or jelly, giving the product less eye appeal. It has excellent flavor as a preserve, however, and at least one of the country's major preservers uses Sumner in a seed-free raspberry jam.

Puyallup is a very large, acid, medium red, fairly firm berry. It has good quality for freezing.

Canby is a large, light red, mildly acid, fairly firm berry. After a rainy period during harvest the berry becomes very soft and is rated as only a fair to good freezing berry because of this. During a harvest season when good weather prevails, Canby has better quality and is a good to very good berry for freezing.

Fairview has good color and flavor as a frozen berry. The berries are medium large and are fairly firm. This is the newest of the commercially grown varieties in the Pacific Northwest.

Washington for 20 years was the mainstay of the Pacific Northwest frozen raspberry industry. In recent years it has become susceptible to root rot and no longer yields well enough to be competitive with varieties like Willamette, Sumner, or Puyallup. It is an excellent berry for freezing and preserving, as it has bright red raspberry color and excellent flavor. However, the berries have decreased in size to the point that it is difficult to get the crop harvested.

Several varieties are listed as good for freezing preservation in the East and Midwest. These include:

Chief, the standard variety for the upper Mississippi Valley. Chief is an early variety of medium size and firmness and good quality.

Latham, a late variety with large medium-red, firm but often crumbly berries. The variety is rated as a good freezing berry, although the quality is not high.

Melton, a very late, large, medium red, firm, good-flavored berry. Melton is grown where viruses are especially troublesome to other varieties.

Newbergh, a late, large-fruited, bright red, firm berry of good quality. This is also grown on limited acreage in the Pacific Northwest, where it is considered to be of marginal quality when compared with Canby, Willamette, and the ther varieties grown in the region.

Sunrise is an early berry of medium size, bright red color, firm texture, and good flavor.

Taylor is the leading New York variety. It is a late variety, medium to large size, deep red color, and high quality.

The following red raspberries are reported to be the best suited for freezing when grown under New York conditions: Willamette, Melton, Taylor, and Newbergh.

Table 4.1 summarizes main raspberry characteristics for the six varieties grown in the Pacific Northwest. While differences show among the varieties for all characteristics, the time of harvest also makes a difference. This is shown in Table 4.2.

Investigations into the seed and other characteristics of the raspberry showed that differences within a variety between early and late harvest times were often greater than intervarietal differences of berries har-

TABLE 4.1

RASPBERRY CHARACTERISTICS FOR SIX VARIETIES

	Canby	Fair-view	Puyallup	Sumner	Wash-ington	Willa-mette
Berry weight, gm.	3.0	2.8	3.3	2.8	1.9	3.6
Number of drupelets	77	68	70	75	52	88
Weight of drupelets, mg.	39	42	46	38	37	40
Seed length, mm.	2.4	2.5	2.5	2.3	2.2	2.6
Weight of seed, mg.	1.3	1.3	1.6	1.3	1.3	1.4
Seed weight, % of drupelet weight	3.3	3.2	3.4	3.5	3.7	3.3

TABLE 4.2

CHANGES IN RASPBERRY CHARACTERISTICS AS THE HARVEST SEASON PROGRESSES; ALL-VARIETY AVERAGE

	Berry Weight (Gm.)	Drupelet Count	Drupelet Weight (Mg.)	Seed Weight (Mg.)	% Seed	Seed Length (Mm.)
1	3.34	75	44	1.8	4.1	2.6
2	3.33	79	42	1.6	3.8	2.5
3	3.28	80	41	1.5	3.7	2.5
4	3.02	75	40	1.4	3.5	2.5
5	2.83	77	37	1.3	3.5	2.4
6	2.95	70	42	1.3	3.1	2.4
7	2.57	66	39	1.3	3.3	2.3
8	2.67	65	41	1.2	2.9	2.3
9	2.57	64	40	1.3	3.3	2.3
10	2.48	61	40	1.2	3.0	2.3

vested at the same time. If small seeds are desired, using late season fruit is one way of obtaining this characteristic.

Processing.—Red raspberries are picked into hallocks measuring about 5 × 5 × 2 in. deep. They are delivered to the plant in flats of 12 hallocks each. In some areas in Oregon hallocks of 4 × 4 × 3 in. dimensions are used. The berries are delivered to the freezing plant as soon after picking as practical, usually as soon as a pick-up load of flats has been filled.

A recent innovation in preparation of berries before freezing is the use of the air cleaner. This move was prompted by tightening requirements of industrial buyers of frozen berries for cleaner, foreign-body-free fruit. In this operation (Fig. 4.9) the berries roll down a sloping grid having the slats close enough together that the berries do not fall through. A strong updraft of air blows up through the grid. The air flow may be regulated to the point where sound berries will go through, but partially dried berries will be lifted from the flow of the product. The first operation using the air cleaner was initiated in 1966 in Woodburn, Ore. It was the experience of this plant that twigs, leaves, and shriveled fruit, as well as spiders, thrips, and other insects were almost entirely eliminated from the berries

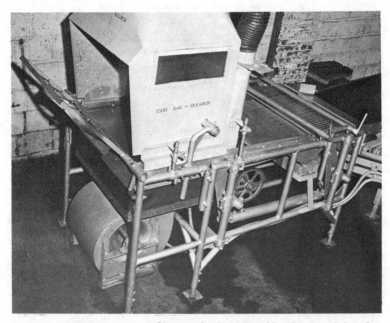

Courtesy of A.B. McLaughlan Co., Inc., Salem, Ore.

FIG. 4.9. MCLAUGHLAN BERRY MULTI-CLEANER

as they went through the air cleaner. The operation was so successful that a large proportion of the Northwest's berry processors have installed air cleaners.

In the past, removal of thrips from cane berries has been a difficult chore, especially in seasons of high infestation. Prior to the use of the air cleaner, the most satisfactory way to remove the insects was by use of detergent washing. There has been some concern about detergent residue in the berries, and this curtailed the use of detergent cleaning to some extent. With the air cleaner doing even a better job than detergent washing, the thrips problem appears to have been solved.

In the plants the flats are emptied into the air cleaner if used. From the air cleaner the berries go to the berry washer and then go over a dewatering shaker and onto an inspection belt. For the retail trade, 6 oz. of sound berries are weighed into a Sefton-type container. Four ounces of 60% sucrose syrup is added to the container and it is sealed. When larger containers are used, the amounts of berries and syrup in the 6:4 ratio are adjusted to make the declared weight. Freezing may be done by traying the sealed containers and placing the filled trays on buggies which are pushed into the freezing tunnel. As each cart is placed in the tunnel, the carts preceding it are moved along through the tunnel. In some plants, especially those plants not physically tied in with the freezing facilities, the berries may be cased, palleted, and frozen by the pallet load.

Raspberries for ice cream, baking, and preserving trades are packed in 10- or 30-lb. slip-covered cans or in plastic-lined steel drums of 55-gal. capacity. Washed raspberries are filled into these larger containers without added sugar, unless the buyer specifies that sugar be added.

Raspberries, Black.—The black raspberry (blackcap) pack is about a quarter as large as the red raspberry pack. Between 2 and 3% of the U.S. pack is put up in the Northeast, 45% in Oregon and Washington, and the balance in the Midwest.

Desirable varieties have minimum seediness, large, plump, juicy-fleshed berries, and deep dark color. Bristol is considered to be the best blackcap for freezing under New York conditions, while in the Pacific Northwest the Munger is the principal variety grown. Munger is superior to other varieties in yield, fruit quality, and plant characteristics.

Black raspberries are handled in the same manner as are red raspberries. Blackcaps are packed in 30-lb. cans or larger containers for the bakery, confectionery, preserve, and other remanufacture trades.

Raspberries, Purple.—Desirable characteristics in this berry are, except for color, the same as in the red varieties. The Columbian, Marion, and Sodus varieties are for freezing. Sodus yields well, and freezes better than the red varieties.

Rhubarb.—This delicious vegetable stem is mentioned here under preparation of fruits for freezing because its principal use is for a fruit dessert or sauce. The frozen pack for the United States has averaged 5 to 6 million pounds per year in the past 10 years or so. About 80% of this pack has been put up in the Western states, with the rest scattered throughout the East and the Midwest.

Three varieties dominate frozen rhubarb production—Crimson, German Wine, and Victoria. Varieties such as Cherry, MacDonald, Strawberry, and Valentine also freeze well, but do not constitute a very large percentage of the frozen market. Recently, hothouse-grown rhubarb has been frozen in increasingly large quantities in Washington, reaching something like 1 million pounds a year, and hothouse-grown frozen rhubarb has also increased markedly in Michigan.

For freezing, the trimmed stalks are cut into one-inch pieces by stationary knives, then washed thoroughly in tanks and by sprays, inspected, and packed into retail and institutional containers. The fruit is mixed with sugar at a ratio of 6:1. For a somewhat longer storage life, the rhubarb may be blanched in steam.

Strawberries.—The strawberry is the most widely grown as well as the most widely frozen small fruit in the United States. Of the total frozen strawberries, 88% are frozen in the West, 8% in the Midwest and 4% in the South.

Varieties.—The strawberry varieties grown and processed vary from region to region. In the United States, the top 6 of the 24 planted and the dates of their introduction are as follows: Northwest, 18%, 1949; Blakemore, 14%, 1929; Robinson, 8%, 1948; Shasta, 6%, 1945; Headliner, 1957 and Tennessee Beauty, 1943, each 5%. Of these varieties the Northwest is grown in the Pacific Northwest States of Washington and Oregon. Blakemore is grown from Georgia and South Carolina in the East westward to Texas, Oklahoma, and Kansas. Robinson is grown in parts of New York, Pennsylvania, Ohio, Michigan, and Illinois. Shasta is a California berry. Tennessee Beauty can be found from Colorado eastward to Maryland while Headliner is grown in the Gulf Coast states. Of these six varieties, most processors rate Northwest and Tennessee Beauty as very good, Blakemore, Headliner, and Shasta as good, and Robinson as poor for freezing.

The varietal picture is one which is slowly but constantly changing. Plant breeders are working on the improvement of the strawberry in many parts of the world. The USDA, at Beltsville, Md., has observed over 1,000,000 strawberry seedlings since 1920. Of these only 1 in 37,000 has been named as a variety. Other plant breeders have tested large numbers of seedlings in their search for better strawberries. Of the many varieties introduced, a few have persisted for many years, but others have disappeared after but short periods in the trade. Susceptibility to disease plays an important part in the life of a strawberry variety.

Some of the strawberries presently grown in the Pacific Northwest and their characteristics are:

Northwest, a high-yielding subacid berry. Many growers get yields of 5 to 10 tons per acre. Northwest has a good red color, medium to firm flesh, and a good flavor, although it is not highly aromatic in character. In recent years it has been the leading variety grown for freezing. Northwest is not a good berry for the preserve trade, as it lacks the desired aromatic flavor after cooking. Also, preserves made from this variety have a limited shelf-life due to their tendency to darken in storage.

The *Marshall* or *Banner* strawberry is an aromatic berry with good red color in the outer flesh and lighter flesh inside. It has tender skin and is at times irregular in shape. It is an aromatic berry with finely balanced flavor which persists after heating, making it a desirable variety for preserving. Much of the early frozen strawberry business in the Pacific Northwest was built around this fine-flavored variety and it still is in highest demand by the preserving trade. Unfortunately, the variety is subject to virus diseases and cannot now be grown in many areas where it once flourished.

The *Hood* strawberry is most apt to replace the Marshall for preserving in the Pacific Northwest. This is, to date, the newest processing variety from the Pacific Northwest, having been introduced to the trade in 1965. It is a fine, aromatic-flavored variety which is well adapted to the preserving trade. It has good color, but tends to be darker when grown in the Puget Sound area than when grown in Oregon's Willamette Valley. In 1966 it outyielded Northwest in number of experimental plots in the Pacific Northwest. It freezes well, although the flesh tends to be soft.

Puget Beauty is an early variety well adapted to growth in the Northern part of Washington State. It is a sweet, highly aromatic berry which can be used to advantage by the preserve trade. Several ice cream manufacturers regard this berry with favor because of its highly aromatic character. Puget Beauty is the earliest-ripening berry in the region, and is grown by some strawberry men to help them line up their picking crews.

The 3 to 4 days by which this variety precedes Northwest will give the grower an advantage in attracting the school children who make up much of the harvest crews. Puget Beauty must be harvested before it is fully ripe, because it is very soft when overripe. The earliness of the variety does not favor its use in Oregon, since it may ripen before school is out for the summer and thus be ready before harvest crews are available.

Siletz is a berry which can be grown in heavy soil where red stele-susceptible varieties do not produce. Although it is rated as being very good for freezing, not many strawberry freezers consider it to be a good berry for the trade. It is a low-flavored berry with soft flesh and variable color when grown in some areas. Siletz often brings one cent per pound less than other varieties.

Molalla is a deep-colored, firm-fleshed berry. It often produces odd shaped berries. Flavor varies from fair to good. The flesh characteristics of frozen Molalla are good.

In California, the Shasta has long been one of the leading varieties for freezing, although other varieties, such as the *Tioga,* may in time take over the top spot in strawberry freezing in this state. The Shasta is light colored, soft, and lacks the flavor found in berries grown in the Pacific Northwest. When it is grown in Oregon, where nights are cooler during harvest season, Shasta has more flavor than when it is grown in California.

The following are varieties grown in other parts of the country and listed as being very good freezing berries.

Tennessee Beauty: Medium size conic. Medium to deep red color. Firm flesh, good dessert quality.

Sparkle: Short, blunt conic to oblate. Glossy rich red color, medium size. Mildly subacid. Good dessert quality.

Pocahontas: Large, attractive, bluntly conic berries. Medium firmness of flesh. Subacid. Good dessert quality.

Midway: Long conic, firm fleshed with tough surface and glossy rich red color. Medium to large size. Subacid. Good dessert quality.

Dixiland: Large, long blunt conic, very firm and attractive berries. Skin and flesh color are bright red. Acid. Fair dessert quality.

Processing of Strawberries.—Harvesting.—Strawberries packed on the Pacific Coast, where more than $^5/_6$ of the frozen strawberries in the United States are packed, have the caps removed at picking. In most other parts of the country caps are removed in the freezing plants. The picking of the berry without hull is accomplished by grasping the stem and cap with one hand pulling with a slight twist with the other.

The berries are picked into flats of 12 or 15 hallocks in the Pacific Northwest. Each hallock holds approximately one pound of berries. In California, shallow trays about 14 × 18 in. in dimension are used instead of flats of hallocks. Hallocks were once made from thin spruce veneer, but in recent years the lack of suitable logs for hallock shook has resulted in the widespread use of plastic-coated paper hallocks. Divided flats made of high-impact plastic are presently being tried on the West Coast.

When berries are delivered to the plants with caps attached, capping machines such as the Morgan strawberry capper or the Gaddie strawberry capper are used for removing caps.

The berries are normally held in the shade until they are trucked to the freezing plant. In some California locations portable field refrigeration units are used. This takes much of the field heat from the fruits and puts a better quality raw product into the plant.

Cleaning.—On delivery to the plant, the berries in the flats are dumped into a washer. The most commonly used washer in the Pacific Northwest is the McLaughlan washer (Fig. 4.10). The washer consists of a shallow pan with water running in it and sharp sprays playing onto the fruit from above. The pan is continuously in sharp reciprocal motion, thus moving the fruit through the washer. As the berries move forward in the pan they are subjected to sharp sprays from above which knock sand, dirt, leaves, and other foreign material from the berries. After washing, the berries pass over a dewatering shaker and onto inspection belts where women remove foreign material, rots, and defects from the berries. When sliced, sugared berries are being packed, the berries are cut into slices of $\frac{1}{4}$ to $\frac{7}{16}$ in. thickness (Fig. 4.11). The trend is toward thicker slices. Slicing is done mechanically in high-speed circular blade slicers separated by proper spaces to give the desired thickness of slice (McLaughlan slicer), or by slow-speed centrifugal slicer (Urschel).

Sweetening.—The sliced berries then go into a mixer where one part sugar by weight is mixed with four parts of sliced berries. Mixing is by ribbon or screw type agitators. The length of time the berries are in the mixer varies a great deal from plant to plant. Sometimes, when tie-ups

Courtesy of A.B. McLaughlan Co., Inc., Salem, Ore.

FIG. 4.10. MCLAUGHLAN BERRY WASHER

Courtesy of A.B. McLaughlan Co., Inc., Salem, Ore.

FIG. 4.11. MCLAUGHLAN STRAWBERRY HALF-SLICER

occur in other parts of the line, the berries may be mixed for prolonged periods. This should be avoided because mixing longer than is necessary to thoroughly distribute the sugar with the berries tends to destroy the slices and results in a product which should more properly be labeled "crushed strawberries." In the ribbon-type mixer, sliced berries and sugar are weighed into one end of a stainless-steel, U-shaped tank from 3 to 5 ft. long and 3 ft. wide. The tank is equipped with slowly revolving stainless steel ribbon mixers, which, as they slowly revolve, mix the sugar with the berries, and move the berries and sugar to the discharge end of the tank. The sugar dissolves during the mixing and a slurry of slices, crushed berries and syrup results.

The screw-type mixer consists of a hopper at the feed end into which the berries and sugar are metered continuously. From the hopper the berries and sugar pass into a cylindrical stainless steel tube or U-trough about 8 in. in diameter and several feet long. Inside the tube, or trough, a slowly turning screw slightly smaller than the inner diameter of the tube propels the berry-sugar mixture toward the discharge end. By the time the discharge is reached the sugar is dissolved. A third way of sweetening berries is performed in two stages. The sliced berries are filled volumetrically into the packages and the partially filled containers then go to a syruper which adds the correct amount of high-density syrup. The containers are then capped, trayed, placed on trucks, and frozen. The greatest loss of character (mushing of slices) comes from the screw-type mixer. The two-stage filling produces the least damage, while the ribbon-type mixer is intermediate in its breaking up of slices.

The Northwest variety breaks down less than Marshall, Siletz, or Puget Beauty. Early-season berries have less tendency to break up, probably due to lower maturity than the berries harvested later in the season. Prolonged holding of berries after harvest results in softer fruit, and prolonged mixing decreases the amount of intact slices.

Freezing.—After mixing, the sweetened berries go to the filler. In many plants a piston filler is used to fill the 10-oz. or 1-lb. retail package. The most commonly used container is a fiber-bodied carton with metal ends, the Sefton container. The sealed containers may be frozen singly on a belt moving through the freezing tunnel, tray-frozen on carts which are moved through the tunnel or cased, palleted, and frozen in a sharp room or in a tunnel. Case freezing is all too common and is not the preferred method. An investigation made by scientists of the Western Regional Research Laboratory showed that the palletized case takes much too long to freeze. In some instances it took nearly a week to bring the center temperature to 0° F. (−18° C.).

Half-sliced strawberries are handled in much the same way as the regular sliced berries except that the berries are cut in a half-berry slicer. This is a machine equipped with a series of parallel Vee-troughs which move back and forth with a reciprocating action. As the berries go down the troughs, they tend to line up with the tip down and are cut in half by rotating circular knives located at the discharge end of the Vee shaped troughs. Mixing with sugar is carried out as in the regular sliced berries.

A very recent development is the packing of half-sliced berries with syrup or liquid sugar in pouches or in shallow plastic boxes which can be sealed shut. These berries, not having gone through the churning sugar mixing process, retain their slice character. By immersing the frozen pouches in cool running water, they can be thawed in a very short time and the resulting product is of excellent quality. Strawberries packed this way are sold as gourmet foods.

Another retail pack is made with whole berries packed in syrup. Washed and drained fruit is packed in a sugar syrup in metal cans which are frozen. If the cans are thawed in running water shortly before the berries are to be opened and consumed, a very attractive product can be obtained.

The third way that strawberries are frozen for the retail trade is individually quick-frozen (IQF). IQF berries are washed, drained, and placed in the freezing tunnel or on wire mesh belts which travel through the freezing tunnel. IQF strawberries are usually retailed in polyethylene bags, 20 to 32 oz. per bag. For institutional use they are packed either in enameled slipcover cans or in plastic-lined fiberboard boxes.

Cryogenic Freezing.—A recent development in IQF strawberries is the cryogenic freezing of the fruit. When immersion freezing in liquid nitrogen (LN) is used, the berries are immersed long enough to freeze a shell equal in thickness to 0.5 to 0.6 the radius of the cross section of the berry. The rest of the freezing is accomplished in the cold gas above the LN, or in conventional 0° to −10° F. (−18° to −24° C.) storage. Prolonged immersion in LN results in the freezing of a hard outer shell which becomes stronger at low temperatures. As freezing continues, internal pressures build up. The faster the freezing, the greater the pressure build-up. Relief of the pressure comes about by expansion of the frozen exterior layers, resulting in cracking or shattering. At any rate, prolonged immersion in LN results in shattered berries.

LNF strawberries do not leak as badly on defrosting as do conventionally IQF berries. In one experiment thawed IQF berries lost ⅓ of their weight in the first 1½ hr., while the LNF berries lost only 6 to 8% in the same time.

An Oregon processor has been LN freezing strawberries for several years. A large hotel chain has bought his pack because they have found the LNF berries can be sliced while still frozen without shattering. The berries are used for dessert in the hotel's carriage trade. The appearance of the berries handled in this way has made the fruit very acceptable to a discriminating trade.

Bulk Frozen Strawberries.—When frozen for remanufacturing, the berries are frozen in accordance with the requirements of the remanufacturer. During the past five years, the cost of 30-lb. containers has increased so much that a number of freezers have returned to the use of steel drums. The berries are usually sugared. Four-to-one ratio of berries to sugar is the most common proportion used. In the past, some fruit was put up at a 3 + 1 level, but it is not economical to ship sugar across the continent. Preservers can add sugar in their plants for less than it would cost when shipped with the berries. The berries are frozen at 0° F. (−18° C.) with minimum delay between packing and start of freezing. They are stored at 0° F. (−18° C.).

For the Bakery and Ice Cream Trade.—Strawberries for the bakery and ice cream trade are packed in barrels, 30-lb. tins, 6½ lb. cans (No. 10 can, double seamed) and 10-lb. tins. Most of the sugaring is at the 4 + 1 level, but some 3 + 1 pack is sold to the ice cream trade. As with strawberries packed for the preserve trade, the fruit and sugar are layered alternately into the container, although some packers will mix the fruit mechanically, and fill containers with the mixture. When cans are used, the containers are normally in a 0° F. (−18° C.) room within an hour after filling.

FRUIT JUICES AND PURÉES

The preparation of juices and concentrates for freezing is similar to preparation of the same products for preservation by other techniques. Fruit juices and fruit purées may be protected from microbiological attack and breakdown by a number of methods. These include the use of such chemical preservatives as sulfur dioxide and salts of benzoic acid and sorbic acid; conventional heat processing after canning; aseptic packing using short-time high-temperature sterilization and sterilized containers; concentration of juice to 70% solids or higher without further processing; and freezing. Of all these methods, freezing is regarded as best for retaining the delicate flavor and characteristic color of fruit juice or concentrate. Frozen orange juice concentrate has become successful primarily because no other means of preservation gives a product of equal quality. The high quality of this product accounts for its tremendous growth during the past 20 years. In other juices, the differences in quality are not quite so clear-cut and although a concentrated frozen apple juice of high quality is available, it is still a poor second to canned and bottled processed apple juice in volume marketed. The same is true for concentrated pineapple juice and grape juice. For jellies or other remanufactured products, juice concentrate preserved by heat or high-solids concentration gives a product which is apparently "good enough" for most purposes. Large quantities of apple, prune, and grape juice are concentrated to 70° Brix or higher and packed in 50-gal. drums. Products at this concentration require neither heating nor storage at low temperatures, because the high osmotic pressure at these concentrations prevents the growth of microorganisms. In recent years the development of aseptic canning processes—where purees, such as apricot puree, are concentrated to 20° to 25° Brix and heated at high temperatures for a short time, then rapidly cooled and aseptically filled into sterile 50-gal. drums—has given a product of good quality which may be stored without freezing. For this reason there is little if any frozen concentrated apricot purée, since the product produced as described above is satisfactory for its most important end products, nectars and jams.

One advantage of the freezing of juices and juice concentrates is that it permits the use of large containers. While slow thawing (as much as 24 hr.) is a disadvantage of frozen juices or concentrate packed in 50-gal. drums, special equipment has been designed to reduce bulk-packed frozen single-strength juice or pulp back to liquid form in 3 min. using a combination of chopping and heat. The juice or concentrate can be slush-frozen with a Votator or similar type of heat exchanger, and filled into 30-lb. tins, 55-gal. drums, or even a 10,000-gal. refrigerated tank car. In California, tomato paste is sometimes slush-frozen and shipped by truck to large industrial remanufacturers of such items as soups and catsup. In Florida, slush-frozen single-strength orange juice of high quality has been shipped in refrigerated tank trucks and even in tankers (ships carrying large stainless steel refrigerated tanks). This is a simpler process than aseptic filling. The chilled or semifrozen juice is unloaded at the destination and repackaged at that time. It has been estimated that by 1970, 50% of all the orange juice will be sold as single-strength.

Preparation

Liquid products derived from fruits may be roughly classified as: clear juices, such as those prepared from cherries and apples; cloudy juices containing a quantity of insoluble solids, such as pear pulp and pineapple juice; and pulps or purées containing fibrous and other insoluble matter, such as those obtained from apricots and peaches. Some products are available in both forms; for example, strawberries may be crushed and used directly as a pulp in such items as jam, or the seeds and insoluble materials may be removed by various means and the resulting clear liquid used for strawberry jelly. This chapter will try to use those terms most commonly found in commercial trade descriptions. The preparation of juices, pulps, and purées of all kinds consists primarily of removing all, or a part, of the insoluble solids from the whole fruit. To accomplish this, there is ingenious equipment of many kinds, and it would be impossible to cover them all in this chapter.

The purée of such fruits as peaches, apricots, pears, and plums is usually made commercially by washing the fruits, cooking them in a jacketed hollow screw heat exchanger, and pulping. The cooking of the fruits has two functions: (1) enzymes which may cause discoloration are destroyed, and (2) the fruit is softened sufficiently so that it is possible to separate pits and fibrous materials from pulp. This is usually accomplished by pumping the hot mixture of pulp, pits, skin, etc., to pulpers and finishers, machines in which paddles revolving at high speed force the crushed, softened fruit through stainless steel screens. This is a two-step operation; the initial screen may have relatively large, 0.25 in., perforations to remove pits, while the second or finisher screen with 0.020-in. holes screens out pit fragments, skin, seeds, etc. Where excessive pit fragments are a problem, paddles may be replaced by special nylon brushes which force the material through the screen without breaking the pits. In order to eliminate the gritty texture of stone cells in pear pulp or the pit fragments from either peach or apricot purée, a disintegrator which grinds these materials into very fine particles has been successfully used.

Fresh or frozen berries are made into a coarse purée by a comminuting machine or paddle finisher having a screen of ¼-in. holes. For juice the purée is treated with 0.2 to 0.5% of a pectin hydrolyzing enzyme preparation and held at 75° F. (24° C.) for 3 hr. Diatomaceous earth filter-aid is added (2 to 4% by weight) and the resulting mixture is separated in a hydraulic press. After a second addition of 0.25 to 0.5% of filter aid, the cloudy press juice is filtered in a pressure filter to obtain a brilliantly clear juice. Gelation is frequently a problem in strawberry juice as well as other fruit juices held in frozen storage. It is caused by activity of pectinesterase enzyme and can be prevented if the enzyme is inactivated by a heat treatment of 185° F. (85° C.) for 2 min. In the production of some juice concentrates enzyme treatments which eliminate the possibility of gelation are used and heat inactivation of pectinesterase is not necessary.

Although some processors heat berries before pressing because the heating reportedly solubilizes color, destroys enzymes, and increases the yield, it has been observed that cold pressing yields a juice having both excellent color and flavor. The yield of juice is about 70 to 75% of the weight of berries juiced—whether cold pressing or hot pressing is used.

One important use of frozen purées and pulps is flavoring for various ice creams. They are supplied either as a straight single-strength pack or a pack of four parts of fruit plus one of sugar. Some of the fruits available are apricot, blackberry, blueberry, boysenberry, cherry, grape, nectarine, peach, plum, raspberry, and strawberry. This material is ordinarily used as a flavoring rather than as material for "ripple-style" ice cream, which uses a heat-sterilized jam-like product. Frozen purées are frequently packed in relatively small operations wherein the fruit is pulped, with or without pre-heating, through a 0.020-in. screen, filled into 30-lb. cans, and frozen.

Enzymes.—Pectic enzymes have been used for many years in the fruit juice and fruit juice concentrate industries. They serve many functions: (1) in the manufacture of grape juice the pectic enzymes, or pectinases, break down the grape pectin and thus increase the capacity of the press as well as the yield of juice; (2) in apple juice, the breakdown of the pectins makes possible the production of clear apple juice which will not develop a cloudy appearance during storage; (3) eliminating much of the burning-on or fouling of heat-transfer surfaces during concentration and permitting higher concentrations to be reached; (4) elimination of gelatin in juice concentrates during storage. For example, it is not possible to concentrate single-strength strawberry juice without occurrence of gelation in storage if the natural pectins are not destroyed. In addition, if pectin is present nondispersible gels develop and cause a marked cloudiness in jelly and other products made from the juice. It is frequently desirable in jelly manufacture to destroy the last trace of native pectin in a fruit juice or concentrate so that a standard quality of pectin can be added to the jelly without having to allow for the natural pectin of the fruit juice; thus a uniform product which will not vary in consistency can be produced.

Pectic enzymes come as liquids of various enzymatic strengths or as a dry powder mixed with filter aid, dextrose, or gelating. The amount of enzyme to be added, as well as the temperature and the length of holding time, varies with different fruits; most enzyme suppliers include specific directions for the use of their preparations with each kind of fruit. For example, for grape juice, a level of 6 lb. of a commonly used liquid enzyme per ton of grapes will cause complete depectinization in about 90 min. at about 135° F. (57° C.).

Concentration

It is expensive to package, store, and ship fruit solids as single-strength juice. Concentrates offer advantages of cost and space-saving both to the remanufacturer and to the consumer. In addition to their use as a beverage, frozen fruit juices and concentrates play an important role in fruit ice creams, ices, sherbets, sundae sauces, fruit syrups, and other soda fountain supplies. Fruit juice concentrate offers a special advantage to jelly manufacturers. By using concentrated fruit juice for making jelly, it is possible to obtain better color and flavor than by employing the conventional cooking of single-strength juice. FDA Standards of Definition require that specific quantities of fruit juice solids be used for various fruit jellies. Since there is a large variation in the solids content of single-strength fruit juices made from the same kind of fruit, it is frequently necessary to vary the formulation with each batch of jelly. Moreover, the use of a single-strength juice requires jelly processors to evaporate large quantities of water, either in an open kettle or in a vacuum pan, to bring the mix up to the required legal percentage of solids. In addition, the use of fruit juice concentrates not only eliminates much of the evaporation required at the jelly plant, but permits the continuous manufacturing of jelly by combining metered streams of fruit juice concentrate, sugar syrup, and pectin solution. In this process the sugar syrup is heated separately to a high temperature and the flavor and color of the concentrate are protected from heat damage by confining heat treatment to a minimum; no evaporation is required. Frozen concentrate plus essence is particularly adapted to this continuous jelly process and produces a jelly of excellent quality. The same economies of packaging and storage which make concentrates appealing to remanufacturers also apply to retail marketing, and the consumer now has the choice of a wide variety of such concentrates (Fig. 4.12).

FIG. 4.12. A SAMPLE OF THE CONCENTRATED FRUIT JUICES FOUND IN
A TYPICAL CALIFORNIA SUPERMARKET

Because most fruit juices are heat-sensitive and their color and flavor
deteriorate under prolonged heating at boiling temperatures, it is customary to concentrate fruit juices in vacuum evaporators. Vacuum concentration has several advantages: it makes possible the low-temperature
concentration of heat-sensitive foods such as orange juice; it permits a
large temperature difference between the boiling temperature of the
product and the heating medium while it maintains a low evaporation
temperature; and it makes possible the use of multiple-effect systems
which result in steam economies.

The delicate flavor of most fruit juices is adversely affected by heat. The
rate of this deterioration is markedly greater the higher the temperature.
In the early days of fruit concentration, it was believed that fruit juices had
to be concentrated at low temperatures to assure high quality. The
maximum allowable temperature for concentration of strawberry juice is
100° F. (38° C.) and in early development of frozen orange concentrates
special low-temperature evaporators which would concentrate juice at

temperatures at about 70° F. (21° C.) were designed. Evaporators operating at 120° F. (49° C.) were felt to produce an inferior product. Evaporation at these low temperatures required strict observance of sanitation, since microbial growth was possible at the concentrating temperature. A heating step was also required to destroy pectinesterase, otherwise gelation would take place in the concentrate during storage. Many of the older low-temperature evaporators comprised a relatively large mass of liquid boiling at a low temperature for periods ranging up to several hours and had the disadvantages that capacity was low and microorganisms could grow at the temperature of the boiling liquid so that thorough cleaning at intervals ranging from 2 to 7 days was necessary. More modern evaporators, on the other hand, employ a small amount of liquid passing rapidly over a large area of heating surface held at a temperature which may be as high as 245° F. (118° C.). While the heating temperature is high, residence time of the liquid in the heating chamber is only a few seconds. This reduced holding time, even at a higher temperature, results in better color and flavor. Such evaporators have higher capacity per square foot of heating surface than the older low-temperature evaporators and are rapidly becoming common in commercial production of orange juice concentrate and pineapple juice. Rapid heating rates are usually obtained by a high ratio of heating surface to holding volume.

Color loss during concentration of fruit juices is related to flavor loss and is much easier to measure objectively. Studies of color loss in strawberry, boysenberry, and grape juices held in sealed tubes under various time-temperature conditions indicate that the logarithm of the time required for a 10% decrease of optical density at the wavelength of maximum absorption is directly proportional to the holding temperature. A 10% loss in color density is selected because it is just noticeable to the eye when the heated juice is compared with its unheated control. It can be seen from these curves that fruit juices can withstand deterioration caused by high evaporation temperatures. Older evaporators used in the food industry are now being replaced by more advanced types. One type of evaporator consists of a vertical short-tube evaporator with a central down-take as shown in Fig. 4.13.

Most evaporators today are constructed in successive multistage units. Each time juice is heated in a chamber in such equipment the operation is known as a stage; the heating medium for each stage may be either steam coming from a central boiler or water vapor coming from a previous stage. When water vapor from boiling juice is used as the heating medium to evaporate juice boiling at a lower temperature, the operation is known as an effect. Many of the new orange juice concentrators in Florida are 7-stage, 4-effect, high-temperature, short-time evaporators. The juice

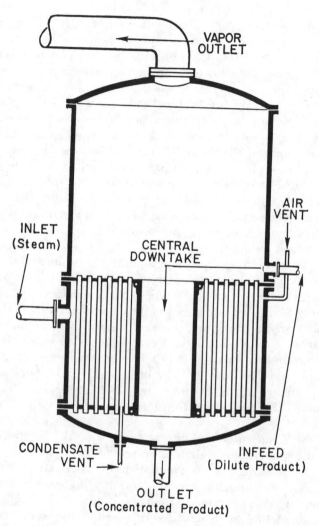

FIG. 4.13. DIAGRAM SHOWING DESIGN OF CALANDRIA EVAPORATOR

makes a single pass through each stage. A vacuum operation is also sometimes used not only to remove water, but to reduce the temperature of the concentrate. For this purpose there is no heating stage and the liquid is flashed off to the final concentration by means of vacuum. The sensible heat of the concentrate is utilized as the latent heat required to boil off liquid and so reduce its temperature to a point corresponding to boiling at the pressure in this final step.

Evaporators.—*Rising-film, Falling-film, and RFC Evaporators*.—
Evaporators are frequently classified either as *natural circulation,* where
the circulation of the product results from the reduction in density of the
solution on heating and from pressure generated by vapor evolved at the
heat exchange surface, and *forced circulation* evaporators, where a circulating pump is used to insure high velocity across the heating surfaces.
Similarly, evaporators are sometimes classified as rising-film or falling-
film, depending upon the movement of the liquid in the tubes. In a
rising-film evaporator, the dilute liquid is fed into the chamber below a
tube-sheet and rises in the tubes. As steam is admitted to the steam chest,
the liquid reaches the boiling point and bubbles of vapor are formed in
the column of liquid. As the bubbles rise they expand and push the liquid
forward with increasing velocity. The mixture of vapor and liquid exits
from the tubes into a vapor separator. Evaporators of this kind are widely
used in the food industry for concentration of clear solutions of low to
moderate viscosity. In a *falling-film* evaporator the dilute liquid is introduced into the chamber above the upper tube-sheet and flows as a thin
film down the inner surface of the tubes. Only a small amount of liquid is
in the falling-film tubes, compared to the column of liquid in the rising-
film evaporator. The *rising-falling* concentrator (RFC) combines both of
the principles. This is a single-pass evaporator where the feed juice is
introduced at the bottom of one bundle of tubes at a temperature above its
boiling point, and the discharged mixture of liquid and vapor is separated
at the top and redistributed over the down-flow pass. This type of
evaporator has been used for concentrating orange, grape, and apple
juices.

   Plate-Type Evaporators.—The plate-type or AVP evaporator, developed in England, is a single-pass evaporator using specially modified
stainless steel plates as heat exchange surfaces. There are more than 100
installations throughout the world, concentrating orange juice, apple
juice, grape juice, ice cream mix, milk, and other liquids. High-quality
tomato paste of 40% solids has been reportedly produced in pilot plant
tests. In one Florida orange juice concentration installation, a 12,500
(lb.) per (hr.) double-effect, plate evaporator and auxiliaries were installed in
an 18 × 27-ft. room with a 9-ft. ceiling. This is considerably more compact
than conventional vacuum evaporators. One of the advantages claimed
for this method of evaporation is the small amount of product in the
system at any one time. In the concentrator described above, there is less
than five gallons of product in each effect at any one time. The system
may be described as a rising and falling film evaporator with liquid fed
from the bottom, rising and descending through alternate plates (Fig.
4.14). The Florida installation described above uses 2 effects to produce
orange concentrate at 65° Brix, with a total retention time of about 1 min.

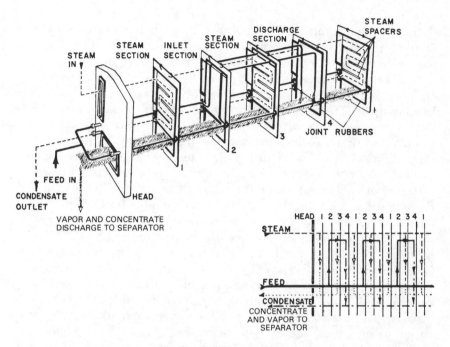

FIG. 4.14. DIAGRAMMATIC ARRANGEMENT OF PLATES IN A PLATE EVAPORATOR
This action occurs simultaneously within each group of plates as shown.

Rotary Steam-Coil Vacuum Evaporator.—Although rotating steam coils have been used for many years for evaporating liquid foodstuffs in open kettles, it is only recently that such a coil has been adapted to vacuum operations. The equipment consists primarily of a rapidly whirling coil completely submerged in the boiling liquid under vacuum. Because of the movement of the coil, fouling is minimized and quality in the product is preserved. Evaporation rates of 55 to 65 (lb.)/(ft.)^2/(hr.) of water have been obtained with such viscous materials as 50% solids cold-break tomato paste, compared to 35 lb. of water evaporated from similar feed material and the same temperature difference, using a swept-film evaporator. This evaporator should be useful for small concentrators of fruit juices or purées, since it reportedly combines high output with low capital cost.

Mechanically-Induced Film Evaporators.—This is a mechanical evaporator in which the liquid to be evaporated flows down the heated walls of a cylinder; a thin film is mechanically induced by rapidly rotating vanes which both agitate and spread the film. Heat for evaporation is transferred through the jacketed cylinder wall. Different manufacturers use various descriptive terms such as "agitated film," "swept-surface," "wiped-film," etc., for their particular models. The vapor mixture produced during heating may be separated in an adjoining vapor separator or in an integral separator contained in the evaporator. The design is most effective for concentrates where high viscosities are encountered.

TASTE Evaporator.—The Thermally Accelerated Short-Time Evaporator (TASTE) is a high-temperature evaporator with very short holding times. The total residence-time during evaporation may be less than a min. and time in individual tube bundles only a fraction of a second. The short holding time permits high operating temperatures. In evaporation of citrus juices, the juice is heated in the first stage to approximately 195° F. (91° C.) and passes through a pressure release valve where it flashes off under a pressure of 19 in. of mercury. The heating and flashing off at a low pressure is repeated through a series of successive tube bundles, each flashing off at a lower pressure than the one preceding. A 4-effect, 7-stage evaporator of this type, having an evaporative capacity of 40,000 lb. per hr., is now being used by many orange juice concentrating plants in Florida.

Vapor Separation.—Evaporation produces a mixture of vapors and concentrated liquid which must be separated rapidly and thoroughly. Vapor separators are essentially cylinders constructed to effect the elimination of entrained liquid from the vapor. The separated vapors are frequently used as a heating medium for an additional effect in another evaporator.

Essence Recovery.—A substantial portion of the volatile components of fruits, which are important contributors to their characteristic flavor, are lost during concentration by evaporation. Many of the liquid flavor components are vaporized along with the water. Various techniques have been used to reclaim these volatile flavor components, and the process is commonly known as essence recovery. Adding back this essence to the concentrated fruit juice restores much of the natural flavor and aroma. Essences, which are mainly dilute water solutions of the volatile flavor components, are now produced and sold by several processors. One may buy on the commercial market strawberry essence, Concord grape essence, apple essence and others. While concentration ratios as high as 1,000-fold have been obtained, most commercial essences are in the 100 to 150 range. The concentration of these essences is usually expressed in terms of "fold." For example, 1 lb. of essence reclaimed from 100 lb. of juice will be labeled 100-fold essence. This does not mean that all of the volatile flavor components in the original 100 lb. of juice are now in the 1 lb. of essence. The flavor "potency" of a 100-fold essence as sold by different processors will vary with the efficiency of their essence recovery units. Buyers have long sought for some more satisfactory objective means of measuring the amount of flavor components than the simple statement of the quantity of fruit from which it was obtained. In the case of one fruit juice, Concord grape, one volatile flavoring component, methyl anthranilate, has been generally accepted as an easily measured component which will reasonably well represent the concentration of the total volatile flavor components. The USDA Standard for Frozen Concentrated Sweetened Grape Juice states that the minimum methyl anthranilate content of a one plus three (4-fold) frozen concentrate should be 1.2 mg. per liter. Analysis of several commercial 150-fold Concord grape essences used to restore aroma to concentrate shows that only about 50% of the original methyl anthranilate in the feed juice is recovered by the essence recovery equipment available. In the case of concentrated pineapple juice, ethyl acetate is used as the stated flavor component to measure the level of volatile flavors, even though its odor is not pineapple-like. Sufficient essence is added to the concentrate so that when diluted back to single-strength level the juice will contain 20 p.p.m. ethyl acetate.

Most volatiles boil off early in the distillation process and the greater part of the volatile flavors may be stripped and recovered from many fruit juices by evaporating only a small fraction of the water present, for example, 8 to 10% of apple juice, 20% of strawberry juice, 30% of blackberry juice, and 40% of Montmorency cherry juice. This vapor fraction is refluxed at atmospheric pressure or under vacuum to concentrate the volatile materials.

Essence recovery, in which both the noncondensables from the evaporating fruit juice, and volatile aromas stripped from the condensate by an inert gas, are compressed and absorbed in a liquid-sealed vacuum pump, has successfully produced high-quality aroma solutions (essences) from orange juice, apple juice, peach purée, and apricot purée.

Freeze Concentration.—Distillation or evaporation by addition of heat is not the only method of removing water from fruit juices. Water may also be removed from juices by freezing out as crystals of solid ice. This process, called freeze concentration, has been commercially used for producing orange juice concentrate.

Freeze concentration is not a new process. Zane Grey, in his book "Betty Zane," describes how more than 200 years ago the American Indians made maple syrup from sap taken from maple trees; "If the Indians had no kettles, they made the frost take the place of heat in preparing the sugar. They used shallow vessels made of bark and these were filled with water and the maple sap. It was left to freeze overnight and in the morning the ice was broken and thrown away. The sugar did not freeze. When this process had been repeated several times the residue was very good maple sugar." A U.S. patent issued in 1884 describes a process for concentrating milk using a scraped freezing cylinder and a centrifuge to separate the concentrated liquid from the ice.

A freeze concentration plant consists of three fundamental elements: (1) a crystallizer or "freezer" which produces a slurry of ice crystals; (2) a separation device where the ice crystals are separated from the mother liquid, this can be either a centrifuge, wash column, or filter press; and (3) a refrigeration unit to cool the liquid and remove the heat of fusion and the frictional heat resulting from hydraulic flow, wall scraping, and agitation of the slurry. With equipment available today it is possible to freeze concentrate most juices to approximately 50° Brix. Orange juice freeze-concentrated to 44.8° Brix need not be cut back with straight juice as is necessary in conventional concentration by evaporation. It should be pointed out that while freeze concentration can produce a frozen juice concentrate of superior quality, as has been demonstrated both with orange juice and apple juice, no frozen juice concentrate is made by freeze concentration commercially. This is due to the fact that juice processors believe that even a slight increase in cost, such as one cent more per can for concentrated juice, will not find a market among economy-minded consumers. It is estimated that the unit cost for freeze concentrating orange juice would be approximately one-half cent per pound of juice processed. According to experts, concentration by evaporation has the following disadvantages:

"In evaporating water from aqueous solutions the volatiles responsible for aroma and fragrance are also driven off. While equipment to condense and return these desirable volatiles has been developed, the recombined end product is inferior to the starting material.

"Distillation takes place with the addition of heat; this heat brings about a breakdown in the chemical structure of the food liquid causing a change in flavor, a diminution of vitamin contents and other nutritive properties. Even when essence recovery systems are employed and all volatiles condensed and added back to the concentrate, the original flavor is not restored."

Freeze concentration is free of the drawbacks associated with the evaporation. It is capable of concentrating fruit juices without appreciable loss in taste, aroma, color, or nutritive value. The refinement of the freeze concentration process has speeded up considerably in recent years with the effort to explore this technique as a possible economic means for desalination of sea water. While in the case of desalination melted ice is the desired end product and the concentrated brine is discarded, freeze concentration of juices seeks to reclaim the concentrated liquid and discards the ice. Basically however, freeze concentration and freeze desalination are the same process. The principal disadvantage of freeze concentration of fruit juices is its present high capital cost. A few years ago the Minute Maid Division of Coca Cola Corp. installed at Plymouth, Fla., 2 plants, including 1 with a rated capacity of 5,000 gal. per day of fresh orange juice. The concentrate from these plants was blended with a more concentrated evaporator effluent to produce a 42° Brix product. This plant is no longer in operation, presumably because of low production capacity. The two most important "bugs" in freeze concentration are: (1) inability to control ice crystal growth over a period of time, and (2) excessive solids loss due to liquid entrapped in the ice crystals.

Since alcohol freezes at a temperature well below the freezing point of water, freeze concentration is particularly applicable to wines. There are more than 100 freeze concentration plants for wine now operating in France. These have water (ice) removing capacities of up to 1,000 kg. per hour. The wine mass comes from the freezer as a crystalline paste and the frozen water is separated from the concentrated wine by decantation, centrifuging, or pressing.

Reverse Osmosis.—Osmosis is a phenomenon that occurs naturally in many biological systems whenever a dilute liquid and a concentrated liquid are separated by a semipermeable material—one which selectively permits one kind of molecule to pass through but holds back other kinds. Under ordinary conditions water passes from the dilute liquid to the concentrated liquid. By applying pressure on the more concentrated liquid, however, it is possible to reverse the flow and force water molecules through the membrane while other molecules (such as sugar, acids, flavors, etc.) are held back. The process may be thought of as molecular filtration.

In plant equipment, apple and orange concentrates up to 40° Brix have been made from their respective juices using a pressure of 2,500 p.s.i.g. Other tests on maple sap indicate that the reverse osmosis technique may be an economically feasible method for concentration of maple syrup.

This process is still in its infancy, but like freeze concentration it is also benefiting from a major research effort on its application to desalination, even though in desalination the water is the desired end product and not the concentrate.

Freezing Methods.—Fruit juices, concentrates, and purées have been frozen by several methods. These include (1) rotating the filled containers in a refrigerated liquid such as glycol, (2) packing the juice in small containers and freezing while in contact with refrigerated metallic plates, (3) freezing in an air-blast tunnel at temperatures of approximately −30° F. (−34° C.), and (4) slush-freezing the juice before packaging, an increasingly popular method because it permits freezing juices in containers of all sizes. While small containers are preferable for blast freezing, it is not uncommon for small jelly processors or food processors to pack and freeze single-strength juices in 30-lb. tins by slush freezing.

The Votator is widely used for slush-freezing both single-strength and concentrated juices. A thin stream of juice is forced at about 50 p.s.i. pressure into the space between the center shaft and the heat transfer tube. Just as rapid evaporation requires a large area of refrigerated heat transfer surface per unit of juice, so does the Votator freeze rapidly by causing a thin layer of juice to pass over a large area of refrigerated heat transfer surface. Floating scraper blades affixed to the rapidly revolving center shaft are forced outward against the heat transfer surface and automatically remove the product film which would otherwise accumulate (Fig. 4.15). Ammonia is ordinarily used as the refrigerant in this equipment, although Freon or brine have also been used as the heat transfer medium. An outlet temperature below 27.5° F. (−2.5° C.) is not recommended for single-strength orange juice, since cooling below this point results in such a stiff slush that it cannot be handled by an automatic filler. By contrast, orange concentrate is commonly chilled at 20° F. (−6.7° C.) because it is still quite fluid at this temperature. After orange juice concentrate is slush frozen in a Votator, the slush is packed into cans and hardened by passage through a low-temperature blast tunnel at −20° to −30° F. (−29° to −34° C.).

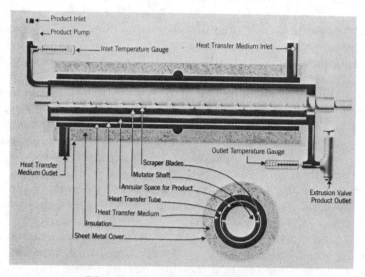

FIG. 4.15. VOTATOR SLUSH-FREEZING UNIT

This cross-sectional drawing depicts the closed-system heat transfer mechanism. A large volume of juice per unit of heat transfer surface is frozen. A clean heat transfer surface is maintained and intense agitation provided.

The Processing of Various Fruits for Juice and Concentrates

Apple Juice.—It is difficult to give a short description of the processing of apple juice because this item is available in several different styles, including: *natural,* a light-colored juice made by adding ascorbic acid during or soon after pressing; *crushed,* a thick, pulpy juice sometimes known as "liquid apple;" *clarified,* a centrifuged juice from which most of the larger apple particles have been removed; *clear,* a filtered, sparkling clear juice; and *opalescent,* a juice with various amounts of suspended insoluble material left to produce a cloudy product. There are regional preferences for each type of juice noted above; for example, most of the juice consumed in the United States is of the clear variety, whereas in British Columbia about ⅓ of the apple juice pack is of the opalescent style. There are also regional preferences for different Brix-acid ratios. Apple juice that is acceptable in one area would be considered poor in quality in another. Western consumers appear to like a juice that is sweet, while those in the eastern states like a rather acid juice. The blending of juices from the many varieties of apple is still somewhat of an art. In general, the blend should include enough acid varieties to give the product an acidity of between 0.40 and 0.50% as malic acid and a sugar content of about 12.5° Brix, or a sugar-acid ratio between 31 and 25. It is desirable to include sufficient of the aromatic varieties of apples such as McIntosh and Golden Delicious to contribute bouquet. Apples also have a wide range of pressing characteristics, from mealy and soft to crisp and juicy, due to differences in variety, maturity, and length of storage.

Until relatively recently most apple juice was made by washing and sorting apples, grinding them in a hammer mill or disintegrator, and pressing out the juice in a hydraulic rack and frame press (Fig. 4.16, 4.17, and 4.18). In this technique ground apple pulp is loaded onto a cloth, preferably nylon, and evenly spread; the pulp is then wrapped in the cloth by folding the corners; a wooden rack is placed on the filled cloth and the process is repeated until a sufficient number of "cheeses" have been made for the capacity of the press. The assembly of racks and cloth is then compressed in a hydraulic or mechanical screw-type press to express the juice from the pulp. One ton of raw apples pressed in such equipment will yield approximately 1,400 lb. or 160 gal., of single-strength juice. This method leaves much to be desired from the standpoint of sanitation, and requires a considerable investment in labor, time, cloths, and racks. There are several modern techniques for juice extraction in the apple industry. These include the Willmes Press, which is essentially a large inflatable rubber tube inside a horizontal cylindrical screen lined with press cloth; several types of centrifuges; continuous screw-presses; and vacuum fil-

FIG. 4.16. LOADING APPLE PULP ONTO PRESS CLOTH

A measured quantity of apple pulp has just been delivered from an automatic measuring box. This pulp will be spread and folded into the cloth.

ters. Most of these depend upon the use of some kind of filter aid such as
rice hulls or shredded cellulose.

In a unique two-stage thick-cake dejuicing system, the apples are
ground to produce a coarse pulp; shredded cellulose is added as a filter
aid, and the mixture is fed into a basket- or sugar-centrifuge where the
free-run juice is extracted. The partially dejuiced pomace leaving the
centrifuge is fed into a vertical screw-press over a small vibrating screen
which serves to remove course fibrous material. The term "thick-cake"
refers to the formation of a press-cake about 4½ in. thick on the centrifuge wall during the first extraction (Fig. 4.18).

FIG. 4.17. PRESSING APPLE PULP

Concentration.—The commercial procedure for producing a frozen apple concentrate is shown in Fig. 4.18 and 4.19.

Clear or opalescent-type apple juice is fed to an essence recovery unit; after removal of the volatile flavor components, the stripped juice is concentrated under vacuum to 45° Brix (4-fold). The concentrated es-

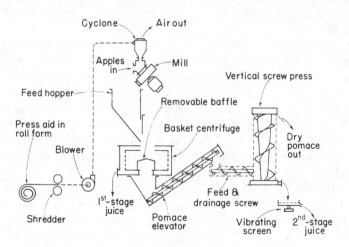

FIG. 4.18. SCHEMATIC DIAGRAM OF TWO STAGE THICK-CAKE APPLE JUICE EXTRACTION PROCESS

sence is added to the cool concentrate, which is packed in 6-oz. cans, frozen, and stored at −10° F. (−23° C.). Since the aromatic components of apple juice are more stable and more easily recovered than those of citrus juices, this added essence technique is used for apples rather than the "cut-back" process used for orange juice concentrate.

In addition to 4-fold frozen apple concentrate, large quantities of a 70°−75° Brix concentrate are made and sold for remanufacturing into such products as jellies. The high solids apple concentrate is marketed, without refrigeration, in 50-gal. drums, tank trucks, and 46-oz. cans for institutional beverage use.

Citrus Juices.—*Orange Juice*.—Frozen concentrated orange juice was the first frozen fruit juice to be produced in large commercial quantities, and continues to be the leader among frozen juice concentrates. Commercial production of frozen concentrated orange juice began in 1945–1946, using a low temperature during evaporation and the addition of fresh juice to the concentrate for the enhancement of flavor.

In the U.S., Florida orange groves produce over ¾ of the U.S. oranges.

In terms of the usual unit for expressing capacity of concentrate evaporators, the total evaporative capacity of the major plants in the Florida citrus area is in excess of 2 million pounds per hour of water.

The tremendous capacity available in Florida for producing and processing oranges obviously makes orange concentrate a big business, and this is further evidenced by the fact that frozen orange juice concentrate is the only frozen fruit which is sold on the futures market of the New York Cotton Exchange. Millions of gallons of frozen orange concentrate are exported from the United States, and Canada is one of the most important purchasers.

Converting fruit into concentrate is a highly mechanized, closely controlled process (Fig. 4.19). The oranges—or other citrus fruit—are handsorted and spray-rinsed. The fruit is then either stored—maximum 24 hr.—or conveyed to the extracting floor. In Florida, an automatically proportioned sample of the fruit drops off the conveyor and goes to a Florida State Dept. of Agr. laboratory, where it is analyzed to insure that state requirements are met. The only important variety of oranges for juice in California is Valencia; several varieties are used in Florida, including the Hamlin, Valencia, Pineapple, and Jaffa.

Conveyor belts carry scrubbed fruit past chutes which lead to a series of extractors. The chutes are proportioned in size to individual extractors. A coring device opens the bottom of the orange as matching metal fingers squeeze the fruit to remove all the juice (Fig. 4.20). The squeezing fingers also expel peel oil onto the surface of the skin, from which it is washed with a stream of water. Thus, none of the bitter orange oil spoils the taste

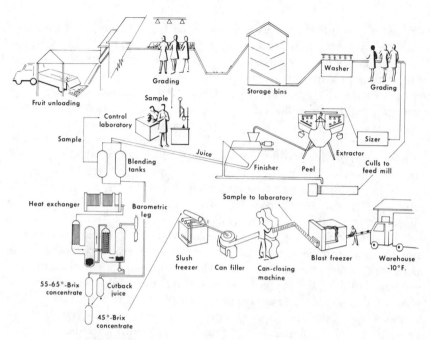

FIG. 4.19. STEPS IN PRODUCTION OF FROZEN CONCENTRATED ORANGE JUICE

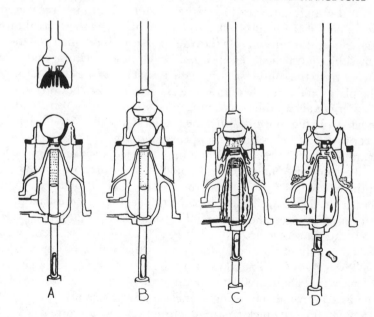

FIG. 4.20. SCHEMATIC REPRESENTATION OF OPERATION OF FMC IN-LINE EXTRACTOR

of the juice—and the oil itself is available for further processing into valuable products. Extracted juice is immediately strained free of pulp—an important step, since incidence of gelation in concentrate is directly proportional to the time the pulp stays in contact with the fresh juice at temperatures above 35° F. (2° C.). Centrifuging removes essentially all the remaining solid pulp.

In one Florida plant, TASTE evaporators utilize a flash concentration at a temperature above 190° F. (88° C.) and require less than 8 min. to convert orange juice from approximately 12 to 65% solids. Evaporated juice is cooled to about 60° F. (15° C.) and mixed with freshly squeezed pulpy juice. This step adds volatiles which restore the fresh juice taste unavoidably lost during concentration, and dilutes the concentrate to market specifications for soluble solids.

The "cut-back" juice is sometimes prepared in separate extractors operating at lower pressure, and larger screen openings are used in the finisher so as to produce large particles of the pulp. The pulp and juice are separated by centrifuging or screening. The pulp, which contains most of the pectinesterase enzyme, is pasteurized to inactivate the enzymes and is then recombined with the juice for use as "cut-back" to restore fresh flavor to the concentrate. Thus, the only unpasteurized part of the concentrate is the fluid part of the "cut-back" juice.

Several orange juice processors have attempted to use recovered essence in lieu of fresh add-back juice. Some of the methods used include freeze concentration, wherein the original aroma is retained and the juice is not subjected to heat; essence recovery under vacuum; and essence recovery from the condensate from the first stage of the evaporator. Most of the plants in Florida do not restore essence because they do not believe the benefits warrant the cost. Orange essence recovered by refluxing the distillate from the concentrating operation is not stable, and the beneficial effect of such added essence disappears on storage. Orange concentrates with added essence recovered in a liquid sealed vacuum pump system, described earlier, retain excellent quality when stored for more than six months at 0° F. (−18° C.). Most of the orange flavor in today's concentrate comes from the naturally accompanying peel oil plus the flavors in the fresh cut-back juice. The mixture of fresh unprocessed orange juice plus concentrate is cooled to about 40° F. (4° C.) in cold wall-tanks and is passed through refrigerated Votators to further lower the temperature to about 25° F. (−4° C.). The concentrate at 25° F. (−4° C.) is filled into 6-oz. cans at a rate of about 1,000 cans per minute per machine. Sealed cans are then passed through a freezing tunnel for approximately 1 hr. at −50° F. (−45° C.) to insure quick-freezing before they go into storage at −10° to −15° F. (−23° to −26° C.).

There are many by-products from the manufacture of orange juice concentrate. These include the peels, pulp, and seeds, which are pressed and dried for cattle feed. The press liquid yields citrus molasses, which is also used for cattle feed, as well as terpene oils for paints and plastics. Orange oil goes to the food, beverage, pharmaceutical, perfume, and soap industries. One producer of concentrate also obtains vitamin P as a by-product from the orange juice concentrate.

Florida's Citrus Commission makes a strong effort to maintain high quality in frozen orange concentrate by prohibiting the washing of citrus pulp and by increasing the minimum fruit solids content from 42° to 45° Brix. For many years the yield of 4-fold frozen orange concentrate was in excess of 1.5 gal. per 90-lb. box of fruit. Under the new quality measures the yield is approximately 1.25 gal. per box.

The Technion of Israel holds patents for an interesting process for the production of highly concentrated citrus juice with a density of up to 83° Brix. This 8-to-1 concentrate reportedly utilizes ultrasonic waves in the final stages of concentration and does not require refrigeration.

Awake.—The effect of adverse weather on citrus production and on the production of orange juice concentrate is dramatically illustrated by the production statistics of two successive seasons in Florida. In 1961–62, more than 116 million gal. of orange concentrate were packed. In the following season, as a result of a severe freeze in the orange-growing area of Florida, less than 52 million gal. were packed.

The shortage of frozen concentrate resulting from the Florida freeze of December 1962 led to the development of several substitute frozen concentrated "orange" drinks in 1963, among them the synthetic frozen concentrated drink, AWAKE. This synthetic concentrate is made of sugar, syrup, water, corn syrup, orange pulp and rind, citric acid, gum arabic, vegetable oil, cellulose gum, potassium citrate, calcium phosphate, vitamin C (ascorbic acid), natural and artificial flavors, vitamin A, artificial color, and vitamin B. By 1965 the product had obtained approximately 10% of the concentrated orange juice market.

The capital requirements for plant and equipment required to produce a given volume of synthetic frozen concentrated orange juice are estimated to be approximately $^1/_5$ of the amount that would be required to produce an equivalent volume of natural frozen orange concentrate. The cost of ingredients for 24 9-oz. cans of synthetic orange concentrate are estimated to be approximately one-seventh the raw product cost for an equivalent reconstituted volume of natural frozen concentrate. These cost advantages enable the manufacturer of AWAKE to advertise and market the product aggressively.

Chilled Citrus Juice.—Chilled citrus juice is made continuously throughout the year from fresh citrus fruit and moves rapidly into consumption outlets after manufacture. In addition, substantial quantities of chilled orange and grapefruit juice are prepared by reprocessing single-strength bulk juice and reconstituting bulk-frozen concentrate.

Citrus Purée.—Frozen fruit purées made from whole citrus fruit are useful ingredients in the commercial preparation of frozen desserts, baked goods, and beverages. Orange, tangerine, lemon, and lime purées are all processed in the same manner. Sound, mature fruit is thoroughly washed with detergent in water and rinsed well. After washing, the fruit is trimmed so that the stem ends and discolored spots are removed. If Washington navel oranges are used, the navel end is removed. The trimmed fruit is either crushed or sliced and is then put through a rotary or tapered screw press fitted with stainless steel screens having 0.027- to 0.044-in. perforations, depending upon the final end-use for the purée. Yield of purée is approximately 50 to 60% of the weight of the whole fruit; it should contain 0.40 to 0.75% of peel oil, depending on the kind of fruit being crushed. The oil content of purée may be adjusted by grating off the skin before crushing, or by adding sufficient single-strength juice to bring the peel oil content down. Purées are packed both with or without the addition of sugar. Sweet purées are usually mixed with one part sugar to five parts purée. The purée may be either directly filled into enameled tin containers or slush-frozen and then filled. Storage should be at 0° to −10° F. (−18° to −23° C.).

Lemon Concentrates.—Most of the lemons in the United States are grown in California but an increasing proportion comes from Arizona. More than 40% of the U.S. lemon crop is processed, primarily into juice and frozen lemonade concentrate.

The commercial production of lemon products in the United States is confined almost exclusively to California, where the Eureka lemon, the principal variety grown, comprises approximately 88% of the total production. The lemon industry is unique in that the fruit is picked for size only and held under controlled storage conditions to mature from green to yellow instead of being left on the trees to color. The operations in processing lemons—inspection, washing, sizing, and extraction of juice—are the same as those earlier described for oranges. After the juice is extracted from the fruit and screened to remove rag and seeds, it is held in brine-jacketed tanks to chill. If necessary the juice can be deaerated in these tanks by applying a vacuum of 20 to 25 in. for about 30 min. Lemon juice is frozen as single-strength juice, concentrate, or lemonade concentrate. For canned frozen single-strength juice, the chilled juice is drawn from the cold holding tanks and further cooled to 30° F. (−1° C.) by passing through a heat exchanger. It is then filled into enamel-lined cans, sealed, frozen, cased, and stored at 0° to −10° F. (−18° to −23° C.). Lemon juice is concentrated in the same evaporation equipment used for oranges except that cut-back juice is not used and the lemon juice is concentrated to 43° Brix or approximately 5¾ to 1.

Frozen Concentrate for Lemonade.—Frozen concentrate for lemonade is second only to orange juice concentrate in production. This product is primarily single-strength lemon juice with sugar added. A small amount (approximately 10%) of concentrated lemon juice is added to give the proper balance of sugar and citric acid. A typical frozen concentrate for lemonade would be prepared as follows: Add sufficient concentrated lemon juice to a mixture of 280 gal. of single-strength lemon juice plus 2,800 lb. sucrose so that the citric acid level of the final product will be from 3 to 3.5%. This will make approximately 500 gal. of 55° Brix concentrate. To give the characteristic appearance of fresh lemonade, some of the juice cells which are screened from the lemon juice after extraction are returned to the concentrate. Unlike most other juice concentrates, which are 4-fold (3 parts water to 1 part concentrate), lemonade concentrate is reconstituted to lemonade by adding 4 to 4½ vol. of water to each volume of concentrate.

Frozen Concentrated Grapefruit Juice.—Frozen concentrated grapefruit juice is processed essentially the same way as frozen orange concentrate. This juice has a tendency to gel because of enzyme activity, so it is heated to 150° to 180° F. (65° to 82° C.) for a few seconds before it goes to the evaporator.

Frozen Concentrate for Limeade.—Frozen concentrate for limeade is prepared by adding enough sucrose to single-strength lime juice to raise the Brix to about 48°. Sufficient lime purée is added to give an oil content of 0.003% in the reconstituted limeade.

Frozen Concentrated Tangerine Juice.—Tangerine juice is extracted in a manner similar to that used for oranges. To avoid off-flavors care should be taken during extraction to avoid incorporation of excessive oil from pulp and peel; the juice should contain not more than 0.020% of recoverable oil and not more than 7% of free and suspended pulp. Unlike orange juice, frozen concentrated tangerine juice shows little tendency to gel and it is unnecessary to heat-inactivate enzymes prior to concentrating.

Grape Juice.—Grape juice can proudly claim to be the first of the heat-processed fruit juices; its history goes back to 1869, when Dr. Thomas B. Welch prepared juice from his own Concord grapes, filtered it, filled it into bottles, and pasteurized it in hot water long enough to kill the yeast. The juice so prepared was used for sacramental purposes at the local Methodist church. Today the yearly production of processed grape juice and grape drinks is commercially significant. By far the most important variety of grapes used for making unfermented grape products is the Concord. Because of its attractive color and characteristic flavor, the grape juice industry is built around this variety.

Processing.—Juice may be expressed from grapes by hot-pressing, i.e., heating and then pressing, or cold-pressing, grinding without heating. Concord grapes are almost always heated to extract color from the skin.

The fully ripened grapes are washed and crushed and the grape mass is heated in a heat exchanger to 140° F. (60° C.). Grapes may be heated to higher temperatures for longer times if additional color extraction is desired. After crushing, the grapes are treated with enzymes to break down the pectin. The juice may be pressed either in a hydraulic rack and frame press similar to those described earlier for apple juice, or by more modern equipment using a Garolla, Zenith, Willmes, Vincent, or other type of press. Practically all of the latter pressing techniques require the addition of a filter aid. Stainless steel screw-presses have been successfully used commercially and are more sanitary and require less labor than the hydraulic rack and frame press. The juice yield per ton of grapes is 175–185 gal.

The extracted juice is flash-heated in a plate-type or tubular heat exchanger to 175° to 185° F. (80° to 85° C.) for pasteurization. It is then cooled to 32° F. (0° C.) and pumped into storage tanks, glass carboys, or wood barrels, where it is held at cool temperatures for 1 to 6 months to allow the mixture of potassium bitartrate, tannins, and other substances commonly known as *argols* to settle.

Detartrated grape juice is concentrated under vacuum by equipment very similar to that used for concentrating apple juice. For the 6-oz., 4-fold retail size, juice is concentrated and sweetened to 48° Brix, and essence is added before packing and freezing.

Buyers of bulk concentrated grape juice must decide whether they wish a high-quality concentrate of 48° Brix which is packed in 50-gal. drums and frozen, or a lower quality concentrate of 72° Brix which can be preserved without either heat processing or freezing. This heavy-density concentrate is usually used in manufacturing jellies, while the frozen 48° Brix concentrate is used primarily for beverages. The grape juice sold as a retail item is a sweetened product and is usually sold at about 48° Brix, the solids being composed of 39% from the fruit and 9% from sucrose.

Guava.—Guava purée is a component in many of the tropical fruit juice drinks which have become so popular in recent years. Although guavas are processed in Australia and Africa, most of the guava purée used in the United States comes from Hawaii. Guavas there are obtained both from wild trees and from cultivated orchards. The guava fruit has a rough-textured yellow skin and varies in shape from round to pear-shaped.

Fruits from wild trees vary from 1 to 3 in. in diameter, but under cultivation the size can be increased to 5 in., and weights up to 1½ lb. per fruit. The color of the inner flesh varies from white to deep pink to salmon red. Those fruits with a thick outer flesh and small seed cavity are considered most desirable for processing because they yield more purée per unit weight than thin-flesh types. Guavas are hand picked. When they have been harvested at prime maturity they will not keep well, so speed and careful handling are necessary in getting them to the processing plant. Small wooden boxes are preferred for carrying the fruit because of the ease with which the ripe fruits are crushed or bruised. Damaged fruit deteriorates very rapidly and starts to ferment; it must then usually be discarded at the plant. Some processors harvest only those fruits which are firm and slightly underripe, and finish off the ripening under controlled conditions at the plant.

Processing.—After an inspection and sorting operation the guava fruits are dumped onto an inspection belt for removal of spoiled or badly damaged fruit, and for hand-trimming of fruits which have only small areas of damage. The sound fruits are then washed in a tank with a dilute detergent solution, mechanically or manually agitated. The fruit comes out of the bath on an elevator and passes under a clear water spray to rinse detergents from the skin. Guavas are one of the easiest fruits to process, since the whole fruit, without peeling or coring, is fed into a paddle-pulper which crushes it into a purée. If the fruit is too firm to pulp by this method it may be necessary to pass it first through a chopper or a slicer. The seeds, fibrous pieces, and skin tissue are screened out with a 0.033 or 0.045-in. perforated screen. The outer flesh of most guava fruits contains a considerable number of hard stone cells. These are removed either by passing the purée through a paddle finisher with a 0.020-in. screen or through a disintegrator which pulverizes the cells. This latter method reduces the graininess but results in a purée whose color is inferior. After removal of the stone cells the purée is passed through a slush freezer and is filled into containers. It is advisable to use plastic-lined enameled containers because of the high acidity of guava purée.

Passion Fruit.—Passion fruits are oval in shape and approximately 2 to 3 in. in their greatest diameter. Two varieties are commercially used—the purple and the yellow. The yellow variety, the common commercial variety in Hawaii, yields about 10 tons per acre, while the purple variety, extensively cultivated in other countries, yields only 1 to 2 tons. Passion fruit juice is golden yellow in color; its sharply acid taste, combined with a distinctive flavor and aroma, makes it very useful for combining with other fruits in juice blends and tropical fruit mixes. In commercial processing approximately ⅓ of the weight of the fruit as delivered is recovered as juice. The pulp must be removed from the tough rind. This is done in Africa by hand-scooping the halved fruit with spoons. In Hawaii, the fruits are sliced into rings of ⅝ in. thickness and then spun rapidly in a perforated centrifuge basket with sloping sides. Speed is regulated to make the slices climb the wall slowly while the juice and seeds escape through the perforated wall.

In Australia the fruit is fed between two revolving cones fitted to the ends of inclined shafts. As the fruit rotates between the spinning cones the skin bursts and discharges the contents. The skin and seeds are then separated from the resulting pulp. Seeds are removed from the pulp in a brush finisher (or a paddle finisher whose paddles are faced with neoprene) having a screen with 0.033-in. holes.

The extraction method serves only to remove the pulp from the rind; separation of the juice from the pulp is accomplished in a two-stage process which employs either a brush finisher, or a paddle finisher with the paddles faced with neoprene. In the first stage, the pulp passes through a stainless steel screen with 0.033 in. holes, and this is followed by a finishing operation with a screen of 60–80 mesh stainless steel to remove broken seed fragments. Because of the importance of the volatile components to the flavor of passion fruit juice, concentrates have not been particularly successful, although some experimental packs have been made with added essence. Although passion fruit juice can be preserved by heating, some flavor deterioration occurs during storage; freezing is therefore the preferred method of preservation.

Pineapple Juice.—Frozen pineapple concentrate is used both as a beverage base in 6-oz. cans or as an ingredient in the manufacture of blended canned fruit drinks, such as pineapple-grape, pineapple-grapefruit, and other fruit drinks. Pineapple juice comes from several sources. These include the juice obtained by pressing the shell scrapings from the Ginaca machine which both peels the pineapple and forms the cylinder from which the slices are made; cores; trimmings; broken pieces from the canning lines; fruit too small for canning; and the juice drained from crushed pineapple preparation, the peeled cylinders, trimming tables, and the slicing operation. The solid material is shredded by various machines and filter aid such as infusorial earth is mixed with the finely ground material before it is fed to a hydraulic press. The liquid material is heated to coagulate some of the solids and the resulting thin slurry is passed through a continuous centrifuge which removes most of the suspended solids, including the fibers and other coarse small pieces. Pineapples too small for canning may be peeled and pressed for juice by machines similar to those used for oranges, or they may simply be crushed, heated, and pressed. Pineapple juice is sometimes homogenized to stabilize the slightly cloudy appearance, since it is believed that the finely divided solids give the juice a better flavor.

For pineapple juice concentrate the juice is concentrated in multiple-effect vacuum pans. Where the method of evaporation is such that the juice must be heated for a long period—say an hour—it is necessary to use pans which operate at temperatures below 140° F. (60° C.). Where short-time evaporators are used, temperatures as high as 180° F. (82° C.) have been successfully used. The fresh pineapple juice at about 12° Brix is concentrated to 60°–65° Brix for remanufacturing purposes, or to 45° Brix for packing in retail-size 6-oz. cans.

Essence is added and mixed with the pineapple concentrate which is then slush-frozen and packed either into 6-oz. metal containers at 45° Brix or in 75-lb., 3-ml. polyethylene-lined fiberboard containers at 65° Brix. Freezing is completed by holding at −10° F. (−23° C.).

Prune Juice.—Prune juice is not a fruit juice in the usual meaning of the term, but rather a water extract of dried prunes. Because it is rich in mineral salts and acts as a mild laxative, prune juice has become a popular breakfast drink in the United States. The leading production area is California, but a large quantity of prune juice concentrate is shipped to eastern states, where it is reconstituted into the single-strength juice. Most prune juice and prune juice concentrate are now being preserved by heat processing. Although Cruess reported as long ago as 1953 that experimental batches of frozen prune juice concentrate were superior in color, aroma, and flavor to most of the heat-pasteurized, canned and bottled juices then found in retail markets, it was not until 1966 that a frozen concentrated prune juice packed in 6-oz. cans was marketed.

Prune juice is made either by the *diffusion* method or the *disintegration* method. In the diffusion method the soluble substances are extracted from the dried prunes by repeated steeping in hot water, followed each time by draining off the liquid extract. After a thorough washing, the prunes are dumped into wooden or stainless steel tanks. About 25 gal. of water per 100 lb. of fruit is added and the mixture is heated to 185° F. (85° C.) by steam coils. The prunes are steeped in the hot water for about 2 to 4 hr. and the surrounding liquid (which now contains much of the soluble solids of the dried prunes) is drained and stored. The process is repeated again with fresh water except that about 15 gal. of water are used for each 100 lb. of fruit in the second extraction. The liquid from the second extraction is drained and combined with the liquid from the first extraction. The process is repeated a third time, with 10 gal. of fresh water being used for each of the original 100 lb. of dried prunes. The three extracts obtained in this manner are mixed together and the vats are emptied of the residue of extracted prunes, which is now practically free of soluble substances and is discarded as waste. The concentration of single-strength commercial prune juice runs from 19° to 21° Brix. If the concentration of the blend of the three extracts is too low, it is increased either by evaporation or by extracting a fresh batch of dried prunes.

A second commercially used method of making prune juice is extraction by disintegration. Several hundred gallons of water are added to 1,200 lb. of washed prunes. The water is brought to a boil and the fruit is cooked for 60 to 80 min. under constant agitation until the prunes are thoroughly disintegrated. Pressure cookers may be used so that the cooking time can be reduced to as little as 10 min. The disintegrated prune mass is dropped from the cooker onto a cloth and the wrapped mass is placed in a hydraulic press. The liquid portion is then extracted under pressure. The extract obtained by this method usually has a density of about 9° to 11° Brix. It can be clarified either by allowing the sediment to settle and siphoning off the clear juice or by filtering through a filter press, using 1% infusorial earth as a filter aid. The resulting extract is a clear liquid which may then be evaporated in open vats or a vacuum evaporator to bring the final concentration up to the desired 19° to 21° Brix before final packing. For making a prune juice concentrate, the juice must be depectinized by addition of 2½ lb. of pectic enzyme per 100 gal. of juice. The enzyme is slowly mixed with the juice and the mixture is allowed to stand overnight. The juice is then filtered and concentrated to approximately 60° Brix at a temperature of not more than 120° F. (49° C.).

Strawberry Concentrate.—Approximately 10% of the sound strawberries delivered to the packing houses are unsuitable for use in fresh and frozen packs because of their size, shape, or minor blemishes. Such berries may, however, be perfectly suitable for such byproducts as strawberry juice. Frozen strawberry juice concentrate is a useful product for jelly manufacture. Concentrates are not only a more economical means to store and ship strawberry solids as compared with frozen single-strength juice, but also have certain advantages in jelly manufacture. Food and Drug Standards require that formulation of strawberry jelly be based on a juice of 8% solids. Natural variation of the soluble solids content (from 5 to 10%) of strawberries results in variable raw material costs for different lots of juice and necessitates changes from lot to lot in jelly formulation. Some lots of single-strength juice tend to form more or less permanent gels during storage, and these may be difficult to disperse in making jelly; in concentrates such gels are enzymatically prevented during processing. Good quality concentrate of density as high as 73° Brix (12 to 1 by volume) has been commercially prepared.

The juice is prepared by pulping cold, sound, whole berries through a ¼-in. screen in a hammer mill. Filter aid is added to the chopped berries in amounts of 3 to 10% and the slurry is pressed in a bag-type press. Depending upon variety and maturity, a yield of cloudy juice ranging from 70 to 80% of the original weight of the berries will be obtained.

The cloudy press juice is immediately treated with pectic enzymes to degrade the pectic substances. This permits production of a clear stable juice and prevents gelling in the concentrated product. The treatment consists of adding pectic enzyme to the juice in a concentration of 0.5% and holding for 3 hr. at 75° F. (24° C.). Filter-aid is added to a concentration of about 0.25% and the juice is clarified in a plate filter.

The clear depectinized juice should be concentrated as rapidly as possible to avoid deterioration of its color and flavor. When natural-circulation evaporators, using either calandria heaters or external heat exchangers, are used, with resulting long evaporation times, the maximum boiling temperature should not be more than 100° F. (38° C.) or a maximum boiling pressure of 1.9 in. mercury to avoid flavor and aroma deterioration. Vacuum concentration of strawberry juice can result in removal of practically all volatile flavor constituents from the concentrate. For a high-quality product these essences may be removed in essence recovery units as described earlier in this chapter. The recovered essence can be mixed into the concentrate to make a "full flavor" product, or can be packaged separately. The latter procedure may be advantageous because a jelly manufacturer may wish to incorporate the essence into the completed jelly just prior to the filling operation, thereby eliminating volatilization of much of the essence during the heating of the ingredients. Both concentrated juice and concentrated essence should be packed in enameled plastic-lined bag containers, frozen and stored at 0° F. (−18° C.) or lower.

Packaging of Fruit Juice Concentrate

The apparently simple act of opening a 6-oz. can of frozen orange juice concentrate actually was followed by unexpected problems. When part of the industry switched to a spiral fiber-foil can with aluminum ends, the consumers found that the magnet on the can opener could not hold the aluminum top when it was removed, and they had to fish for the top when it fell into the concentrate. Can manufacturers tried to solve this problem by making one end of aluminum and the other of steel with directions on the container to open the other end, i.e., the steel end. By 1965, various easy-open containers were being used. These consisted of fiber or metal cans with either a device on the lid such as a key-shaped or ring-shaped piece of metal enabling the consumer to lift and pull, or a variety of tear strips, for which the user pulls a tape or string to remove the top portion of the can.

In the early years of the retail frozen concentrate industry most products were packed in 6-oz. metal cans. An increasing proportion of concentrates are now packed in a spirally-wound fiber-foil body with metal ends. The ends are made of either tinplate or aluminum and are lacquer- or enamel-coated inside. The fiber body is usually a 4-ply composite, composed of 2 plies of natural kraft liner-board sandwiched between 2 thin sheets of aluminum foil. Both metal and fiber-body cans come in a variety of sizes, 6-oz. cans being most popular at the retail level, although there is an increasing proportion of 8-oz. and 12-oz. cans. For institutional use, 32 fl. oz. of concentrate is packaged in either 401 by 509, 401 by 510, or 404 by 414 cans.

Several experimental containers other than the conventional metal or fiber-foil cans are being test marketed. These include 8-oz. rectangular blocks of frozen orange juice concentrate encased in plastic material and packed in a carton; pouch freezing of orange concentrate, miniature milk cartons with aluminum foil liner and tear tab top, and containers with reclosable plastic lids.

Nomenclature of Fruit Drinks and Juices

In recent years there has been a proliferation of various combinations of juices, artificial flavors, and water offered for sale as beverages. The U.S. FDA, and various states, have set up broad standards of content for fruit drinks. These standards usually specify various quantities of color, preservatives, acids, sweeteners, and other ingredients. For example, under current FDA classifications, orange juices and drinks are listed in Table 4.3.

PRODUCT STABILITY

Some of the factors involved in quality changes of fruits from harvest through frozen storage have been discussed from a general viewpoint.

TABLE 4.3

FOOD AND DRUG ADMINISTRATION CLASSIFICATIONS
ORANGE JUICES AND DRINKS

	% Single-Strength Juice
Orange juice	100
Fruit drinks	
Orange juice drink	Not less than 50
Orange ade	Not less than 25
Orange drink	Not less than 10
Orange soda	No juice required—must be true fruit flavor
Imitation orange	No juice required

The following section will be concerned with more specific changes in various products as affected by handling practices, and ways of measuring these changes.

Product Quality Changes Resulting from Processing

The preparation of fruits for freezing preservation is calculated to preserve the fresh quality as much as possible until the fruit is consumed. This period includes the stages of prefreezing, processing, freezing, frozen storage, thawing, and perhaps thawed storage. Physiological changes which can lower quality can go on at any of these stages, and processing must take the subsequent changes into account. The changes occurring before and during freezing have been discussed previously; these proceed relatively slowly and decrease in rate as temperature is lowered. However, the changes occurring upon thawing can be profound and very rapid, owing to the favorable environment. Mechanical and chemical disruption of cells, usually resulting in their death, greatly increases permeability of membranes and allows mixing of cell contents which would otherwise be separated. As ice first begins to melt, cell contents, including enzymes and their substrates, are mixed in a concentrated solution and at a temperature favorable to reaction. The result is often a rapid degradation in quality unless steps are taken in processing to prevent such reactions. To the uninitiated who freeze a whole apple or pear, watching the apparently normal fruit turn brown and rotten-looking after a few minutes of thawing can be a spectacular and impressive demonstration of enzymic reactions. To prevent such reactions the fruit must be blanched or treated chemically as described in Volume 3. Unfortunately, any such prefreezing treatment has some undesirable effects. Blanching in steam or hot water causes severe softening and loss of soluble solids, while treatment with sulfur dioxide to inactivate enzymes has a deleterious effect on flavor. Adding ascorbic acid or other acids may protect the fruit sufficiently for some uses although the protection is incomplete. Packing the fruit in syrup has a beneficial effect on color, flavor, and texture. It not only excludes air but has an inhibiting effect on enzymic browning.

This effect is demonstrated in Fig. 4.21. Although a syrup of about 30% sugar concentration is beneficial to texture as well as to flavor of most fruits, a higher concentration can cause dehydration by osmosis, with consequent shriveling and toughening. Furthermore, color may bleed from the fruit along with water, and this is not reversible.

Blanching fruit in syrup has not been used to any great extent, but offers some possibilities for protecting quality if used correctly. As noted above, increasing the soluble solids within the fruit lessens freezing damage, and blanching in syrup of the right concentration could aid in accomplishing this, as could vacuum infiltration of syrup.

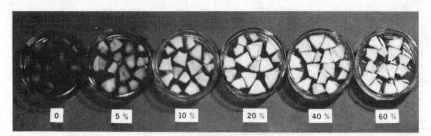

FIG. 4.21. INHIBITION OF ENZYMATIC BROWNING OF APPLE SLICES IN SUGAR SOLUTIONS OF CONCENTRATIONS SHOWN

Delay in processing raw fruit results in defects in the final product, either from microbial growth or by enzyme action. For example, bruises in fruit will darken from polyphenol oxidase activity. If the fruit is rapidly processed after harvest the bruises may be hardly noticeable in the final product, but if there is a period of several hours or longer for enzymic oxidation to proceed before the fruit is frozen, the bruises may become quite dark and obvious. This is one argument for decreasing the time between harvest and freezing by development of mobile processing equipment which can be moved to the field. If delay is unavoidable, the fruit should at least be cooled as quickly as possible; for example, in the way that tank trucks of cold water are used for transporting and cooling red sour cherries.

Packaging is of course important in protecting quality of frozen fruit. Only one point of interest will be mentioned here—namely, the use of liquid-in-bag type packaging for liquid nitrogen-frozen tomatoes. By freezing slices in small liquid-tight pouches, juice which exudes from the tomatoes during freezing surrounds the slices during frozen storage and thawing. This partially prevents dehydration during storage and allows maximum rehydration upon thawing.

Subjective Quality Changes During Storage and Distribution

Frozen fruits are not all affected in the same way by conditions conducive to deterioration of quality. Assessment of overall quality in some is related mostly to color, in others more to texture or flavor. Thus, subjective quality evaluations of frozen fruit by a trained panel can be used to detect changes in color, flavor, and texture long before the product becomes commercially unacceptable. Laboratory panels found significant flavor differences in about 90% of 55 lots of commercially-frozen strawberries within two weeks of storage at 20° F. (−6.7° C.); but quality control experts from frozen food plants indicated that it might take 4 to 6 weeks at 20° F. (−6.7° C.) to cause deteriorative changes which would be serious enough to cause consumer complaints. Overall quality of frozen strawberries was found to be related to retention of red color and ascorbic acid along with flavor.

Raspberries showed flavor differences in two weeks at 20° F. (−6.7° C.), but preferences of trained and untrained panels were about evenly divided between these and control samples stored at −20° F. (−29° C.). The principal reason why the samples at 20° F. (−6.7° C.) could quickly be distinguished from the controls was the difference in the migration of acidity from berries to syrup, and not necessarily the development of objectionable flavor. Color likewise migrated from raspberries, also from boysenberries, to syrup at storage temperatures of 10° F. (−12° C.) or higher, but only slightly at 0° F. (−18° C.) or lower.

Laboratory panel data on 52 different lots of commercially frozen peaches demonstrated that overall quality is directly related to the extent of browning, and the single most important factor in determining the extent of browning is the degree of container fill in the 10° to 25° F. (−12° to −4° C.) temperature range. Peaches in hermetically sealed tin containers were found to retain color and ascorbic acid better than those in composite containers. Color retention was further improved if the fruit was packed under 15 to 25 in. vacuum. The accessibility of atmospheric oxygen was, of course, the direct factor involved.

Red sour pitted cherries stored at 20° F. (−6.7° C.) showed texture and flavor differences to a trained panel in 3 to 4 weeks. Color differences in baked pie did not change appreciably for 6 to 8 weeks. In thawed unbaked cherries, browning was readily observed at 20° F. (−6.7° C.) in the fruit exposed to the headspace. Firmness was found to increase with storage time. As with peaches, sealed tin cans were superior to composite containers from the standpoint of preventing enzymatic discoloration at elevated temperatures.

Apricots are similar to cherries and peaches in the way their quality deteriorates under adverse conditions.

Fluctuating storage temperatures did not have any special effect on cherries or berries. Storage under fluctuating temperatures and simulated distribution patterns indicated that changes which occurred under variable temperatures were similar to those which occurred at an equivalent steady temperature.

Orange juice is unique inasmuch as the consumer's opinion of "quality" depends to a large extent on retention of a stable "cloud", which is a colloid consisting mostly of pectin.

Coagulation and precipitation of this cloud is unsightly and degrades the textural or "mouth feel" quality of the juice.

The recovery of the volatile aromas which are distilled off during concentration of fruit juices has been described. These water-soluble aromas are utilized as dilute water solutions and are known as essences. These include alcohols, esters, acetals, aldehydes, furfural, methyl furfural, aromatic aldehydes, ketones, esters, terpenes, and aromatic hydrocarbons. The stability of essences in frozen storage appears to vary with the method of manufacture. For example, a high-quality water-soluble essence of orange juice distilled under vacuum at temperatures of 110° to 115° F. (43.3° to 46.1° C.) did not noticeably deteriorate at this temperature, and the addition of the recovered water-soluble essence to the concentrate contributed to the characteristic odor of fresh orange juice. Unfortunately, the effect of this essence disappeared after six months at 0° F. (−18° C.) storage. By contrast, an "aroma solution" from fresh orange juice, recovered by a previously described vacuum stripping technique known as WURVAC, was added to orange concentrates and maintained a product quality equal or superior to commercial cut-back orange concentrate after storage for more than six months at 0° F. (−18° C.).

The flavor of fourfold apple juice concentrate with added essence was stable for approximately 2 years, 1 year and 2 to 4 months at 0°, 10° to 20°, and 30° F. (−18°, −12° to −7°, and −1° C.), respectively.

Most juice concentrate used in remanufacturing, as contrasted to retail consumption, is not frozen but is sold as a high-density concentrate and kept at room temperature; the high concentration prevents spoilage. It is best to store apple and grape concentrates at 35° F. (2° C.). Although commercial storage of concentrates at room temperature is not uncommon, definite browning occurs in sixfold (approximately 70° Brix) boysenberry juice concentrate held for 6 months at 70° F. (21° C.); some browning occurs even at 40° F. (4° C.) in 6 months. The color changes were attributed to a decrease in anthocyanin pigments accompanied by an increase in brown products of one or more deteriorative reactions.

Chemical and Physical Changes and Objective Methods of Analysis

Evaluation of frozen fruit products by trained panels, based as it is on statistical processes, is slow and requires many people. Since it is not suited to quality control or other quick applications, there is a demand for objective tests which will give equally good results in a shorter time and with fewer people. Objective tests which have been used for evaluation of frozen fruits are discussed below.

Color.—Two kinds of color measurements are used to measure quality loss in frozen fruits. One is the loss in natural fruit color and the other is an increase in browning. Combinations of these measurements can sometimes be used to advantage, or the ratio of color in fruit and syrup.

Browning.—It has been mentioned above that overall quality of frozen peaches stored in the 10° to 25° F. (−12° to −4° C.) range was directly related to extent of browning, and that both extent and intensity of browning are directly related to ascorbic acid content. In this case, the number of browned slices in a package was counted as a measure of extent of browning, and intensity of browning was measured by reflectance of slices in a Hunter Color and Color-Difference Meter. As slices become darker, the "R_d" and "b" values of this instrument decrease in essentially linear fashion; the first is a measure of total reflectance and the second of yellowness, which gradually decreases as it becomes masked by increasing brown pigment from enzymic oxidation of phenolic compounds.

Red sour cherries are subject to browning in the same way as peaches, and this can also be measured by reflectance.

Loss of Red Color.—In addition to browning, cherries lose their natural red anthocyanin skin color. This can be measured by a decrease in the "a" value of the reflectance meter. At the same time the syrup becomes increasingly redder, so that the ratio of red color in syrup to that in the fruit (color index) increases during storage at temperatures above freezing. This increase parallels loss of quality and can be used to measure it. Figure 4.22 illustrates the transfer of color in cherries stored at 20° F. (−7° C.).

Raspberries are similar to cherries in that they lose red color to the

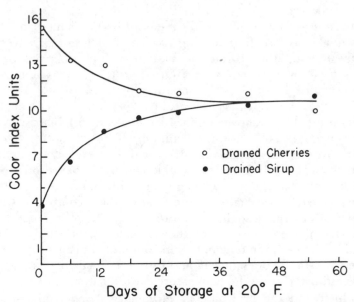

FIG. 4.22. EFFECT OF STORAGE TIME AT 20° F. (−7° C.) ON COLOR DISTRIBU-TION IN FROZEN SYRUP-PACKED CHERRIES

syrup during storage, but they do not brown. Therefore the red color in the syrup is purer and more soluble: it can be used alone as a measure of quality deterioration if the fruit is uniform, but since this is not always the case, a measurement of the ratio of red color in fruit to that in the syrup is more useful. During storage at 20°–25° F. (−7° to −4° C.) expression of fruit and syrup color as a ratio tends to eliminate variations due to original color and processing conditions, and thus permits evaluation of temperature history irrespective of these variables. Since raspberry color is soluble, it can be extracted from the fruit by blending in water or acid solution and filtering, then measuring absorbancy of an aliquot with a colorimeter at 515 mμ. Syrup can be merely diluted appropriately for color measurement. During storage at 20° F. (−7° C.) the ratio of color in fruit to color in syrup decreases from about 3.5 to 1.5 or less in 30 days. However, after the color ratio reaches 1.0 there is little further change.

Red strawberry color can be measured satisfactorily in packaged berries by the "a" value of the Hunter Color and Color-Difference meter. It was found that this value best represented the major color changes in strawberries; the decrease in "a" value per day varies exponentially with temperature in the 10° to 30° F. (−12° to −1° C.) range.

Boysenberry color is similar to that of raspberries. It leaks from berries to syrup during storage, and the ratio of absorbancy in fruit and syrup at 515 mμ gives a good measure of quality deterioration in boysenberries, as it does in raspberries.

Ascorbic Acid.—The ascorbic acid content of berries is high and decreases as quality is lost, thus making it a good indicator of adverse handling treatment. Little ascorbic acid is lost in fresh strawberries stored at 35° F. (1.7° C.) or in frozen berries stored at 0° F. (−18° C.), but in composite containers stored at temperatures between 10° and 30° F. (−12° and −1° C.) the loss in ascorbic acid per day varied exponentially with temperature. This is shown in Fig. 4.23. Since the original ascorbic acid content is not usually known, the amount of loss cannot be determined by measuring only reduced ascorbic acid. However, it was found that by measuring both the reduced ascorbic acid and its oxidation products, dehydro-ascorbic acid and diketogulonic acid, an indication of the storage history could be obtained. The total of these components remained relatively constant, so that by knowing the average rate of loss of ascorbic acid at a given temperature, the original content could be estimated, and from this the amount of loss and amount of quality change in that particular sample. Boysenberry and raspberry quality changes can likewise be estimated from the total ascorbic acid analysis.

Fruits which undergo browning by enzyme-catalyzed oxidation of

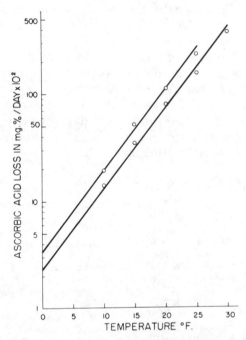

FIG. 4.23. EFFECT OF STORAGE TEMPERATURE ON RATE OF ASCORBIC
ACID OXIDATION IN TWO DIFFERENT LOTS OF STRAWBERRIES

phenols, including peaches, cherries, apples, and apricots, lose ascorbic acid through secondary oxidation by the oxidized phenols. Furthermore, these fruits have a relatively low initial ascorbic acid content compared to berries. Therefore, ascorbic acid analysis is not as good a measure of quality in these fruits, although in peaches packed with added ascorbic acid, the latter has been shown to decrease with loss in quality.

Soluble Solids.—Raspberry soluble solids have been shown to increase in the fruit and decrease in the syrup during storage above 20° F. (−7° C.). At 30° F. (−1° C.) a soluble solids ratio of 1.0 was obtained in 1 to 2 weeks. This ratio is obtained only under complete thawing conditions and therefore is a good "thaw index." The soluble solids ratio reaches 1.0 at 30° F. (−1° C.) in syrup packed cherries in 8 to 10 days and in syrup packed peaches in 6 to 10 days. A ratio of 0.9 is reached in less than 5 days, indicating rapid exchange of solids and water. On the other hand, at 20° F. (−7° C.) the ratio increases very slowly until an apparent equilibrium value of 0.8 is reached. Therefore, when values of 0.9 to 1.0 are found, it is virtually certain that the product has been completely thawed at some time. The soluble solids ratio is thus useful in indicating the history of many syrup-packed fruits.

Total Acid.—The ratio of total acid in fruit to that in syrup decreases markedly in raspberries in 1 to 2 weeks of storage at 20° F. (−7° C.). This ratio is interesting in that it starts in the freshly packed fruit at about 3.0, decreases to 1.0 in about 7 days, and then slowly decreases further to an equilibrium value of 0.5 to 0.7 in 40 to 80 days. Thus a total acid ratio of 1.0 or less would show a considerable change. It appears that this easily measured ratio is a sensitive quality index.

ADDITIONAL READING

DeFIGUEIREDO, M. P., and SPLTTSTOESSER, D. F. 1976. Food Microbiology: Public Health and Spoilage Aspects. AVI Publishing Co., Westport, Conn.

GOULD, W. A. 1977. Food Quality Assurance. AVI Publishing Co., Westport, Conn.

HAARD, N. F., and SALUNKHE, D. K. 1975. Postharvest Biology and Handling of Fruits and Vegetables. AVI Publishing Co., Westport, Conn.

HARRIS, R. S., and KARMAS, E. 1975. Nutritional Evaluation of Food Processing, 2nd Edition. AVI Publishing Co., Westport, Conn.

JOHNSON, A. H., and PETERSON, M. S. 1974. Encyclopedia of Food Technology. AVI Publishing Co., Westport, Conn.

KRAMER, A., and TWIGG, B. A. 1970, 1973. Quality Control for the Food Industry, 3rd Edition. Vol. 1. Fundamentals. Vol. 2. Applications. AVI Publishing Co., Westport, Conn.

NAGY, S., SHAW, P.E., and VELDHUIS, M.K. 1977. Citrus Science and Technology. Vol. 1. Nutrition, Anatomy, Chemical Composition and Bioregulation. Vol. 2. Production, Processing, Products and Personnel Management. AVI Publishing Co., Westport, Conn.

PANTASTICO, E. B. 1975. Postharvest Physiology, Handling and Utilization of Tropical and Subtropical Fruits and Vegetables. AVI Publishing Co., Westport, Conn.

PENTZER, W. T. 1973. Progress in Refrigeration Science and Technology, Vol. 1–4. AVI Publishing Co., Westport, Conn.

PETERSON, M. S., and JOHNSON, A. H. 1977. Encyclopedia of Food Science. AVI Publishing Co., Westport, Conn.

RYALL, A. L., and PENTZER, W. T. 1974. Handling, Transportation and Storage of Fruits and Vegetables. Vol. 2. Fruits and Tree Nuts. AVI Publishing Co., Westport, Conn.

SACHAROW, S. 1976. Handbook of Package Materials. AVI Publishing Co., Westport, Conn.

TRESSLER, D. K., VAN ARSDEL, W. B., and COPLEY, M. J. 1968. The Freezing Preservation of Foods, 4th Edition. Vol. 2. Factors Affecting Quality in Frozen Foods. Vol. 3. Commercial Freezing Operations—Fresh Foods. Vol. 4. Freezing of Precooked and Prepared Foods. AVI Publishing Co., Westport, Conn.

TRESSLER, D. K., and WOODROOF, J. G. 1976. Food Products Formulary. Vol. 3. Fruit, Vegetable and Nut Products. AVI Publishing Co., Westport, Conn.

UMLAUF, L. D. 1973. The frozen food industry in the United States—Its origin, development and future. American Frozen Food Institute, Washington, D.C.

WILLIAMS, E. W. 1976A. Frozen foods in America. Quick Frozen Foods 17, No. 5, 16–40.

WILLIAMS, E. W. 1976B. European frozen food growth. Quick Frozen Foods, 17, No. 5, 73–105.

WOODROOF, J. G., and LUH, B. S. 1975. Commercial Fruit Processing. AVI Publishing Co., Westport, Conn.

WOOLRICH, W. R., and HALLOWELL, E. R. 1970. Cold and Freezer Storage Manual. AVI Publishing Co., Westport, Conn.

5

Freezing Meats

L. J. Bratzler, A. M. Gaddis
and William L. Sulzbacher

M eat is a highly nutritious but extremely perishable food, and its preservation has always been of vital importance to mankind. This is due to the cyclic variation in the abundance of its supply and the necessity to maintain its quality during the periods of shortage and during its distribution and marketing. In order to maintain some degree of palatability and wholesomeness, it must be subjected to some type of processing very soon after slaughter. Freezing and storage at low temperatures are the means by which the highly prized qualities of fresh meat can best be maintained. Similarly, desirable characteristics of special meat products can be frequently maintained satisfactorily.

The full potential of commercial freezing of red meats has not been realized. The meat industry has for some years smoothed out seasonal variations by freezer storage of pork cuts for curing as required. A major area for development, the distribution of frozen cuts at retail levels, has made little progress. The reasons for this are manifold and complex. To clear the way for such application, the following must be achieved: marketing reorganization with centralized packaging of retail cuts; adjustment of labor-capital relationships; full acceptance by the consumer of packaged, frozen cuts; and improvement in technology. Meat storage in home freezers and the wide prevalence of locker plants for local patrons have increased during the past several decades. Consumers are well aware of the excellence possible with frozen meat, but the frozen product in many instances cannot compete with the fresh product. This is partly due to a distrust of commercial, frozen meat, the ready availability of fresh cuts, comparative prices, and the relatively poor color of frozen cuts.

FACTORS PRIOR TO FREEZING THAT AFFECT MEAT QUALITY

What Is Good Quality?

Excellence in meat includes a number of characteristics; these are mainly flavor, aroma, juiciness, color, and tenderness. There is a great deal of interdependence between most of these quality factors.

Flavor and Aroma.—Flavor and aroma are closely related and difficult to define. The flavor of cooked meat arises from water- or fat-soluble precursors. Water extracts of raw meat produce a meaty flavor on heating. Results suggest interaction of meat juices and fibrillar elements during cooking. A purified water extract of raw beef has been found to contain inosine, an inorganic phosphate, and glycoprotein. The glycoprotein, in addition to glucose, contains the amino acids serine, glutamic acid, glycine, alanine, isoleucine, leucine, β-alanine, and proline. Mixtures of these amino acids heated with glucose, inosine, and phosphate produce meaty odors and flavors in fat or water. The opinion has for some time been held by the Japanese that mononucleotides are largely responsible for meat flavor.

Glutamic acid, inosine, and hypoxanthine can be used for flavoring. This suggests that the increase in flavor of meat during aging may be related to nucleotide breakdown, with the formation of inosine, and ribose and hypoxanthine. However, there are reports of artificial production of various meat flavors by heating a pentose with cysteine and other amino acids in water. This suggests that the odor and taste of beef and pork might be simulated by water-soluble constituents. Other research indicates that the components of meat flavor may be divided into two groups.

There is a basic meat flavor common to all species which is developed in the lean by heat. The precursors of this can be removed from the muscle tissue with water. Flavors characteristic of species are developed by heating the fatty tissue. There are considerable differences between species in the fatty tissue. However, there may be some doubt that the flavor differences are due basically to variations in unsaturated fatty acid composition. Fat has a definite effect on flavor, and an adequate finish is needed for the best flavor.

Variation in flavor has been noted in sheep and cattle, and there is evidence that this may be inherited. It is well known that older animals have more flavor than immature animals. There are also notable differences in the flavor of different muscles. Also, the biochemical condition of a muscle may affect its flavor. Muscles with high ultimate pH[1] possibly have lower intensities of flavor.

Juiciness.—Juiciness is generally considered in terms of richness and amount of juice. The sensation is influenced by smoothness and a lasting fluidity. This quality characteristic is frequently found to be related to tenderness and flavor. There is a direct association with the amount of intramuscular fat which disappears at higher fat levels. Juiciness is also related to the water-holding capacity of the muscle.

Color.—Color is important to the overall impression of the quality of the meat. Myoglobin, a respiratory pigment, is mainly responsible for the color, and the appearance of the meat surface is due to the quantity of the myoglobin molecules, its chemical state, and the biochemical condition of the muscle. In fresh meat, the most important chemical form is oxymyoglobin, which has the bright red color desired by the consumer. Poor color may be due to a number of factors, some of which may not be related to poor eating quality. If the ultimate pH is high the color will be dark, due to the presence of the purplish-red myoglobin and a closed structure of the muscle. Low ultimate pH may promote a very pale color due to an open meat structure, and oxidation of the myoglobin to the brown pigment, metmyoglobin. The formation of metmyoglobin frequently foretells general autoxidative deterioration due to coupled reactions with unsaturated fatty acids.

Tenderness.—Tenderness or toughness is a quality representing the summation of properties of the various protein structures of skeletal muscles. The degree of tenderness can be related to three categories of protein in muscle. These are stroma protein (connective tissue), myofibrillar (actin, myosin, and tropomyosin) proteins, and sarcoplasmic proteins and reticulum; and the importance of their relative contribution depends on circumstances. Tenderness varies in the muscles of an animal and in a given muscle.

Tenderness is probably the most important single characteristic of meat because it determines the ease with which it can be chewed and swallowed. Many factors influence the quality of meat. These are broadly: the nature of meat, genetic characteristics, feeding and handling of the animal, classification, inspection practices and standards, and other antemortem and postmortem conditions.

The Nature of Meat

The composition of lean meat may be approximated as 75% water, 18% protein, 4.0% soluble nonprotein substances including mineral components, and 3% fat. The essential unit of muscle tissue is the long multinucleate fiber. The diameters of the muscle fibers vary within the muscle and with the age of the animal and degree of activity. The fiber consists of formed protein elements, the myofibrils, between which is a solution, the sarcoplasm, and a fine network of tubules, the sarcoplasmic reticulum. The fiber is bounded by a very thin membrane, the sarcolemma, to which connective tissue is attached on the outside. Each fiber is composed of many myofibrils, a variable number of nuclei, and inclusions such as mitochondria, glycogen granules, and liposomes or fat droplets embedded in the sarcoplasm of the cell. Table 5.1 defines the 3 protein classes and their characteristics.

Myofibril Proteins.—The myofibrillar proteins are myosin, tropomyosin, and actin. These proteins amount to 10% of the muscle and are important in the functional properties of meat. Such proteins undergo changes during rigor mortis that are related to the tenderness and other important properties of muscle. They are the contractile proteins and together with connective tissue constitute the structure of meat. The myofibrillar proteins possess a high degree of the water-holding capacity which is one of meat's most important physical properties.

TABLE 5.1

PROTEIN COMPOSITION OF VERTEBRATE MUSCLE

Protein Class	Definition
Sarcoplasmic proteins	Those proteins soluble at ionic strengths of 0.1 or less at neutral pH. Constitute 30–35% of total protein in skeletal muscle and slightly more than this in cardiac muscle. Contains at least 100–200 different proteins. Sometimes called myogen.
Myofibrillar proteins	Those proteins that constitute the myofibril. Make up 52–56% of total protein in skeletal muscle but only 45–50% of total protein in cardiac muscle. Although high ionic strength is required to disrupt the myofibril, many of the myofibrillar proteins are soluble in H_2O once they have been extracted from the myofibril.
Stroma proteins	Those proteins insoluble in neutral aqueous solvents. Constitute 10–15% of total protein in skeletal muscle and slightly more than this in cardiac muscle. Includes lipoproteins and muco-proteins from cell membranes and surfaces as well as connective tissue proteins. Although exact percentage composition can vary widely depending on source of the muscle, collagen frequently makes up 40–60% of total stroma protein and elastin may make up 10–20% of total stroma protein.

Source: Whitaker and Tannenbaum (1977).

Sarcoplasmic Proteins.—The sarcoplasmic protein fraction contains most of the enzymatic activities and therefore also influences the functional properties of meat. Myogen, globulins, myoglobin, metmyoglobin, hemoglobin, myoalbumin, creatine kinase, phosphoglyceride dehydrogenase, and pyruvate kinase are components that have been identified in sarcoplasmic extracts. There are said to be at least 50 components, many of which are enzymes of the glycolytic cycle.

Stroma Proteins.—The remaining structures in meat are the mitochondria containing the insoluble enzymes responsible for respiration and oxidative phosphorylation, the muscle membrane or sarcolemma, and collagen, reticulin, and elastin fibers of the connective tissue. It appears well established that the collagen content has an appreciable effect on tenderness and that the state of the myofibrillar proteins affects both tenderness and water-holding capacity.

Non-protein Substances.—There is present in meat about 4.0% of soluble, nonprotein substances. Among these there are nitrogenous compounds such as creatine, inosine monophosphate; di- and triphosphopyridine nucleotides, amino acids, carnosine, and anserine. Carbohydrates are present, including glycogen, glucose, and glucose-6-phosphate. Inorganic constituents such as phosphorus, potassium, sodium, magnesium, calcium, and zinc are concerned mainly with osmotic pressure and electrolyte balance inside and outside the cell. These factors are highly important in muscular contraction and relaxation during life, and in postmortem muscle they have a large effect on tenderness and water-holding capacity.

Intramuscular fat is an important part of muscle structure, and it influences the characteristics of meat. In addition to triglycerides, there is a considerable content of phospholipids and unsaponifiable constituents such as cholesterol. There are also small but important amounts of vitamins A, B, C, D, E, and K.

Genetic Characteristics

There is a rough relationship between breed of animal and meat quality. Over years of selection, strains of animals have been developed that tend to yield the best meat. However, a great deal of variability in meat characteristics exists among animals of the same strain. Much of the selection in the development of strains may have been on the basis of rate of gain, grade, and fat thickness, and not on specific meat quality factors. Tenderness and some aspects of meat flavor have a fairly high degree of heritability. It would seem that much may yet be possible in the improvement of meat quality through effective breeding and selection programs. Excessive fatness in meat animals is a great source of waste today. Intramuscular fat is frequently much higher in the muscles than is necessary for the best quality. Meat cuts destined for freezer storage benefit in stability with a minimum amount of fat. In recent years there has been some improvement in the development of meat-type hogs, and recently more attention is being paid to meatiness in cattle. Intramuscular fat contributes to tenderness, juiciness, and flavor, but does not exert increasing influence beyond a certain modest concentration. Much of the fat added by fast gain is superfluous and wasteful.

Feeding and Management

Some changes or variation in meat characteristics can be obtained by nutrition and handling. Fatness can be somewhat varied by the level of nutrition. This affects tenderness, juiciness, and possibly flavor. However, in general, lowering the plane of nutrition seems to decrease to a disproportionate degree the marbling or intramuscular fat. This lowers the quality of the meat. In hogs, the use of large amounts of roughage produces a little higher proportion of lean cuts without affecting the intramuscular fat. Feeding provides little leeway in the variation of meat quality. Even use of a low level of nutrition encounters the difficult-to-solve problem of putting the right amount of fat on in the proper location. No single nutrient known will give a consistent and pronounced effect on meat characteristics. The absence or excess of certain substances can cause abnormal characteristics. However, this is a nutritional deficiency condition. Tocopherol or vitamin E, when fed, will tend to increase the stability of fatty tissue. However, this has not been considered to be a practical measure for improving meat. In hogs, the feeding of unsaturated fatty acids causes their deposition in the meat tissue, with the net result of reduced storage life. This may also cause a vitamin E deficiency. Various hormones serve as growth promoting agents. They may increase rate of gain and increase muscle growth with less fat deposition. Tenderness decreases with age. However, flavor increases with age, and also the amount of fat. Differences have been noted between the flavor of grass-fed and dry lot-fed cattle.

Classification of Meat

Cattle over six months of age produce beef, while veal is produced by very young cattle, or calves. The meat from hogs is known as pork, irrespective of animal age. Sheep under 14 to 16 months of age produce lamb, while mutton comes from sheep older than 14 to 16 months.

A further classification of cattle can be made, such as: steer, male castrated before sexual maturity; heifer, immature female; cow, mature female; bull, mature male; and stag, male castrated after reaching sexual maturity. There is no difference in palatability characteristics between steer and heifer beef of the same quality grade. Beef from the remaining classes may show considerable variation due to animal age or maturity. Corresponding classes of hogs are barrow, gilt, sow, boar, and stag. No differentiation is made between pork from barrows and gilts. Pork from the three remaining classes is used mainly for processing or sausage manufacturing purposes. Some boars produce meat that is inferior because of the sex odor that may be present. In sheep, the comparable classes are wether, ewe lamb, ewe, ram, and stag. Lamb and mutton from the respective classes find different uses like the swine classes do, although the sex odor incidence in rams is less than in boars.

All meat-producing animals can be further subdivided into breeds. However, research data do not indicate palatability differences attributable solely to breed. Trimmed retail-cut yield differences are mainly a function of internal and external carcass fat content, and the highest yields are found in those carcasses with the least fat. Degree of animal fatness and resulting carcass fatness can be regulated largely by nutrition or feeding and management practices. Intensive genetic selection for certain traits, such as leanness, or "meat-type" in hogs, has markedly reduced the relative amount of pork fat marketed as live hogs. It is often stated that carcass attributes are more variable between individuals of the same breed than between the averages of breeds of the same species.

Preslaughter Conditions

Fresh meat of a given species is far from uniform. This is indicated by wide differences in time of keeping in freezer storage. There are inherent biological variations that are responsible for differences in characteristics, quality, and stability of the fresh meat. Also, much lack of uniformity may be due to inability or failure to control conditions in the interval preceding slaughter. Stress exerted on the animals at that time has a considerable physiological effect. Exhaustion, fasting, and excitement tend to cause glycogen loss. However, associated problems range from too much glycogen, with rapid and extensive glycolysis postmortem, to low glycogen, with slow, insufficient glycolysis.

The pig possesses the greatest variation in postmortem changes and differences in muscle characteristics. Pork, not entirely coincidentally, presents the biggest problem of stability in freezer storage. Due to genetic makeup and sensitivity to antemortem conditions, a considerable amount of pork has a rapid glycolysis and is pale and watery with low ultimate pH. The incidence of this condition averages about 18% and may approach 40 to 50% in the summer. Some strains of hogs are predisposed to this condition, and there may be a genetically controlled excess or imbalance of glycolytic enzymes.

Beef and lamb appear less susceptible to conditions immediately preslaughter. However, dark cutting beef is not infrequently encountered. This condition is due to glycogen depletion, with resultant high ultimate pH. There is a considerable difference of opinion concerning the effects of natural and artificial (or induced) glycogen depletion preslaughter. At the least, it causes considerable variation in meat properties. Cattle, lambs, and pigs, when subjected to stress conditions for several hours preslaughter, usually have lower muscle glycogen, higher postmortem muscle pH, darker color, and improved muscle tenderness and juiciness. Under such conditions there have been reports of less flavor or decreased desirability in flavor. The nature and extent of this possible (controversial) flavor change is not known. Normal glycogen reserves at slaughter produce meat that is bright in color, low in pH, of an open structure, with minimum waterholding capacity, and greater stability microbiologically.

Meat Inspection

Companies doing interstate or foreign meat business are compelled to have publicly supported federal meat inspection. This service has been in operation in the United States since 1906. At present it is under the direction of the Meat Inspection Program, USDA, Washington, D.C. There are many services and purposes but its principal duties are related to product healthfulness, such as the removal of bad or diseased meat from the food supply, enforcement of sanitation in meat dressing, handling, and processing procedures; correct and informative product labeling; and the prevention of adding harmful ingredients to meat and meat products. Antemortem and postmortem, processing, labeling, worker health, and sanitation are examples of inspection services performed. Meat packing companies that have only an intrastate operation may come under either state or local governmental jurisdiction for inspection procedures. Livestock slaughtering is becoming decentralized in that the new dressing facilities are being built in or near the areas of meat animal feeding and production. Often these new plants dress only one species of livestock and may do little or no meat processing.

Slaughtering Meat Animals

Beef.—Most of the cattle are transported to the slaughterhouse by truck according to plant requirements. Thus, extended holding periods are eliminated, although it is general practice to rest the animals prior to

FIG. 5.1. SHROUDING BEEF CARCASSES

Beef carcasses are shrouded with a muslin cloth as they leave dressing floor on way to the cooler. The shroud helps to smooth and whiten the fat.

FIG. 5.2. BEEF AGING COOLER

Beef cooler in the Chicago plant of Armour and Co., where sides of beef are
chilled and aged before being made into wholesale cuts.

slaughter. Research has shown that extreme antemortem stress may seriously affect carcass characteristics, and may bring about such defects as dark-cutting beef.

Since the passage of the "humane slaughter" law in 1958, stunning with a sledge hammer has been replaced by stunning or penetrating devices driven either by air or by a gunpowder cartridge. Koshering, or slaughter according to Jewish precepts, is the only exception. Specialized beef slaughter plants utilize an "on the rail" automated procedure that includes hide pullers and other labor saving devices. After the carcasses have been split, the halves are washed, shrouded, weighed, and transferred to the chill cooler and finally to the holding cooler (Fig. 5.1 and 5.2). USDA Choice steers will yield about 60% of live weight as carcass: lower-grading cattle will yield less.

Pork.—Recent research has shown that preslaughter treatment of hogs may have a marked effect on carcass muscle quality. The incidence of pale, soft, and watery pork may be increased if the animals are severely stressed prior to slaughter. Electrical stunning or carbon dioxide immobilization satisfy the humane slaughter requirements. Hogs are bled, scalded, dehaired, cleaned, eviscerated and split in a rapidly moving conveyor-type operation. Carcass yield will range from 70 to 75% of live weight.

Lamb and Mutton.—Sheep and lambs are generally stunned electrically, although their size makes CO_2 immobilization practical. Similarly to hogs, the slaughter operation is done on a conveyor disassembly line. The carcasses are not split; they average about 50% of live weight.

Postmortem Treatment

Improper procedures can materially affect the quality of the meat and its stability potentialities. With death, the cytochrome oxidase system becomes inactive due to lack of oxygen, and the adenosine triphosphate (ATP) is depleted. The inorganic phosphates formed stimulate the breakdown of glycogen to lactic acid, which lowers the pH. As ATP decreases, actomyosin is formed and rigor mortis sets in. Depending on extent and rapidity of lowering of pH, denaturation of proteins sets in and this lowers the water-holding capacity and causes water loss or drip. Denaturation of the sarcoplasmic proteins makes them more susceptible to proteolysis. Rapid chilling, besides its inhibiting effect on proteolysis and microorganisms, is extremely important for its slowing effect on pH drop, lessening of the amount of denaturation of sarcoplasmic proteins, slowing loss of water-holding capacity, and lessening the shortening or contraction of the muscles and actomyosin formation. This sequence of events makes a more tender product.

Beef.—The beef carcass halves, or sides, will have an internal temperature of about 104° F. (40° C.) when they leave the kill floor and are moved into the chill cooler. Rapid cooling of the sides is desirable in order to reduce the incidence of deep seated incipient spoilage. A high relative humidity in the chill cooler is maintained in order to control carcass weight and surface dehydration. Cooling units may be of many kinds but overhead fin types are very popular at present. Achieving an internal round temperature of 50° F. (10° C.) in 600-lb. cattle within 20 hr. or less after slaughter is desirable. Weight loss during this period is assumed to be about 2.5%, but actual weighings indicate a range down to 0.5%.

Following the initial chilling period of 18 to 24 hr., the beef is moved to the storage or holding coolers. These are operated at temperatures of 32° to 36° F. (0° to 2° C.). Relative humidity is best controlled by maintaining the correct temperature differential between the cooling surface and the ambient air. A compromise between low shrinkage in weight and little bacterial and mold growth generally determines storage cooler operational conditions. The beef is shipped from the storage coolers and the total elapsed time from slaughter to shipment may range from 24 to 72 hr.

Some of the beef that is used for sausage manufacturing purposes may be boned and used immediately. This "hot beef" that has not passed through complete rigor mortis has superior water holding or binding characteristics. Similarly, if this hot beef is frozen before developing complete rigor, water-holding capacity is superior to that of beef that is normally cooled, boned, and stored after slaughter. The boning and freezing operations must be done expeditiously if increased water-binding capacity is desired.

Pork.—The split pork carcasses are generally cut into wholesale cuts after the carcasses have been chilled for 20 to 24 hr. following slaughter. Shipment of the cuts is made shortly after cutting and packaging or crating. There has been some interest in boning heavy hog carcasses immediately after dressing so as to take advantage of the prerigor characteristics discussed in connection with hot beef. Also, the curing and processing of unchilled pork cuts, such as hams, is being studied.

Lamb, Mutton, and Veal.—Freshly dressed whole lamb, mutton, and veal carcasses cool quickly because of their small size. They move into the distributive trade channels soon after chilling. Many of the mutton carcasses are boned and used for sausage making purposes.

Aging.—The practice of allowing meat to hang at cooler temperatures (32° to 38° F.; 0° to 3° C.), or aging, for tenderizing effect is not generally done by meat packing companies. The institutional trade may require aged beef cuts such as loins and ribs. The hotel supply section or division of the parent company may provide this service or it is supplied by specialized hotel supply meat purveyors. Before freezing, care is exercised in trimming excessively aged meat to eliminate all moldy, slimy, and discolored areas that are objectionable from an esthetic or flavor standpoint. It is customary to age beef carcasses. Lamb requires little aging, and pork, because of its lipid instability, will not tolerate such processing.

The tenderizing effect of aging is a function of time and temperature. The degree of aging increases with higher temperatures and in a geometric, not straight-line, progression (Fig. 5.3).

During this ripening period there is a tendency toward increase in pH and osmotic pressure, proteolysis sets in, and intramolecular rearrangement occurs. Sodium and calcium ions are released from the muscle proteins and potassium ions are absorbed. These changes increase the waterholding capacity. During aging there is an increase in water-soluble nitrogen which is believed to arise mostly from sarcoplasmic proteins, but proteolysis is not extensive. The degree of protein hydrolysis is less at higher ultimate pH values.

The first 7 to 10 days of aging has the most tenderizing effect. Beyond one or two weeks changes in flavor may also occur. There is little known concerning the reasons for improvement in flavor usually attained in aging. The inosinic acid that is found may have a function in the basic meat flavor. However, extended aging results in loss of flavor. Prolonged aging is not advisable for meat intended for freezer storage since much of the stability of the lipids may be lost.

PROTECTION OF FROZEN MEAT QUALITY

Packaging

Prevention of desiccation, or freezer burn, is perhaps the most important function of packaging materials for frozen meat. This characteristic, or moisture-vapor proofness, is particularly important when frozen meat is stored in freezers of low relative humidity and/or rapid air movement.

There are a number of means by which deleterious changes of meat in freezer storage may be minimized. Meat cuts should be wrapped in a good type of locker paper with adequate moisture-vapor and air barrier characteristics.

There are many films, laminates, foils, paper, etc., with excellent vapor proofness that are available, and are constantly being improved. Such characteristics as transparency, durability, and flexibility are also important and are considered when packaging materials are selected.

Good moisture-vapor barrier properties are essential even when long storage is not required. Desiccation or freezer burn can take place rapidly. This results in considerable deterioration in appearance and color. Dehydration in the freezer greatly speeds up the rate of autoxidation, since it causes protein denaturation and allows greater contact and penetration of the air. A locker paper with fairly good barrier properties will permit storage at 0° F. (−18° C.) for 9 to 12 months depending on the kind of meat. Pork will not keep as long as beef and lamb. The ideal method of protection is vacuum or inert gas packing in cans. This type of packaging is so effective that temperature of freezer storage is no longer critical as long as it is no higher than 15° F. (−9° C.). Such storage is expensive and frequently not practical. There are now films available which can be used effectively for vacuum packaging meat. In selecting methods of packaging, consideration should be given to the time in storage required for maintenance of good quality. Oxidation of the fat in meat cuts is largely a surface action and penetration is not deep. The relationship between the surface area and weight of the cut is a factor to be considered in selection of storage conditions.

Retail Cuts.—One of the national packers developed a rather complete line of frozen retail meat cuts in the 1930's and again in the mid-1950's. The first venture failed because of inadequate technology and frozen display facilities. A second effect revealed additional problem areas, such as unrealistic in-store profit margins, high packaging costs, higher plant labor costs, and consumer resistance to frozen packaged meats. However, users of institutional meat cuts do not object to frozen meat cuts, many of which are portioned and ready for cookery.

At the present time most of the frozen meat retail cuts can be classified as "specialty items." These include the ground meat patties, breaded, and similar items that are often found in delicatessen stores. In addition to consumer resistance to frozen packaged meats, such technical problems as label attachment and the surface color degradation that occurs when frozen meat in a transparent wrapper is exposed to light must be solved. Many meat industry executives feel that centralized meat packaging is

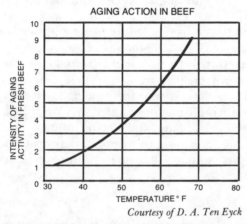

Courtesy of D. A. Ten Eyck

FIG. 5.3. EFFECT OF ACCELERATED TEMPERATURE ON AGING IN BEEF

necessary for increased efficiency and lower marketing costs. This centralized meat packaging operation would have fewer problems with frozen packaged retail cuts than with fresh packaged retail cuts.

Wholesale Beef Cuts.—In the domestic U.S. trade very few sides or quarters are frozen. Because of size and shape, these units are difficult to package economically. The amounts that are frozen are determined by inventory position. Generally, no packaging is practical except from a sanitation aspect; the material may be of a heavy muslin cloth or similar material.

Wholesale cuts such as rounds, loins, etc., are normally partially or entirely boned to allow for packaging in waxed paperboard containers or cartons. Added product protection may be supplied with carton liners or individual cut wrappers. The largest tonnage of frozen beef is of the processing kind that is used in sausage production. This is boneless and may be packaged and frozen in 50- or 100-lb. waxed cartons.

Wholesale Pork Cuts.—Inventory position and expected future demand generally govern the quantity of wholesale pork cuts frozen and stored. Hams and bellies are two items commonly frozen and stored for future special promotions, such as Easter hams, for example. The product may be frozen fresh, and after storage be defrosted and then cured and processed. Glazing with water may be used to minimize freezer burn during bulk freezer storage. Cured pork items are seldom frozen and stored. The salt added during curing accelerates fat oxidation and resultant rancidity even if the meat is stored at freezer temperatures.

Some fresh pork sausage is frozen for distributive advantages, but storage is relatively short. Some of the salt producers have developed an antioxidant that is incorporated into the salt to give protection against rancidity development. This has particular merit in connection with fresh pork sausage that is normally seasoned with salt, pepper, and other spices. Fresh boneless pork trimmings for future processing needs are generally packaged and frozen in 50- or 100-lb. waxed boxes.

Wholesale Lamb Cuts.—New Zealand has taken advantage of the freezing process to promote the export lamb and mutton trade. Acceptability of the frozen product in carcass form has been excellent in England. Legs, for example, are individually wrapped in plastic film and packaged in cartons for freezing, shipping, and storing. Suitable moisture-vapor-proof wrapping material is used to maintain desirable external carcass appearance. Boneless mutton and lamb for sausage production are handled like beef and pork.

Antioxidants

There is considerable appropriate interest in the use of antioxidants in meat products. Antioxidants are substances that are capable at low concentrations of slowing the rate of oxidation of lipids or other oxidizable substances. Except for products such as lard or sausage, the use of antioxidants is not approved for meat cuts and meat products. Antioxidants generally used are butylated hydroxyanisole, butylated hydroxytoluene, η-propyl gallate, and citric acid. Combinations are used which have effects greater than their performance singly. This behavior is called synergism. Acid compounds, themselves ineffective, have a synergistic action in the presence of phenolic antioxidants and are also able to chelate trace metal autoxidation catalysts. An effective antioxidant for meat cuts would have great advantage for some uses, and it is probable that a suitable one will one day be approved.

The addition of a stabilizing agent to animal tissues constitutes a difficult problem. An antioxidant for meat cuts must meet certain standards. It must be capable of interrupting the fat-oxygen reaction chain. Its molecules must be small enough to pass readily through animal tissue cell walls. It must have a reasonable solubility range so as to enter both aqueous and fat phases. The stabilizer may not react with the aqueous phase, but if soluble in the aqueous phase the antioxidant will be able to travel through the aqueous channels to reach isolated fatty portions in the heterogenous product. It must be soluble in an edible medium for dispersion through meat tissue. The antioxidant must have a high inhibitory efficiency so that low concentrations may be used. There must be imparted no flavor, color, odor, or toxicity to the product stabilized and the cost must be reasonable. The mechanism of antioxidant action is not completely understood and there is much leeway for the development of more suitable and effective antioxidants. A number of phenolic antioxidants will retard autoxidation catalyzed by heme pigments. However, the best phenolic inhibitors are practically insoluble in water. In contrast, water-soluble synergistic antioxidants with the exception of ascorbic acid have no effect on heme pigment catalyzed oxidation. Synergistic antioxidants which have been approved for use in lard are citric acid, phosphoric acid, thiodipropionic acid and its esters, and lecithin. Many of the normal constituents of meat, such as amino acids, nicotinic acid, and para-amino benzoic acid, have synergistic activity. The manner in which synergists act is not completely known and probably varies. A number have the property of combining with and inactivating pro-oxidant trace metals. Some, as in the case of ascorbic acid, may reduce oxidized primary antioxidants. Others, such as phosphoric acid and organic acids, may form fat-soluble complexes with primary phenolic antioxidants. Phosphoric acid will form complexes with fat hydroperoxides.

In meat cuts, color loss due to oxidation of myoglobin is usually coupled with rancidity or lipid autoxidation. Apparently, oxidation of one will initiate the autoxidation of the other, but in the normal course of events it is probable that pigment oxidation triggers the process. An antioxidant may be successful in protecting against lipid oxidation but still may not insure color stability. Phenolic antioxidants will not reduce metmyoglobin, and their quinone oxidation products catalyze oxidation of myoglobin. Increased color deterioration due to metmyoglobin formation has been frequently observed in frozen fresh meats treated with phenolic antioxidants. Ascorbic acid under certain conditions will protect meat color. It will reduce metmyoglobin but is not generally suitable, since it accelerates oxidation in some frozen fresh meats and inhibits it in others. Retardation by ascorbic acid is probably dependent on the level of tocopherol present in the meat. The effectiveness of ascorbic acid is increased by the presence of compounds such as ethylenediaminetetraacetic acid (EDTA) or polyphosphates and sufficient amounts of phenolic antioxidants. However, proper distribution of phenolic antioxidants in meat cuts is presently impossible. Ascorbic acid combined with liquid smoke has given excellent stability to frozen pork, and it is exceedingly valuable in the fixation and stabilization of color in cured meat.

Aside from difficulties in distribution of antioxidants through meat cuts, antioxidants available generally fall short of being entirely adequate. In the case of uncured, cooked meat, oxidation during freezing and thawing may be controlled by tripolyphosphate alone or with ascorbate. No satisfactory combination is available for freezer-stored, cured meats. Tripolyphosphate and ascorbic acid will give excellent protection for a time, but then extremely rapid oxidation follows. BHA has proved to be reasonably effective with cured pork, but could not prevent development of off-odor. Frozen, fresh ground meat can be stabilized by any of the approved antioxidants.

Further study on antioxidants is greatly needed. The investigation of naturally occurring materials of potential antioxidant activity appears to be a useful line of research. Among these, the polyhydroxy flavones present in plants are particularly potent and have dual functions of both chain breaking and metal deactivation in the molecule. Many of the flavanoids possess the necessary water solubility because of natural combinations with sugars.

Tocopherol.—The natural antioxidant, tocopherol, is well distributed and reasonably effective in meat fat, but is stored in rather low amounts in animal tissues. If the amount of tocopherol could be increased the problem of lipid stability would be greatly diminished. There have been attempts made at deposition of antioxidants in animal tissues. A large number of antioxidants have been fed to animals but only tocopherols were stored in adipose tissue. A considerable amount of work has been done on the feeding of tocopherol to animals. Increases in body tissues were achieved which gave improved stability. The efficiency of tocopherol storage seemed poor, however, and the process was considered wasteful. This work should be reviewed, and possibly repeated, since much of the evaluation was done on rendered fat and there was no thorough testing of the meat tissues. Tocopherol administration needs further examination in the light of new knowledge of the antioxidant activity of vitamin E, other metabolites, and ubiquinone. Ubiquinone is present in probably much greater concentration in animal tissues than tocopherol. Earlier evaluations of tocopherol feeding were based on the increased content in the rendered fat. Stability cannot be exactly determined without consideration of the tissue as a whole. Much of the pro- and antioxidant activity occurs at interfaces between the fat and the lean. Also, color and phospholipid stability must be considered in addition to the triglycerides.

Smoke, Spices, and MSG

Some material, added to meat to improve or add distinction to the flavor, also will improve the stability in freezer storage. One of the oldest and most effective is the smoking process. Smoke contains a number of phenols and other classes of compounds. There is much interest in the process of smoking and the composition of smoke, and progress has been made toward identification of the compounds present. Smoke penetration is only superficial, most of it accumulating on the surface. So far as effect on stability is concerned, the problem is as in the case of antioxidants and their dispersion. There are a number of liquid smoke and synthetic smoke preparations available. These formulations vary greatly in their antioxidant properties and contribution to flavor. There are commercial curing mixtures that contain smoke ingredients for the purpose of obtaining uniform distribution through the meat. However, there is little information on the antioxidant properties supplied. Smoking of cured cuts adds greatly to their stability in freezer storage. However, because of the accelerating effect of sodium chloride on rancidity, the freezer storage of cured meats is not generally recommended.

Many of the spices have antioxidant properties, and among them sage is probably the most effective. They are very useful in increasing the stability of freezer-stored pork sausage. However, this is not powerful enough to completely counteract the pro-oxidant action of the sodium chloride.

Monosodium glutamate (MSG) which has the property of intensifying flavors has been known for some time to have properties approaching those of an antioxidant. Experience has indicated considerable stabilizing effect on lipids and color on many types of freezer-stored meat and meat products. The compound not only stabilizes fresh frozen meat cuts and ground meat, but it is strikingly effective with freezer-stored, cured meat and sausage which otherwise deteriorate so fast. Samples treated with glutamate have been found to have greater stability in the freezer than BHA-treated meat. Little appears to have been done to extend the use of glutamate and evaluate it from a basic standpoint. Apparently it penetrates meat tissue fairly rapidly. According to recent research, the compound increases the reducing activity of meat. There appears to be little known about glutamate's capabilities when it is combined with other antioxidants and synergists.

FACTORS THAT AFFECT QUALITY OF FROZEN MEAT

Freezer storage is a highly effective means of preservation. The main functions are the inhibition of microorganisms and the checking of proteolytic, hydrolytic, and lipolytic activities. Oxidative processes are slowed, but in the case of some products autoxidation may be accelerated during freezer storage.

Drip Upon Thawing

Changes in the muscle protein take place in freezer storage and there is no way of preventing this. Protein denaturation can only be lessened by the freezing technique or by selection of meat with the proper characteristics. One of the most insurmountable disadvantages of freezer stored meat is the exudation of fluid (drip) on thawing. This liquid contains proteins, peptides, amino acids, lactic acid, purines, vitamins of B complex, and various salts. The amount of such constituents is probably related to the degree of cell damage received during freezing and storage. Two factors determine the amount of drip. One controls the extent to which the fluid, once formed, will drain from the meat. Such variables as the size and shape of the cut, ratio of cut surface, and amount of large blood vessels are significant. The second factor is related to the water-holding capacity of the muscle proteins. These properties being constant, the amount of drip depends on the rapidity of freezing and the resulting size of the ice crystals. The explanation of this is fairly simple. At very fast rates of freezing, tiny ice crystals are formed in the cells, leaving the meat virtually unchanged structurally. At slow freezing rates, extracellular ice crystals are formed that are quite large. These large crystals distort muscle fibers and damage the sarcolemma. As this proceeds, the remaining extracellular fluid increases in ionic strength and by osmotic pressure is able to draw water from the cell interiors of the muscle. The structure is not only distorted by the large ice crystals, but of far more importance is the denaturation of proteins by the high ionic strength of the extracellular fluid. With denaturation the proteins lose their waterholding capacity. Protein damage is a function of time and temperature of freezing. Thus, the quantity of drip will tend to increase with time in storage. The rapid rate of freezing necessary is frequently impossible to achieve commercially because of the high thermal inertia of thick cuts. Drip can be minimized by freezing carcasses immediately after slaughter. Aging before freezing tends to diminish drip to some extent. This is believed due to alterations in ion-protein relationships in which sodium and calcium are released, and potassium ions are absorbed by the myofibrillar proteins. pH has a profound effect on drip. Drip is greater, the lower the pH. At a high ultimate pH, even relatively slow rates of freezing result in virtually no drip. The pH of so-called normal meat is close to the isoelectric point of the proteins. At minimum pH values, the water-holding capacity is low and the proteins are more readily denatured. Even at similar pH's, different muscles vary in amount of drip, and therefore differ in susceptibility to damage. Proteins are far more stable at higher pH's.

Changes in Tenderness and Juiciness

Freezing tends to improve tenderness through a physical action on the tissue. Tenderness appears to hold up well during freezer storage, even though some protein damage occurs progressively. If desiccation is allowed to take place through poor protection, considerable loss in tenderness and juiciness will occur early in storage. Desiccation also favors lipid autoxidation and the development of off-flavors. Even under normal conditions and with adequate protection, there is a tendency toward some loss in juiciness as the result of progressive protein damage. Decrease in water-holding capacity occurs, resulting in loss of fluid (drip) upon thawing, thereby affecting the juiciness of the cooked meat. Loss in water-holding capacity as the result of freezing and subsequent storage is a change that is very difficult to prevent.

Autoxidation of Lipids

The most serious change that takes place in freezer-stored meat is the autoxidation of the lipids. Such deterioration is a problem in all types of freezer-stored meat and meat products, fresh and cooked meat, cooked and uncooked cured meats, and other meat preparations. Autoxidation of meats is dependent upon availability of and contact with oxygen. Very small amounts of lipid autoxidation are sufficient to produce off-flavors which render the product unpalatable and which may in extreme cases introduce a factor of diminished nutritive properties. The source of most of the off-flavors stems from the oxidative cleavage of the unsaturated fatty acids. A variety of complex reactions occurs and compounds of a volatile and odoriferous nature are formed. The principal class of compounds responsible for rancid flavors and odors seems to be the aldehydes. It is not known whether the basic flavor of meat is affected by the oxidative processes taking place or if the change is due entirely to the added effect of the aldehydes. There are isolated cases where flavor in food has been observed to decrease in the very early stages of autoxidation. However, too little is yet known about meat flavor to definitely determine this influence. Flavors developed by lipid autoxidation can be desirable and characteristic in some special meat products.

The fatty acids in meat are present as part of two different kinds of lipids. These are the triglycerides and the phospholipids. As an integral part of the meat these two lipid classes do not oxidize in a way that is directly related to their unsaturated acid composition. The triglycerides, which are the major lipid, are variable in amount and contain relatively small proportions of polyunsaturated acids. The phospholipids are a fairly constant component of meat. This rather complex and heterogeneous class of compounds contains large proportions of unsaturated acids, and a significant amount of C_{20} or higher unsaturated acids. Isolated in the pure state the phospholipids oxidize with great rapidity. However, the phospholipids seem quite stable in their natural state in uncooked meat. The triglycerides, either separately or as part of the meat cut, autoxidize readily enough. The initiation of such change or reaction of the triglycerides is either preceded or followed by oxidation of the heme pigment, myoglobin, to metmyoglobin. The heme pigments are powerful pro-oxidant catalysts, and it is probable that the ferric pigment, metmyoglobin, accelerates the triglyceride autoxidation. The apparent stability of the very high unsaturated phospholipids at the same time is remarkable since the heme pigments are in virtually the same medium. These appear to be a protective mechanism for the intimately related phospholipids.

The triglycerides, even at low temperature, oxidize readily to form hydroperoxides and their aldehyde scission products. Autoxidation of unsaturated fatty acids has been widely studied. The aldehydes formed from oleate, linoleate, linolenate, and arachidonate have been isolated and identified. Each unsaturated acid forms a characteristic group of aldehydes. Linoleate and arachidonate are similar in some respects in the products formed.

Temperature and Length of Storage

Prior to World War II, 10° to 15° F. (−12° to −9° C.) temperatures were regarded as satisfactory for frozen meat storage. Industry and Armed Forces research results indicated that meat quality changes were less if storage was done at 0° F. (−18° C.). Since that time most of the recommended storage periods are based on this temperature. It is realized that even lower temperatures would result in longer stable storage, but the higher cost of maintaining these lower temperature conditions may not be warranted by the resulting increase in value based on these quality factors.

Another factor entering into the frozen meat storage operation is that of temperature fluctuation. This should be kept to a minimum even if the average temperature is maintained at 0° F. (−18° C.). Extreme temperature fluctuations that may occur during defrosting, for example, should be avoided. This is one factor that contributes to the relatively short storage life of frozen packaged meat held or displayed in self-service cases.

As to length of storage, it should be always kept in mind that freezing meat and its subsequent frozen storage do not improve the product. Some segments of the meat industry erroneously believe that frozen meat has unlimited frozen storage life. Based on commercial 0° F. (−18° C.) storage conditions, one can expect only imperceptible quality changes in beef and lamb to take place in 6 months. This assumes proper handling prior to freezing, use of correct packaging materials, and good commercial freezing operations. While research results do not all agree, there are indications that minor quality changes, especially in flavor, begin to occur after six months of storage. Comparable storage periods for pork and veal are about four months. Fat in pork cuts lacks the stability of beef and lamb fat, and may develop a slightly stale or rancid flavor; thus, the shorter maximum frozen storage period is recommended (Table 5.2).

Freezer Storage of Cured Meats

Cured meats autoxidize in the frozen state much more rapidly than uncured or fresh meats. This seeming pro-oxidant effect is due to the presence of sodium chloride, in spite of the fact that salt, except under

TABLE 5.2

STORAGE LIFE OF FROZEN MEATS (IN MONTHS)

	10° F.	0° F.	−10° F.	−20° F.
Beef[1]	4	6	12	12+
Lamb	3	6	12	12+
Veal	3	4	8	12
Pork[2]	2	4	8	10

[1] Diced products have shorter life.
[2] Cured products such as ham and bacon can be stored a few weeks only.

unusual conditions, is apparently not a pro-oxidant. It will not cause the oxidation of methyl linoleate in an emulsion. Nevertheless, in an interface junction between solid sodium chloride and lard, accelerated oxidation has been observed to take place. This observation has been aptly compared to the physical condition existing in frozen cured meat. It has been found that bacon stored at higher levels of freezer storage temperature kept better than bacon stored at 0° F. (−18° C.) or lower. Also, possibly related to this observation is the fact that oxymyoglobin solutions oxidize more rapidly at lower freezer storage temperatures. However, freezer-stored fresh meat does not seem to show such an effect since its stability increases as temperatures are lowered. There may be an indirect relationship involved in this. Sodium chloride has a powerful influence on the protein of meat. It acts on the respiratory enzymes and inhibits most of the meat's native reducing activity. The discoloring effect of sodium chloride on red meat is well known. Upon the addition of salt, the myoglobin is rapidly oxidized to metmyoglobin. The strong action of sodium chloride may set free powerful hematin catalysts, and the ferric are more active than the ferrous heme compounds. In view of these facts, it is not surprising that the fatty acids are exposed to strong oxidative attack.

The removal of protective enzyme action and formation of active catalysts by sodium chloride does not explain the tendency toward greater oxidation rates of bacon sides as freezer storage temperatures are lowered. This might well be related to the effect of myoglobin solutions and influence of contact of solid sodium chloride with triglycerides. The effect of freezing on oxidation rates possibly is due to marked changes in pH. Precipitation of ice and salts causes large decreases in the pH of frozen, cured meats. Lowered pH increases the rate of autoxidation of lipids and pigments. The use of nitrite with sodium chloride and heat has been observed to reduce the amount of autoxidation due to the formation of catalytically inactive ferrous nitric oxide hemochromogen (nitrosohemochromogen). This appears to show that sodium chloride owes its activity to at least two different factors. Increasing the pH or use of abundant quantity of ascorbic acid will decrease the rate of autoxidation.

Cooking of cured meat has been reported to inhibit sodium chloride promoted autoxidation in the freezer. In this case, possibly catalysts are rendered inactive or reducing compounds are formed that protect the triglycerides.

Freezer Storage of Cooked Meat Products

Cooked fresh or uncured meat and meat products are frequently very unstable. While oxidation is scarcely discernible at the usual freezer storage temperatures, it is extremely rapid during freezing or thawing. There is no induction period and no hydroperoxide accumulation. This deterioration is due to phospholipid oxidation. The mechanism by which the phospholipids are preferentially oxidized is not clear. It is not known whether the sensitivity of the phospholipids is due to heat destruction of protective systems or to the formation of active catalysts. The phospholipids are integral parts of the muscle tissue, and are closely associated with the muscle proteins and pigments. The heat of cooking inactivates enzymes and denatures proteins. Phospholipid oxidation under these conditions is apparently a heme catalyzed reaction. Meats free of heme pigments, such as crab and shrimp, do not show this type of autoxidation. Denatured heme pigments may be more active catalysts. In the case of phospholipid oxidation, acceleration is believed to be brought about by the ferric cooked meat pigment. This deterioration can be prevented by the addition of nitrite, which converts the meat pigments to nitrosomyoglobin and the catalytically inactive ferrous nitric oxide hemochromogen. This indicates that heme compounds are involved; it also shows the catalytic power of the cooked meat pigment. The lipids of meat are therefore composed of two classes which are distributed in the meat tissues differently and appear capable of oxidizing independently. However, the mechanism involved and reasons for the selective action are not clear. The significant action of both salt and heat might be to destroy protective enzyme systems and bring about the formation of more active catalysts through protein denaturation. Sodium chloride apparently causes oxidation by several joint mechanisms. It is not controlled very well by lowering the temperature of freezer storage, and there are no entirely satisfactory antioxidant combinations. The phospholipid oxidation can be controlled by freezer storage and by antioxidants.

The Microbiology of Frozen Meat

So long as meat is held at a temperature lower than 15° F. (−9° C.) there is no microbial growth, and all bacteria, yeasts, and molds which may be present are held in a dormant state. At such freezing temperatures, the direct chemical activities of the microorganisms cease and they are unable to produce toxins or enzymes.

One must, however, consider these questions with respect to the microbiology of frozen meat. These are: (1) Can the microorganisms survive freezing? (2) How well can the microorganisms grow after the meat is thawed? and (3) Will extracellular microbial enzymes released before freezing continue to have an effect?

Early workers investigating the survival of microorganisms in frozen meat probably incubated their cultures at 99° F. (37° C.), the temperature usually used in public health laboratories. Counting at this temperature, they noted a decrease in numbers of microorganisms that was correlated with time in freezer storage. When, however, counts are obtained by incubating cultures at 68° F. (20° C.) very little decrease is found with time in the freezer, and, where the meat samples are well wrapped to protect the meat from freezer burn and oxidation, there is no significant change in the numbers of microorganisms. Usually, in fact, there will be a slight increase in numbers of microorganisms as compared to the unfrozen controls. This increase is probably only apparent, and due to the effect of freezing in breaking up clumps of organisms on or in the original meat.

Fortunately, however, it has also been demonstrated that potentially dangerous organisms, such as staphylococci and salmonellae, will not grow in ground meat held at 45° F. (7° C.) or below. All refrigerators should, and many do, operate below this temperature.

When we consider our second question, how well can the microorganisms grow after thawing, the facts are even more reassuring. Experimental evidence indicates that microorganisms grow about at the same rate, or even a bit slower, after thawing than they do in unfrozen controls. Furthermore, freezing and thawing result in a prolonged lag phase before the microorganisms begin to grow at all. Thus, it may be said that thawed meat is no more perishable than unfrozen meat.

Many of the microorganisms commonly found on meat excrete extracellular enzymes. These enzymes may be excreted in sufficient quantities prior to freezing to exert a significant effect on meat quality during freezer storage, even though the original microorganisms are completely dormant. The phenomenon has been demonstrated in the laboratory for lipases which are released by Pseudomonas cultures and can act on fats at −20° F. (−29° C.) in two weeks. To what extent this may apply to other systems is not yet known, but it emphasizes the fact that strict sanitary precautions must be taken at all stages of the freezing process if high quality is to be assured.

A final precaution should be sounded with respect to precooked, frozen meat dishes. Here the natural spoilage flora have been reduced or completely eliminated by the cooking process and since the food could be contaminated with other microorganisms, such as spores of *Clostridium botulinum* or *C. perfringens*, special care should be exercised to see that the temperature of the food is kept lower than 38° F. (3° C.) prior to cooking.

TAILORING MEAT FOR A PURPOSE

One of the most important properties of meat is its ultimate pH. This may vary from about 5.2 to 6.6 due to preslaughter conditions and environment, which were earlier discussed. If the pH is lowered within the range of normal meat, fresh meat will be discolored, cured meat color will be improved, the oxidation of the fat of raw meat will be accelerated, and the stability of the fat of cooked meat will be unchanged. Variations through this range of pH can have tremendous bearing on the properties of the meat to be freezer-stored. Fresh meat with pH at the upper end of the normal pH scale will show a much more stable frozen product from the standpoint of lipids, color, and waterholding capacity. Also, there should be little drip on thawing. Indeed, selection or adjustment of the pH of meat to be used for freezer-storage could solve most of the problems involved in such processing.

As already indicated, it is possible to tailor meat to the desired properties by preslaughter treatment. Much improvement might be attained in freezer-stored meat by selection of fresh meat with pH in the upper range. According to the knowledge already available but incompletely tested, there should be no great difficulty in adjusting the ultimate pH to the desired level. Summarizing all information on the effect of different kinds of stress on animals before slaughter, evidently a high ultimate pH can be attained with little adverse effect on palatability. The following important characteristics are obtained: the muscle tenderness is increased, juiciness is higher, and color is more stable, although somewhat darker. Even at relatively slow rates of freezing, drip is small due to the greater waterholding capacity and stability of the proteins. Research appears to indicate little significant change in flavor. Such meat is more susceptible to bacterial attack, and absorbs curing salts less rapidly. This appears to be the greatest fault of meat with a high ultimate pH. The color of the meat is darker, but it is far more stable and in any case color darkens upon freezing. The desired conditions for preslaughter adjustment may be obtained by administration of adrenaline, neopyrithiamine, epinephrine, or sodium iodoacetate. Freezing soon after slaughter slows down the rate of ATP disappearance. One pertinent recommendation for tailoring of meat to desired properties for freezer storage is the injection of magnesium sulfate before slaughter as a relaxant to slow down the subsequent rate of decrease of ATP, followed by freezing postmortem in the course of five hours.

Control of meat properties according to the various means discussed may seem not practical at this time. This is to some extent true, and certainly more study is needed of such processes. However, since it is probable that most meat in the future will be marketed as prepackaged units, and since a large part of the meat may be freezer-stored until purchased, uniformity of the meat supply will be essential. The adjustments possible by the antemortem and postmortem treatments will tend to improve stability and color, increase the effectiveness of antioxidants, and give better overall palatability.

ADDITIONAL READING

ASHRAE. 1974. Refrigeration Applications. American Society of Heating, Refrigeration and Air-Conditioning Engineers, New York.

BAUMAN, H. E. 1974. The HACCP concept and microbiological hazard categories. Food Technol. 28, No. 9, 30–34, 70.

COLE, D. J. A., and LAWRIE, R. A. 1975. Meat. AVI Publishing Co., Westport, Conn.

DeFIGUEIREDO, M. P., and SPLITTSTOESSER, D. F. 1976. Food Microbiology: Public Health and Spoilage Aspects. AVI Publishing Co., Westport, Conn.

FABRICANTE, T., and SULTAN, W. J. 1975. Practical Meat Cutting and Merchandising. Vol. 1. Beef. Vol. 2. Pork, Lamb and Veal. AVI Publishing Co., Westport, Conn.

GOULD, W. A. 1977. Food Quality Assurance. AVI Publishing Co., Westport, Conn.

GUTHRIE, R. K. 1972. Food Sanitation. AVI Publishing Co., Westport, Conn.

HARRIS, R. S., and KARMAS, E. 1975. Nutritional Evaluation of Food Processing. 2nd Edition. AVI Publishing Co., Westport, Conn.

JOHNSON, A. H., and PETERSON, M. S. 1974. Encyclopedia of Food Technology. AVI Publishing Co., Westport, Conn.

KOMARIK, S. L., TRESSLER, D. K., and LONG, L. 1974. Food Products Formulary. Vol. 1. Meats, Poultry, Fish and Shellfish. AVI Publishing Co., Westport, Conn.

KRAMER, A., and TWIGG, B. A. 1970, 1973. Quality Control for the Food Industry, 3rd Edition. Vol. 1. Fundamentals. Vol. 2. Applications. AVI Publishing Co., Westport, Conn.

KRAMLICH, W. E., PEARSON, A. M., and TAUBER, W. F. 1973. Processed Meats. AVI Publishing Co., Westport, Conn.

LEVIE, A. 1970. The Meat Handbook, 3rd Edition. AVI Publishing Co., Westport, Conn.

PETERSON, M. S., and JOHNSON, A. H. 1977. Encyclopedia of Food Science. AVI Publishing Co., Westport, Conn.

SACHAROW, S. 1976. Handbook of Package Materials. AVI Publishing Co., Westport, Conn.

TRESSLER, D. K., VAN ARSDEL, W. B., and COPLEY, M. J. 1968. The Freezing Preservation of Foods, 4th Edition. Vol. 2. Factors Affecting Quality in Frozen Foods. Vol. 3. Commercial Freezing Operations—Fresh Foods. Vol. 4. Freezing of Precooked and Prepared Foods. AVI Publishing Co., Westport, Conn.

UMLAUF, L. D. 1973. The Frozen Food Industry in the United States—Its Origin, Development and Future. American Frozen Food Institute, Washington, D.C.

WHITAKER, J. R., and TANNENBAUM, S. R. 1977. Food Proteins. AVI Publishing Co., Westport, Conn.

WILLIAMS, E. W. 1976A. Frozen foods in America. Quick Frozen Foods 17, No. 5, 16–40.

WILLIAMS, E. W. 1976B. European frozen food growth. Quick Frozen Foods 17, No. 5, 73–105.

WOOLRICH, W. R., and HALLOWELL, E. R. 1970. Cold and Freezer Storage Manual. AVI Publishing Co., Westport, Conn.

6

Freezing Poultry

Donald deFremery, Alvin A. Klose
and Robert N. Sayre

Marketing poultry as a frozen product has many advantages. Consequently, freezing has been an essential part of the great expansion in the production of poultry products. About 30% of uncooked chicken, 80% of turkeys, and almost all of further processed poultry meat products are marketed in the frozen form. Freezing has permitted the leveling out of seasonal fluctuations in production to meet current demands; it has provided a means of eliminating marketing losses due to microbial spoilage and rancidification; and it has enabled overseas shipment of poultry products. To the consumer, freezing and frozen storage under proper conditions have brought assurance of wholesomeness, optimum appearance and eating quality, and convenience. The large proportion of fryer chickens that are marketed in a chilled but unfrozen condition represent a challenge to the food freezing industry. Chilled poultry has obvious limitations in shelf-life due to microbial growth; however, it is preferred over frozen poultry in many cases because of some or all of the following factors: (1) consumers' belief of superior quality of chilled product, which may be attributed to past experiences with poorly processed, packaged, and stored frozen poultry or to supposed present inferiority of frozen products; (2) methods of pricing the products at the retail level, which favor the chilled product; and (3) inadequate processing and freezing methods for retention of optimum flavor, tenderness, and juiciness. Since we may expect an acceleration in the present trend toward further processing of poultry and its sale in cut-up, fabricated, and precooked form, the proportion of total poultry production that is frozen should necessarily increase.

MAGNITUDE AND CHARACTER OF THE INDUSTRY

Over the past 30 years poultry production has increased percentage wise more than any other meat, and freezing and frozen storage have played an important role in this increase. More than 80% of the production is eviscerated under federal inspection, with 12% of the young chickens, 34% of the mature chickens, and 86% of the turkeys marketed as a frozen product. A substantial part of this production is processed by relatively few firms. For example, the 20 largest firms accounted for 44% of the young chickens and 51% of the turkeys. This concentration in the industry, together with the significant economies of large scale in poultry processing, has resulted in a trend to larger unit operations with processing rates as high as 10,000 chickens per hour. A rate of 10,000 chickens per hour would amount to about 70 million pounds live weight per year on a full-time one-shift basis. Over 90% of the tonnage of young chickens are processed in plants with capacities of at least 2,500 chickens per hour, and 50% of the weight of turkeys was processed in plants operating at a rate of at least 2,500 units per hour. This processing plant size has an obviously large refrigerating requirement, in all cases for chilling purposes, and in many cases for freezing and frozen storage. For example, for a large plant processing and freezing 400,000 lb. of poultry per day, a total of 70 million B.t.u. would be removed, requiring about 250 tons of refrigeration.

FACTORS PRIOR TO FREEZING
THAT AFFECT PRODUCT QUALITY

Quality of frozen poultry, as related to the palatability of the cooked product, is the sum of many individual characteristics, e.g., texture, juiciness, flavor, odor, color and microbial contamination. Another way to consider overall quality, a way which is often overlooked by those concerned with frozen foods, is to realize that frozen quality represents the nature of the product *before* freezing as well as the treatment the product receives *after* freezing. Phases of poultry processing essential to the economical production of optimum quality frozen poultry begin with selection of high-quality live birds, and include slaughtering, defeathering, evisceration, chilling, cutting up, further processing, packaging, and freezing. For each of these phases, the technical requirements and standard commercial practices will be described. Without exception, frozen storage cannot upgrade the quality of frozen poultry. The best that frozen storage can do (and all it is really expected to do) is to maintain initial quality.

In some instances, particularly in phases of rapid and continual change, the reader will be directed to sources of information that are periodically brought up to date. Alternative types of equipment or processes will be compared and their advantages and limitations will be listed, but it should be remembered that choice of a particular process or equipment will often be dictated by the size of the operation and local conditions.

Selection of Raw Material

Use of high-quality, fresh raw material is the first prerequisite for the successful production of any frozen food product, and it is especially essential for poultry products. The poor regard for frozen poultry held by some consumers undoubtedly arises in part from the use of material that was substandard before freezing. Any practice of freezing chilled poultry that has been held in an unfrozen state to the point of incipient spoilage should be rigorously discouraged.

United States standards for quality of individual live birds, A, B, and C, are assigned on the basis of health, feathering, and absence of excessive numbers of pin feathers, conformation, fleshing, fat covering or finish, and freedom from defects such as skin bruises, flesh bruises, and breast blisters. United States standards for quality of dressed poultry and ready-to-cook poultry, A, B, C, are assigned on the basis of conformation; fleshing; fat covering in the skin layer, generally termed finish; freedom from pin feathers and hair; freedom from exposed flesh resulting from cuts, tears, and missing skin; freedom from disjointed and broken bones and missing parts; freedom from discoloration of the skin and flesh; and freedom from freezing defects. Freezing defects in the frozen packaged product include darkening over the back and drumsticks, pockmarks due to dehydration, and thin layers of clear or pinkish colored ice.

It is clear from the above that in order to end with an optimum frozen product, birds must be well fleshed, well finished, have a good conformation, and be free from bruises and mechanical damage that may be incurred during catching, crating, and processing. Grade A quality material should be used for production of frozen whole birds. Grade B and C birds, in which the meat is as wholesome and as nutritious and tasty as Grade A material, should be used preferably for the production of further processed items.

Age.—The chronological age of the bird influences palatability primarily through its well-known effect on tenderness.

The USDA has established classes, standards, and grades for poultry, and maintains a grading service to implement the use of these aids to uniform quality. Kinds of poultry include chickens, turkeys, ducks, geese, guineas, and pigeons. Classes of chickens are Cornish game hen, usually 5 to 7 weeks of age and not more than 2 lb. ready-to-cook weight; broilers or fryers, usually 9 to 12 weeks of age and making up the major part of chicken production; roasters, usually 3 to 5 months of age; capons; stags, intermediate in maturity between roaster and rooster; stewing hens, a by-product of egg laying flocks and usually 1 to 2 years of age; and mature roosters. Studies with chickens (Rhode Island Red X Brown Leghorn) that were 2, 6, 12, or 18 months of age demonstrated that tenderness decreased in both breast and leg meat as the age of the bird increased. A study of immature meat-type chickens of a much narrower age range (6, 10, or 14 weeks) revealed chronological age had no bearing on meat tenderness.

Turkeys are classified into fryer-roasters, usually under 16 weeks of age; young hens and young toms, both as a rule 5 to 6 months of age; yearling hens and yearling toms, both usually under 15 months of age; and mature turkeys, customarily in excess of 15 months of age.

The effect of age on tenderness of young turkeys appears to be minor. Turkeys (Broad Breasted Bronze) were less tender at 22 weeks of age than they were at 9 weeks, although the difference was only marginal. In contrast, there is an *increase* in tenderness with increasing age (12, 18, or 24 weeks) in thigh muscles and in the outer layers of the breast muscle. Experimenters commented that birds used with essentially insignificant development of connective tissue would be classified as immature. In parallel with the findings on chickens, mature turkeys would be expected to be tougher than immature turkeys. The other kinds of poultry (e.g., ducks, geese) are also classified by age.

Experimental studies by several groups have shown that, in general, age has a marked influence on flavor intensity only when mature birds are compared to immature birds. For example, 19-month-old White Leghorn hens have more flavor in leg meat (but not breast meat) than 13-week-old females of the same breed. Similar results are reported studying immature chickens (6, 10, or 14 weeks old, Vantress X White Rock). The youngest group had slightly less intense chicken flavor than either of the two older groups, which were indistinguishable from each other.

Genetic Strain.—Within a given class of poultry, genetic strains do not have any effect on the flavor, tenderness, or juiciness of the cooked meat. Thus, growers should select their stock on the basis of other factors, such as feed conversion or rate of growth. In a study of 8 breeds and crosses of chicken broilers grown to 12 weeks of age, researchers were not able to detect differences between breeds with respect to flavor and tenderness. Similar conclusions were made regarding flavor, tenderness, and juiciness in turkeys. Differences in genetic strain fail to influence tenderness in 5 or 6 selected strains of turkeys. Such studies have used closely related varieties of birds. Results of experiments on two lines of chickens selected for difference in growth rate, White Rock (heavy, fast-growing, meat-type), and Brown Leghorn (light, slow-growing, egg-type) indicate the meat-type birds are generally more tender. In other studies, fast- and slow-growing meat-type chickens, regardless of breed or diet, were indistinguishable with regard to flavor or tenderness. Thus, in most cases, genetic strain plays a negligible role in cooked meat quality.

Diet.—Diet, in general, does not affect the quality of poultry. The only exception to this statement (and it is a major exception) is the detrimental effect that fish oils have on flavor and aroma. This observation was first reported more than 40 years ago and it has since been examined in detail by many researchers.

Studies on chickens demonstrate that diets containing 2% cod liver oil contribute a definite "fishy" flavor to the cooked meat. If this oil is removed from the diet two weeks prior to slaughter, the off-flavor is largely eliminated.

Turkeys are apparently more susceptible to this defect than chickens. More than four weeks on a fish oil-free diet are required for turkeys to lose their fishy flavor and aroma. Actually, the highly unsaturated fatty acids present in fish oil are the cause of the undesirable flavor and odor. This has been demonstrated by careful work in which highly unsaturated linseed oil imparted fishy flavors and odors just as effectively as fish oils.

When the fat component is more saturated, the source of the fat (beef, corn, or soybean) and its amount (up to 26% of the diet) play no role in the flavor and aroma of the cooked product. Varying the grain in practical-type diets also has no effect on flavor, aroma, tenderness, or juiciness.

With the marked advances in poultry genetics and feeding practices that have occurred during the past 30 years, many growers have expressed concern over the quality of modern, fast-growing strains fed on

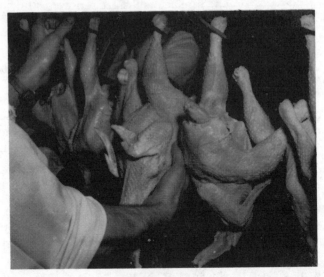

Courtesy U.S. Dept. Agr.

FIG. 6.1. POULTRY SUSPENDED ON CONVEYOR LINE
As the birds move past, the U.S. Dept. Agr. Grader examines and evaluates the poultry, keeping in mind each of the quality factors.

the new high-energy diets, as compared with older strains on older types of diet. Their worries are intensified when they hear statements like "Chicken doesn't taste as good as it did when I was a child." In studies of diets representative of those in use either in 1930 or recently and in diets fed to both the 1930 and the recent type broilers, no quality factor was affected, either by breed or diet. Quality factors of broilers are in general unaffected when practical type diets are replaced with semi-synthetic diets.

One final dietary factor should be mentioned, and that is the effect of a brief period of fasting prior to slaughter. Starving for 16 to 24 hr. before slaughter has no influence on cooked aroma or tenderness. Additional processing requirements for optimum frozen quality will be emphasized as commercial processing steps are discussed in their chronological sequence below.

Slaughtering

The main effect that variation in the slaughter conditions has on poultry quality is the completeness of bleeding and the resulting appearance of the carcass. Most researchers agree that the appearance of the dressed carcasses is acceptable, regardless of method of slaughter. In the overall dressing operation birds are suspended by their feet on shackles attached to an overhead conveyor line which moves birds past the stationed operations at rates as high as 90 birds per minute (Fig. 6.1). The general standard method of slaughter is bleeding the bird over a two minute period by severing the arteries and veins at the junction of head and neck by an outside cut. While equipment has been developed in the past for mechanical bleeding, the operation is now a manual one, with one operator handling not more than about 45 birds per minute. Inadequate bleeeding results in an objectionable red appearance in the skin layer and consequently a lower grade for the carcass. An electric shock is used immediately prior to slaughter in some processing plants, particularly for turkeys, to stun the bird and minimize violent struggling during the bleeding period which can cause bruised and broken wings. However, electric stunning has been discontinued in some cases because it occasionally causes muscle spasms which break the rib cage of the bird.

Feather Removal

Scalding.—Feather removal is one of the most important processing operations from the standpoint of the final appearance and eating quality of the frozen product. Feathers are loosened by immersion in or spraying of water at controlled temperature(s) and time(s).

The most detrimental effect that harsh scalding treatments have is an increased toughness in the cooked meat. Although there is some disagreement as to which time-temperature combinations lead to significant toughening, in general, water temperatures fall into 3 ranges: 125° to 130° F. (51° to 54.4° C.), for 60 sec. or more, which is satisfactory for most of the feathers on broiler chickens; 138° to 140° F. (59° to 60° C.), for about 45 sec., which is necessary for turkeys in order to obtain a completely feather-free carcass without excessive hand labor; and about 160° F. (71° C.), for short periods to loosen the relatively stubborn hock and neck feathers. Turkeys appear somewhat more resistant to scalding effects than young chickens. This may be related to the fact that turkeys generally have a greater insulating layer of feathers, skin, and fat to protect the muscle. Skinned turkeys are much more susceptible to toughening by scalding than unskinned controls. The observation that fowl can withstand relatively harsh scalding conditions (compared to young chickens) is pertinent in this regard.

Another major defect possessed by birds that have undergone severe scalding treatments is an increased tendency to lose body moisture. Moisture loss is appreciable only when birds are chilled in air (rather than ice slush). This has been shown both for turkeys and for chickens. The increased loss of moisture is due to the fact that the outer layer of skin or cuticle is removed when birds are scalded at 140° F. (60° C.) or above to provide a feather-free product. This removal results in a skin surface much more susceptible to darkening and dehydration if special precautions are not taken in chilling, packaging, and freezing. Although the cooked aroma and flavor of these birds are unaffected, excessive scalding temperatures and times are to be avoided because they lead to poor appearance and toughening of the underlying flesh. In order to prevent scald water from being introduced into the lungs and air sacs, regulations require that breathing of the bird must have stopped prior to scalding.

This preparation of poultry feather-removal is unfortunate since the scalding treatment has no beneficial effect on poultry quality beyond its feather-loosening ability, and excessive scalding has a toughening effect. Consequently, the poultry processor must continually balance poor feather removal (inadequate scalding) against lower organoleptic quality (excessive scalding) by adjusting the scalding conditions.

Feather-Picking.—The interrelationship between scalding and feather-picking means that gentle scald conditions must be followed by severe picking conditions to effect removal of the major portion of the bird's feathers. Removal of feathers from the scalded birds may be accomplished by a variety of power driven, mechanical feather pickers (Fig. 6.2). All of the common commercial feather picking machines depend on the rubbing, stripping, and beating action of flexible rubber fingers generally mounted on two rotating double drums between which the suspended bird is conveyed. Proper clearance between the drums and the bird is essential to prevent abrasion and resulting defects in appearance. If the mechanical pickers are unduly harsh, abrasion or tearing of the skin may result. Excessive beating of the bird by the rubber fingers will also lead to toughness in the flesh of the cooked bird. Several types of double drum picking machines are available, with special designs for removing body feathers, wing feathers, neck feathers, and hock feathers. In addition to double drum pickers, there are several free-floating or cyclic pickers in which the birds are not suspended, but are impelled against stationary rubber fingers by moving surfaces studded with stiff rubber fingers.

Eviscerating

Poultry is normally eviscerated as soon as possible following feather removal to inhibit bacterial growth and to prevent the transfer of visceral flavors to the meat. Although the holding of New York dressed (uneviscerated) poultry at 32° to 40° F. (0° to 4° C.) for 24 hr. before evisceration appears to have only a negligible effect on off-flavor development, the quality of these birds following subsequent frozen storage is inferior to warm-eviscerated, frozen controls. If New York dressed birds are held longer than one day before evisceration, or if they are frozen and stored

Courtesy Gordon Johnson Co.

FIG. 6.2. AN AUTOMATIC, ON-THE-LINE RUBBER-FINGERED FEATHER PICKER

before evisceration, definite visceral flavors are imparted to the meat. In general, the best meat flavor is achieved by minimizing the period of time in which viscera remain within the carcass. While commercial evisceration carried out with the birds suspended on a conveyor line is mainly a manual operation, many mechanical aids (e.g., for stripping linings from gizzards, removal of feet, heads, and lungs) have been developed that make the operation very rapid. Inspection for wholesomeness is accomplished on the evisceration line, and then the bird is ready for chilling.

Chilling Operations

After warm evisceration, when the internal carcass temperature may range from 90° to 100° F. (32° to 38° C.), it is desirable to lower the temperature rapidly in order to inhibit bacterial growth and other deteriorative changes. Over ⅓ of the total heat to be removed to attain a desired 0° F. (−18° C.) frozen storage temperature is removed during the washing and chilling process. USDA regulations for ice and water chilling require that poultry carcasses shall be chilled to 40° F. (4° C.) or lower within 4 hr. for weights under 4 lb., 6 hr. for 4- to 8-lb. weights, and 8 hr. for carcasses over 8 lb. For air chilling of ready-to-cook poultry, air movement and temperature should be such that the internal temperatures of the carcasses are reduced to 40° F. (4° C.) or lower within 16 hr. There is practically no commercial large-scale air chilling of poultry employed at the present time because of the slower cooling rate of 30° to 35° F. (−1° to 2° C.) air compared to ice and water, the loss of weight experienced in air chilling, and the deterioration in surface appearance which is particularly marked with high-temperature scalded birds.

Slush Ice Chilling.—While most poultry is chilled by combinations of cold water and ice, the particular procedures and types of equipment vary widely. Slush ice chilling can result in an excellent product, since even cold-tolerant, psychrophilic microorganisms grow only at an extremely slow rate at ice temperature; the liquid medium also keeps the carcass surface moist and attractive in appearance. Excessively long periods in ice slush, i.e., longer than necessary to reduce internal temperature below 40° F. (4° C.) and to adjust to continuous schedules of processing and freezing, should be avoided, because they result in unnecessarily large amounts of water absorption by the meat, and in extreme cases by the leaching of flavor constituents or precursors from the meat. USDA regulations permit ready-to-cook poultry to be held in water-saturated ice for 24 hr., but beyond this time any further holding must be in fresh, continually drained ice. Tolerances for moisture absorption during ice slush chilling, which can be extensive with procedures involving agitation, have been set as follows: for ready-to-cook poultry that is to be consumer-packaged and/or frozen, maximum percentage moisture absorption of the drained product shall be 4½% for turkeys 20 lb. and over, 6% for turkeys 10 to 20 lb., 8% for turkeys under 10 lb. and for chickens 5 lb. and under, and 6% for all other kinds and weights of poultry.

The oldest and simplest method of ice chilling is to place birds and flake or crushed ice in alternate layers in tanks, using ½ to 1 lb. of ice per pound of poultry, and then filling all voids with water. The water can be agitated or circulated by bubbling air through the tank or by a circulating pump. Relative amounts of ice, water, and birds depend on initial carcass temperature and temperature of the water. Some plants provide refrigerated water, and in some cases a preliminary chilling with water precedes the introduction of ice slush. Of course, the chilling water and the water used to make ice must come from a potable source, and the ice must be handled in a sanitary manner. Ice may be procured in block form and be crushed at the processing plant, it may be produced at the plant in flake form, or it may be produced at the plant in the form of a free-flowing slush ice that can be pumped from a central supply to the tanks. Flake ice or subcooled ice is made by scraping thin ice shells from refrigerated cylinders, with the refrigeration system working at relatively low suction pressures. Clear ice or nonsubcooled ice may be made by refrigeration equipment working at considerably higher suction pressures and is harvested periodically from vertical plates or tubes. Slush ice or snow ice machines, working at about the same suction pressures as the clear ice makers, and coupled with a storage tank and piping system, provide a low labor cost operation. However, proper adjustment of ice crystal size and flow characteristics is necessary to avoid plugging of lines and valves by separation of ice and water.

Advantages of ice slush chilling in tanks are its simplicity, relatively low cost of equipment, and opportunity for assurance that maximum and optimum tenderization is accomplished before freezing. Disadvantages of tank chilling are the large floor space required, the considerable amount of hand labor involved in placing the birds in tanks and removing them, and the slow rate of cooling compared to continuous, mechanical, agitated slush ice chilling systems. While air chilling results in 1 to 2% moisture loss, tank chilling in ice slush can produce a percentage moisture absorption correlated directly with chilling time and inversely with size of bird, and can approach or exceed 8% in some cases. Maximum tolerances established by the USDA have been listed previously. The factors of moisture absorption and required chilling (aging) times for optimum tenderization will be discussed further in the following description and evaluation of mechanical, continuous chilling devices.

Mechanical Chillers.—Increased rates of chilling over that accomplished by static ice slush can be achieved by substituting a lower temperature chilling medium such as brine (although this is not practiced commercially) or preferably by introducing a mechanically activated movement or agitation of carcasses in the chilling medium. By greatly accelerating chilling rates, mechanical chillers provide an efficient continuous operation in contrast to the batch-type static tank chilling. An illustrative but not exhaustive list of mechanical poultry chillers include the types described briefly in the following paragraphs. Statements of rate, capacity, and ice requirements are indicative, but could vary widely from actual working characteristics due to variations in plant size, water temperature, and entrance and exit carcass temperature.

A parallel-flow tumble system, in which birds float through horizontally rotating drums suspended in long tanks of chilled water or ice slush, provides forward motion of the birds by pumping and recirculating the chilling medium from one end of the tank to the other, and a sideways tumbling motion by the rotation of the drum. Such a system may contain two tanks in series, the first tank using cold water and the second tank ice slush at the rate of 0.5 to 1.0 lb. of ice per pound of carcass. A unit occupying about 200 sq. ft. of floor space can chill over 30 birds per minute with an immersion time in the chiller of about 20 min.

A counter-flow tumble system is also available. Again there are two cylindrical drums, but the forward motion of the birds is controlled by a helical screw mounted inside of the drums, with partitions fastened to the screw in order to lift the birds somewhat above the water level before releasing them as the drum continues to revolve and push the birds forward. Poultry is chilled in the first drum by overflow water at 34° F. (1° C.) from the second drum which uses ice at about 0.5 lb. per pound of carcass. Immersion times are about 30 min. with rates up to 6,000 birds per hour.

A slightly different principle of agitation is employed in the oscillating vat or rocker system, which consists of two tanks that are mounted on eccentric rollers and rock from side to side as the birds are conveyed through by the regulated flow of the circulated cooling media. Cold tap water is used in the first tank and ice and water at around 34° F. (1° C.) in the second tank. Rates of 2,500 birds per hour with an immersion time of 40 min. can be achieved with this equipment.

While the three mechanical chillers described above are of the free floating type, the fourth chiller to be described is a continuous drag chiller in which the birds are suspended by the hocks on shackles and pulled through two tanks in series. The first tank contains chilled tap water at about 40° F. (4° C.) and the second tank ice slush at about 33° F. (1° C.). A total immersion path of 50 ft. of chilled water and 200 ft. of ice slush requires about 1 hr. The equipment can handle 3,000 birds per hour.

Mechanical chillers increase cooling rates 2- to 4-fold over those of static tank chilling and hence are desirable for optimum processing efficiency. However, rapid, agitated chilling introduces problems of excessive moisture absorption and inadequate aging for which allowances must be made and precautions taken. A comparative study of mechanical chillers and tank chilling for their effects on chilling rates and moisture absorption in eviscerated chicken broilers reveals moisture uptake of 2 to 3% from washing of the carcasses prior to chilling, and a total uptake due to washing and chilling ranging from 7% for 4-hr. static tank chilling to 18% for some mechanical chilling conditions. For any particular chilling system, moisture absorption increased about 50% by extending the evisceration cut to the thigh area or by leaving the neck attached to the carcass. Much of the absorbed water was held in the superficial layers of skin and between skin and muscle and was lost rapidly during draining and subsequent holding. When this loosely bound water is not allowed to drain away before freezing, an undesirable amount of water collects in the package on thawing. Studies have shown that the muscle tissue absorbs very little water during the chilling process, probably less than 2%, while the skin and surrounding fatty and connective tissue absorbs as much as 12%. Weight loss during drainage is rapid for the first 15 min., moderate for the next 45 min., and very slow thereafter. Many processing operations are limited to a 10- to 15-min. draining period before packaging.

Salts such as sodium chloride and sodium polyphosphates are potentially useful in controlling moisture absorption and retention in poultry meat, but they have not received acceptance in most commercial products. The addition of 0.5 and 1.0% sodium chloride to ice slush reduces the 24-hr. moisture uptake by about 10%. A rapid, mechanically agitated, 15-min. ice slush chilling condition results in a 40% reduction in moisture uptake by the addition of 2% sodium chloride to the ice slush. Comparable percentage reductions in moisture uptake were obtained by 2% salt addition to static ice slush for a 15-min. chilling period. Regulations generally set a tolerance for sodium chloride in chilling media, e.g., 70 lb. to 10,000 gal. of water, above which approval must be obtained and the product appropriately labeled. Low sodium chloride concentrations have been used to control the flow characteristics of slush ice.

Sodium polyphosphates, principally mixtures of sodium pyrophosphate and sodium tripolyphosphate, are permitted in further processed products such as turkey rolls. Effects of polyphosphates on processing characteristics have been studied extensively, and may be summarized as follows. Addition of polyphosphates to the chilling water reduces water absorption during chilling, reduces water lost as drip during refrigerated storage, and reduces water lost during thawing and cooking of the bird. Polyphosphates also have a detectable antioxidant effect.

When poultry is marketed in an unfrozen condition, the time delay between slaughter and cooking is great enough that toughness, caused by lack of aging, is never a problem. However, when birds are frozen at the processing plant, adequate aging becomes extremely important. Assurance of an optimally tender product depends on the allowance of a sufficient aging period, above freezing temperatures, between slaughter and freezing for adequate tenderization to take place. Tenderization reactions proceed at a negligibly slow rate at temperatures below 25° F. (−4° C.). Aging for tenderization can, of course, occur after thawing and before cooking, as well as after slaughter and before freezing. Since the holding of poultry in chill tanks involves considerable expense and an interruption of the smooth flow of the processing line, the poultry processor needs to know how much time should be allowed for aging. A considerable amount of work has been reported on the time necessary for optimum tenderization of poultry meat. Young chickens require 8 to 12 hr. to become tender. Studies on young turkeys indicate freezing should not commence before 12 to 16 hr. postmortem if birds are to reach an acceptable level of tenderness.

It should be emphasized that the tenderizing times for turkeys mentioned above refer only to fryer-roaster turkeys (12 to 18 weeks old). Older turkeys appear to require less time before freezing to become tender. Turkeys that are 22 to 27 weeks old (15 to 25 lb. live weight) are reported to be adequately tender with only 1 to 2 hr. of postmortem aging. Presumably this difference is due to the longer time required to freeze the muscle tissue in the larger birds.

Although tenderization does not occur at the usual temperature of frozen storage, inadequately aged birds become significantly more tender if they are held for 2 to 14 days at 25° to 27° F. (−4° to −3° C.). Tenderization can also take place following thawing; under these conditions, it is just as rapid as tenderization before freezing. The most important determinant of tenderness is the total elapsed time which a bird spends in the unfrozen state between slaughter and cooking; it is immaterial whether that time is spent before freezing or after thawing.

The studies reported above on the relation between aging time and tenderness have utilized either mechanical devices or trained taste panels to evaluate tenderness. Two excellent reports have appeared which measured the reactions of a *consumer-type* (untrained) taste panel to poultry that had received various aging treatments. The experiments on turkeys indicated that the consumer will comment on toughness if birds are aged for less than eight hours. A four-hour aging period is required for chicken fryers to reach an acceptable level of tenderness.

The muscle that gives rise to the greatest number of toughness comments in unaged poultry is the *pectoralis superficialis,* the large breast muscle. Consequently, this is the muscle that is used as an index of proper tenderization. In contrast, the thigh and leg muscles are generally more tender, although it takes them a longer time to become optimally tender. From 2 to 4 days are required for tenderness changes to be completed in these muscles.

Product Protection

For successful marketing of frozen poultry products, adequate protection and attractive packaging are key factors.

Packaging materials should have several characteristics, and two of the most important are relative impermeability to moisture and to oxygen. Moisture loss is not only a direct loss of product weight but also results in unsightly freezer burn areas on the product surface.

Oxygen availability, in addition to factors mentioned earlier, greatly accelerates rancidity development. Chickens in oxygen-permeable packaging have less than one-third the storage life of chickens in packaging that is relatively impermeable to oxygen.

Organoleptic deterioration and chemical changes involving oxygen consumption are independent of moisture loss and are dependent only on storage temperature and partial pressure of oxygen. Best results are obtained by storing the product under nitrogen with complete exclusion of oxygen. Carbon dioxide is produced from samples stored in a nitrogen atmosphere, indicating that anaerobic reactions are involved in some of the deteriorative changes taking place during frozen storage.

Oxygen contributes as much to deterioration as a substantial rise in storage temperature. Improved packaging was found to increase the storage life as much as a 20° F. (11° C.) drop in temperature.

Packaging.—Whole, ready-to-cook birds are trussed and formed into a compact attractive shape, often by inserting the legs under a specially cut strip of skin, or into a formed wire retainer. Giblets wrapped in parchment paper are generally inserted in the crop cavity, and the neck in the visceral cavity. The birds are then packaged in form-fitting plastic bags that are translucent, fairly tough, and reasonably impermeable to moisture and air.

There is an increasing trend, especially with chickens, to cut up the bird at the processing plant and package the complete bird, or separate and package by parts. This operation exposes much more surface for possible moisture loss and rancidification by atmospheric oxygen. Consequently, the overall package, which may include inner plastic film liner, cardboard carton, and possibly a sealed overwrap, should be compact and completely filled, and reasonably impermeable to moisture loss and air exposure.

Whole, plastic-bagged birds such as turkeys are packed in fiberboard cartons or similar containers for handling in frozen storage and shipment to retail markets. These cartons should be rectangular, of a shape to facilitate palletizing, and strong enough to withstand stacking loads 16 ft. high in refrigerated warehouses.

Materials.—Attempts have been made to develop edible coatings which could be placed on birds by dipping or spraying to prevent moisture and gas transfer. These have not been particularly successful to date. Corn syrup reduces the rate of fat oxidation, increases the brown color of cooked chicken, and reduces moisture loss through an ice glaze. Starch coatings are permeable to water and cause a white color that persists after cooking. An acetylated monoglyceride coating was the best moisture barrier of the three mentioned. However, the acetylated monoglycerides are not as effective as cellophane or Cryovac bags for controlling dehydration.

Several kinds of plastic films are presently in use for the storage of frozen poultry. Some of the desirable features of these films are impermeability to moisture and oxygen, transparency, heat sealability, heat shrinkability, toughness, and flexibility at low temperatures. Different films have different properties that can be combined by forming laminates from two or more different films. Heat shrinkability can be attained by stretch-orienting the molecular configuration of the film during manufacture; upon heating, the molecules will realign, resulting in shrinkage of the film. Form fitting is achieved in most cases by the ability of the plastic film to shrink when the sealed bag is immersed for a few seconds in water somewhat below the boiling point (Fig. 6.3).

A heat-sealable film in packaging frozen poultry is desirable. Polyethylene film has the advantage of being heat-sealable and pliable at low temperatures, but it does not provide a good barrier to oxygen. As a result, polyethylene is often combined with another film as a laminate. Polyester is one of the films used in this type of combination. It provides low permeability to water and gases as well as being heat-shrinkable and flexible at low temperatures. Oriented polypropylene is another film often used in combination with polyethylene to provide clarity and improved barrier qualities. Polyvinylidene chloride is a transparent, heat-shrinkable film which is impermeable to both water and oxygen. Vacuum packaging in one of these films, followed by heat shrinking, provides a very satisfactory wrap for frozen poultry.

FREEZING AND FROZEN STORAGE

Preservation of poultry meat by freezing and frozen storage is done with the hope of maintaining the characteristics of the thawed product at or near those of the product prior to freezing. The effect of freezing and frozen storage on poultry appearance and palatability is influenced by

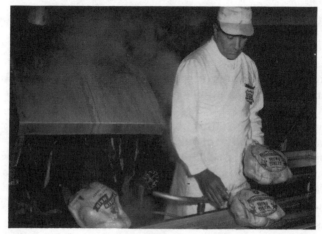

Courtesy U.S. Dept. Agr.

FIG. 6.3. A MOMENTARY DIP IN HOT WATER SHRINKS THE PLASTIC FILM SURROUNDING TURKEYS TO REMOVE AIR POCKETS AND WRINKLES

rate of freezing, temperature and duration of storage, packaging, and handling during and after thawing.

As muscle is chilled, biochemical reaction rates are diminished. This preserving influence becomes increasingly effective as the temperature is lowered, until ice crystals begin to form. Freezing is essentially a form of low temperature dehydration. During the early stages of freezing, ice crystals are composed of relatively pure water, and the unfrozen solution becomes progressively more concentrated as the temperature is lowered.

This concentrated salt solution, particularly at temperatures slightly below the initiation of freezing, is the cause of much of the loss in product quality that can be associated with freezing and frozen storage.

Ice Formation

Ice crystals start to form in poultry muscle at about 27° F. (−3° C.) and grow rapidly in size until the temperature falls to 20° F. (−7° C.). With a given capacity for heat removal, the rate of cooling decreases when ice formation starts. This is due to the latent heat of fusion (106.0 B.t.u. per lb.-° F.). The specific heat of unfrozen poultry meat is 0.7 B.t.u. per lb.-° F.; after freezing it drops to 0.37 B.t.u. per lb.-° F. Also, the heat conductivity of ice is four times that of water. These two factors combine to increase the rate of cooling once the freezing plateau of a portion of the muscle has been passed.

Since the water in muscle contains many solutes, is physically isolated in small channels, and is hydrogen bonded into protein molecules, it does not all freeze at any particular temperature. The proportion of frozen water increases as the temperature is lowered; 74, 83, 88, and 89% of the water in lean meat is frozen at 23° (−5°), 14° (−10°), −4° (−20°), and −22° F. (−30° C.), respectively. The final cryohydric point or eutectic point of meat is between −58° (−50°) and −76° F. (−60° C.). Further lowering of the temperature beyond this point will not cause any additional solidification of water or solutes. However, there is still unfrozen water tightly bound to the muscle proteins, and this constitutes 8 to 10% of the water in muscle.

As various solutes reach saturation due to the transition of solvent water to ice, they start to crystallize out of solution. Thus, both the concentration and the composition of the cell fluid change upon freezing. Studies on the pH of frozen poultry muscle show that upon freezing there is an initial increase of 0.3 pH unit from pH 5.7 to pH 6.0. This shift is attributed to precipitation of about 30% of the total phosphates in muscle.

Water translocation and resultant solute concentration during ice crystal formation is dependent both on the physiological condition of the muscle and the rate of cooling. Since almost all poultry handled in commercial processing is either in rigor mortis or in a postrigor condition at the time of freezing, the location of ice crystals is dependent on the rate of freezing.

Slow heat removal from postrigor muscle results in the extracellular nucleation of a relatively few ice crystals which then grow to a large size by pulling water from within the fibers. The extensive dislocation of water in this type of freezing may cause irreversible changes in the normal water-solute relationship. Injury to proteins within the fibers results from high solute concentration. This slow freezing causes great histological distortion of the fibers and their contents, with complete disappearance of the muscle striations.

However, upon thawing, the fibers regain an appearance identical to unfrozen fibers. The passive collapse of cells during slow freezing does not result in any appreciable mechanical injury, and any damage is due to the high solute concentration.

Rapid freezing, accomplished by a large temperature differential and efficient heat transfer, causes intracellular ice formation. Supercooling of the muscle increases the probability of ice crystal nucleation, and ice formation will start before water can diffuse out of the cell. As the rate of cooling increases, the probability of nucleation increases, and a greater number of small ice crystals are formed with less translocation of water. Ice crystallizes within muscle fibers as spears which rapidly extend parallel to the long axis of the fibers. The spears grow in diameter as well as in length, pushing aside the structural material within the cell. As the rate of freezing increases, more and more small spears form within a fiber. When freezing is accomplished within hundredths of a second, small ice rodlets form within single sarcomeres. These small ice crystals are unstable, and the beneficial effects of minimal water dislocation accomplished by rapid freezing can easily be lost during frozen storage or thawing.

One of the major considerations in choosing the freezing rate for poultry is the desired appearance of the frozen product. Fast freezing has been shown to have only small effects on either palatability or drip formation. However, color of the frozen bird is greatly influenced by rate of freezing. Alteration of the freezing rate can produce any color from dark red to chalky white. Large ice crystals allow incident light to penetrate the skin and muscle so that the dark red deoxygenated myoglobin can be seen. As the ice crystals become smaller due to more rapid freezing rates, more light is refelected from the surface, giving a white appearance. Optical changes take place in both skin and muscle. The white appearance is due to opacity of the skin and is not really a change in color. Rapid freezing for the production of a light color is particularly important for birds scalded at 140° F. (60° C.) where the cuticle of the skin has been removed. Young birds with little fat under the skin will also be dark colored unless they are frozen rapidly.

Chemical Alterations

Various chemical reactions take place during frozen storage. Ribose increases during a storage period of 149 days at −4° F. (−20° C.). Amino nitrogen increases fivefold and there is a marked increase of anserine and carnosine in the water extract. The lactic acid content of chicken meat declines during storage at 14° F. (−10° C.) resulting in a gradual rise of pH. Proteolysis in muscle held at −5° F. (−21° C.) is indicated by an increase in water-soluble nitrogen and nonprotein nitrogen. A decrease in amino nitrogen indicates that amino acids also are degraded.

One of the more important changes taking place during frozen storage is the loss of protein solubility in salt solution and the resultant loss of water-holding capacity and meat tenderness. The loss of protein solubility, i.e., denaturation, is probably a manifestation of many types of chemical reactions taking place in the frozen condition. Decreased solubility of the myofibrillar proteins is the major cause of lowered protein solubility. The number of sulfhydryl groups and the ATP-ase activity decline with storage time and free amino acids increase in the nonprotein nitrogen fraction. Storage at −112° F. (−80° C.) for two years causes no appreciable change in protein solubility whereas a storage temperature of 0° F. (−18° C.) is sufficient to stabilize the product for one year. However, the above mentioned indicators of protein solubility denaturation begin to appear after 15 to 20 weeks at 14° F. (−10° C.).

Phospholipid (60% of the muscle lipid) decreases during storage at 14° F. (−10° C.); free fatty acids increase. About 70% of the free fatty acids come from phospholipid and the remainder from triglyceride.

A theory for protein denaturation in frozen muscle states that the concentration of tissue salts results in the release of free fatty acids which in turn react with actomyosin to render it insoluble.

Rancidity and development of off-odors and off-flavors are major factors in determining storage life. Turkey fat is less stable to oxidative deterioration during frozen storage than chicken fat. Dietary tocopherol is deposited much more efficiently in chicken fat than in turkey fat. Since there was no difference in fatty acid content of body fat between chickens and turkeys, it was concluded that tocopherols are important in stabilizing poultry fat. Delays in chilling or extended holding in the unfrozen condition greatly increase fat deterioration during frozen storage.

Physical Alterations

Recrystallization is an important physical change which may take place during frozen storage, resulting in translocation of water and consolidation of the unfrozen solution. Small ice crystals are thermodynamically less stable than large crystals, and water molecules tend to migrate from small to large crystals. This migration is promoted by fluctuating storage temperatures. Repeated temperature fluctuation between 7° (−14°) and 20° F. (−7° C.) causes ice crystal growth. However, no measurable redistribution of ice is found in muscle after 180 weeks of storage at a constant 7° F. (−14° C.). The major effect of temperature fluctuation on frozen turkeys is excessive frost accumulation inside the package. Fluctuating temperatures result in moisture loss and quality deterioration only slightly greater than that found in samples held at a constant temperature equivalent to the mean of the fluctuating temperatures.

Storage temperature influences both the rate of desiccation and the development of freezer burn. Dehydration increases with increasing storage temperature. There is a similar temperature effect on the degree of freezer burn of poultry skin. Freezer burn is an extension of desiccation resulting in irreversible protein denaturation. Freezer burn of liver is most severe in rapidly frozen samples: Freezer burn follows the pattern of ice cavities caused by sublimation of the ice crystals. An explanation for more severe burn in rapidly frozen muscle is the increased surface area of cavities left by smaller ice crystals.

Biological changes are essentially stopped at −112° F. (−80° C.). However, this temperature is not possible or practical in most commercial storage facilities. In addition, such a low temperature would not be necessary for quality maintenance during the normal length of commercial storage. Temperatures of 0° F. (−18° C.) or below have generally been recommended for storage times up to one year. Poultry stored at

0° F. (−18° C.) maintains desirable quality for more than twice as long as that stored at 10° F. (−12° C.).

Commerical Freezing Practices

General Considerations and Requirements.—Since poultry to be frozen is almost always adequately protected against moisture loss or shrinkage by tight moisture-proof packaging, comparisons of freezing methods or systems on the basis of moisture transfer from the product are not pertinent. Also, if the products are adequately chilled before placing in the freezing system, no system can be considered to be a potential hazard for development of an unwholesome or rancid product. Requirements will therefore be discussed in terms of refrigerating efficiency, rate of freezing, and appearance of the frozen product.

Studies by various workers have established the correlation between rapid freezing rates and small ice crystal size with resulting pleasing light frozen appearance. Given present commercial practices in production and processing, rapid freezing is necessary to obtain an optimum, light appearance in the frozen product. High-temperature scalded turkeys, lacking the outer surface cuticle, have an objectionably dark surface if not frozen rapidly. Poorly finished birds with little fat under the skin also benefit greatly in appearance from faster freezing. Since it is only the outer 2 to 3 mm. of the surface layer in which ice crystal size, and hence rate of freezing, influence appearance, some processors have introduced a rapid crust-freezing, followed by a slower freezing of the inner portion of the carcass in a cold storage warehouse. Rates of freezing are greatly reduced by placing packaged birds in cartons before freezing, so the alternative of freezing packaged birds on open shelves is widely practiced.

Obviously critical factors in establishing sufficiently rapid freezing are temperature, velocity, and type of the cooling medium. Some have observed a marked increase in freezing rate of turkeys, and improved appearance, by decreasing air-blast temperature from −12° to −21° F. (−24° to −29° C.), but very little additional effect by reducing the air temperature to −31° F. (−35° C.) (Fig. 6.4 and 6.5). Increasing air velocity beyond 600 f.p.m. had a small beneficial effect. The largest difference in freezing rate was noted between birds frozen on an open shelf and birds frozen after being packed in a carton.

Freezing packaged poultry by low-temperature brines or glycols is being practiced to an appreciable extent, although air-blast freezing is still the predominant method. Reported times required for −20° F. (−29° C.) calcium chloride brine to lower the internal temperature of warm eviscerated, packaged birds to 15° F. (−9° C.) are 1½ hr. for broilers, 5 hr. for

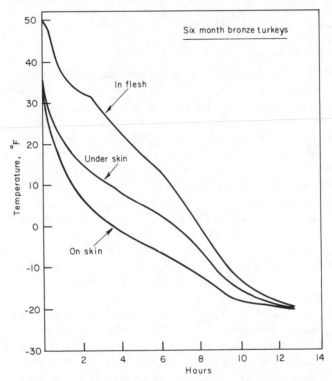

FIG. 6.4. TEMPERATURES AT THREE LOCATIONS DURING FREEZING OF PACKAGED READY-TO-COOK BRONZE TOM TURKEYS OF 21-LB. WEIGHT

Thermocouples were inserted under packaging film and on skin, under skin and on top of flesh, and at a one-inch depth in the flesh, all three being in the breast region.

12-lb. turkeys, and 7 hr. for 25-lb. turkeys. Comparative studies of liquid immersion and air-blast freezing in relation to cooling rates and appearance reveal that the air-blast temperatures needed are much lower than −40° F. (−40° C.), e.g., around −100° F. (−73° C.), in order to duplicate the very light skin appearance developed by −20° F. (−29° C.) liquid immersion freezing. Figure 6.6 illustrates the very rapid rate of temperature decrease at various depths in 15-lb. turkeys immersed in −20° F. (−29° C.) liquid. These rates are in sharp contrast with those shown in Fig. 6.4 and 6.5 for freezing by means of air blast in the same temperature range. The importance of air temperature and air velocity in air-blast freezing is seen in Fig. 6.7 and 6.8. Reduction of air temperatures below −40° F. (−40° C.) appears to have very little additional beneficial effect, and increasing air velocities above 1,200 f.p.m. seems to be of little value.

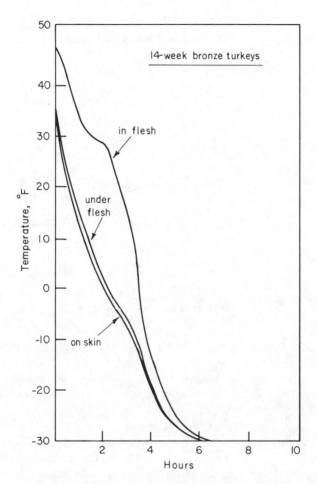

FIG. 6.5. TEMPERATURES AT THREE LOCATIONS DESCRIBED IN FIG. 6.4 DURING FREEZING OF PACKAGED, READY-TO-COOK BRONZE TOM TUR-KEYS OF 7-LB. WEIGHT

Air-Blast Freezing.—The major part of commercially frozen poultry is produced in air-blast systems which vary widely in design, capacity, and operating characteristics. Air-blast systems are adaptable to almost all types of products, so that public cold storage warehouses invariably use such systems for the extensive custom freezing that they conduct for poultry processors who have only inadequate freezing capacity in their own plant. For a freezing operation in the poultry processing plant, air-blast systems have much to offer. They are relatively inexpensive in construction, trouble-free and clean in their operation, and adaptable to all sizes of product unit, packages, and cartons. Product movement in and

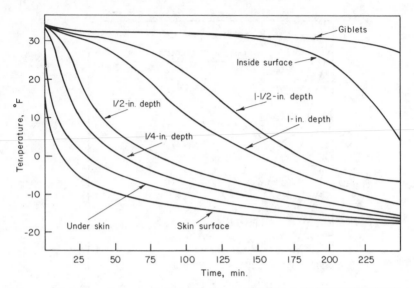

FIG. 6.6. TEMPERATURES AT VARIOUS DEPTHS IN BREAST OF 15-LB. TURKEYS DURING IMMERSION FREEZING AT −20° F. (−29° C.)

out can be accomplished by multi-shelved racks on wheels, by pallets and fork lift trucks, or by conveyor belts activated by gravity or power.

Air-blast freezers may be designed as insulated tunnels or as uninsulated cells installed within a large cold storage room maintained in the 0° to −10° F. (−18° to −23° C.) range. Cell sizes vary; a typical one is about 20 ft. wide by 25 ft. long. The cell-within-storage-room design is obviously efficient in reducing construction and insulation costs in walls and doors, in minimizing heat gain and frosting from the ambient atmosphere, and in minimizing trucking distance from freezer to cold storage space.

As mentioned in the previous section, high air velocities across the product, up to 1,500 f.p.m., are recommended to give rapid freezing and optimum appearance of the poultry surface. Fortunately this air velocity will also provide optimum freezer coil operation. Air temperature should be in the range of −30° to −40° F. (−34° to −40° C.) which can be provided by a two-stage ammonia refrigeration system or by alternative systems.

In the design and routine loading of the air-blast cell, it is desirable to have short air paths in order to assure a low pressure drop through the product load and small differential of temperature between air entering and air leaving the load. It is generally good practice to stack the product full width in the direction of the air flow, proceeding from back to front.

Plastic-bagged poultry may be placed on open tiered racks or may be

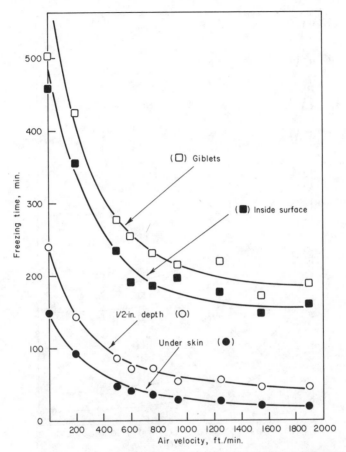

FIG. 6.7. RELATION BETWEEN FREEZING TIME (TIME FOR TEMPERATURE TO FALL FROM 32° to 25° F.; 0° to – 4° C.) AND AIR-BLAST TEMPERATURE IN 5- TO 8-LB. CHICKENS WITH INITIAL TEMPERATURE OF 32° TO 35° F. (0° TO 2° C.) AND WITH AIR VELOCITY OF 450 TO 550 F.P.M.

placed in cartons, with provision for adequate air flow around the bird, and the cartons then may be placed on racks. Conventional freezing racks are about 3 × 5 × 6 ft. high and can be moved on casters or by fork lift truck. Often the carton tops are left off until after freezing. Some telescoping cartons have been constructed with cut-outs in the side walls to provide optional air flow. By introducing spacers between layers of cartons and spaces between adjoining cartons in the same layer, palleted loads of some products can be frozen satisfactorily. The importance of adequate air velocity at the product surface (or nearest accessible surface to the product surface in a cut-up, packaged product) cannot be over-emphasized.

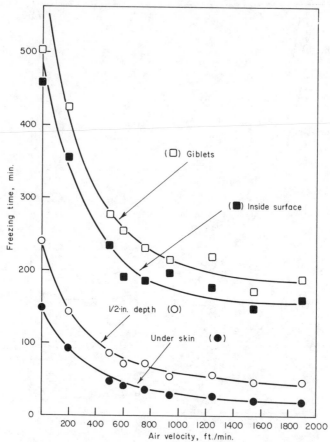

FIG. 6.8. RELATION BETWEEN FREEZING TIME (TIME FOR TEMPERATURE
TO FALL FROM 32° TO 25° F.; 0° TO −4° C.) AND AIR-BLAST VELOCITY IN 5-
TO 8-LB. CHICKENS WITH INITIAL TEMPERATURE OF 32° TO 35° F. (0° TO
2° C.) AND WITH AIR-BLAST TEMPERATURE OF −20° F. (−29° C.)

Essential to an efficient air-blast freezing operation are properly designed automatically operating outer frozen storage freezer doors, with
air curtains, to provide minimum refrigeration loss and condensation
between atmospheric air temperature areas and refrigerated areas of the
plant.

Liquid Immersion Freezing.—Freezing of plastic-bagged whole
ready-to-cook poultry in low temperature (20° to −20° F. (−7° to
−29° C.)) liquids has had limited commercial application in comparison
with air-blast and liquid spray systems. While higher temperatures can be
used than in air-blast systems, and hence more efficient, higher refrigera-

tion suction pressures are possible, a liquid immersion system has many practical disadvantages. Liquid immersion is adapted to very few products. Any break, however small, in the plastic bag enclosing the bird will result in liquid leaking into the product and will necessitate repackaging. If salt brines are used in conventional metal equipment, corrosion problems can develop. Despite these limitations, if birds are completely immersed in their passage through the low temperature liquid, very light, uniform frozen surface appearance can be developed in the product.

Appreciable commercial use of liquid immersion freezing of poultry was first introduced in Canada. The ideal liquid for this operation should be nontoxic, noncorrosive, and inexpensive, and should have low viscosity and freezing point and high thermal conductivity. Of liquids that can be used, calcium chloride and sodium chloride brines are inexpensive, of low viscosity, and have good thermal properties, but they are corrosive and calcium chloride could be considered toxic. Sodium chloride brines are used extensively. Alternative liquids are the glycols, such as glycerol and propylene glycol. These are nontoxic, noncorrosive, and have acceptable although not optimum viscosity and thermal properties. Propylene glycol finds extensive use, but primarily in a liquid spray process that is described below.

Liquid immersion systems are varied in design. Most of them require a large, shallow, insulated tank of refrigerated liquid equipped with ammonia or Freon refrigeration coils and some means for orderly conveying the packaged birds through the tank within a given time interval. Trough-like channels, each about 14 in. wide, may be built into the tank to facilitate movement. The birds may be flumed through the tank by recirculating the freezing liquid, or some mechanical means of conveying them through the tank may be employed. It is essential to reduce the temperature of the entire surface of the bird uniformly, including that part of the surface normally floating above the liquid level. It is possible to introduce baffles at the liquid level to hold the birds completely submerged, but this is a potential source of damage to the package. A better method is to provide a continuous spray of the freezing liquid over the exposed surfaces of the birds as they float through the tank. Another requirement for uniform freezing rate and uniform color over the entire surface is that the packaging film be skin tight at all places. Gaps or bridges between film and skin will lead to slower freezing rates and variegated color.

As discussed in an earlier paragraph, the general aim in liquid immersion freezing is to freeze an outer shell of the bird rapidly, so as to develop a desirable light surface appearance, and then to finish freezing of the whole carcass by moving the birds to a blast freezer or to a frozen storage room with reasonably good air movement. The depth of the frozen shell will vary with each particular operation; the shell must be thick enough to maintain the surface layer in a completely frozen state throughout the remaining transfer and freezing operation. A thickness of ½ to ¾ in. is generally considered adequate. Birds may be removed at the discharge end of the freezing tank by a continuous motor-operated stainless steel wire belt. Total immersion time can be controlled by rate of removal.

A final step of a liquid immersion freezing operation is a rinsing of the bag with cold water to remove the surface film of liquid refrigerant. For relatively high-priced liquids such as propylene glycol, the glycol can be recovered from the rinse water by fractional distillation.

Selection of the operating liquid temperature represents a balancing of the effect of liquid temperature on appearance, freezing rate, and refrigerating efficiency. Hence no general recommendation can be made, although temperatures in the range of 0° to 10° F. ($-18°$ to $-12°$ C.) may optimize most of the above factors.

It is of obvious economic importance to know how long an immersion time (or how thick a frozen crust) is required to preserve the light frozen appearance during subsequent completion of freezing in air blast and storage at commercial cold storage temperatures. Pilot runs under the planned commercial operating conditions may often be necessary. Under controlled experimental conditions, using an equal-volume mixture of methanol and water as the immersion liquid and an initial carcass temperature of 32° to 35° F. (0° to 2° C.), 4- to 6-lb. chickens need a minimum immersion time of 20 min. in $-20°$ F. ($-29°$ C.) liquid to provide a crust sufficient to maintain a maximum white appearance during subsequent completion of the freezing on open shelves in an air blast at $-20°$ F. ($-29°$ C.) and 300 to 500 f.p.m. For 14- to 15-lb. turkeys this minimum immersion time was 40 min. The chalky white appearance was stable for at least 20 weeks in $-20°$ F. ($-29°$ C.) storage, but darkened noticeably after 5 to 6 weeks at 0° F. ($-18°$ C.) or after 2 weeks at 20° F. ($-7°$ C.). Creamy, more natural skin appearance can be established with shorter immersion times, and can be maintained with higher blast freezing and storage temperatures.

Adequate agitation of the liquid about the carcass surface is an important factor, but only limited quantitative data are available. Using gelatin gel models to simulate a poultry carcass, researchers are able to reduce freezing times for a ½-in. outer shell about 50% by increasing agitation in a 50-50 (volume) mixture of propylene glycol and water, and about 35% in a 29% calcium brine. Liquid-film heat transfer coefficients, (B.t.u.)/(hr.)/(ft.)^2/(° F.), vary widely with the type of freezing liquid and the level of agitation, and range from 13 for a viscous glycerol mixture at a low level of agitation to 167 for calcium chloride brine at a high level of agitation.

Also, an increase in agitation of brine solution from a still condition to 84 f.p.m. flow reduces freezing time by 40% with a 0° F. (−18° C.) brine and by 60% with a 10° F. (−12° C.) brine. Due to viscosity differences, a 0° F. (−18° C.) salt brine solution was as efficient in freezing rate as a −10° to −15° F. (−23° to −26° C.) glycol solution.

Liquid Spray Freezing.—Freezing packaged, whole, ready-to-cook poultry by spraying cold liquid over it has received wide commercial adoption, primarily through the development of special commercial freezing processes and equipment. This type of equipment may consist of a 10-, 20-, 30-, or 40-ft. metal conveyor belt, operating in a cabinet and over a tank so that a continuous spray of 15° F. (−9° C.) or lower temperature propylene glycol solution (about 46% glycol and 54% water by weight) can be directed from above on the surface of the birds as they are conveyed through the cabinet. Equipment for a 40-ft. unit may include two pumps for recirculating the glycol solution at about 1,400 g.p.m., refrigeration coils and attendant equipment of about 60 tons capacity, main conveyor belt and motor, and a short conveyor belt with motor at the discharge end which picks up the birds at the end of the spray section and conveys them through a water spray to remove residual glycol. This glycol rinse can be concentrated by distillation and added back, so that actual consumption of glycol is minimal and has been estimated at 15 cents worth of glycol per 1,000 lb. of product.

Usually a liquid spray freezing system is designed and operated to freeze only an outer shell of the bird so as to provide an acceptably light final frozen appearance, after which the partially frozen bird is moved immediately into an air-blast freezer or cold storage room where the remainder of the freezing takes place. The fraction of the total refrigerating load required to bring the product from chill temperature of 40° F. (4° C.) to cold storage temperature of 0° F. (−18° C.) that is borne by the spray system may range from 20% for 20-lb. turkeys to 40% for 2-lb. fryers. At these percentages, commercial liquid spray freezing units are able to handle 14,000 lb. of turkeys per hr. and 5,000 lb. of fryers. As in all liquid freezing systems, compared to air-blast systems, the refrigeration cycle is more efficient with about a 0° F. (−18° C.) refrigerant evaporator temperature.

Using a brine spray, crust freezing to a ½-in. depth required 25 min. at a −3° F. (−19.4°C.) brine temperature, 30 min. at 0° F. (−18° C.), 45 min. at 5° F. (−15.0° C.), 90 min. at 10° F. (−12.2° C.), and 4 hr. at 20° F. (−7° C.).

Liquid Nitrogen and Carbon Dioxide Freezing.—A review of the comparative value of liquid nitrogen and carbon dioxide systems and older systems utilizing air blast or refrigerated liquids, shows six commercial systems using either liquid nitrogen as a spray or carbon dioxide as a spray or pulverized solid. These systems have been developed by the large commercial producers of liquefied gases, and the freezing costs and ultimate usefulness of the systems depend to a large extent on the cost of the liquefied gases in large quantities.

Direct immersion in liquid nitrogen results in shattering or cracking of large pieces of meat, so that for this and other considerations involving refrigerating efficiency, present systems involve evaporating the liquid nitrogen in the freezing chamber and utilizing the refrigerating capacity held in the latent heat of vaporization of liquid nitrogen (85 (B.t.u.) per lb.)) plus the capacity represented by the rise in temperature of the nitrogen gas from its boiling point ($-320°$ F.; $-196°$ C.) to the desired final temperature of the product ($0°$ F.; $-18°$ C.).

Equipment and methods for freezing food by a direct spray of liquid carbon dioxide have been patented. Food of discrete particle size is conveyed through a freezing tunnel either by a screw conveyor or by a horizontal conveyor belt. This type of operation is well adapted to diced poultry meat. A freezer utilizing solid particles of carbon dioxide in an inclined rotating cylinder has been developed.

A liquid nitrogen flash freezer which consists of a conveyor constructed within a double-walled, vacuum-insulated cylinder is also available.

Liquid nitrogen or carbon dioxide freezing has found greatest application and value in freezing products in small particles such as diced, cooked, deboned chicken meat.

Plate Freezing.—Plate freezers have been applied to the freezing of poultry meat products that can be formed or packaged into rectangular shapes that have flat surfaces and lend themselves to close packing for good thermal contact between plate surface and product. They have also been used for rapid chilling of flat trays of precooked, deboned poultry meat.

Special Systems.—The demand for continuous, labor-saving freezer systems has resulted in the development of several almost completely automated units. Such systems may be designed to handle packages, cartons of birds, or unwrapped pieces of chicken or turkey. Product may be transported through the freezing chamber on belts or trays.

The Greer system, pictured in Fig. 6.9 and 6.10, is adaptable to all sizes of whole birds, packages, or cartons. The system automates the processing operation from the point where birds are placed in the bottom halves of cartons until they are frozen and ready for a carton top to be put on. A typical design handles about 50 birds per min. with about 150,000 lb. total capacity. Refrigeraton coils and fans are located at the side of the machine so as to give a high-velocity two-pass air flow that applies the coldest air to the warmest product. Frost or ice build-up is minimized since the trays' shelves never come outside the freezer.

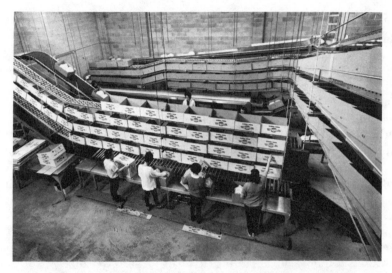

Courtesy of Joy Manufacturing Co., Greer Division, Pittsburgh, Pa.

FIG. 6.9. LOADING AND UNLOADING GREER MULTI-TRAY FREEZER, AN AUTOMATIC
POULTRY FREEZING SYSTEM

In background, operators are placing unfrozen birds in cartons and sliding cartons on feed
conveyors for freezer. In foreground, lids are being placed on cartons containing completely frozen birds.

Courtesy of Joy Manufacturing Co., Greer Division, Pittsburgh, Pa.

FIG. 6.10. LOADING BIRDS IN CARTONS ONTO FREEZER TRAYS IN GREER MULTI-TRAY
FREEZER, AND AUTOMATIC CONVEYORIZED AIR-BLAST FREEZING TUNNEL

Proper number of cartons for each tray are automatically positioned and counted, and
then gently pushed onto the tray by a sweeping arm.

Thawing

Rate.—Thawing conditions are usually not given much consideration in the handling of frozen poultry since, in most cases, thawing takes place after the product is sold to the consumer. As with freezing rates, the rates of thawing have been shown not to have a significant effect on the palatability of the frozen product. Likewise, the rate of thawing has no effect on the tenderness of properly aged birds.

Drip Loss.—Drip loss from thawed poultry is usually small, particularly from whole birds. From 0.5 to 1.5% of the unfrozen weight of chickens is lost as drip after freezing. A small but significantly increased drip loss results from either slow freezing or slow thawing procedures (1.5% drip loss for slowly frozen birds compared to 0.5% loss for those rapidly frozen).

Two processing practices can cause appreciable increases in drip loss. A large proportion of the water taken up during processing is lost upon thawing. Another factor is cutting up birds prior to freezing. Frozen and thawed cut-up chicken can lose 10% of its weight as drip.

Bone Darkening.—Bone darkening is a condition which develops when young poultry is frozen and thawed. Organoleptic properties of the muscle are not affected, but the brown to black appearance of the long bones and the surrounding muscle results from leaching of hemoglobin out of the bone marrow. Subsequent oxidation of the red hemoglobin to methemoglobin produces the dark color. Bone darkening is a problem only in young birds for two reasons. First, more hemoglobin is present in the bone marrow of young, rapidly growing birds. And second, incomplete calcification of the bones allows the hemoglobin to escape from the marrow cavity. Aging prior to freezing, rate of freezing, and freezing in an inert atmosphere have no effect on the extent of bone darkening. Genetic background, extent of bleeding, and processing conditions have no influence on bone darkening. Bone marrow is the only source of the pigment that darkens thawed bone.

Microbiology.—Questions often arise regarding the advisability of refreezing previously frozen and thawed birds. Frozen and thawed poultry spoils at the same rate as similar material which has not been frozen. Repeated freezing and thawing has no effect on bacterial activity. Meat does not become more perishable after freezing and thawing, but nutrient and palatability losses as drip might be greater from thawed meat than from unfrozen meat. One aspect that should be considered if thawed poultry is to be refrozen while still in commercial channels is the color of the frozen product. As with fresh poultry, thawed birds if they are refrozen slowly will appear dark due to formation of large ice crystals.

CONSUMER ACCEPTANCE

The ultimate consideration regarding freezing preservation of poultry is its effect on palatability and consumer acceptance. Investigators do not agree completely upon this effect, but any differences in organoleptic quality have generally been small. A trained taste panel reported that unfrozen broilers were preferred to frozen birds for all palatability factors after 51 days of storage at 14° F. (−10° C.). Differences were small. No significant differences in organoleptic qualities are found between frozen birds and ice-packed birds throughout 14 days of storage. Likewise, there is usually no difference in flavor between unfrozen and thawed poultry muscle. A slight taste panel preference exists for unfrozen samples based on small differences in tenderness and juiciness. Generally, consumer acceptance is as high for frozen fryers as for unfrozen birds. Extensive consumer surveys indicate that ¾ of consumers preferred fresh to frozen fryers at the retail market, but ⅔ of these consumers froze the birds at home. Another study showed that 63% of consumers froze poultry after purchase and before cooking, even if stored for only a few days.

Properly packaged poultry is stable at this temperature for a year and possibly for two years. Thawing is best accomplished rapidly with care to prevent excessive microbial growth.

ADDITIONAL READING

ASHRAE. 1974. Refrigeration Applications. American Society of Heating, Refrigeration and Air-Conditioning Engineers, New York.

BAUMAN, H. E. 1974. The HACCP concept and microbiological hazard categories. Food Technol. *28*, No. 9, 30–34, 70.

DeFIGUEIREDO, M. P., and SPLITTSTOESSER, D. F. 1976. Food Microbiology: Public Health and Spoilage Aspects. AVI Publishing Co., Westport, Conn.

GUTHRIE, R. K. 1972. Food Sanitation. AVI Publishing Co., Westport, Conn.

HARRIS, R. S., and KARMAS, E. 1975. Nutritional Evaluation of Food Processing, 2nd Edition. AVI Publishing Co., Westport, Conn.

JOHNSON, A. H., and PETERSON, M. S. 1974. Encyclopedia of Food Technology. AVI Publishing Co., Westport, Conn.

KOMARIK, S. L., TRESSLER, D. K., and LONG, L. 1974. Food Products Formulary. Vol. 1. Meats, Poultry, Fish and Shellfish. AVI Publishing Co., Westport, Conn.

KRAMER, A., and TWIGG, B. A. 1970, 1973. Quality Control for the Food Industry, 3rd Edition. Vol. 1. Fundamentals. Vol. 2. Applications. AVI Publishing Co., Westport, Conn.

MOUNTNEY, G. J. 1976. Poultry Products Technology, 2nd Edition. AVI Publishing Co., Westport, Conn.

PETERSON, M. S., and JOHNSON, A. H. 1977. Encyclopedia of Food Science. AVI Publishing Co., Westport, Conn.

RAPHAEL, H. J., and OLSSON, D. L. 1976. Package Production Management, 2nd Edition. AVI Publishing Co., Westport, Conn.

SACHAROW, S. 1976. Handbook of Package Materials. AVI Publishing Co., Westport, Conn.

STADELMAN, W. J., and COTTERILL, O. J. 1973. Egg Science and Technology. AVI Publishing Co., Westport, Conn.

TRESSLER, D. K., VAN ARSDEL, W. B., and COPLEY, M. J. 1968. The Freezing Preservation of Foods, 4th Edition. Vol 2. Factors Affecting Quality in Frozen Foods. Vol. 3. Commercial Freezing Operations—Fresh Foods. Vol. 4. Freezing of Precooked and Prepared Foods. AVI Publishing Co., Westport, Conn.

WEISER, H. H., MOUNTNEY, G. J., and GOULD, W. A. 1971. Practical Food Microbiology and Technology, 2nd Edition. AVI Publishing Co., Westport, Conn.

WILLIAMS, E. W. 1976A. Frozen foods in America. Quick Frozen Foods 17, No. 5, 16–40.

WILLIAMS, E. W. 1976B. European frozen food growth. Quick Frozen Foods 17, No. 5, 73–105.

WOOLRICH, W. R., and HALLOWELL, E. R. 1970. Cold and Freezer Storage Manual. AVI Publishing Co., Westport, Conn.

7

Freezing Fish

A. Banks, John A. Dassow, Ernest A. Feiger,
Arthur F. Novak, John A. Peters,
Joseph W. Slavin and J. J. Waterman

Freezing is applied to many widely different products made from species of fish that vary in chemical and physical composition and in initial quality, and that are subjected to widely different processing techniques. Freezing is used on the vessel to permit the landing of high-quality pelagic fish, such as tuna, which are thawed for canning, or a high-quality groundfish, such as haddock, which are thawed, filleted, and refrozen. In the production of breaded and precooked convenience seafoods, such as fish sticks and fish portions, freezing is employed in processing the blocks of raw fish-fillets as well as in processing the final product. Another common application is in the freezing of fish fillets or fish steaks made from fish that have been stored under refrigeration at above-freezing temperatures for varying periods of time.

The beneficial effects realized from the application of the freezing process to a fishery product vary with the basic characteristics of the product and with the manner in which the product is treated throughout all phases of handling prior to, during, and after freezing. An understanding of the effects of handling, processing, and storage on the quality of various types of frozen fish and frozen fishery products is essential for the production and distribution of high-quality seafoods to the American consumer.

This chapter discusses the freezing of fish and fish products. Initially, the nature of the quality changes that take place prior to and after freezing are described; then the many complex factors that influence the quality of the products are discussed. Later sections cover commercial fish freezing in the United States and Europe.

THE QUALITY OF FROZEN FISH

Quality Changes in the Unfrozen Product

When fish die, they tend to undergo chemical and physical changes that adversely influence their quality in the fresh state as well as their suitability for subsequent freezing and frozen storage. This deterioration is due primarily to bacterial and autolytic changes.

Fish muscle is essentially sterile when the fish are landed on the vessel; but, soon after death, psychrophilic or cold-loving bacteria, predominantly of the pseudomonas species, multiply and secrete enzymes that act on the food and produce spoilage. This spoilage is reflected in the development of objectionable off-odors and flavors and of a soft, watery-textured flesh.

The rate at which spoilage occurs varies with the specific characteristics of the product and with the sanitation, handling, and storage methods used. Of the various factors that influence the quality and the shelf-life of the unfrozen fish, temperature of storage is the most important. For example, cod or haddock fillets, which have a maximum shelf-life of 12 to 14 days when stored at 32° F. (0° C), spoil twice as fast when stored at 42° F. (5.6° C.). The effect of temperature on quality has been measured quantitatively. Data suggest that, for cod, $Q_{10} = 3$—that is, for each 10° C. (18° F.) reduction in temperature between 77° F. (25° C.) and 32° F. (0° C.)—the shelf-life of the fish is tripled.

Prior to the freezing of fish, autolysis resulting from the activity of the enzymes naturally present in the fish (as contrasted with those produced by bacteria) takes place relatively rapidly. The rate at which these changes occur varies with the different species of fish and with the treatment employed in handling and storage. If the fish are not eviscerated, autolysis will occur at a much faster rate, particularly in those species that have a large visceral cavity and that are particularly subject to enzymatic activity. In general, fatty species, that is, species having a high lipid content—such as mackerel, herring, and tuna—are more prone to enzymatic activity than are those having a low lipid content—such as haddock and flounder. Autolytic changes are reflected in a marked breakdown of texture, excessive amounts of drip, and rancid-type odors. With the lowering of temperature, these effects of enzymatic action decrease.

Quality Changes in the Frozen Product

Fish in frozen storage undergo changes in flavor, odor, texture, and color. The rate at which these changes occur varies with the handling and processing techniques employed prior to, during, and after freezing.

Bacterial growth in frozen fish is not a problem because, at temperatures below 15° F. (−9° C.), the activity of marine bacteria is largely inhibited and they are essentially dormant. The influence of freezing on the enzymes secreted from marine bacteria, however, has not been thoroughly investigated; these enzymes may adversely affect the stability of the frozen product.

The development of toughness can be a significant problem in fish held in frozen storage. The mechanism of textural changes is not completely understood, but the available evidence indicates that the changes in texture may be associated with the denaturation of proteins.

In a review of studies on fish protein, two principal theories have been set forth to explain the cause of protein denaturation during frozen storage: (1) concentration of solutes, especially inorganic salts, to a point during freezing where they damage the protein in the stored fish, or (2) lipid hydrolysis.

The salt-denaturation theory assumes that exposure to salts, which become concentrated as water is progressively frozen out during the lowering of storage temperature below 32° F. (0° C.), causes direct damage to the protein. Although it is true that maximum damage to protein occurs at temperatures just below freezing, sufficient evidence is not available to indicate that the concentration of salt *per se* is of primary importance.

The lipid-hydrolysis theory comes from an observation that during frozen storage the increase in free fatty-acid content parallels protein damage as measured by inextractability. In pursuing this theory, fish actomyosin in solution is rendered increasingly insoluble during storage at 32° F. (0° C.) in the presence of increasing amounts of dispersed linolenic and linoleic acid. In a more recent study, a working hypothesis has been formulated relating lipid hydrolysis to denaturation of fish-muscle protein induced by frozen storage. The findings from studies of model systems by Anderson and co-workers suggest that electrostatic interaction between protein and fatty acid is the primary factor in protein inextractability.

The work carried out on fish proteins thus indicates that no simple explanation will be found for the changes that take place during freezing and frozen storage and for the relation of these changes to textural changes. Protein damage may be due to a number of different interrelated chemical and physical changes that vary with species and methods of processing, storage, and handling.

Color and flavor changes resulting from the oxidation of fish oils and pigments are also important in fish held in frozen storage, particularly in the fattier species of fish. Atmospheric oxidation of fish oils increases the formation and decomposition of peroxides, which produce acid and carbonyl compounds, many of which have a very unpleasant flavor or odor. These deteriorative changes can be measured chemically as well as organoleptically.

The oxidation of the fish oils in fish flesh varies with the quantity and the type of oil, the highly unsaturated types being less stable than are the others. It has also been reported that oils undergo more pronounced and more rapid deterioration when in the flesh than when extracted from the flesh. These findings, which have been verified by studies on herring and sardine oil, indicate that the mechanism of deterioration in the extracted oil may be quite different from that in the unextracted oil in the flesh.

Although the rate of deterioration of fish oil varies with many different factors, the stability of the oil increases with a decrease in storage temperature. Since low temperature reduces the rate of chemical reaction, the role of low temperature in decreasing the rate of deterioration of oils is understandable. In addition to control of temperature, however, control of atmospheric oxygen is essential to minimize oxidative rancidity. The use of inert atmospheres, vacuum packing, or suitable protective coverings markedly inhibits rancidity development.

Factors Influencing Quality

The quality of frozen fishery products is influenced by many factors. It can be affected by the condition of the living fish, the composition of the fish or fishery product, and the condition of the raw material used for freezing. Freezing considerations are also important. Very slow freezing can result in excessive drip and in an end product of lower quality. Postfreezing handling and treatment also have a marked effect on quality, and care must be taken to minimize oxidative rancidity and dehydration during frozen storage.

This section discusses the more important factors in prefreezing, freezing, and postfreezing treatment that affect the quality of frozen fish products. Particular attention is given to the effect of the basic biochemical properties of the raw material and treatments during handling and storage on the stability and the shelf-life of the frozen product.

Prefreezing Considerations.—The stability of the frozen product is markedly influenced by the biological composition of the product and by the quality or condition of the raw material prior to freezing. Although little documentary evidence is available on the precise role of these factors, recent studies indicate that they are perhaps most important in relation to product stability and shelf-life.

Biological Composition.—To a great extent, the basic biological properties of a particular species of fish determine whether it can be frozen successfully and also determine the amount of protection required to minimize adverse changes in quality during freezing and frozen storage. The deterioration that cod muscle undergoes in frozen storage is influenced by the condition of the living fish; the condition of the fish deteriorates at the spawning season and also deteriorates increasingly as the size of the fish increases. Such findings, which are based on the cell-fragility method, may also have some bearing on organoleptic changes in fish held in frozen storage.

Lipid content markedly influences the storage stability of the frozen product. Rancidity caused by oxidation of fish oils and pigments is a more serious problem with some species of fish than with others. In general, fish with a high oil content are more susceptible to oxidative color and flavor changes than are those with a low oil content. However, the type of oil is also important. Pink salmon containing only 6% oil developed rancid odors and flavors much quicker than do king salmon, red salmon, and coho salmon, which contain about 16, 11, and 8% oil, respectively. Another example is that of sablefish, which develop rancidity at a much slower rate than do lake chub, which contain about the same quantity of oil.

In addition to lipids, the composition and structure of the proteins in fish muscle also influence the stability of the frozen product. The proteins of some species may be more susceptible to protein-fatty acid interaction and to denaturation than are those of other species. Unfortunately, however, little is known of the biochemical properties of fish and how these properties influence their storage stability. A complete study relating storage capability to biochemical composition may provide insight into some of the factors governing the frozen shelf-life of fish.

The relative suitability of different fishes for freezing and frozen storage is shown in Table 7.2. As is indicated by the table the frozen-storage stability of different species varies considerably. Obviously, fish that have low or moderate storage stability require more protection and better treatment during frozen storage than do those that have a high degree of stability.

Condition of Raw Material.—The condition of the raw material to be frozen significantly affects the quality of the frozen product. Particular attention has been directed toward the state of rigor and the organoleptic quality of the fresh product.

State of Rigor.—Considerable attention has been directed toward the state of rigor because of its possible importance in the freezing of fish at sea. Recent reviews indicate that although there may be some differences in the quality of the product, depending on whether the fish are frozen prerigor or postrigor, in general, this factor does not appear to be greatly significant in the freezing of whole fish at sea.

The state of rigor may, however, be more important in the freezing of fillets than in the freezing of whole fish because of the absence of the structural frame for support. Studies indicate that freezing during rigor mortis can adversely affect the quality of frozen-thawed fillets. Cod fillets frozen just after entering rigor were discolored when thawed, and ex-

TABLE 7.1

RELATIVE SUITABILITY OF FISHERY PRODUCTS FOR FREEZING AND FROZEN STORAGE

High Suitability	Medium Suitability	Low Suitability
Haddock	Ocean perch	Mackerel
Cod	Whiting	Tuna
Flounder	King, red, or coho salmon	Catfish
Shrimp	Hake	Sea herring
Halibut	Lake herring	Spanish mackerel
King crab	Red snapper	Pacific sardines
Pollock	Crawfish	Smelt
Scallops	Rockfish	Clams
	Dungeness crab	Chub
	Carp	Chum or keta salmon
	Buffalofish	Whale meat
	Blue pike	
	Yellow perch	
	Swordfish	
	Pacific oysters	
	Alewives	
	White bass	

hibited ragged edges and poor texture. These findings indicate that fish fillets should be frozen prior to, or at the end of, rigor mortis.

Quality of the Raw Material.—A decrease in the quality of raw material used for freezing results in a disproportional decrease in the storage life of the frozen product. If low-quality fish are used for freezing, the initial quality of the frozen product will be similarly low, and the shelf-life of the frozen product will be considerably reduced. In early studies, the quality of fish at the time of freezing had a marked effect on the shelf-life and quality of the frozen product. These findings were confirmed on frozen whiting fillets. Results of study on pollock fillets indicate that the stability of these species in frozen storage drops rapidly as the duration of pre-freezing holding in ice increases. The adverse effect of more than very brief prefreezing holding is proportionally greater in frozen fish that are stored at very low temperatures than in frozen fish stored at higher temperatures. Figure 7.1 shows graphically the great effect of length of preliminary iced storage of pollock fillets upon the allowable subsequent length of frozen storage at +20°, 0°, and −20° F. (−7°, −18°, and −29° C.). The product can be satisfactorily held for well over two years at −20° F. (−29° C.) if the iced holding time was no more than 2 or 3 days; on the other hand the maximum shelf-life was only 8 to 10 months when the iced storage period was as long as 12 to 14 days. The advantage of very low freezer storage temperature is realized only if iced storage is kept to a very few days. At a freezer storage temperature of 0° F. (−18° C.) the shelf-life is in all cases very much shorter than at −20° F. (−29° C.), but at

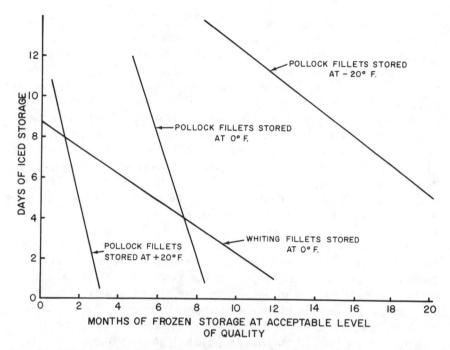

FIG. 7.1. EFFECT OF LENGTH OF PRELIMINARY ICED STORAGE OF POLLOCK FILLETS UPON ALLOWABLE SUBSEQUENT LENGTH OF FROZEN STORAGE

the higher temperature there is proportionately less advantage in keeping the iced storage period very short.

Whiting fillets are even more sensitive to brief iced storage holding than are pollock fillets. Frozen storage quality can be maintained satisfactorily at 0° F. (−18° C.) for a year only if iced storage has lasted no more than a day and whiting fillets held for 8 to 10 days on ice are spoiled and not suitable for freezing.

Freezing Considerations.—The relation of freezing to changes in quality reveals fast freezing produces very small crystals of ice in the muscle, whereas slow freezing produces large crystals. In the early 1900's, the theory was proposed that the large crystals formed in slow freezing mechanically damaged and ruptured the cells in fish muscle and caused increased drip in the thawed product. This theory, which led to considerable speculation of the benefits of fast freezing, is not supported by those who find no evidence of punctured cells during slow freezing. Investigations based on the premise that the rupture of cells will liberate desoxyribonucleic acid (DNA) from the nucleus of the cell showed some correlation of DNA with cell damage, but did not conclusively demonstrate the effects on cell damage of slow or fast freezing.

Work conducted to date has been concerned primarily with the influence of the rate of freezing on biochemical changes in the muscle tissue; very few investigations have been conducted to determine if the rate of freezing affects the quality of the product as judged by organoleptic examinations. A review of the information available indicates, however, that with raw material of high initial quality, a freezing time of several hours to as long as 26 hr. does not significantly influence the quality of the product. Fast freezing is used widely because of its practical advantages in mass producing frozen seafood. Slow freezing is still used, however, for large whole fish such as halibut, tuna, or salmon; and no significant detrimental effects have been noted that are attributable to the rate of freezing.

The method of freezing may affect the appearance or quality of the frozen product. For example, in sharp freezing, a process in which the product is frozen on refrigerated grids or plates, bulging or voids may occur because of the lack of outside pressure to control the expansion of the product. Freezing in an air blast can result in excessive "freezer burn," or dehydration, because of low relative humidity or of the use of air velocities of over 500 f.p.m. with inadequately packaged products. Also, the package will be distorted unless devices are used to control expansion.

In immersion freezing, the solution used must meet FDA approval and must not adversely affect the quality of the product. The temperature of the freezing medium and the length of time that the product is immersed must be precisely controlled; otherwise the solution will penetrate into the fish and may adversely affect the quality of the product.

In immersion or spray freezing with liquid nitrogen or Freon, care must be taken to prevent whitening of the surface of the product, a phenomenon of these refrigerants caused by the low temperatures and very fast removal of heat from the product. The weight loss of fish properly frozen in liquid nitrogen is 50 to 75% less for those frozen in moving air. Ultra-fast freezing to cryogenic temperatures does not, however, appear to have such a beneficial effect on the quality of the frozen fish as is found with certain fruits and vegetables. In fact, freezing to very low temperature ($-290°$ F., $-179°$ C.) can cause irreversible loosening of bound water and denaturation which can influence the quality of the product when subjected to conventional frozen temperatures.

Postfreezing Considerations.—*Temperature*.—In addition to affecting the quality of the product during the freezing process, temperature and time are the most important factors influencing the shelf-life and quality of fish products held in frozen storage. An increase in storage temperature markedly reduces the shelf-life of the product. The effect of various temperatures on the maximum frozen-storage life of pollock and haddock fillets suggests that the maximum shelf-life is about doubled with each 10° F. (6° C.) lowering of storage temperature.

The effects of temperature may differ with different species of fish. With fatty species—such as mackerel and herring, which are especially susceptible to oxidative rancidity—storage at −20° F. (−29° C.) or lower is necessary to obtain a satisfactory marketing shelf-life; whereas with lean species—such as haddock and cod, which are not especially susceptible to rancidity—storage at 0° F. (−18° C.) or −10° F. (−23° C.) is adequate. A time-temperature coordinate system can estimate mathematically the loss in quality of frozen fruits and vegetables subjected to a number of different temperatures for various periods of time. This tool shows considerable promise for use in the estimation of the loss of quality in frozen fishery products. With a knowledge of the rates of quality loss of products at different temperatures, the loss of quality that would result when the product is subjected to certain conditions of time and temperature can be calculated. A typical example in the application of the time-temperature system follows.

Using the maximum storage-life information for pollock fillets shown in Table 7.3, one can calculate the relative rate at which quality is lost at different temperatures. Figure 7.2, which gives the best-fitting curve of the data presented in Table 7.3, shows that as the temperature of storage increases, the relative rate of loss in quality increases significantly. In fact, with each 10° F. (6° C.) rise in temperature of the product, its shelf-life is halved; in the usual notation, $q_{10} = 2$.

The second step in applying the time-temperature system is to construct a diagram such as is shown in Fig. 7.3. The lightly shaded area in Fig. 7.3 reflects the maximum storage shelf-life of pollock fillets under the temperature and time conditions shown. The general concept here is that maximum storage life can be integrated so as to represent a total area as

TABLE 7.2

MAXIMUM FROZEN STORAGE LIFE OF POLLOCK AND HADDOCK FILLETS AT VARIOUS TEMPERATURES

Product[1]	Shelf-life at Storage Temperatures of				
	+20°F. (−7°C.)	+10°F. (−12°C.)	0°F. (−18°C.)	−10°F. (−23°C.)	−20°F. (−29°C.)
	(Weeks)				
Pollock fillets	5	14	32	43	102
Haddock fillets	12	28	37	70	

[1] Fillets were prepared from fish held 1 day or less on ice.   Longer iced-storage will shorten the frozen storage life appreciably.

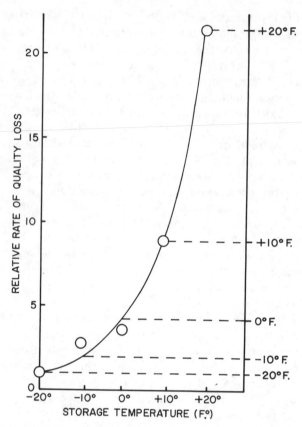

FIG. 7.2. RELATIVE RATES OF QUALITY LOSS IN FROZEN POLLOCK
FILLETS AS A FUNCTION OF STORAGE TEMPERATURE

shown in the diagram. This area remains constant. As the temperature becomes lower the shelf-life increases; conversely, as the temperature becomes higher, the shelf-life decreases. The relation of the area estimated from the known time-temperature history of the product to that for the maximum shelf-life of the product is indicative of the amount of quality lost. For example, if the area under the temperature-distribution curve is 50% of the maximum area for the storage quality, then the product can be assumed to have lost 50% of its quality.

A chart that facilitates conversion of the effect of storage time at a particular temperature to the equivalent effect at another temperature for products have $Q_{10} = 2.0$ is shown in Fig. 7.4. Reference to the figure shows that the loss of quality in a product held for 40 weeks at, say, $-10°$ F. ($-23°$ C.) is no more than the loss of quality in a product held for only 10

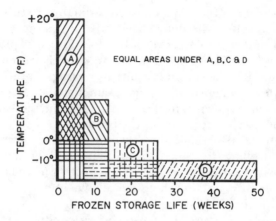

FIG. 7.3. MAXIMUM STORAGE LIFE OF POLLOCK FILLETS AT VARIOUS
TEMPERATURES

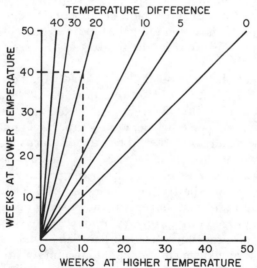

FIG. 7.4. CHART SHOWING EQUIVALENT STORAGE TIMES AT DIFFER-
ENT TEMPERATURES, FOR $Q^{10} = 2$

weeks at $+10°$ F. $(-12°$ C.). Conversion charts for other values of Q_{10} can
be constructed easily.

Protection from Moisture Loss and Oxidative Rancidity.—Frozen fishery
products undergo adverse changes in quality if subjected to contact with
air or to loss of moisture. The air surrounding the frozen product is
usually at a lower moisture-vapor pressure than is the product and there-

fore acts as a sponge in removing moisture from the product. This loss of moisture may result in the dehydration of the flesh to a point where chalky and fibrous texture develops, discoloration takes place, and off-odors and off-flavors develop. Contact of air with frozen fish causes oxidation of the oil in the fish and results in "rusting," or discoloration, of the flesh and development of rancid odors and flavors. These changes are mainly due to a breakdown of the lipids in the fish and they vary with type of fats and chemical composition of the product.

The storage life of frozen fishery products can be increased significantly by controlling the microclimate surrounding the product to minimize dehydration and oxidation. This protection can be obtained by packaging or glazing, by controlling the relative humidity of the storage room, or by combining these methods.

Packaging.—Packages used for frozen fishery products must have a low permeability to moisture vapor, a low rate of transmission of oxygen, and resistance to the absorption of oil and of water. They must also fit the product tightly to minimize air spaces, or voids. Products in a loosely fitting package will lose quality rapidly because of (1) oxidation due to the excessive amount of air surrounding the food, and (2) dehydration due to the migration of moisture from the product to the inside surface of the package.

Ordinarily, fatty fish lose quality more rapidly than do lean fish, and they therefore must receive added protection against the adverse effects of oxidation and rancidity. Conventional waxed cartons overwrapped with films containing polyethylene, waxed paper, cellophane, or combinations of these materials do not protect the product adequately.

Fatty fish can be protected by use of a vacuum-pack with heat-shrinkable bags; the air is removed from the package, and the film is shrunk tightly around the fish by immersing it briefly in hot water or passing it through a steam chamber. Instead of being vacuum packed, the package can be purged with nitrogen to remove the oxygen. Pouches and bags made of polyester films coated with, or laminated to, polyethylene, cellophane, polyvinylidene chloride, or aluminum foil are quite satisfactory as are also those made of certain combinations of these materials.

Another method of packaging fatty fish is to dip them into a protective alginate or other solution, package them, fill all voids with the dipping solution, overwrap the carton, and freeze the product. An alternative method consists of individually freezing the product, packaging it, flooding the cartons with a glazing solution, overwrapping the cartons, and refreezing the product. The above methods have been used successfully for packaging frozen mackerel, smelt, and herring.

The so-called nonfatty products such as haddock, cod, certain shellfish, and precooked fish, must also be protected—principally against the loss of moisture and to some extent against oxidative rancidity. Since the packaging requirements for the lean fish and shellfish are less demanding than are those for fatty fish, conventional waxed chipboard cartons and overwrapping films can ordinarily be used satisfactorily. For best results, fish fillets, breaded convenience items, or steaks should be packaged in tightly fitted waxed cartons overwrapped with a highly moisture-vapor proof laminate made up of combinations of waxed paper, polyethylene, aluminum foil, or cellophane. Individual wrapping of the product with cellophane or polyethylene prior to packaging will also add greatly to shelf-life, at a very low cost. The new cook-in-the-bag package being used for shrimp, fillets, and other heat-and-serve seafoods offers considerable promise. Materials such as polyethylene and combinations of foil, polyethylene, and paper are being used satisfactorily for this purpose.

Very little quantitative information is available on the precise role of packaging in extending the shelf-life of frozen fish. The literature indicates, however, that a good moisture-vapor proof package can double the frozen shelf-life of fish and shellfish that are particularly sensitive to oxidative rancidity.

Glazing.—The function of a glaze is to provide a continuous film or coating that will adhere to the product and retard both the loss of moisture and the rate of oxidation. Glazes are, in the main, applied to large whole or eviscerated fish, although they are sometimes used in conjunction with packaging. The glaze is applied by dipping or spraying.

Although many different types of glazes have been introduced, ice is still the only glaze of any considerable commercial importance.

Many patents have been issued describing the addition of various chemicals to water to reduce the brittleness or the rate of evaporation of the ice glaze. Colloids and thickeners such as Irish-moss extractives, cellulose gum, and pectinates have been used to improve the effectiveness of the glaze. An alginate referred to as "Protan" has been used successfully in the glazing of packaged mackerel fillets.

A glaze containing corn-syrup solids was effective in increasing the shelf-life of frozen salmon steaks. However, the writer has found that water glazes containing sodium alginate or carboxymethylcellulose were not any more effective than was a plain ice glaze in extending the shelf-life of frozen packaged whiting and round haddock.

Antioxidants have been incorporated into dips or glazes in an attempt to protect fatty fish from rancidity. Ascorbic acid and ascorbates, different gallates, and other antioxidants retard the development of rancidity in different species of fish, but the effect varies greatly and is generally unreliable. Sprats glazed with water containing ascorbic acid developed a foreign taste and odor and kept no better than did water-glazed fish. However, the development of rancidity in the fat of frozen rainbow trout can be delayed by dipping them in a weak solution of ascorbic acid.

The present trend is to use packaging materials wherever possible to protect the product from loss of moisture and from oxidation. In fact, pallet-size loads of whole frozen fish are being packaged with large plastic bags to minimize the deterioration in quality. Glazes are still used for some whole halibut, salmon, or fresh-water fish and for fish steaks or portions readily susceptible to rancidity.

Relative Humidity.—Frozen fishery products contain about 80% water and therefore have a relatively high moisture-vapor pressure. As the relative humidity of the storage environment is decreased, the moisture-vapor pressure of the air decreases proportionally, causing a greater difference between the moisture-vapor pressure of the product and that of the air surrounding it. Moisture then migrates from the product to the air until equilibrium is established. The rate at which the moisture is removed from the product is a direct function of the difference in vapor pressure between the air and the product.

Fishery products are stored, transported, and handled under conditions in which both the temperature and the relative humidity may vary considerably. Temperature has a greater effect on the quality of the product than does relative humidity. However, the detrimental effects of a low relative humidity can be important, particularly when unpackaged frozen seafoods are being stored. Whole tuna stored at 0° F. ($-18°$ C.) and at a relative humidity of 70 to 80% for 10½ months lost 20 times as much weight as did tuna stored similarly except at the higher relative humidity of 90 to 95%. Results of tests show that unpackaged seafoods may lose weight at a significant rate when stored at a low relative humidity.

It is therefore important that cold-storage plants used for long-term storage of frozen seafoods be designed to maintain a relative humidity of 90% or higher. The use of a low storage temperature and a high relative humidity will significantly extend the quality of unpackaged and packaged frozen seafoods.

Thawing and Refreezing.—Double freezing—that is, thawing a frozen product and subsequently refreezing it—may also influence the quality of the product in its final frozen form. In comparing refrozen fillets of sea trout with singly frozen fillets, after 10 months of storage at 0° F. ($-18°$ C.) the refrozen fillets exuded twice as much drip as did the singly frozen fillets. However, the palatability of the singly and doubly frozen fish did not differ appreciably.

Refrozen fillets cut from stored, frozen, round, trap-caught cod were equal in quality to the round fish when initially examined prior to frozen storage. During storage at 0° F. (−18° C.), however, the refrozen fillets deteriorated more rapidly than did the singly frozen fish.

Refrozen stored fillets cut from whole haddock that had been frozen prerigor in brine and postrigor in cold air had a shorter frozen-storage life than did the whole stored fish, but the differences were minor.

Refrozen haddock fillets cut from fish stored partially frozen at 28° F. (−2° C.) for 12, 20, and 30 days were of unacceptable quality after 1 month of storage at 0° F. (−18° C.). The refrozen fillets were judged to be tough, dry, rancid, decomposed, ammoniacal, and musty.

The above indicate that if good freezing and thawing procedures are used, the refrozen fish fillets will lose quality at a slightly faster rate than will singly frozen fish. However, if poor freezing, thawing, or storage conditions are used during the initial freezing operations, these adverse changes in quality will be reflected in a significantly shorter shelf-life of the refrozen product. It is therefore particularly important to employ extra care when fish that are to be refrozen are being handled.

Shelf-Life.—The quality and shelf-life of frozen fish varies with basic biological factors inherent in each different species and the manner in which the fish are handled from the time of capture until the time of delivery to the consumer. The freezing time and the freezing method used are of secondary importance, especially if reasonably good commercial methods are used.

With advancements in technology, it has been possible to double-freeze products, thereby producing raw material that is suitable for refreezing into the final processed form. Extra care must be taken in handling these products, since they are more susceptible to damage than are single frozen products.

Quantitatively assessing the role of each factor that affects the shelf-life of frozen-stored fish is an exceedingly complex task because of the many variables involved. Some limited information indicates that the composition of the species, the quality of the fish prior to freezing, the frozen storage temperature used, and the packaging method used are perhaps the most important factors influencing the shelf-life of frozen fish. Of secondary importance are biological considerations within a particular species—for example, environment, age, or method of catching or killing.

The complexity of evaluating the effect of the many different variables influencing quality is further increased by the lack of an objective means of measuring the deterioration of quality in the product. The literature is difficult to evaluate because of the widely different subjective techniques used to determine organoleptic changes. Also, it is almost impossible to rule out preferences of individual taste panels for certain types of fish and fish products.

In some cases, panelists use terms such as "high-quality shelf-life" or "maximum shelf-life" as criteria for establishing recommended storage periods. High-quality shelf-life is usually the period before the product becomes significantly different from the frozen control, whereas the maximum shelf-life is the period before it becomes inedible.

Although the information available is limited, it does nonetheless provide a basis for estimating shelf-life of some of the more important commercial species. The summary presented in Table 7.4 shows approximately what can be expected for the storage life of these products when they are kept at 0° F. (−18° C.).

COMMERCIAL FREEZING OF FISH IN THE UNITED STATES

From the data available on a world-wide basis, production of fish and shellfish exceeds 100 billion pounds, of which the United States produces a tenth on a live weight basis. In terms of the world commercial production of fish and shellfish, the United States is in fifth place. The major producers, in decreasing order, are Peru, Japan, China, and the U.S.S.R.

TABLE 7.3
APPROXIMATE STORAGE TIMES FOR PACKAGED AND GLAZED FISH AND SHELLFISH[1]

Product[2]	Storage Time in Months at 0°F. (−18°C.)	
	(A)[3]	(B)[4]
Fatty fish		
Mackerel	2–3	4–6
Salmon	2–3	4–6
Sea herring	2–3	4–6
Smelt	2–3	4–6
Sprat	2–3	4–6
Trout	2–3	4–6
Lean and medium fatty fish		
Cod fillets	3–4	7–10
Haddock fillets	3–4	7–10
Fish sticks	3–4	7–10
Flounder fillets	3–4	7–10
Ocean perch fillets	3–4	7–10
Plaice	3–4	7–10
Pollock fillets	3–4	7–10
Sole	3–4	7–10
Shellfish		
Shrimp	3–4	6–8
Scallops	3–4	6–8
Clams	2–3	4–6
Lobster (cooked)	2–3	4–6
Oysters	2–3	4–6

[1] The storage times at a designated temperature will vary with the quality of the fish prior to freezing.
[2] Values for fish up to three days in ice before freezing.
[3] (A) Hardly detectable changes in quality occur; product is still of good acceptability.
[4] (B) Very significant changes in quality occur, and product is of low acceptability.

Over the past years, the per capita consumption of fish and shellfish in the United States has remained relatively stable as has the total catch of fish and shellfish. Thus, with increasing population, it has been necessary to rely heavily on imports to meet the demand created by a rising population. About 60% of the fish consumed in the United States consists of products imported from foreign countries. The commercial catch of the United States goes into various marketing forms, such as canned, frozen, fresh, etc. It is interesting to note that about 35% of the domestic landings of fish go into production of fish meal and oil. A rough estimate indicates that over one billion pounds of fish and fishery products are subject to refrigeration during some phase of handling, processing, and distribution. This includes fish that are iced or frozen on the vessel for further processing ashore, and fish and shellfish products processed and distributed in the fresh or frozen condition.

Fishing Areas

The commercial fisheries of North America vary widely with geographical location as to types of products produced and processing methods employed. Each area has a particular specialty for which it is noted. For example, New England is famous for its production of groundfish fillets from haddock, cod, flounders, and ocean perch. Other products include sea scallops, which are marketed in the fresh and frozen state, and canned Maine sardines. Hard and soft shell clams and live lobsters are also specialties of New England.

The Middle Atlantic and South Atlantic are noted for production of food items such as blue crabs, red snapper, oysters, and scallops. The largest single item from this area is menhaden, which is manufactured into fish meal and oil.

The Gulf area of the United States is the heart of the shrimp industry.

TABLE 7.4

1975 TOTAL WORLD CATCH ALL EDIBLE FISH—65 MILLION TONS

	Catch (Million Tons)
Asia	30.0
Russian	9.0
Africa	5.0
U.S. and Canada	5.0
Others	3.5
Total European catch	
EEC countries	5.3
Non EEC countries	7.2

Source: Williams (1976B).

Of secondary importance are red snapper, mullet, oysters, and blue crab.

The Great Lakes region is noted for its yellow pike, chubs, carp, catfish, and other fresh water species. Production of fresh water trout by fish farming methods is a rising business in the Midwest, and many of the fish are sold live or as fresh or frozen product.

Southern California is noted for its landing of tuna, an item which has achieved prominence as a canned food. Halibut, salmon, groundfish, and dungeness crab come from the Pacific Northwest.

Alaska is famous for its king crab and salmon. Untapped groundfish resources and large unexploited populations of small shrimp are just part of the vast potential of the Alaskan fisheries.

A large number of diversified fishery products is produced in the United States for human consumption. These include fresh and frozen fish fillets, dressed fish, raw breaded and precooked fish sticks and portions, fish pies and fish dinners, breaded or raw shrimp, and salmon and halibut fish steaks; frozen, fresh or canned crab meat; cured and smoked fish; canned shrimp, sardines, tuna, salmon and other canned items; and shellfish such as oysters and lobsters, either whole or in varying product forms. In recent years some attention has been given to production of convenience items, includng freeze-dried fishery products, frozen fish dinners, and specially packaged heat-and-serve products such as shrimp in a boil-in-a-pouch package. Hopefully, in the future more attention will be given to producing and marketing fish in new, different, and more appealing forms.

Catching, Handling, and Freezing Fish on the Vessel

Fishing vessels used in the United States include Pacific trawlers and combination vessels, halibut schooners, tuna clippers and seiners, Atlantic beam trawlers, and Gulf shrimper and snapper vessels. These fishing craft may range in length from 25 to 200 ft.; the tuna seiners are the largest in operation for the United States. In Canada several larger freezing trawlers, over 200 ft. in length, are employed to freeze and process groundfish at sea. These trawlers are similar to those used in Europe for freezing the catch at sea.

Methods of Catching Fish.—In the American fisheries fish are caught by hook and line, by purse seine, or by trawling, either on the bottom or in the midwater area. Pound nets, drift nets, and gill nets are also used to catch some species of fish. Hook and line fishing is practiced commercially in the Gulf of Mexico red snapper fishery and in some of the smaller inshore fisheries. Pole and line are used to catch tuna and mackerel-like fishes, but in recent years hook and line techniques used in the tuna fishery have been replaced with purse seines for the most part. Trolling with hook and line is used in the Pacific salmon fishery. Another type of hook and line is the longline,which consists of a line to which a number of smaller lines carrying baited hooks are attached. Longlining was developed by the Japanese, and is now used extensively for harvesting halibut and Atlantic tuna. The purse seine is used in salmon, herring, menhaden, and tuna fisheries. Seines carried on some of these vessels may be extremely large, ranging to 500 fathoms in length and 45 fathoms in depth. One seine may contain as much as 80,000 lb. of fish.

Trawling is used to harvest groundfish such as haddock, cod, flounder, pollock, and ocean perch. The otter trawl, commonly found in New England and Canada, consists of a flattened conical bag made of netting. The mouth of the bag is held open horizontally by two large doors which are attached to the mouth by chain bridles. Floats hold the mouth of the net open and wooden rollers keep the bottom of the net slightly off the ocean floor. Trawling is conducted in shallow areas and on the continental shelf and slopes to depths of about 500 fathoms. The trawl is towed behind the vessel at a speed of 2 or 3 m.p.h., for 1 or 2 hr., the length of time depending on the fishery; then the net is hauled up and the fish are dumped on deck. The midwater trawl has been used in recent years to harvest schooling fish, such as Pacific hake, which are available in the midwater areas. A telemetering device is used to measure the height of the trawl off the bottom. The depth of the trawl can be controlled by varying the speed of the vessel and the length of the trawl wire.

Maintaining the Quality of Fish on the Vessel.—As discussed previously, postmortem changes in fish caused by enzymatic and bacterial activity will proceed at a fast rate unless the product is refrigerated as soon as it is landed on the vessel. It is generally known that chilling delays and minimizes spoilage and that the ideal chilling system cools fish rapidly to wet ice temperatures. In addition it is essential to minimize bacterial contamination of the fish during all stages of handling on the vessel. Dirt should be washed off the fish as landed and the surfaces with which the fish come in contact be maintained in a clean condition. Care must also be taken to wash the fish after gutting it, and to use clean ice.

Icing of Fish.—Ice is used to chill groundfish, halibut, snapper, and in some cases, tuna on the vessel. Fish properly iced will cool rapidly and will retain quality for 1 to 2 weeks, depending on the species. When adequately iced, medium-sized cod and haddock can be chilled to 31° F. to 32° F. (−0.6° C. to 0° C.) in 3 to 7 hr. Since spoilage may be advanced twice as fast at a temperature of 37° F. (3° C.) as at 31.5° F. (−0.3° C.), it is essential that the fish be cooled to ice storage temperatures quickly.

Two types of ice are used in the United States and Canada—flake ice, which is rather finely ground, and crushed-block ice, which consists of large irregular shaped pieces. Fishing boats operating out of New England use crushed-block ice almost exclusively. In recent years, Canadian boats operating out of the Maritime Provinces have turned from crushed-block ice to flake ice. The melting rate of flake ice is higher than that of crushed ice because of its increased surface area per pound of ice. It is claimed by some that this offsets the disadvantage of rupture of the fish due to the coarser pieces of crushed ice.

Ice to be effective must be clean when it is used aboard the vessel. Some bacteriological tests on ice in the hold of a fishing vessel show bacteria counts to be as high as 5 million bacteria per gram of ice. These findings indicate that chlorinated or potable water should be used in making the ice at the ice plant and the ice should be stored under sanitary conditions. It is important to discard unused ice from the vessel at the end of each trip.

In New England a ratio of 1 part ice to 3 parts fish is commonly used, whereas in Canada the ratio is 1 to 2. Groundfish, such as haddock and cod, are eviscerated and washed with running sea water before being iced in the hold. In summer months the gills are traditionally removed to delay spoilage. Smaller fish, such as ocen perch and flounder, are not eviscerated but instead are rinsed with a hose and iced directly in the hold of the vessel. Methods of icing vary with construction of the fish hold and the pen layout for the particular vessel. On most vessels employing bulk storage the following method of icing will result in proper maintenance of quality. The floor of each pen is first covered with a layer of ice 8 to 12 in. deep and a similar quantity is placed along the sides. A layer of fish not over 4 to 6 in. deep is then placed on the bed of ice and covered by an 8-in. layer of ice. Layers of ice and fish are built up in the same manner until they reach a height of four feet. At this point, a shelf consisting of aluminum or wooden pen boards is installed to prevent crushing of the bottom layers of fish. The layers of ice and fish are then built up on the shelf as previously described. Properly iced groundfish may be kept for as long as 12 to 14 days.

Work conducted in Europe indicates that the quality of iced fish can be extended by approximately three days by storing the fish in individual boxes with sufficient quantities of ice. This containerization minimizes bruising, eliminates excessive weight on the fish and facilitates unloading of the catch. Fish are stored in boxes on some small draggers operating out of New England and on vessels operating in the Great Lakes region; however, this method is not in general use.

In icing halibut, care should be taken to pack both the belly and gill cavities with ice. This will permit the water from the melting ice to flow away from the fish. If the drain water is allowed to accumulate around the fish the halibut will smell sour. Refrigerated coils located under the deck of the fish hold may be used to keep the ice from melting while the vessel is enroute to the fishing grounds. Very few vessels in the United States are outfitted with such coils because of the short distance to the fishing grounds and unsatisfactory results in maintaining the quality of the catch. Attempts to use mechanical refrigeration to supplement ice have sometimes resulted in spoilage of the entire catch of fish, apparently because the ice melting-rate was not sufficient to cause melt water to flow over the fish at a rate high enough to wash bacteria away, and to cool the fish adequately.

Storage of Fish in Refrigerated Seawater.—Research findings indicate that halibut, groundfish, shrimp, and herring spoil just about as fast in refrigerated sea water as in ice. Ease of handling, reduction of weight losses, and elimination of bruising are advantages to be considered in short-term storage in refrigerated seawater. In all cases it is important that adequate seawater circulation be maintained and that the temperature be kept at 30° F. (−1.0° C.). Also, since bacterial build-up in tanks and in connecting pipelines can be a problem, it is essential that a refrigerated seawater system be designed for easy cleaning. Chlorine should be used to clean the entire system after each trip.

It should be noted that refrigerated seawater is commercially employed in the United States for preserving fish that are to be used as canned seafood or in industrial products. It is used on salmon trollers and for storing and transporting large quantities of salmon aboard barges or cannery tenders where previously little or no refrigeraion was employed. Many trollers have reconverted to icing because of some industry dissatisfaction with chill water systems. Problems arose from inadequately designed units or from improper operation of refrigeration equipment.

Refrigerated seawater installations for chilling menhaden and other industrial fish on the vessel are increasing in number. Previously these fish have been handled without refrigeration and were usually processed within a day of capture. With the advent of longer trips, however, it is now necessary to chill the catch; refrigerated sea water is an ideal medium for cooling large quantities of fish and for efficiently unloading these fish. It is usually a simple matter to pump the fish from the sea water tanks into the plant. Refrigerated seawater has also been used by Canadian firms for storing halibut on the vessel. This practice is being discontinued because of quality problems.

Freezing Fish at Sea.—Technical requirements have been established for the freezing of fish at sea and for thawing these fish for reprocessing ashore. Methods in use include moving air, contact-plate, and immersion. Types of fish and shellfish frozen at sea include: groundfish (such as haddock and cod), salmon, tuna, king crab, shrimp, sea bream, and other species. Commercial application of freezing at sea has proceeded faster in Europe than in North America. The Soviet Union, the United Kingdom, West Germany, Poland, and other nations have large numbers of vessels equipped (1) to freeze whole fish for later thawing ashore, and (2) to process and freeze the catch at sea. The modern factory vessel contains fish processing equipment for filleting, packaging, and freezing the fish, as well as fish reduction facilities. A typical factory vessel is outfitted with stern trawling gear for harvesting the catch, and has a storage capacity of over one million pounds of frozen fish.

In the United States freezing at sea is limited to those fisheries where long trips make preservation by other means difficult or impossible. Considerable work has been done by researchers in the United States and in Canada on freezing and thawing of groundfish at sea but commercial application is limited. In recent years, Canadians have built several trawlers designed to freeze the catch at sea. In the United States two large factory vessels are being built to process and freeze fish at sea. Results of these commercial endeavors will to a great extent determine the course of future action in freezing of groundfish at sea on the North American continent.

In the United States, freezing at sea is used for tuna, salmon, king crab, and more recently shrimp. Tuna freezing is the largest single application, as evidenced by the fact that over 250 million pounds of tuna are frozen at sea aboard U.S. fishing vessels. Freezing of tuna in brine began in the mid-1930's when it was found that vessels could not carry sufficient crushed ice to land fish in high-quality condition. Commercial application of freezing at sea increased because of the need for the tuna industry to go farther and stay at sea longer to land a profitable catch.

Methods used to freeze tuna at sea involve typical freezing systems. The tuna are frozen in brine wells lined with galvanized pipe coils on the sides and top. A large vessel may contain 12 or 14 brine wells with a total fish capacity as high as 450 tons. In handling the catch, the fish are loaded into the wells through a deck hatch or manhole and stored in chilled seawater until the well is full. Storage density varies from 45 to 50 lb. of fish per cubic foot of space. The dirty seawater is pumped overboard and a fresh solution pumped into the well. Salt is then added gradually and mixed by the brine circulating pumps until a concentration of 10 to 15% is attained.

The time required to freeze tuna varies considerably depending on the size of the fish, the capacity of the refrigeration system, and the manner in which the brine system is operated. In general practice, freezing times range from 48 to 72 hr. or even longer for large-size fish or poorly operated brine systems.

After the fish are frozen to a temperature of 15° F. to 20° F. (−9° to −7° C.) the brine is pumped overboard and the tuna kept in the storage wells, which are maintained at approximately 10° F. (−12° C.) by direct expansion of ammonia in coils lining the inside of the wells. Prior to unloading at the cannery, the fish are partially thawed by adding seawater to which salt is added to prevent freezing. This brine solution is recirculated over the fish to permit thawing. In freezing tuna, the fish may be held in the wells for excessive periods of time prior to freezing, or be frozen at a very slow rate, causing the small fish to absorb excessive amounts of salt. During recent years, the quantity of small tuna such as skipjack has increased considerably. It is essential that proper freezing and handling techniques be employed with these smaller fish because they will not withstand the abuse that the larger size fish can take without excessive damage.

A brine spray system is used on some of the smaller tuna vessels. In one system, the fish hold is lined with refrigeration coils which are cooled by direct expansion of ammonia. Brine is then sprayed on these coils until a layer of ice about one foot thick is built up. When the fish are put into the tanks the compartment is flooded with brine which is cooled by the coils and recirculated by a pump. In a similar system used by the Canadians, the brine is pumped from the bottom of the hold through a heat exchanger and sprayed over the fish. The brine is recirculated until the fish have been cooled to their freezing point. In general, the above systems are not very satisfactory because of difficulty in cooling the fish uniformly and rapidly. For example, the fish are usually piled one on top of another, thus making cooling by spray very slow. It should be noted that these systems are used on vessels that make only short trips and are not used on vessels that stay at sea for periods of 1 to 3 months.

Freezing salmon at sea was adopted on a commercial basis on vessels after World War II so the vessels could make longer trips and land higher quality fish. Fish were frozen in tanks or wells in a sodium chloride brine. Usually, the tanks were located on the decks of the vessels and after freezing the brine would be pumped to another tank, and the fish unloaded through the manholes in the tanks, and manually transferred to the refrigerated hold. With introduction of refrigerated seawater, however, few if any salmon are frozen in brine on vessels today.

No substantial quantity of trawl or bottomfish is frozen at sea in the United States or Canada. However, results of research by American and Canadian investigators show that groundfish can be satisfactorily frozen at sea by brine immersion or by contact-plate. In the *Delaware* experiment, it was found that cod and haddock could be satisfactorily frozen in a 23% sodium chloride brine solution, thawed in water, filleted and refrozen, and marketed as packaged fillets. The final product was quite satisfactory and differed very little from singly frozen fish. Experiments by Canadian and American workers indicate there is little difference between the quality of cod frozen in brine or with contact plates and subsequently thawed in water or by microwave energy. It was noted, however, that some loss in quality did take place due to the second freezing. Also, improperly freezing the product initially at sea results in an inferior product for thawing and refreezing. It is, therefore, essential in freezing fish at sea that the freezing be carried out under carefully controlled conditions which permit maintenance of the highest quality product.

Preparation and Freezing of Selected Fishery Products

The major raw fish products produced under refrigeration include fish fillets, fish steaks, and dressed fish. Following is a discussion of methods of preparing these products and of equipment used for freezing.

Production of Fish Fillets.—Filleting of fish was introduced in 1921 by Dana Ward in New England. Since that time, the industry of producing fresh and frozen fillets has grown significantly. Annually over 60 million pounds of fillets are produced from groundfish such as cod, haddock, flounder, sole, and ocean perch.

A fillet is a piece of flesh cut away from either side of the fish along the backbone behind the visceral fin down to the tail section of the fish. If that portion of the flesh that lines the visceral cavity remains with the fillet, the fillet is called a full nape fillet. Most high-quality fillets do not contain the nape.

Fresh fillets are marketed with skin on or off, or as a butterfly fillet in which the fillets adhere by the upper tough skin of the belly as in whiting.

Fish used in the manufacture of fillets such as haddock are eviscerated after capture, on the vessel, and are stored in ice in the fish hold. When the boats dock the fish are unloaded from the hold in canvas baskets which hold approximately 100 lb. of fish. Methods of handling vary depending on the port and plant in question. In Boston, where considerable quantities of fish fillets are produced, the fish are emptied from the unloading baskets to weigh boxes located on platform scales at the dock. After the fish have been weighed, they are pushed from the weigh box into 500-lb. capacity tote boxes for transport to the processing plant. At other ports, fish may be emptied onto a conveyor where the ice is removed and the fish are conveyed to a weigh box on a scale. When the proper weight is reached the weigh box door is opened manually and the fish are emptied onto another conveyor for transport into the plant. This type of arrangement exists at the larger plants which have dock facilities where the vessels can be unloaded.

Smaller species of fish such as ocean perch are unloaded from the vessel into cylindrical de-icing devices made of heavy metal mesh. As the fish and ice tumble toward the lower end of the unit, the ice is washed through the sides of the machine, the de-iced fish then fall onto a conveyor and are conveyed to weighing boxes. After being weighed, the fish are conveyed by belt directly into the storage bins in the shore processing plant where they are iced.

The processing operations involved in production of chilled and frozen fish fillets consist of washing, scaling, sorting, inspecting, filleting, skinning, brining, packaging, and weighing. In a small plant many of these operations are performed by hand. In larger plants equipment is available for scaling, filleting, skinning, and weighing.

In a typical fish filleting plant, haddock or cod are dumped into a tank of running chlorinated water in which they are washed to remove slime, blood, ice, etc. An automatic conveyor removes the fish from the wash tank and transfers them to a hinged wooden box which holds approximately 500 lb. of fish. A layer of crushed ice is placed on the bottom of the box; another is added when the box is one-half filled; and, finally, when the box is full, the fish are covered with ice. Then the boxes are trucked to a refrigerated room where they are held until they can be prepared for freezing.

When required for filleting, the boxes of similar sized fish are removed from the refrigerated room, and the fish are emptied through the hinged door in the side of the box into the hopper of a scaling machine in which the scales are removed and the fish is washed with seawater. From the scaler, the fish pass through wash tanks and thence to manual filleters or to the filleting machines. The waste is conveyed to the reduction hopper. The fillets are conveyed to skinning machines and then to a brine tank where they are immersed for a short period of time in a solution of sodium chloride containing a few parts per million of sodium hypochlorite.

The brined fillets drain as they are conveyed to plastic or stainless steel pans in which they are carried to packing tables; there they are packaged.

The kind of scaler used will depend upon the type of fish. Scalers for ocean perch and whiting consist of a revolving expanded metal or wire mesh drum; they are capable of processing as much as 15,000 lb. of fish per hour. Species such as cod and haddock are scaled in a device in which the fish are drawn head first against a high-speed rough metal wheel. This scaler will handle about 4,000 lb. of fish per hour. Both types of scaler require sufficient water to wash away the loosened scales.

Machines are available for filleting haddock, cod, pollock, ocean perch, and herring. In the Baader machine, which is used for haddock and cod, the fish are grasped by the tail mechanically and are guided past two rows of knives. As the fish pass across these knives, the fillets are removed. Most filleting machines can be adjusted so as to cut the fillets either with the nape on or with it off.

Skinning machines are in common use in most large fish processing plants. The skin-on-fillet is usually conveyed to the machine where the skin is removed by a rotating series of small knives. The cut can be adjusted to control yield of the fillet.

Brine is used in some commercial plants to assist in lowering the drip of fish fillets. Although this method has been used for many years, it does not appear to have any marked advantage. In fact, in many cases, the brine may serve as a source of bacterial contamination of a fish fillet. Also, brine dipping time is usually difficult to control in commercial tanks.

Fish fillets for distribution as fresh, unfrozen, are packed in 10-, 20-, or 30-lb. capacity metal tins. The fillets are wrapped in 1- or 2-lb. lots with cellophane and placed in the tins, or are just laid in the tin and covered with cellophane on top.

Fillets for freezing are usually packed in 1-, 5-, or 10-lb. packages. In the 1-lb. pack, the fillets are placed into waxed chipboard cartons which are overwrapped in a suitable moisture-vapor proof material. Fillets to be packed in larger lots are wrapped in 1- or 2-lb. lots with cellophane and placed into waxed chipboard cartons.

In recent years, considerable quantities of fish have been used in production of fish blocks. In preparing these blocks, boneless and skinless fillets are placed in waxed chipboard containers, either parallel to or perpendicular to the length of the container. The thick portion of the fillet is usually placed adjacent to an edge of the container and the thin portion placed in the center, which is built up with fillets until the desired weight is obtained. Most processors add an extra ¼ lb. of fillets over and above the intended net weight to assure against deformation during freezing. The fillet blocks are frozen in multi-plate compression freezers, using spacers ³/₃₂-in. smaller in depth than the height of the container. This results in a compact, uniformly made fish block suitable for production of fish sticks and portions.

Production of individually quick-frozen fillets is a more recent development in fish processing. These fillets are frozen individually in cold air on a conveyor belt or in liquid nitrogen. After freezing, the fillets may be glazed and packaged in 5- or 10-lb. packages for distribution to the market. Individually quick-frozen fillets have the advantage that the entire pack does not have to be thawed for serving individual portions.

Preparation of Frozen Steaks.—Halibut and salmon, which are caught mainly in the Pacific Northwest and in Alaska, are trimmed and washed thoroughly after being landed, and they are frozen and stored as whole glazed fish. These whole fish are withdrawn at intervals as orders are received. Steaks or fillets are cut from either the hard-frozen fish or the partially-thawed fish, and the cut product is packaged, refrozen if necessary, and returned to cold storage.

Steaking Frozen Fish.—In steaking frozen halibut, the usual starting material is dressed head-off fish. After the halibut are removed from frozen storage, the dorsal and the ventral fins are shaved away with a large, sharp knife. The halibut are then steaked one at a time with a hand saw. With the initial cut, 2 or 3 in. of gristle is removed at the nape of the neck. The second cut removes belly and nape as one unit, and the third cut separates this unit into two pieces. The belly piece is conveyed or placed into a scrap box; the remainder is cut into steaks ¾ or ⅞ in. thick.

If three sawyers are working together, a second sawyer receives the belly piece, trims away the fins and the thin belly wall that is less than one inch thick, and cuts steaks from the remainder. Steaks from the first and second sawyer are passed along to a third sawyer who dices or cuts them into serving-size pieces suitable for packaging. If there are only two sawyers, the second sawyer trims and steaks the belly piece and dices all of the steaks. Sometimes a number of belly pieces are accumulated, trimmed, and steaked by the first sawyer. Trimmings and sawdust from the steaking operation are used for pet food or for mink food. The yield of salable halibut steaks from dressed frozen halibut is about 70% of the weight of the dressed halibut.

Halibut Packaging.—After the steaks have been cut into serving-size pieces, they are dropped into a glazing tank containing cold water, from which they are lifted by an inclined mesh belt and delivered to the packing table. Some glazing operations employ a spray of water above and below the mesh belt to obtain a thicker glaze.

Waxed cartons for packaging the steaks come in flat cut blanks and are formed by hand, by hand-operated machines, or by automatic machines. They then are either carried in large boxes to the packers or are conveyed along the packing table on a supply belt. The packer takes cartons from the supply belt, places them on the scale pan, fills them with pieces of halibut steaks to a weight of one pound, and then places them on a conveyor belt running to the lid-closing table, where they are closed by hand. These cartons of necessity must be large enough to hold odd-shaped frozen pieces with crowding, consequently they contain a considerable amount of air space. The closed cartons then are fed either automatically or by hand to the machine that overwraps them with printed waxed paper and heat-seals the paper. Wrapped cartons are packed in fiberboard shipping cases and are returned to storage at 0° F. (−18° C.) or lower. Institutional-size packs of steaks are packed in fiberboard boxes that have a capacity of 15 lb. These boxes or cases are lined with heavy waxed paper, and a sheet of this paper is placed on top of each layer of steaks. When the box is full, the lining paper is folded over the top layer of steaks, and the box is glued shut. If the box is of two-piece construction it is strapped shut.

Salmon Steaking.—Salmon are steaked with the same type of equipment used for halibut and are handled in much the same manner. In steaking salmon, the first saw is used to trim away the fins, cut off the head behind the gills, and cut off the collar or nape piece. The second saw is used to cut steaks right down to the tail piece. Tail steaks too small for a single small serving are used in animal food. For institutional packs, however, the tail piece is filleted and is packed with the steaks. Salmon steaks ordinarily are not large enough to require dicing or cutting into individual portions.

Salmon are packed in retail-size and institutional-size cartons in the same way as are halibut steaks. The yield of salable steaks ordinarily is about 70% of the weight of the dressed heads-on frozen salmon, but it may exceed this value.

Steaking Thawed Fish.—It is the practice in some plants to pack new supplies of steaks by partially thawing frozen dressed halibut that have been held in storage at 0° F. (−18° C.) or colder. The frozen halibut are brought out of storage; the dorsal and ventral fins are shaved off with a sharp knife; and the halibut are thawed in air, on the floor of the fish house, or in circulating water.

The first steps in preparing slacked halibut for steaking are washing and trimming. The halibut is scraped inside and then is scrubbed inside and out with water. The thin belly flaps are cut away, and the nape is cut off back to where a good slice can be obtained. The washed and trimmed halibut then are transported to the steaking plant or room.

Slices ⅞ in. thick are cut one at a time on a specially constructed slicing machine. The tail portion of the halibut is collected, along with the nape pieces from the small halibut, filleted, and packed in fillet cartons for freezing.

As the steaks leave the slicing machine, they fall onto a conveyor and are washed under a heavy spray of water. Continuing on the conveyor, they are allowed to drain before they fall from the end of the conveyor into a stainless-steel rotary briner. Here they are brined, like fillets, as they travel around a complete circle. After the steaks leave the briner, they are conveyed to the dicing table where they are cut into pieces for packing into cartons of one pound capacity. Only steaks from large halibut require extensive cutting to obtain pieces of proper size.

The cut steaks then are conveyed to the inspection table, where each steak is candled over a frosted glass plate illuminated from below by strong lights; this helps to disclose any imperfections or parasites present in the flesh. The imperfections are trimmed away with a pair of shears, and the inspected steaks are put into pans and are trucked to the packing tables where the steaks are packaged in waxed chipboard cartons. Freezing under pressure causes the tightly packed halibut steaks to fill the voids in the carton, thus eliminating air pockets. The resulting product is a completely filled carton having flat surfaces.

Commercial Methods of Freezing Fishery Products

Refrigeration machines used for freezing fishery products in the United States vary in design from rather simple batch-freezing units which require considerable labor in product handling, to automatic loading and unloading units that utilize mechanical or electronic controls to regulate operation in accordance with the requirements of the product. These machines usually are classified according to their general physical characteristics as being of the sharp, air-blast, contact-plate, or immersion types. A brief discussion of representative commercial types of freezing equipment and typical product freezing rates follow.

The Sharp Freezer.—Fish frozen by this method are placed directly on the shelves or on aluminum pans or plates covering the pipe coils.

The sharp freezer is presently limited to round or dressed fish such as halibut or salmon; panned fish such as whiting, mackerel, or herring or in some cases institutional 5- and 10-lb. packages of fillets or steaks. Consumer-size packages of fish fillets undergo bulging because of the absence of a method of controlling expansion during freezing.

The rate of freezing in commercial installations is quite low. At evaporator temperatures of $-5°$ F. to $-20°$ F., $(-20.5°$ C. to $-29°$ C.) 14 and 16.5 hr. are required to cool 2- and 2½-in. thick packages of fish fillets, respectively, from 50° F. to 0° F. (10° C. to $-18°$ C.). Faster freezing can be accomplished by using lower evaporator temperatures or by circulating cold air over the products.

Excessive handling of the product is another disadvantage of the use of sharp freezers. As much as 3 hr. are required for loading and 3 hr. for unloading 40,000 lb. of fish fillets. Infiltration of air and accumulation of frost on the shelf coils during loading and unloading are problems. In a New England plant, handling requirements and air infiltration have been reduced by utilizing conveyors to carry the product from the processing room into the freezer, and from the freezer to the cold-storage room. Some labor is still required, however, to transport the product from the conveyor to the freezer shelves and from the shelves to the conveyor.

Air-Blast Freezers.—The freezers are usually fully loaded at one time by rolling or pulling a rack of shelves of fish into the insulated room. More recently, conveyors have been used to move fishery products continuously through the blast room or tunnel. Most freezers of this type operate at air temperatures of −30° F. (−34° C.) or lower. The velocity of the air moving over the product generally varies between 500 and 1,000 f.p.m. to give the most economical freezing. Lower air velocities result in slow product freezing, whereas at higher velocities the freezing cost per pound of fish frozen per hour increases considerably.

Blast freezers are used for freezing such fishery products as shrimp, fish fillets, steaks, scallops, or breaded precooked products packed in institutional-size packages; round, dressed, or panned fish; and shrimp, clams, or oysters packed in metal cans. More recently, they have been used for freezing fish fillets wrapped in moisture-vapor proof, heat-sealable packaging films and for freezing unpackaged fillets, shrimp, and precooked fish sticks. Unpackaged fish sticks frozen in a conveyor type freezer are packaged automatically after freezing to further reduce handling cost.

Dehydration of product, or "freezer burn", may occur in freezing unpackaged whole or dressed fish in blast freezers unless the velocity of air is kept to about 500 f.p.m. and the period of exposure to the air is controlled. Consumer-size packages of fish fillets or fish-fillet blocks requiring close dimensional tolerances undergo bulging and distortion during freezing.

The Contact-Plate Freezer.—Contact-plate freezers in commercial use in the United States are of three types: batch, semiautomatic, and automatic. The batch freezer is best suited for an operation where a number of differently sized products are to be handled. The semiautomatic and automatic freezers are used for freezing very similarly sized packages of fish in large quantities. These freezers are customarily used in the production of precooked, specialty-type seafood products (see Chapter 2).

FREEZING FISH IN EUROPE

Because of wide fluctuations in supplies, the fish industries of most major producing countries have always relied heavily upon long-term preservation to even out distribution of their products. It is not surprising, therefore, to find that preservation by freezing and cold storage, in its early stages of development, was tried out as an alternative to traditional forms, since *a priori* it offered a means of preservation without radical changes in basic qualities. In Europe, these trials took the form of brine-freezing certain classes of fish at sea, for example, halibut by British concerns, flat fish and other fish not suitable for salting by French companies fishing the Newfoundland Banks, and cod and other species by Germans. Unfortunately, these pioneering efforts did not meet with lasting success, probably for technico-economic reasons. Attempts in Britain in the early 30's to brine-freeze and cold store part of the prolific catches of herring were also unsuccessful because of the development of rancidity in the products.

Subsequent research in various establishments throughout the world led to the identification of the basic principles of good practice, in particular the need to use a method of freezing other than brine-freezing and the need to employ temperatures of storage well below 0° F. (−18° C.) for proper preservation. Prior to 1939 there was, however, little commercial development and although greatly increased supplies of frozen fish were made available in Europe during the war years, notably from North America, Iceland, and Norway, quality in many cases was adversely affected by delays in transport. Consequently, there was a build-up and strengthening of prejudices against frozen fish which can be encountered even today, and subsequent commercial developments were thereby held up, particularly with regard to the freezing of fish for further processing.

After 1945, the industries in various countries were encouraged to develop the production of frozen fish so as to conserve catches and thus to offset shortages of other foodstuffs. As a result, the freezing of fish assumed some importance in national economies. A noteworthy feature, too, was the adoption of modern techniques in freezing and cold storage, highlighted by earlier research.

Although at first, therefore, the main objectives of commercial developments were to make a fuller use of the prolific catches available at the time, the general pattern of freezing operations soon began to assume the form common in North America, and increasing quantities of consumer packs of fillets and other products are now being produced for sale from refrigerated cabinets.

From 1945 to the present, fillet production in Europe more than doubled, and this refers only to the production of frozen fillets of demersal (i.e., bottom-dwelling) fish. Considerable quantities of other fish and other forms of fish are also frozen. This does not include statistics of the considerable production of frozen fish in the U.S.S.R. However, production in Denmark, Norway, and the U.K. is still increasing, while that in Iceland would seem to have reached a peak. It is noteworthy that the production of frozen fillets on board German trawlers now exceeds production on shore.

General Principles

It is now well established that the better the initial condition of fish, as regards both microbiological condition and freshness, the quicker the freezing, within limits, and the lower the temperature of storage, the better the quality of the frozen products and the longer they can be cold stored. Two types of product can be reçognized: one, frozen for further processing, for example, for thawing, filleting and distribution as wet fish, or for refreezing in some other form; and the other, frozen in a consumer pack which is thawed before or during cooking by the consumer. The first type of product requires more stringent treatment. For example, if firm fillets of good appearance are to be obtained from thawed frozen whole white fish, then the fish must be frozen within 1 to 4 days of catching, depending on species, and kept well iced during this time; in addition, the freezing process must be quick and efficient. High-quality packs of frozen fillets can be obtained from white fish iced for seven days after catching and products that are acceptable in certain countries can be obtained from fish frozen after longer periods in ice than this. Although the effects of rate of freezing are difficult to detect on the basis of the texture of the cooked fish, an unduly long time spent in freezing will clearly affect flavor, and it is thus customary to use an efficient and quick process for the freezing of consumer and industrial packs of fillets. Holding temperature is of critical importance for both types of products; a temperature of $-20°$ F. $(-29°$ C.) is recommended to preserve the texture of frozen whole fish so that they can be satisfactorily processed when thawed, and to preserve the texture and flavor of fillets.

The freezing and cold storage of fish in most European countries is operated under Codes of Practice or in some cases under regulations. These codes may be government-controlled or controlled by trade associations. In adition to specifying good practice for freezing and storage, other factors, for example, pretreatment of the fish, bleeding and icing, and humidity in cold stores, may be covered by the codes. In general, the codes fall short of the recommendations that have been made from time to time by research establishments, but this is only to be expected since time is required for industry to gear itself to the best practice. There are indications that the codes may be modified from time to time to keep in step with industrial developments.

Plant and Equipment

For the most part, the freezing of fish in Europe is based on apparatus specially designed for freezing the products in individual consumer packs or larger industrial packs. Freezing between horizontal refrigerated plates or in an air blast is used for this purpose. In both types of freezer the process is completed within a few hours.

Although earlier forms of plate-freezers employed Dole Vacuum Cold Plates, more modern types make use of extruded aluminum plates with internal channels to carry the refrigerant. It is claimed that freezing times are thereby reduced very considerably. Large installations use ammonia as the refrigerant, the liquid being pumped from a central surge drum to the various freezers. Smaller self-contained units employ R22. Packages of fish may be frozen in open trays fitted with spacer sticks or in metal trays fitted with lids.

Air-blast freezers are usually of the truck-and-tray type employing air at −30° to −40° F. (−34° to −40° C.) moving at about 1,000 f.p.m. The trucks are usually moved through the tunnel mechanically either against the air-flow (counter-current type) or across it (cross-flow type). Completely automated blast freezers are in use in Germany. In most blast freezers, the products are frozen in metal trays fitted with spring-loaded lids to prevent distortion of the packs. Air-blast freezers are popular for whole or gutted fish such as herring and salmon. Herring and other small fish may be frozen in packs in open trays; larger fish are frozen singly.

Of special interest is the vertical plate freezer specially designed in the U.K. for freezing fish at sea (Fig. 7.5) and now widely used in a modified form for this purpose. The freezer employs refrigerated plates to provide a series of vertical compartments or "cans" into which the fish are loaded for freezing. Very good contact between the fish and the plates is obtained

Courtesy of Jackstone Froster, Ltd., Grimsby, U.K.

FIG. 7.5. END-UNLOADING 12-STATION VERTICAL-PLATE FREEZER

in this way. Modern forms employ extruded aluminum plates of the type referred to above and are fitted with means for warming the plates to facilitate release of the blocks at the end of the freezing cycle. The fish are loaded into the freezer at the top and in some models the blocks can be compacted by means of hydraulic rams. Some freezers can be unloaded from the bottom, and others from the side. The blocks obtained are usually 4 in. thick and can weigh up to 100 lb. The plates may be refrigerated by the primary refrigerant (R12) fed from a surge drum through pumps fitted with an induction drive, or by a secondary refrigerant— usually trichlorethylene. The plate temperatures are normally maintained at about −40° F. (−40° C.) and the total freezing time of a 4-in. block under such conditions is about 3½ hr.

In addition to efficient freezers at sea, land freezing of fish, particularly smoked fish, in wooden boxes, is carried out in so-called sharp freezers which consist essentially of a large room equipped with facilities for recirculating air through a low-temperature cooler. Freezers of this kind are mainly to be found in the U.K., while in Norway the Dahl process for freezing fish in boxes with cold brine is still used to prepare herring for bait.

Liquid nitrogen is not as yet used commercially for freezing fish in Europe.

Cold Stores.—Most cold stores, particularly those at the ports close to production centers, have been built since the war and are thus of modern design. There is extensive use of modern expanded plastics for insulating purposes and most of the stores are designed to operate down to −20° F. (−29° C.) or below. With any particular block of stores, the tendency is to provide a small number of large chambers rather than a large number of small ones. Operating costs are reduced by the use of fork-lift trucks for the movement and stacking of products, and individual rooms are made high enough, up to 20 ft., to take full advantage of the mechanical stacking thus provided. Although it is now well known that the refrigeration of cold stores by grids on exposed walls will reduce in-store desiccation, the recent tendency is to refrigerate by unit air coolers, on grounds of economy in costs, particularly as regards defrosting. Since most consumer packs are well protected by wrappers, the consequent drying effect does not matter a great deal, but loss of moisture from the product may be more serious when storing industrial packs, which are not so well protected, and blocks of sea-frozen fish, which in many cases may not even be glazed.

Transport.—The provision of suitable equipment for the transport of frozen fish and other foods in Europe has grown concurrently with developments in freezing, though there was a very considerable lag period at the start. This is understandable, since the development of new equipment takes time and can only occur once a demand has been well established.

In consequence, in the early stages reliance was placed upon insulated containers, sometimes refrigerated with solid carbon dioxide, for the transport of frozen fish. Since the late fifties, however, there have been marked improvements in the standard of equipment provided, particularly for road transport. New and better insulants, mainly expanded plastics, are now in universal use and an increasing proportion of the vehicles are mechanically refrigerated, with gasoline or diesel engines providing the power. The provision of small, efficient diesels has opened the way for these new developments to spread to the railways, who are opposed to the use of gasoline engines. Standards of efficiency of containers and procedures for testing have now been laid down by various European organizations.

European Commercial Practices

Freezing at Sea.—In Europe, the concept that the combination of freezing and cold storage is primarily a process of preservation has been kept well to the fore and applied to the development of freezing at sea those fish caught on grounds too distant for chilling with ice to afford satisfactory preservation. These applications have taken two forms; the freezing of filleted fish, and the freezing of whole gutted fish.

Freezing filleted fish at sea began in Europe in 1954 with the coming into operation of Fairtry I owned by a Leith, Scotland, company. This factory stern trawler has a length of 245 ft. between perpendiculars and measures 2,605 gross tons; she processes only her own catch and does not take fish from other vessels. The factory area is under cover below the fishing deck, and the catch, mainly cod, is gutted, washed, and then filleted and skinned by machine. The fillets are then weighed into 7-, 14- or 28-lb. lots, packed in trays and frozen in either specially-designed air-blast freezers or in horizontal plate freezers having a total capacity of about 30 tons per day. When the blocks are frozen they are removed from the trays, packed in outer cartons, and stored in the low temperature hold at $-5°$ F. ($-21°$ C.). The ship makes about 4 voyages a year, mainly to the fishing grounds in the northwest Atlantic, and can carry up to 600 tons of frozen fillets.

The freezing of filleted fish at sea presents a number of technological problems associated with the physical and biochemical changes that go on when the fish pass into *rigor mortis*. Although the products obtained have an excellent flavor, certain defects in appearance reduce sales-appeal and tend therefore to restrict further development. Similar sister-ships of Fairtry I have, however, been built and have been in operation since about 1960 (Fig. 7.6), working mainly off Newfoundland and Greenland. The annual production of frozen fillets from the trawlers is approximately 2,000 tons each.

The products are used for special outlets—for the preparation of fish fingers and fish portions, and for supplying retail fish fryers, and there is some evidence of a vigorous consumer demand for the products. Recently, other U.K. concerns have shown an interest in the freezing of fillets at sea, as have concerns in other countries, including the U.S.S.R.

Experiences in Germany run somewhat contrary to past experiences in the U.K. Fish frozen at sea has accounted for an increasing part of German production, the amount from all sources expanding from 2,900 tons in 1960 to 10 times that now, mainly fillets; the latter figure corresponds to a live catch weight of over 150,000 tons. At the present time, a third of the German deep sea trawlers have freezing plant on board, including four that freeze the whole of the catch; a third of the freezer-trawlers are equipped with filleting machinery.

Although a few German trawlers have frozen whole fish, usually singly without heads, the emphasis has been on freezing fillets at sea, mainly in

FIG. 7.6. THE FACTORY STERN-TRAWLER "FAIRTRY II"
Owner, Chr. Salvesen, Leith, U.K.

the northwest Atlantic. The average length of voyage for ships carrying freezing plant to these waters has increased significantly from 32 days in 1961 to 50 days now. Some of the largest trawlers that freeze the whole of their catch have undertaken trips of 80 days or more.

A typical German vessel freezing part of its catch as fillets is the stern trawler Othmarschen, built in 1965 (Niegsch 1965). She has a length of 220 ft. b.p. and measures 1,394 gross tons. About ⅔ of the cargo space provides low temperature storage at −22° F. (−30° C.), with room for about 340 tons of frozen fillets; the remainder of the stowage is for fish chilled in crushed ice in the traditional manner. There are three processing lines on the enclosed factory deck for cod, codling, and redfish, respectively. The fish are headed, filleted, and skinned by machine, and the fillets are then packed in aluminum trays and frozen in 11-kg. blocks in two horizontal plate freezers having a total capacity of 26 tons per day.

The stern trawler Bonn, built in 1964 as the first of a fleet of six sister ships, is representative of the vessels freezing all of their catch. With a length of 254 ft. and a gross tonnage of 2,560, she is closely comparable in size with the Fairtry. On this vessel the machine-cut fillets are automatically weighed and packed before being frozen in eight vertical plate freezers with a total capacity of 30 tons a day. Cold storage space at −18.5° F. (−28° C.) is available for about 500 tons of blocks of fillets. The refrigeration plant has been designed for tropical operation, and ships of this class have made exploratory voyages to the hake grounds of the south Atlantic.

Other European countries that have invested in fishing vessels equipped for freezing fish at sea include Spain, Greece, Israel, Italy, Portugal, Norway, and France. Of these, Spain and Greece have by far the largest number of freezer trawlers in service, principally for operation in the central and southern Atlantic.

Very considerable quantities of fish are also frozen at sea by Russian, Polish, and East German companies. For instance, Russia alone has in operation several large factory ships each fed by a number of conventional trawlers and each able to freeze at the rate of about 100 tons of fillets a day, plus 25 freezing trawlers of the Fairtry type, each capable of freezing about 30 tons of fillets a day.

Broadly speaking, the pattern of operations is similar in most of these ships. The fish, landed direct from the trawl or transferred from the attendant trawler, are washed, gutted, filleted, and skinned mechanically and then frozen in an air-blast freezer. The fish are packed into trays fitted with spring-loaded lids to provide a degree of compaction in the blocks. More recently built ships are using plate freezers. Ammonia is used for refrigeration and the frozen products are stored at −5° to

FIG. 7.7. THE FREEZER-TRAWLER "NORTHELLA"

Owner, J. Marr and Son, Hull, U.K.; Builder, Hall, Russell and Co., Aberdeen.

−25° F. (−21° to −32° C.). One of the problems encountered is the variable delay between catching and gutting, which may affect the appearance and texture of the frozen fish, particularly when the fish are exposed to temperatures much above 50°–60° F. (10° to 16° C.) and become subjected to the effect of high-temperature rigor. Facilities are provided on those ships fishing in warmer waters for the fish to be chilled on arrival on board, either by mechanically chilled seawater or by seawater chilled with ice. In colder waters, fish may be kept cool by treatment with ordinary seawater, but at times in the Arctic it is necessary to cover fish exposed on deck to prevent them from freezing prematurely.

Modern additions to the Russian fleet are more ambitious in scope and include canning as the main processing operation. These ships, however, also carry refrigeration equipment and dielectric thawing equipment to preserve glut catches for subsequent canning during periods of light fishing.

Whether or not fish should be filleted before freezing at sea is a debatable point. While it is true that freezing and cold storage space can be saved by freezing fillets, extra space is required for filleting and for additional crew. Requirements for crew can become disproportionally larger if, as is often the case, round trips are far longer than those of a conventional trawler. Fishermen normally work at least a 16-hr. day when wet fishing, but cannot do this for protracted periods on long freezing voyages.

The development of freezing at sea in the U.K. has taken the form of freezing whole gutted fish. These developments are based on earlier experimental work (Fig. 7.7) using vertical plate freezers specially designed for the purpose of freezing cod in the form of compact blocks 4½ in. thick, which could then be thawed out and treated in the same way as iced fish. Subsequently, a large scale commercial experiment, in which about 25 tons of sea-frozen gutted fish were produced on each of 8 voyages, confirmed this earlier work and led ultimately to the building and operation of "Lord Nelson," a stern trawler fitted with freezing equipment. The idea was taken up by trawler owners in the Humber ports and there are now many trawlers in operation freezing whole fish and more are planned.

All of these ships are equipped with vertical plate freezers. A typical representative of the fleet is the trawler Victory, working from the port of Grimsby. Her length is 215 ft. b.p. and she measures about 1,800 gross tons. She has ten 12-station, top loading, side unloading, vertical plate freezers producing blocks of whole frozen white fish 42 in. long × 21 in. wide × 4 in. thick; each block weighs approximately 100 lb. (Fig. 7.8). Total freezing capacity is about 35 tons of fish a day. The primary

Courtesy of Associated Fisheries, Grimsby, U.K.

FIG. 7.8. STOWING BLOCKS OF FROZEN WHOLE COD ABOARD THE FREEZER-TRAWLER "VICTORY"

refrigerant, R22, is used to cool a secondary refrigerant, trichlorethylene, which in turn is pumped through the freezer evaporator plates at a temperature of about −40° F. (−40° C.).

The fish are sorted and gutted by hand at the after end of the covered factory deck, passed through a rotating cylindrical washer and conveyed forward mechanically to storage bins alongside the two rows of freezers running fore and aft on either side of the factory space. The fish are packed neatly by hand between the pairs of freezer plates and are reduced to a temperature of about −5° F. (−21° C.) at the center of the fish in about 3½ hr. Hot trichlorethylene is then circulated through the evaporator plates, partly to help release the blocks from the freezer and partly to ensure that when reloading begins the wet fish do not stick to the plates and so prevent the formation of a compact block of fish. The discharged frozen blocks are fed through insulated hatches in the factory deck to the cold store below, which operates at a temperature of −20° F. (−29° C.) and has a capacity of about 500 tons.

Trawlers of this type make voyages of 30 to 60 days' duration, depending on the catching rate, and work principally off Greenland, Newfoundland, and Labrador. British trawlers freezing whole fish are so far based entirely on the Humber ports of Hull and Grimsby, but they are operating from at least one other British fishing port as well.

Most firms operating these trawlers have interests in port wholesalers and are able to channel supplies of sea-frozen fish through them. On landing, the frozen blocks of fish are not subjected to the usual auction, but are taken direct to cold storage where they are held at −20° F. (−29° C.) until required.

Thawing.—Thawing is an important factor in subsequent handling and, because of the quantities dealt with, special equipment is provided. Two procedures are in use. In the first, the blocks of fish are thawed by dielectric heating, the blocks being conveyed on a rubber belt through a series of six dielectric units. In order to even out the flow of heat, the blocks are first of all immersed in plastic trays of water in order to fill up the voids in the blocks. Blocks of fish 4 in. thick can be thawed by this procedure in about 1 hr. (Fig. 7.9).

The second procedure employs continuous cross-flow air-blast thawers. In this type of plant several conveyors are placed one above the other and so arranged that when blocks of fish are loaded onto the top conveyor they fall at the end of the traverse on to the one immediately below. The fish are transferred from one conveyor to another in this way and finally emerge at the bottom of the defroster on the side opposite to the point of entry. In addition to reducing the length of the defroster required, this arrangement speeds up thawing slightly, since blocks of fish break up as

Courtesy of Radyne, Ltd., Wokingham, U.K.

FIG. 7.9. DIELECTRIC THAWING PLANT FOR BLOCKS OF WHOLE SEA-FROZEN
WHITE FISH

they make each drop. During passage through the defroster, the blocks of fish are subjected to a blast of moist air at about 70° F. (21° C.) and moving at about 1,200 f.p.m., the air being humidified by water sprays as it is circulated. Thawing of 4-in. thick blocks of fish takes 4 to 4½ hr.

This type defroster is somewhat wasteful of fan power because of the large volumes of air that have to be circulated at high speeds. A more economical defroster in this repect has been suggested by Merritt and Banks (1964). This is a parallel-flow defroster in which the air and fish move in the same direction and which therefore allows the use of air at a slightly higher temperature than 70° F. (21° C.) at the entry duct, since it strikes the fish at their lowest temperature and is progressively cooled as it moves down the defroster.

Air defrosters are built with an output of 1 to 2 tons of fish per hour; smaller units of a batch type are also in use. One circulates slowly moving warm air between blocks laid out on racks in a similar fashion to the defroster used in the "Northern Wave" experiment, and another type blows warm moist air between trays of frozen fish on trucks. In the former, thawing may take 12 to 18 hr., but this does not appear to affect quality, provided that the fish are not overheated. The speed of thawing in the second type of batch defroster is about the same as that in the larger, continuous air-blast defrosters.

When the fish are thawed they are treated in exactly the same way as iced fish. They are filleted, usually by hand, and may then be packed in ice boxes for dispatch to the retailer or they may be lightly brined and cold smoked before retail sale. Most of the wet fish are sold through fish fryers.

By and large, quality is high, although difficulties are sometimes encountered which can be related to periods when the trawlers encounter heavy fishing. The aim of the producers is to keep costs down by moving the fish quickly once it is landed and cold stored, and to this end they try t arrange for continuous supplies to be made available at a fixed price to their customers.

Freezing on Shore.—*Whole Fish*.—Freezing and cold storage are particularly useful in spreading the availability of heavily seasonal supplies of such fish as herring and salmon. As a consequence, considerable quantities of herrings are frozen whole in 7-lb., 14-lb., or larger packs for subsequent thawing and processing, including smoking, refreezing after nobbing or filleting, and marinating. Larger packs still are used for the subsequent production of pet foods. Sprats and similar fish are also frozen in bulk for subsequent canning as sardines or brisling.

Research has shown that herring need to be frozen within 24 hr. of catching and kept well chilled during this time if satisfactory products are to be obtained, and the regulations controlling the sale of herring for freezing in the U.K. go part way toward enforcing this principle.

Considerable quantities of salmon are frozen in Europe for export. The products, including others imported from outside Europe, are thawed and then smoked or in some cases used for the manufacture of salmon paste. In the U.K., and in other countries too, some frozen salmon are thawed and sold without further treatment. Relatively small quantities of mackerel are likewise frozen for sale after thawing, without further treatment.

All these fish are fatty fish and can become rancid rapidly in cold store. Glazing and vacuum packing in plastic bags, coupled with storage at −20° F. (−29° C.) are used to control such deterioration.

Occasionally, during times of glut, gutted white fish such as plaice, cod, or haddock caught on short trips will be frozen for short periods of cold storage and then thawed, filleted, and refrozen in commercial or consumer packs. Fish that has been bought for the production of fish sticks is often treated in this way. This procedure is preferred to holding the fish in ice for extended periods.

Fillets.—The procedure employed in all European countries for the production of packs of fish fillets is basically the same, although there may be differences in the initial quality of the fish and in the actual species frozen.

Fish may be caught by trawl, by line, or by seine-net and are usually gutted, washed, and well iced as soon as possible after capture. On landing, fish that has been treated properly at sea will be at 32° F. (0° C.) or a little below. Fish for freezing is bought in the open market, although there is an increasing tendency toward purchase by contract. Once purchased, the fish are transported to the processing factory where they are well washed and then filleted by hand or by machine. The fillets are then packed either into cartons for consumer packs weighing 8 to 14 oz. or into molds for industrial packs weighing from 5 to 14 lb. The industrial packs may or may not be wrapped in parchment or waxed paper; the unwrapped blocks are glazed when frozen, and sometimes the wrapped ones too. The consumer-packs usually have an inner and outer wrap. When frozen, both types of product are packed into larger outer cartons for cold storage and transport.

A wide range of products is produced, including fillets of cod, haddock, whiting, plaice, hake, halibut, and lemon sole.

Of special interest is the use of vacuum packing for such products as frozen smoked salmon, herring, kipper fillets, smoked haddock fillets, and rainbow trout. The value of this procedure for fatty fish has been clearly demonstrated. Some packs of kipper fillets and smoked haddock fillets include a small pat of butter within the pack, which itself has been designed so that the fish can be cooked by plunging the package into boiling water without any prior thawing; this type of pack is known as boil-in-the-bag.

Fish Sticks.—The production of fish sticks is broadly on the lines employed in North America. In the U.K. fish sticks account for about 14% of frozen food products. Cod is the chief ingredient although some products are made with hake, imported frozen from South Africa. The sticks may be cut from blocks of fillets or from blocks of shredded fish. The blocks may be hard frozen and cut with a band-saw, or partially thawed blocks may be guillotined to obtain the portions. The pieces are covered with a batter, coated with bread-crumbs, and then fried in vegetable oil. When cooled the finished products are packed into cartons holding 6 or 10 1-oz. fingers and then frozen and cold stored. The process is fully mechanized up to the packing stage. In all cases precautions are taken to remove bones.

In the U.K., fragments of fish from the cutters are used, together with other trimmings, to manufacture fish cakes; a mixture of shredded fish, a binder, usally rusk, and mashed potato, is shaped into small circular portions or cakes, battered, and fried. Trimmings obtained as a by-product of salmon smoking are used in the same way. Some of these products are also frozen. In Norway such material is often made into fish balls, a typical Norwegian dish, and frozen.

Quality Control.—All the larger firms employ quality control throughout the manufacture of their various products. The raw material is assessed for freshness by taste panel and by an objective test, for example, trimethylamine estimation, and the limits laid down by the individual firms are then applied and low-quality fish rejected. Standards vary between different countries and between different firms, but the general tendency is toward the utilization of only the better quality fish for freezing. Frozen products obtained from other sources, including imported material, are also subjected to quality assessment. Quality control staff are also employed in the larger factories to make routine production line checks, and many companies also check the quality of their products on sale to the public.

Temperature of storage is an important factor affecting quality, for example of frozen material bought for further processing or of consumer packs on sale to the public, and there is therefore considerable interest in the cell fragility test which, it is claimed, can be used to assess the effects of frozen storage treatment on fish.

ADDITIONAL READING

AM. FROZEN FOOD INST. 1971. Good commercial guidelines of sanitation for frozen soft filled bakery products. Tech. Serv. Bull. *74*. American Frozen Food Institute, Washington, D.C.

ASHRAE. 1974. Refrigeration Applications. American Society of Heating, Refrigeration and Air-Conditioning Engineers, New York.

BAUMAN, H. E. 1974. The HACCP concept and microbiological hazard categories. Food Technol. *28*, No. 9, 30–34, 70.

DeFIGUEIREDO, M. P., and SPLITTSTOESSER, D. F. 1976. Food Microbiology: Public Health and Spoilage Aspects. AVI Publishing Co., Westport, Conn.

GUTHRIE, R. K. 1972. Food Sanitation. AVI Publishing Co., Westport, Conn.

HARPER, W. J., and HALL, C. W. 1976. Dairy Technology and Engineering. AVI Publishing Co., Westport, Conn.

HARRIS, R. S., and KARMAS, E. 1975. Nutritional Evaluation of Food Processing, 2nd Edition. AVI Publishing Co., Westport, Conn.

JOHNSON, A. H., and PETERSON, M. S. 1974. Encyclopedia of Food Technology. AVI Publishing Co., Westport, Conn.

KRAMER, A., and TWIGG, B. A. 1970, 1973. Quality Control for the Food Industry, 3rd Edition. Vol. 1. Fundamentals. Vol. 2. Applications. AVI Publishing Co., Westport, Conn.

MOUNTNEY, G. J. 1976. Poultry Products Technology, 2nd Edition. AVI Publishing Co., Westport, Conn.

PETERSON, M. S., and JOHNSON, A. H. 1977. Encyclopedia of Food Science. AVI Publishing Co., Westport, Conn.

RAPHAEL, H. J., and OLSSON, D. L. 1976. Package Production Management, 2nd Edition. AVI Publishing Co., Westport, Conn.

SACHAROW, S. 1976. Handbook of Package Materials. AVI Publishing Co., Westport, Conn.

TRESSLER, D. K., VAN ARSDEL, W. B., and COPLEY, M. J. 1968. The Freezing Preservation of Foods, 4th Edition. Vol. 2. Factors Affecting Quality in Frozen Foods. Vol. 3. Commercial Freezing Operations—Fresh Foods. Vol. 4. Freezing of Precooked and Prepared Foods. AVI Publishing Co., Westport, Conn.

WEISER, H. H., MOUNTNEY, G. J., and GOULD, W. A. 1971. Practical Food Microbiology and Technology, 2nd Edition. AVI Publishing Co., Westport, Conn.

WILLIAMS, E. W. 1976A. Frozen foods in America. Quick Frozen Foods 17, No. 5, 16–40.

WILLIAMS, E. W. 1976B. European frozen food growth. Quick Frozen Foods 17, No. 5, 73–105.

WOOLRICH, W. R., and HALLOWELL, E. R. 1970. Cold and Freezer Storage Manual. AVI Publishing Co., Westport, Conn.

8

Freezing of Shellfish

*A. Banks, John A. Dassow, Ernest A. Feiger,
Arthur F. Novak, John A. Peters,
Joseph W. Slavin and J. J. Waterman*

V ariations in processing methods for shellfish are related to the seasonal characteristics that affect yield, easy of meat removal, and color. Speed in handling the product from time of butchering to freezing is a most important process requirement. Quality control procedures generally emphasize sanitation, food regulatory requirements, and compliance with end-product specifications of the processor.

In this chapter we will explore the freezing of crabs, lobsters, shrimp, oysters, scallops, clams, abalone and related products.

CRABS AND LOBSTERS

Crabs and lobsters are members of the *Crustacea* class and the order *Decapoda*. Common biological characteristics of these invertebrates are a hardened outer shell or exoskeleton, five pairs of jointed legs, gills for respiration, and an open circulatory system in which the blood, or hemolymph, bathes the tissues in open channels or sinuses before returning to the single-chambered heart. The animal grows by successive molts of its shell during its lifetime. The timing of the fisheries harvest and the meat quality and yield are related in part to the molting cycle of crabs and lobsters. These biological characteristics and individual species variations have important effects on the freezing and storage properties of these Crustacea.

Production of Frozen Crab and Lobsters

In comparing the relationships between catch and frozen products of the major species of crabs and lobsters, it should be noted that the amount and value of frozen king crab meat exceed the combined volume and value of frozen products derived from blue and Dungeness crabs, northern lobsters, and spiny lobsters. Blue crab is most important in frozen specialties, such as the much-favored crab cakes, deviled crabs, and stuffed crabs. Frozen Dungeness crab is relatively more important in the frozen sections and whole crabs. The northern lobsters, at least the domestic catch, appear only as a relatively minor frozen item because most of the catch is sold to restaurants and consumers as live lobsters. The small domestic catch of spiny or rock lobsters is greatly exceeded by imports of frozen rock lobsters, which are 25 times the amount of the domestic catch.

Minor species of crabs landed and frozen in part for marketing included 2.5 million pounds of red crab *(Cancer irroratus)* taken mostly in New England, and 0.9 million pounds of stone crab *(Menippe mercenaria)* in Florida. The tanner crab *(Chionectes* spp.) is a relatively large deep-water crab taken in the North Pacific and North Atlantic. At present only small quantities are landed in Alaska and on the Canadian east coast; however, the fishery for this crab, being introduced in Canada as the queen crab, has a great potential if the harvesting and processing economics can be improved.

Characteristics in Relation to Freezing Preservation

The group characteristics of crabs and lobsters have important applications to the handling of the Crustacea and to the processing methods in freezing and storing the meat.

Live Handling.—It is essential that crabs and lobsters be kept alive after capture, and in healthy condition up to the time of processing. If the animals are allowed to die or even become very weak prior to processing, there are irreversible changes in the flesh which cause adverse changes in the texture, appearance, and flavor of the cooked meat. The yield of meat, for example, tends to decrease because the meat tends to stick in the shell and breaks apart as it is removed. The texture becomes chalky or friable and looks stringy. The delicate flavor may be lost and the meat tends to discolor easily during processing or storage. The extent of the changes varies according to species and the degree of poor condition at the time of butchering and processing.

Circulatory System.—One of the significant physiological features of crabs and lobsters, as compared with fish, is the nature of the circulatory system. The crab blood or hemolymph comprises a large percentage of the animal weight and is highly variable in relative volume depending on the condition; varying, for example, in some crab species from 30 to 40% of the weight of the animal. This is 8 to 10 times as high as the relative volume of blood in fish. The hemolymph is relatively colorless and circulates in the peripheral areas in fairly open circulation patterns rather than through an organized network of blood vessels and capillaries as in higher animals. As a result, the yield and quality of the meat of crabs is affected by large losses of the blood if the animal dies slowly out of water or as a result of injury.

Molting Cycle.—Crabs and lobsters grow by passing through a series of molts. Before molting, the tissues of the animal are relatively high in solids and low in moisture. Immediately after molting and before the new shell hardens, the tissues absorb water and swell to allow for new growth. During the immediate postmolting period, the meat yield is significantly lower per unit of body weight, then increases as the tissue protein develops, the shell hardens, and the new growth is consolidated. In king crab, the yield of cooked meat per unit of body weight increases from 14 or 15% a couple of months after the molting period, to 24 or 28% after 7 or 8 months.

Pigments and Meat Color.—One problem encountered in the freezing and storage of crab and lobster meat is the difficulty of retaining the characteristic meat color. Varying from snow white to creamy white, depending on species, the meat may discolor to various degrees, depending on exposure to air and high temperatures during processing and on poor conditions during freezing and storage. Yellowing of the meat usually indicates some degree of oxidation during processing or long cold storage. Fading and discoloration of the red or orange-red carotenoid surface pigments may also take place under the same conditions; however, this is quite variable in any particular species and from one species to another. The bright red pigments of the king crab are quite stable in comparison to the orange-red pigments of the Dungeness crab. Pigments at the joints and in the claw appear to be the most prone toward oxidation and discoloration.

The development of blue or black discolorations, usually simply called "blueing," is one of the most troublesome color problems. These colors may develop to a moderate degree during or shortly after cooking or may appear after freezing and during storage. One type of blueing occasionally appearing on king crab meat does not develop until the meat is thawed and allowed to stand exposed to air a short while. Tests with the cause of blueing in king crab meat indicated that the reaction could be reversed by means of a reducing agent, such as sodium sulfite, and could be inhibited by dipping the meat prior to freezing in a dilute solution of ascorbic acid. On occasions, notably in king crab and spiny lobster tails, a deep bluish-black curd-like discoloration develops in and around the joints and the spaces between the muscle and the shell. The cause of these blueing reactions is not completely understood, but appears to be related to biuret-type reaction(s) between the copper pigments in the blood and the heat-denatured muscle proteins.

Freezing Characteristics.—The freezing and storage characteristics of crabs and lobsters cover a wide range, from very limited keeping quality of only a few weeks in the case of blue crab and northern lobster to good keeping quality up to one year in king crab and spiny lobster, or even longer. Dungeness crab appears to rate between blue crab and king crab. The northern lobster is largely marketed as live lobster because of the difficulty of retaining good quality in the frozen form and the fact that there is little surplus in the domestic catch available for freezing in any case.

In the late 1940's and early 1950's the Alaska king crab fishery was developed when it was shown that the cooked crab meat had superior freezing and storage characteristics in comparison with other crab species.

Texture Changes During Storage.—The texture of fresh cooked crab and lobster meat is moist to juicy, with rather fibrous but tender muscle segments. During freezing and storage the muscle fibers tend to become dry, slightly tough or spongy, and stringy. This is particularly true of blue crab and northern lobster meat, although any of the species will undergo these adverse changes if stored too long or at too high a temperature. Recognition of this important time-temperature variable and the need for proper packaging to exclude air and minimize voids are necessary in planning production of frozen crabs and lobsters.

Comparative tests with frozen Dungeness and king crab meat show that Dungeness crab meat was unpalatable after three months of storage at 0° F. (−18° C.) when packaged in moisture-vapor proof cellophane, but king crab was palatable after one year under similar conditions. However, Dungeness crab meat packed tightly in hermetically-sealed cans and stored at −20° F. (−29° C.) was of good marketable quality after one year.

Lower storage temperatures are of similar benefit for improving the keeping quality of northern lobsters. Cooked lobster meat in cans was of better quality after 18 weeks of storage at −20° F. (−29° C.) than that stored at 0° F. (−18° C.). If lobsters are frozen and stored at −20° F. (−29° C.), they are in good condition after 6 months but when stored at −5° F. (−21° C.) they are at the storage limit after 3 months. Crab meat to be held up to 1 year should be stored at −20° F. (−29° C.).

Flavor and Odor Changes During Storage.—The flavor and odor of crabs and lobster, although fairly distinctive, are quite mild and sweet, with a pleasant aftertaste. When stored at too high a temperature or in packages with air voids, the mild sweet flavor is quickly lost, for example, in the cases of northern lobster and blue crab meat, in just 2 or 3 weeks. Initially the flavor changes to one which is flat and without character. After this phase a hay-like, slightly acrid taste develops, in company with off-odors of a similar nature. These off-flavors appear to be oxidative in character, particularly those noted in the surface pigment layer or in the meat of the joints or claws. In king crab the surface meat of the shoulder joint is quite apt to develop a strong, almost rancid, bitter taste with an oily, persistent character.

Such flavor and odor changes are accompanied usually with adverse texture changes. Presently, few chemical data are available on changes that occur in freezing and storing crab and lobster meat. Generally adverse changes are minimized if the crabs and lobsters are processed, packaged, and frozen quickly with minimum exposure to air and if stored at −20° F. (−29° C.) in packaging materials that have very low or zero permeability to oxygen.

Composition and Food Value.—Crab and lobster meats are quite high in protein, from 16 to 20%, and low in fat, from 0.8 to 1.6% of the weight of the cooked meat ready to eat. The total mineral content, exclusive of sodium chloride added during processing, is about the same in crab but is more variable than in most fish species, from 1.2 to 1.6% by weight of the cooked crab meat. The natural sodium content is much higher than in most fish, about 200 mg. per 100 g. of cooked crab meat as compared to 60 or 70 mg. per 100 g. of fish flesh. In marketed king crab the salt content of the frozen meat is from 1.0 to 1.5%, a level acceptable to most palates. Frozen Dungeness crab meat tends to be slightly higher and more variable in salt content, owing to the use of brine in processing.

Crab and lobster meats contain about 125 and 200 mg. % of cholesterol respectively, a value about twice or more as high as that of fish fillets or chicken meat, a factor important in some restricted diets. The content of calcium, magnesium, phosphorus, and iron is about the same for crab and lobster meat as that for lean fish fillets. The protein, as in all fish and shellfish muscle protein, is complete, with the needed amount and ratio of the essential amino acids; therefore, even in small quantities the inclusion of crab and lobster in the diet frequently can be important for protein balance as well as for appetite appeal. The food energy value is low, from 80 to 90 Cal. per 100 g. of the cooked meat, similar to the energy value of lean fish fillets.

Species Characteristics

The factors more or less common to crabs and lobsters have been considered in relation to freezing preservation and quality. In the following sections, considerations peculiar to the five main species of the group are discussed. Of interest here are descriptive and habitat characteristics which affect the handling and processing methods, keeping in mind the general group characteristics.

King Crab

Ranking first in the group with respect to utilization in the frozen form, the Alaska king crab *(Paralithodes camschatica)* is probably the most familiar species in the retail market because of the rapid growth of the industry in just the last ten years. Since 1953, the landings of king crab in Alaska increased from 4.6 million pounds to 86.7 million pounds in 1964, 126 million pounds in 1965, and 160 million pounds currently.

King crabs are the largest of the commercial species of crabs and range up to 24 lb. each, although the usual commercial range in weight is from 8 to 12 lb. The yield of the meat varies most commonly during the harvesting season from 18 to 26% of the landed weight and averages 20%.

King crab meat is about 70% leg meat, 15% shoulder meat, and 15% body meat. The yield is apt to be low at the beginning of the season, usually in early fall, and improves as the season progresses (Fig. 8.1).

Courtesy of Bureau of Commercial Fisheries, U.S. Dept. of Interior

FIG. 8.1. A KING CRAB VESSEL IN KODIAK HARBOR, ALASKA

King crabs are taken over a wide area and during a long season, from early fall to late spring; therefore the sorting of crabs for condition and in some cases the modification of process details according to the condition are important. For example, when sections or the legs in the shell are being frozen, the processor demands well-filled legs, with a bright, clean, outer shell. Crabs that have molted too recently or, conversely, those older crabs which have skipped a molting period are not acceptable. The former are apt to be light in weight and the latter have discolored, scarred shells that are unsightly.

Depending on condition and the end product in mind, the time of the cooking operation may be increased. Many operators lengthen the cook if they notice that the meat tends to have a blue-gray discoloration in the coagulated protein. The variations in the raw material and the necessary adjustments for process control are notable parts of the king crab processing which require experience and good judgment.

Blue Crab.—In 1964, the blue crab (*Callinectes sapidus*) was the most important species landed in the United States but accounted for less than 5% of the commercial production of frozen crab meat. The demand for chilled, cooked, blue crab meat and the fact that blue crab has not been satisfactory for freezing and storing are the main reasons for the small volume of frozen meat produced. However, the use of the meat in frozen specialties and the marketing of frozen soft crabs have been highly successful. Therefore, the value of frozen blue crabs with specialties is more than 25% of the total value of crabs and lobsters.

The blue crab is a small crab. It ranges in weight from 2 to 3 crabs per pound and is from 3 to 6 in. in width across the carapace. There is very little leg meat in the blue crab, and in contrast to king and Dungeness crabs, the premium meat is the body meat. The "lump" meat that forms the muscles which operate the swimming legs is considered the choice portion. Next in value is the white or flake meat of the remaining muscles of the body. Claw meat brings the lowest price. The total meat yield is somewhat low, only 14%.

Dungeness Crab.—The Dungeness crab (*Cancer magister*) is the common shore crab of the Pacific coast from California to the Aleutian Islands in Alaska. It is a fairly large crab; a common market-size crab measures from 6½ to 9 in. across the carapace and weighs two pounds or more. The most common market form is fresh, cooked, whole crab. Approximately half the total meat is leg and claw meat and half is body meat. Meat yield is 25% of the landed weight on the average, and the leg meat is the premium portion. Most frequently, chilled, frozen, and canned Dungeness crab meat are marketed as mixed leg and body meat.

Dungeness crab is caught in bays, inlets, and in ocean areas along the continental shelf area in waters up to 40 fathoms in depth. For reasons not completely clear the abundance of Dungeness crab tends to be cyclic. Two-to four-year periods of relative abundance are usually followed by similar periods of poorer catches. As a partial result of this cycle, product diversification for frozen Dungeness crab and development of frozen specialties have been minimal (Fig. 8.2).

The frozen production is dominated by the production of frozen, whole and eviscerated cooked crabs because the item may be thawed and sold for "fresh" cooked crab or may be used by the restaurants when the fresh, live crabs are not available. Chilled, cooked, Dungeness crab meat is a major market item, but because of its limited shelf-life there has long been an interest in some form of pasteurization to extend the shelf-life, much like the heat pasteurization process popular for blue crab meat. Currently both this procedure and the use of low-level gamma radiation pasteurization are under test and show promise. It is unlikely that a diversified, frozen Dungeness crab industry will develop until the annual

Courtesy of Bureau of Commercial Fisheries, U.S. Dept. of Interior

FIG. 8.2. HAULING DUNGENESS CRAB TRAP INTO FISHING BOAT

production can be both increased and stabilized, possibly by the development of the extensive fishing areas in Alaska.

Northern Lobster.—The northern, or true lobster *(Homerus americanus)* is one of the most valuable fishery commodities landed, a circumstance determined by the high demand for this notable delicacy when it is purchased live and cooked and served promptly. There is almost never a surplus of the domestic catch of the inshore lobsters. These vary in size from ½ to 2 lb. each and are the traditional fishery in Maine, where over 80% of the domestic lobster catch is landed. Frozen lobster meat does not store well at the usual temperatures and since little meat has been available in past years, the production of frozen lobster meat has been very small in relation to the catch (Fig. 8.3).

In recent years a fishery has developed for the offshore deep-sea lobsters, many of which weigh more than two pounds. These larger lobsters are more difficult to market alive and therefore are more suitable for meat production. A substantial amount of this meat production is frozen for later use in frozen lobster specialties. Imported lobster meat from Canada has also been important in the domestic supply. The Bureau

Courtesy of Bureau of Commercial Fisheries, U.S. Dept. of Interior

FIG. 8.3. REMOVING LOBSTERS FROM TRAP

of Commercial Fisheries estimates that of the 1.5 million pounds of lobster meat imported from Canada, one million pounds was frozen. Both fresh and frozen meat have been used in the production of a half million pounds of frozen lobster specialty items like lobster Newburg, lobster soup, stuffed lobster, and lobster pie. Small but increasing amounts of frozen whole lobsters are being marketed as the newer immersion and flash-freezing methods, such as liquid nitrogen freezing, are developed.

Spiny Lobster

The spiny lobsters or rock lobsters (Palinuridae) are known also as lobster tails, southern lobsters, and in Australia as marine crayfish. The spiny lobsters lack the large claws of the northern lobsters, have long spiny antennae or feelers, and have a flexible rather than a stiff tail-fan like the northern lobsters. Most of the meat is in the tail and the domestic and imported production consists primarily of frozen tails. The flavor and the texture differ only slightly from the northern lobster if the tails are well frozen and prepared. There are two common species in the domestic fishery, *Panulirus argus*, found off the Florida west coast and in the Caribbean area, and *P. interruptus* found off Southern California.

The domestic production of spiny lobsters is a little over four million pounds and mostly from the Florida fishery, 6.2 million pounds of spiny lobsters established a record high for the domestic fishery. The high volume of imported frozen tails (over 100 million pounds on the live weight basis dominates the market. Australia, South Africa, and New Zealand provide the bulk of these imports. In Australia, there are nine important species of spiny lobsters in the extensive fisheries, but a single species *(P. cygnus)* from the west coast of Australia dominates the commercial production, most of which is exported to the United States. A small but growing production of spiny lobsters is from the Caribbean area and the Latin American countries. A new commercial spiny lobster fishery in the waters of the Republic of Panama has excellent prospects and recommended fishing methods similar to those used in the Florida fishery.

Commercial Freezing Practices

King Crab.—King crabs are harvested commercially from southeastern Alaska to Kodiak Island and the Aleutian Islands. They are captured by means of large pots consisting of iron frames about 6 ft. square and 3 ft. high covered with wire mesh. The pots are baited with fresh fish and then dropped to the bottom at 10 to 15 fathoms depth with a heavy line and marker buoy. The fisherman lifts his pots every few days and keeps the male crabs having a carapace width of 6½ in. or more. The crabs are held alive on the vessel in a well with circulating seawater. At the dock the live crabs are transferred to seawater tanks to await processing. Weak crabs may be sorted out and processed immediately, but generally all dead and injured animals are discarded.

The crab is butchered by use of a stationary iron blade. The back or carapace is removed and the crab is split into halves or sections. The sections are cleaned and washed to remove viscera and blood.

Cooking.—Two cooking systems are used: a one-stage cook of 20 to 22 min. in seawater at 212° F. (100° C.) and a two-stage cook of 10 min. in fresh water at 160° to 165° F. (71° to 74° C.), followed by removal of the meat and a second cook of the meat for about 10 min. in either fresh water or dilute (3%) brine. The cooked sections are cooled in cold water and divided into the parts consisting of entire legs and the adjacent shoulder sections. The legs are inspected and sorted for freezing in the shell if the shell appearance is satisfactory, otherwise the legs are processed for meat removal.

Meat Removal and Packing.—The shoulder, claw, and body meats are removed by shaking or by blowing with water under pressure. The legs are divided at the joints, in some cases by use of a band saw, and fed into two large rubber rollers. The rollers are adjusted for proper clearance and are rotating so that the shell is squeezed as it passes through and the meat is forced out. Meat yield averages about 20% but may be 26% or more by weight of the mature live crabs (Fig. 8.4).

The meats are washed, sorted, inspected for shell and debris, and packed into cartons or trays for freezing. The packed meats usually consist of 70% leg meats, 15% shoulder, and 15% body meat. The meats are packed in a block mold and frozen, with the leg meats on the outside and the shoulder and body meats in the center. The fill weight of drained meats is checked to assure full net weight after freezing and thawing. Approximately 10% water by weight of the meats is added to the blocks to eliminate voids. Dimensions of the blocks are determined by the sizes of the frozen portions to be cut and packaged later. Typical weights are 15-, 28-, and 30-lb. per block (Fig. 8.5).

Freezing and Storage.—The blocks are quick-frozen under pressure, and if not frozen in a waxed carton are given an ice glaze before packing in the shipping case. Most of the block production is shipped to cold storages in the Pacific Northwest and stored at 0° to −10° F. (−18° to −23° C.).

Frozen King Crab in the Shell.—The cooked legs are sorted for freezing in the shell. The legs having a clean, bright-colored shell and a proper meat fill are washed, trimmed, packed into trays or cartons, and blast frozen. The frozen legs are glazed and packed into 10- or 25-lb. cartons. In later processing, the legs may be cut and split into ready-to-cook

Courtesy of Bureau of Commercial Fisheries, U.S. Dept. of Interior

FIG. 8.4. WHOLE COOKED KING CRAB (EVISCERATED) READY FOR
FREEZING

portions, reglazed, and packaged for retail sale. Much of the frozen crab
in the shell goes to restaurants or to markets where it is thawed just before
use or sale. For special seafood displays it is common to freeze the whole
crab, eviscerated, with the carapace in place. King crab meat frozen in the
shell retains more of the flavor of the fresh cooked king crab than does the
meat frozen in the block. The storage life of the king crab frozen in the
shell and well protected from dehydration is 6 months or more at 0° to
−10° F. (−18° to −23° C.).

Blue Crab.—The blue crab industry is important in ten states along the
southern Atlantic and Gulf states, but is concentrated mostly in Maryland,
Virginia, and Florida. These 3 states produce about 70% of the U.S. catch.

Courtesy of Bureau of Commercial Fisheries, U.S. Dept. of Interior

FIG. 8.5. PACKING KING CRAB MEAT FOR FREEZING IN BLOCK

Harvesting.—The blue crab inhabits the shallow waters and is harvested traditionally by a wide variety of gear. In Chesapeake Bay, where about 50% of the crabs are taken, the major types of gear for hard-shell blue crabs are trotlines, pots, and dredges.

The trotline is a baited, hookless line anchored on the bottom and is used to catch hard crabs when they are actively feeding, usually from April through November. The fisherman lifts the line and places it on a spool projecting from the boat. As the boat moves along the line, he catches the crabs in a dip net as the baits, with any crabs clinging to the bait, are raised to the surface.

The pots used for blue crab are cubical in shape, two feet on each side, and covered with chicken wire; they are divided into an upper and lower compartment. The crabs enter the lower or bait compartment and then pass into the upper or trap compartment through a slit in the partition. The fisherman lifts his pots daily and removes the crabs from the top compartment.

Dredges are used principally in the winter when the crabs are in deeper water and inactive. The dredge has teeth along the bottom bar of a metal frame to dislodge the crabs from the bottom. The crabs pass into a mesh bag that can hold between 3 and 4 bushels of crabs. Two dredges are dragged from a boat and hauled alternately. The crabs are sorted and placed in barrels. To harvest peeler crabs, a lighter form of the dredge without the teeth, called a scrape, is used in the same way.

Cooking.—After being unloaded, the crabs are weighed and dumped into circular iron baskets for cooking in vertical retorts. Cooking conditions vary considerably because of the type of equipment used, but most plants use steam at 250° F. (121° C.) for 3 to 20 min. after temperature and pressure in the cooker are brought up to the desired level. A few plants use boiling water for 15 to 20 min. Many processors deback the cooked crabs, wash, and refrigerate the crabs overnight before picking the meat, a practice that tends to increase the meat yield (Fig. 8.6).

Meat Removal.—The meat is picked primarily by hand. In recent years there have been some favorable results obtained in studies of the mechanization of the industry but at present the development of a completely successful machine to remove blue crab meat has not been achieved. The picked meat is divided into three categories: (1) the "lump" meat, which is the large muscle controlling the swimming legs and is the

Courtesy of Bureau of Commercial Fisheries, U.S. Dept. of Interior

FIG. 8.6. BLUE CRAB PICKING ROOM IN CHESAPEAKE BAY AREA

premium product, (2) the "regular" or flake meat, which is the remainder of the muscles from the body and is the second in value, and (3) "claw meat," which is the lowest in price. About ½ of the meat is flake, about ¼ is lump, and ¼ is claw meat. Total meat yield varies with the season, and in one study varied from 12.4 to 16.3% meat.

Preservation of the Meat.—Most of the meat is packed in hermetically sealed 1-lb. cans and is sold in the fresh chilled form. To extend the marketing period for the chilled product, a heat pasteurization process was developed and has been used successfully for 1-lb. sealed cans of meat which are stored and distributed at 32° F. (0° C.) or as near that as practicable.

Blue crab meat does not freeze and store well; therefore only a relatively small volume of frozen meat is produced. For institutional use and for later processing into crab specialties, the meat is frozen in 5-lb. cans or plastic bags. A small amount is packaged in 12-oz. cartons and frozen for retail sale.

A popular frozen item for many years has been the frozen soft-shell crab, a product common only in the blue crab industry. Soft-shell crabs are crabs that have just molted. They are obtained by holding hard-shell crabs in floats until the molt occurs. The soft crabs are removed from the water within a few hours and are graded for size. The crabs may be held in cool storage for 2 to 3 days before processing, then are killed, eviscerated, washed, wrapped individually with parchment, packed 1 or 2 dozen to a carton, and frozen.

Dungeness Crab.—Dungeness crabs occur along the Pacific Coast from California to the Aleutian Islands in Alaska. They are harvested with circular iron pots, 3 ft. or more in diameter, and constructed with 2 entrance tunnels on the side. Pots are hauled up every day or two, the male crabs of legal size are placed in seawater wells on the boat either dry or flooded, and the crabs are delivered alive at the plant. Only live, vigorous crabs are processed; therefore, if the crabs are held in tanks or in cool storage prior to processing, the condition of the crabs is checked before butchering.

Cooking.—Both whole crabs and butchered crab sections are cooked. The crabs are butchered by use of a fixed iron blade. The carapace is removed and the crab is eviscerated, split into halves, washed, and conveyed to the cooker. The halves or sections are cooked in boiling water 10 to 15 min. in stainless steel batch or continuous cookers. If whole (uneviscerated) crabs are cooked, the cooking time is longer, about 20 to 25 min. The hot crabs are cooled in water prior to meat removal and then are dumped onto a stainless steel table for meat removal.

Packing and Freezing.—The body and leg meats are weighed separately and packed in about equal proportion in No. 10 C-enamel cans holding 5 lb. net weight of drained crab meat. Number 2 cans, holding 1 lb. of meat, are also packed, but in smaller volume. The cans are sealed under a low vacuum, frozen in a sharp or blast freezer, and stored at 0° to −10° F. (−18° to −23° C.). For retail sale the meat is usually thawed and repacked into trays or cartons for display as fresh chilled crab meat. The thawed meat should be used within a few days.

Freezing Dungeness Crab in the Shell.—Cooked whole and eviscerated crabs are prepared and frozen for restaurant use and retail sale. Both blast and brine immersion freezers are used for the whole cooked crabs. For brine freezing, the crabs are placed in metal trays or baskets and lowered into circulating brine of 88° (eutectic) salometer at 0° to 5° F. (−18° to −15° C.) for up to 45 min., then removed, and dipped into fresh cold water to remove excess brine and provide a light ice glaze. The frozen whole crabs are packed in flexible film bags or in shallow cartons for storage and distribution. Owing to the market demand for chilled fresh or thawed crab in the shell, this item comprises the larger part of the frozen Dungeness crab production.

Frozen Dungeness crab meat stores well only if protected by suitable packaging in sealed containers against dehydration and oxidation, and stored at −10° F. (−23° C.) or lower. Under these conditions it stores well for six months.

Northern Lobster.—The inshore lobsters are caught in traps, also called pots. The typical pot consists of an oblong box made of laths spaced to allow many of the undersized lobsters to escape. In search of bait the lobster enters through a funnel-shaped opening. The lobsterman makes his rounds once or twice daily in a small boat, hoisting each pot to the surface, removing the lobsters, returning any undersized lobsters to the sea, and returning the pot to the bottom. In cold weather, the lobsters are found in deeper water, farther from shore. The lobsters are held alive in seawater pounds or tanks until they are sold or processed.

Preservation of Lobster Meat.—The lobsters are cooked in boiling dilute salt brine (3%) for 10 to 20 min., then cooled before the meat is removed manually. The meat averages 22% yield and is mostly in the tail.

Lobster meat is frozen in sealed cans, 14 to 16 oz. each, and is sold mainly to processors for preparation of lobster specialty items. A substantial volume of frozen lobster meat is imported from Canada. For best results the meat should be stored at −10° F. (−23° C.) or lower, and used within 3 months.

Deep-Sea Lobsters.—The deep-sea lobsters are caught by large trawlers off the coast of New England in waters up to 200 fathoms deep. The lobsters are placed in seawater tanks in the vessel. Difficulties are sometimes encountered in keeping offshore lobsters alive on board the vessel. A possible solution is to freeze the whole lobsters aboard the vessel in either a blast or immersion freezer.

Freezing Whole Lobsters.—Early tests of freezing raw or uncooked lobsters were unsuccessful because it proved impracticable to remove the meat without tearing it into small pieces after the lobsters were thawed and cooked. On the other hand, the tests showed that the meat of frozen raw lobsters after thawing and cooking was greatly superior in flavor and texture to that of frozen cooked lobsters. A method which overcomes the difficulty of the meat sticking to the shell of lobsters frozen raw and later cooked involves immersion in boiling water before cooling and freezing. The heating period should only be of sufficient duration to cook the meat next to the shell but not the meat below the surface. Heating in boiling water for 1½ min. is said to be sufficient for a 1-lb. lobster. A small volume of lobster tails and claws, frozen raw or cooked, is vacuum packed in plastic pouches and marketed.

Another method of preparing raw whole lobster for freezing uses electric shock to paralyze or kill the animal, which is then placed in a quick freezer. It is claimed that lobsters frozen raw by this method may be stored for six months before thawing and cooking, with good results. Still another method of extracting meat from spiny lobster uses shell freezing techniques to separate the meat from the outer shell and flash freezing using liquid nitrogen.

Spiny Lobster.—The spiny lobster is caught most successfully with baited traps made of wood slats. In Florida, the traps are 2 by 3 ft. at the base, about 18 in. high, with sloping sides, and with a funnel opening on the top. The trap is fished in the inshore waters about 5 to 10 fathoms deep. The fisherman picks up his traps every day or two and delivers his catch alive to the buyer who transfers them to live tanks or cool storage rooms (35° F; 2° C.). There is an increasing use of circulating seawater tanks on fishing vessels in areas such as Australia where distances to the spiny lobster grounds are greater.

Freezing Spiny Lobsters.—Preparation of the spiny lobster consists of breaking the tail from the body and removing the intestine. The tail is about ⅓ of the weight and is about ⅔ meat. The tails are washed and sorted into 4 sizes, from 6 to 16 oz. each, for freezing.

The raw tails are frozen individually or in blocks, protected with an ice glaze, and packed in waxed cartons. The frozen lobster tails store well if they are protected from dehydration and are stored at −10° F. (−23° C.) or lower.

SHRIMP

Production of Frozen Shrimp

Shrimp is an important food because of its high nutritive value and palatability. As an article of commerce of the state of Louisiana and as a source of income, the shrimp industry is large. Nearly 70% of the U.S. catch comes from Louisiana water around the mouth of the Mississippi River. It is estimated by the Louisiana Wildlife and Fisheries Commission that over 200,000 people are employed in the work of catching shrimp and 125,000 additional persons engaged in the various processing phases of the industry.

In spite of the size of the shrimp industry and value of the commodity, the United States has not been able to match the production of shrimp to the demand.

Although shrimp are highly perishable and are usually caught during certain seasons only, they are preserved in large quantities by freezing. During the past two decades increasing quantities of frozen shrimp have been imported into the United States; at the present time these imports equal or exceed the total production of the continental United States and Alaska.

The domestic shrimp industry is located principally along the coast of the Gulf of Mexico. Some shrimp are caught also along the Atlantic and Pacific coasts and in Alaskan waters. The principal shrimp packing states are Florida, Louisiana, and Texas.

Three different species of shrimp are captured in southern waters in important commercial quantities. These are *Penaeus setiferus*, *Penaeus aztecus*, and *Penaeus brasiliensis*. A fourth species *Hymenopenaeus robustus*, also called royal red shrimp, is only caught in deep water, 200 fathoms or more, in the Gulf of Mexico and off the northeastern coast of Florida. It has not as yet assumed commercial importance. The species taken in Alaskan waters include the following: *Pandalus borealis*, *Pandalus dispar*, *Pandalus goniurus*, *Pandalus platyceros*, and *Cragon franciscorum augustimana*.

Along the coast of the Gulf of Mexico the white shrimp (*Penaeus setiferus*) spawn at sea. The eggs hatch at sea and the young shrimp are carried by wave action into the shallow waters of the streams and bayous along the coast. Under favorable conditions of food, temperature, and salinity they grow very rapidly, attaining marketable size in four to six months. As they grow they gradually move out to sea to complete their life cycle.

Shrimp are caught by towing a trawl net from the stern of the trawlers. After 1 to 3 hr. of trawling, the net is hauled aboard and emptied. Since catches contain many species of sea life besides shrimp, considerable time is required to separate the shrimp from the remainder of the catch.

The Quality of Frozen Shrimp

Influence of Prefreezing History.—It is recognized that high temperatures or lack of refrigeration in the storage of perishable foods results in rapid loss of quality. This is even more true of shellfish than of meat, since the former do not go through an aging process. Deterioration of shellfish quality is considered to result from the action of enzymes, originating both from their tissue and from the contaminating microorganisms originally present on the shrimp or introduced during catching, handling, and processing. Spoilage of the product is believed to be mainly the result of bacterial action, and is caused by the consequent formation of compounds which impart off-odors, off-flavors, and color changes.

Effect of Delayed Handling on Quality.—Freshly caught shrimp allowed to remain 6 hr. at air temperatures prior to icing spoil after 6 days of ice storage, as indicated by organoleptic observation and the indole content. Similarly, shrimp held 2 and 6 hr. prior to heading, washing, and icing, after 6 and 11 days of iced storage have significantly higher bacterial counts, higher tissue pH's, lower organoleptic scores, and greater loss of characteristic sweetness than those iced immediately after catching. These results stress the need for rapid handling of freshly caught shrimp to assure retention of superior quality.

Effect of Length of Ice Storage on Quality.—The use of larger boats which are able to extend offshore fishing areas has resulted in an increase in length of time in ice storage on the trawlers and a longer interval of time between catching the shrimp and their processing and freezing. Also, new species of shrimp which are more susceptible to deterioration in refrigerated storage are entering commercial channels. This has resulted in delivery to processing plants of shrimp with high bacterial counts and appreciable development of "black spots" or melanosis. Furthermore, the progressive changes in ice-stored shrimp can be divided into three phases. During the first phase of 0 to 7 days of ice storage, the shrimp lose their sweet flavor. This is followed by phase two, of 8 to 14 days, and is characterized by tasteless product, and in the last phase (more than 14 days storage) rapid deterioration, accompanied by off-odors and off-flavors.

Chemical and Bacteriological Changes Affecting Quality Prior to Freezing.—*Effect of Bacteriological Condition*.—Since shrimp live only a few minutes after removal from their natural habitat, microbial spoilage starts immediately through marine bacteria on the surface and in the digestive system, and through microorganisms which happen to contaminate the shrimp on the deck, in handling, and from ice used during their storage. Fish and other marine organisms caught with the shrimp may also, chiefly through slime and exuded intestinal contents, contaminate them. Removal of the heads reduces the bacterial count somewhat because the head carries approximately 75% of the bacteria. The bacterial counts of freshly caught headless shrimp are largely determined by the bacteria and debris adhering to the surface. The average bacterial count on shrimp as prepared for icing under commercial conditions, headed and washed by the fishermen, was 7,400 per gram. With expeditious handling and thorough washing under commercial conditions, headless shrimp can evidently be placed in ice storage on board trawlers carrying only a relatively low microbial load.

In commercial practice shrimp are packed in alternate layers of ice and shrimp. Since the melting ice from the upper layers of shrimp washes down over the lower layers, position in the bin influences bacterial count. One test showed that while shrimp from the layer next to the top had only a slight increase in bacterial counts, the bacteria in the lowest layer increased a thousandfold after nine days of storage.

Studies have been made of the types of bacteria initially present in fresh Gulf of Mexico shrimp caught adjacent to the Texas coast. The main groups present were Achromobacter, Bacillus, Micrococcus and Pseudomonas; these made up 78% of the 1,200 isolates. In biochemical characteristics, 62% of the isolates were proteolytic, 35% lipolytic, 18% reduced trimethylamine oxide, and 12% formed indole. During ice storage there was a steady increase in percentage of Achromobacter to a marked domination of the flora (82% after 16 days). The well-established significance of this genus as a cause of spoilage in fresh foods correlated quite well with the known ice storage-life of fresh shrimp.

Black Spot, Cause and Prevention.—The prevention of deterioration in the quality of fresh and ice-stored shrimp involves two main problems, namely, maintaining low microbial counts, and prevention of oxidation, chiefly of phenols, into melanins. This condition is known as "black spot" or melanosis. In general, the discoloration begins to develop in the membrane which connects together the two ends of overlapping shell segments. When the black spot condition is severe the shrimp show pronounced black bands where shell segments overlap, giving a banded or zebra appearance to the tail. The tail and head fins become black and the crawling legs change color, first at the joints and finally over the entire leg. The top and sides of the head are affected, so that the interior becomes a soft mass and the carapace or shell becomes flexible instead of stiff, as it normally is.

The dark color is the result of melanin pigments which form on the internal shell surfaces or, in advanced stage, on the underlying shrimp meat. These pigments are produced by an oxidative reaction of tyrosinase on tyrosine, and the reaction is accelerated by copper and other metallic ions. Black spot has been observed on all species of shrimp taken from waters contiguous to North America. Earlier assumptions that this discoloration was connected with microbial activities are definitely ruled out. A recent comprehensive study confirms the concept of this black spotting as a nonmicrobial phenomenon.

Objective Tests for Quality Change.—Since chemical compounds may decrease or increase in concentration as a result of catalytic action of tissue enzymes or through bacterial action, studies were undertaken to determine whether such changes could be used as objective tests for quality of both ice-stored and frozen shrimp. The results showed that glycogen sugar content increased rapidly during the first three days of ice storage, then decreased slowly until the sixth day, after which time it decreased rapidly to a minimal value. Similar results were obtained for lactic acid content. The acid-soluble orthophosphate content of shrimp tissue decreased very rapidly during the first seven days of storage and then less rapidly throughout the remainder of the storage period. These three tests can therefore be used to indicate changes in prime quality. Tests useful for determining onset of spoilage of ice-stored shrimp are trimethylamine nitrogen, volatile acids, bacterial content, Nessler ammonia, and sulfhydryl groups determined by iodine titration. The first four tests showed low values for the first 10 to 12 days of storage and then increased very rapidly as spoilage occurred. Sulfhydryl groups decreased until the onset of spoilage and then increased very rapidly.

The pH of shrimp tissue is a fairly good indicator of shrimp quality. Freshly caught shrimp has a pH of about 7.2, and this progressively increases to 8.2 or higher after 21 days of storage. A pH value of 7.7 or lower is indicative of prime quality shimp, those having values from 7.7 to 7.95 as poor quality but acceptable, and those having a pH of 7.95 or above as spoiled.

It is postulated that loss of quality during the early period of storage is mainly caused by autolysis and with longer storage spoilage occurs mainly through bacterial action.

Factors Affecting the Quality of Frozen Shrimp.—*Storage Stability.*—Shrimp frozen immediately after catching (within 3 hr. after the net was emptied) maintained the highest quality during a 12-month storage period at 0° F. (−18° C.). Even with proper and adequate icing on the trawler, the quality of the frozen shrimp decreased in direct relation to the length of time of storage in ice. Bacteriological data on these samples paralleled organoleptic results. Frozen storage for 1 day at 0° F. (−18° C.)

caused an average bacterial count reduction of 50% for market shrimp. This increased to 82% after 2 months' storage. Percentage reductions during 2 to 12 month frozen storage of laboratory and commercially frozen products ranged from 48 to 99%.

Effect of Packaging.—The problem of proper packaging of green or headed shrimp has been investigated. Results obtained show that large losses of product moisture can occur through the use of poor moisture-vapor barriers for the containers. Organoleptic and tenderness tests after cooking the shrimp which had been packaged in various types of containers resulted in the following observations: toughness increased directly with loss of moisture from the product during storage, and poor packaging contributed to the development of disagreeable odors and flavors and general loss of quality.

Effect of Temperature.—The quality of frozen shrimp is not only affected by the prefreezing history and the type of packaging, but also by the temperature at which the shrimp are stored. Deterioration occurs in samples stored for 3 months at 10° F. (−12° C.) and at 10 months they are inedible. Shrimp stored at −40° F. (−40° C.) have the appearance of freshly frozen shrimp and were of excellent quality at the end of 12 months' storage. Those samples stored at 0° F. (−18° C.) were of only slightly poorer quality than those stored at temperatures fluctuating between 0° and 10° F. (−18° and −12° C.). Temperatures above 0° F. (−18° C.) are not recommended for storage of frozen shrimp, and it is doubtful if temperatures below 0° F. (−18° C.) will result in a sufficiently superior product to warrant the added expense.

Bacteriological studies of these samples showed that the number of viable bacteria was significantly reduced during freezing at −40° F. (−40° C.). The reduction was greater in peeled shrimp than in unpeeled shrimp. Storage at 10° F. (−12° C.) was more destructive to the bacteria than storage at lower temperatures, while at −40° F. (−40° C.) there was no decrease in numbers during 12 months' storage, and no evidence was obtained that temperature fluctuating between 10° F. and 0° F. (−12° and −18° C.) was more lethal because of the fluctuations.

Commercial Freezing of Shrimp

The U.S. imports of shrimp in large part are a result of the annual loss of shrimp because of spoilage between the time it is caught and the time when the shrimp are ready for shipment to remote markets. This loss is often as high as 10% of the total catch. In addition to this loss, some of the marketable shrimp are lowered in quality. Because of the character of the product and the conditions under which it is caught, some loss is inevitable, but the losses should and can be considerably reduced below the present level. Such reductions would aid in conservation of natural resources, assist the fishermen to obtain a better return for their labors, and result in a higher quality shrimp being received by the consumer.

Studies indicate that the highest quality shrimp are produced if the shrimp are frozen as soon as possible after removal from the water. There are difficulties involved in this procedure. First, freezing facilities are limited even at the shore bases, and the boats do not have large facilities for freezing shrimp.

To preserve the fresh quality of shrimp, better means of refrigerating fresh shrimp from the time they are caught until they are delivered at the shore plants are required. In recent years, the shore plant facilities have been expanded considerably by the enlargement of existing plants and by the construction of new plants at the shrimp landing ports.

The freezing of shrimp on board the trawler is an ideal situation and produces shrimp of the highest quality. There are, however, limitations to the procedure. The equipment needed to outfit a trawler properly is rather costly, and a large crew is necessary to carry out the heading operation if the equipment is lacking.

Thus far, mechanical refrigeration is being introduced, but has not been widely used to preserve shrimp on board the trawler. Ice usually provides the refrigeration.

Harvesting and Handling of Shrimp.—Practically all shrimp are caught by means of a trawl which is towed slowly along the ocean floor. This trawl is a large funnel-shaped bag with a winged bag held open by otter boards so that the shrimp can be entrapped. At intervals, depending upon the volume of the catch in the net, it is hoisted aboard the vessel and emptied on the deck. The net is again thrown overboard for another drag while the crew of the vessel separates the shrimp from the various other marine life which is also caught in the trawl. The shrimp are sorted, washed, and stored with alternating layers of ice in bins in the hold of the vessel. Some crews, while at sea, remove the heads. This saves considerable space in the bins, since the tails of the shrimp are the only edible portion.

On many vessels the shrimp are graded during this operation, the larger and smaller shrimp being placed in separate baskets (Table 8.1).

The first operation aboard the vessel has a particular bearing on the market value of the shrimp. If permitted to lie on the deck of the vessel in the hot sun, shrimp deteriorate rapidly in the course of a few hours, and the quality is reduced materially. No matter whether the shrimp are to be sold fresh, cooked, canned, or frozen, they must be packed in a sufficient amount of ice at the earliest possible time. Care should be exercised in handling them to be sure that the shrimp are not injured, since a bruise

TABLE 8.1

SIZE CLASSIFICATION OF SHRIMP

| Type and Condition of Shrimp | Count-Number of Headless Shrimp per Pound | | | |
| | Not Peeled | | Peeled | |
	Regular	Deveined	Regular	Deveined
Raw, chilled or frozen	15 and less	16 and less	18 and less	19 and less
	16–20	17–21	19–24	20–25
	21–25	22–26	25–30	26–31
	26–30	27–31	31–36	32–38
	31–35	32–36	37–42	39–44
	36–42	37–43	43–50	45–53
	43–50	44–51	51–60	54–63
	51–60	52–61	61–72	64–75
	61 and over	62 and over	73 and over	76 and over

causes more rapid deterioration as a result of enzymatic and bacterial action. The U.S. Dept. of Health, Education, and Welfare has published recommended procedures for the harvesting and sanitary handling of shrimp.

Unloading Cargo.—Upon arrival in port, the shrimp boat proceeds to the raw shrimp plant dock where the catch is unloaded either by basket or hoist by power conveyor. The shrimp are separated from the ice by washing and are then weighed prior to entering the processing plant.

Preparation of Shrimp for Freezing.—The introduction of economically feasible freezing methods caused a major revolution in the shrimp industry. Refrigeration opened new markets for shrimp in inland areas, where they were virtually an unknown commodity. Shrimp are readied for the frozen market in the following forms: (1) frozen headless; (2) frozen peeled and deveined; (3) uncooked frozen and breaded; and (4) headed and unshelled.

Freezing Headless Shrimp.—Methods of handling headless shrimp from the packing and weighing line to the freezer are more or less standard. The principal differences depend upon whether the processing plant operates its own freezer or uses the facilities of a public freezer.

When the shrimp are landed, they are taken to the plant where the heads are removed from any remaining whole shrimp. After being washed, the headless shrimp are inspected to remove defective ones, graded according to size, and either iced for distribution as fresh shrimp or packed for freezing. In addition to shrimp with shells on, peeled and deveined shrimp are also frozen. These operations are done by machinery in many of the plants. Size grading machines are used extensively, and peeling and deveining operations, formerly done largely by hand, may now be done much more rapidly by machine.

Cartons of several sizes are used for frozen shrimp, such as the small 8-,

10-, and 12-oz. consumer-type waxed carton with overwrap, and the larger sizes having a capacity of 2½, 5, or 10 lb. Following the freezing operation, the larger cartons may be opened and the shrimp glazed by spraying cold water on the surface of the frozen block of shrimp or by immersing the product in cold water. About 8 oz. of water are used per 5-lb. carton of shrimp. After the shrimp are sprayed, covers are attached to the carton, and the cartons are turned upside down. Glazing of shrimp in the larger-size cartons is inadequate, however, because the glaze evaporates at the edges of the block during frozen storage and the shrimp then become desiccated. A more recent method that is being used rather extensively, and is the preferable technique, is to omit the glazing entirely and to rely on a moisture-vapor proof overwrap on the carton, with careful packing of the shrimp to minimize voids within the carton; this will go far toward preventing desiccation during prolonged frozen storage.

Plant Procedure.—The procedure used to process the headless shrimp embraces the following steps:

Receiving and Unloading.—Two men working as a team in the hold of the fishing vessel shovel iced headless shrimp into a portable power conveyor. The conveyor elevates the shrimp from the vessel hold and discharges them into a wash vat located on the dock. Here the storage ice is washed away from the shrimp.

Inspecting and Grading.—A conveyor removes the shrimp from the vat and feeds them past a team of seven inspectors. The inspectors remove by hand all extraneous matter and shrimp of inferior quality. The inspected shrimp are discharged into the receiving hopper of a grading machine.

Packing and Weighing Fresh Shrimp for Freezing. Mastering.—The frozen 5-lb. cartons of shrimp are removed from the freezer by 2 teams of 2 men each and placed 10 cartons to the master carton. The master carton is then sealed and placed in cold storage to await shipment.

Freezing Peeled and Deveined Shrimp.—*Receiving and Grading.*— Iced fresh headless shrimp are delivered by truck in 100-lb. boxes. The boxes are unloaded through a wall opening directly into a cold storage room. From the holding room shrimp are emptied into a vat where the storage ice is flushed away and the shrimp are washed. A conveyor belt removes the shrimp from the vat and feeds them past a team of inspectors who manually remove any extraneous matter and damaged shrimp. The inpected shrimp are then fed into the grading machine, which sorts them into size categories and discharges each size through one of several metal chutes. The shrimp are caught in metal containers which when filled are pulled manually along roller tracks to a scale. This weight is checked by a recording clerk and the shrimp are rolled back into the refrigerated holding room to await further processing.

Heading, Peeling, and Deveining of Raw Shrimp.—When the shrimp are landed, they are taken to the plant, where the heads are removed from any remaining whole shrimp. After being washed, the headless shrimp are inspected to remove defective ones, graded according to size, and either iced for distribution as fresh shrimp or packed for freezing. The shrimp with shells on are peeled and deveined by machines. The peeled and deveined shrimp are discharged onto a conveyor feeding to an inspection station, and shells and veins are discharged onto a waste belt.

Inspection.—The peeled and deveined shrimp are fed past a team of inspectors who check for any remaining shell or vein, pulling aside any shrimp needing further cleaning. The cleaned shrimp continue on to the next operation. The incompletely cleaned shrimp are diverted to hand operators who complete the operation and place the shrimp back on the conveyor feeding to the next operation.

Preparation for Freezing.—The peeled and deveined shrimp are discharged by the conveyor to 2 adjoining work stations where 2 teams of 5 workers each place the shrimp on thin aluminum sheets. Each sheet holds approximately 2½ lb. of shrimp, and the shrimp are spaced in such a manner that they do not touch each other. The aluminum sheets are separated by angle irons and placed in 12 stacks of 5 each on a rolling rack; they are manually rolled into the freezing tunnel.

Freezing and Cold Storage.—A temperature of −25° to −40° F. (−32° to −40° C.) is recommended for freezing shrimp. This temperature is obtainable in multiple freezers and in blast freezers, and is low enough to freeze the shrimp so rapidly that the cellular breakdown is kept at a minimum. The temperature of storage of the frozen shrimp should be maintained at 0° F. (−18° C.), and preferably lower, at all times. At the lower storage temperatures, the development of a rancid flavor in the shrimp is minimized.

Immersion Freezers.—The immersion-type freezer is used largely for the freezing of shellfish such as shrimp aboard the fishing vessel. The product is frozen by immersing it in an agitated cold brine solution of a fixed concentration and temperature.

The type of immersion freezer employed depends on the nature of the product being frozen. Although sharp, blast, or plate freezers are somewhat versatile, inasmuch as they will freeze a variety of different fish species in various forms, the application of the immersion freezer is restricted largely to the specific type of fish or shellfish for which it was designed.

The following factors should be considered in the selection of a commercial immersion freezer aboard a fishing vessel: (1) type of products to be frozen, (2) allowable freezing time for products of average weight, (3) handling requirements, (4) source of available power, (5) refrigeration-equipment space requirements, (6) average catch, (7) space required to hold a portion of the fish prior to freezing, (8) cost and dependability of refrigeration equipment, (9) effect of brine on the product, (10) freezing temperature of the brine, and (11) cost of maintaining a clean brine supply. The following describes various types of immersion freezer employed on vessels.

A sodium chloride brine solution—its use having been proved in other types of immersion freezers—was first used. The shrimp frozen in this solution were of good quality; however, they sometimes fused together into a solid mass upon freezing, and subsequently dehydrated in cold storage unless protected by spray glazing. By using a brine solution composed of salt and sugar, the shrimp do not fuse together upon freezing and they can be stored for a period of 60 days or more without dehydration. This solution is presently being employed in the immersion freezers aboard several shrimp fishing vessels.

The freezing is done in a stainless-steel tank approximately 5 ft. wide, 7 ft. long, and 4 ft. high, which is located on the deck of the vessel. The brine temperature is maintained at 0° F. (−18° C.) by refrigerant R-12 circulating through plates placed within the tank. A hydraulically driven propeller inside the tank provides the necessary brine agitation. The frozen shrimp are maintained at 0° F. (−18° C.) by an additional diesel-driven compressor supplying refrigeration to overhead plates located within the holds.

The freezing operation is as follows: shrimp, after being caught, are headed and washed. They are weighed in 50-lb. lots and put into stainless steel wire baskets. These baskets, in turn, are set in the freezing tank. When 15 min. have elapsed the baskets are picked up and the shrimp are dumped into containers, which are then placed in a cold storage room. At the end of each day, the shrimp are put into 50-lb. master cartons. By this method, it is possible to process and freeze more than 3,000 lb. of shrimp in 10 hr.

The advantages of this type of freezer are that it (1) produces a high-quality product, (2) reduces handling costs, (3) eliminates shore processing costs, (4) freezes products quickly and efficiently, (5) permits thawing of individual shrimp rather than 5-lb. fused blocks of shrimp, and (6) provides a protective glaze which reduces dehydration in cold storage.

Applications Unique to Freezing of Shellfish.—Two factors of unique importance to seafoods are: where the freezing is done; and whether the product has been previously frozen. Many tons of shellfish are frozen aboard ships as they are caught. Most are frozen in bulk, to be thawed and reprocessed later. It would seem questionable to use a costly instant freezing method in later processing if the shellfish had previously been frozen by a slower method. If cryogens such as liquid nitrogen are to be used for instant freezing of shellfish on the ship, the ship must have holding capacity for a weight at least equal to the weight of shellfish to be frozen while the ship is at sea. An alternative method would be to supply the cryogen to the trawler from a supply ship.

A factor that can nullify all efforts to improve quality and lower costs of frozen foods is poor handling in distribution and storage. It matters little how a food is frozen originally if the product is allowed to thaw and stand in an overloaded display case or on a loading platform. Much progress has been made in handling practices in recent years, but there is still room for improvement.

Advantages of Rapid Freezing.—Liquid nitrogen as a commercial refrigerant for shrimp is gaining in acceptance as a result of intensive research and development work that has been performed during the past ten years. Experimental evidence obtained from chemical, bacteriological, histological, and organoleptic tests showed that the use of liquid nitrogen as a cryogen for the preservation of foods, especially fishery products, resulted in superior foods when they were compared to those preserved by conventional methods of refrigeration. Comprehensive studies on shellfish showed that the shelf-life of both shrimp and oysters was extended either (1) after rapidly freezing in liquid nitrogen, or (2) after using liquid nitrogen to maintain temperatures for refrigeration of the products until they were brought ashore for processing. It is observed that the shellfish frozen by conventional refrigeration had higher indole and trimethylamine nitrogen content than the samples cooled by liquid nitrogen, at the end of each test period. In addition to this fact, the rate of formation of these compounds was more rapid in the iced controls than in the samples cooled by liquid nitrogen. Histological studies showed that the product frozen rapidly by liquid nitrogen developed smaller ice crystals, and less protein changes than the shellfish frozen by mechanical refrigeration. Formation of smaller ice crystals and less concentration of the protein resulted in less drip-loss upon thawing of the product frozen by liquid nitrogen. Organoleptic test panels conducted with trained individuals further confirm these qualitative results. The taste panels considered odor, appearance, sweetness, flavor, and texture. All of the other food products experimentally frozen by liquid nitrogen were superior in appearance, and had physical qualities identical to those of products frozen by conventional methods.

All of the products frozen by liquid nitrogen had a smoother texture and a more normal, fresh color while in the frozen or thawed state, as against a water-logged and bruised appearance when they are frozen by conventional methods (Fig. 8.7).

Courtesy of Fishing Gazette

FIG. 8.7. INSPECTING IQF SHRIMP FROM LIQUID NITROGEN FREEZER

OYSTERS

Since the record high of almost 170 million pounds of oyster meats harvested at the turn of the century, production has declined steadily, with about 60 million pounds being harvested; of this amount only 2.3 million pounds are frozen.

Reasons for this decline are complex and involve local tradition, political differences, pressures and expediencies, individual selfishness, and ignorance of biological facts and factors. Many elements only indirectly involved with shellfish management and production have complicated the problem.

However, even moderately enlightened management of a resource can accomplish a great deal.

It is estimated that in territorial waters of the United States there are about 1,400,000 acres of bottom designated as oyster-producing areas. Of this entire acreage only about 185,000 acres are privately leased or controlled. Even though a considerable area of these privately controlled grounds is not cultivated, these beds produce, nevertheless, about 50% of the total oyster crop of the United States. There is a great difference in productivity between privately controlled and public oyster beds.

Species of Oysters Harvested

Although more than 100 species of oysters have been described, only three are of economic importance in the United States. The Eastern oyster *(Crassostrea virginica)*, which is found from Maine to Texas, accounts for over 50 million pounds of meats marketed. On a regional basis, the Gulf of Mexico was the leading production area with 23 million pounds, followed closely by Chesapeake Bay with 22 million pounds.

The Pacific oyster *(Crassostrea gigas)* is not native to the United States. Seed oysters imported from Japan are planted on suitable beds located primarily in Washington where they grow to marketable size in about 18 months; in contrast, the Eastern oyster requires 30 or more months to reach marketable size and the Olympia requires 48 to 60 months. National spatfall (reproduction) does occur with the Pacific oyster; but unless summer water temperatures are high enough, natural reproduction is not sufficient to support the industry.

The Olympia oyster *(Ostrea lurida)* is of importance primarily from a historical point of view, as it contributes only 35,000 lb. of meats to the oyster market, as compared with the 700,000 lb. harvested in 1926. This excellently flavored little oyster has not been able to compete with the more vigorous Pacific species.

Oyster Diseases and Predators

Oysters are vulnerable to attack by a large number of enemies. Probably the most spectacular instance in recent years was the virtual destruction of the oyster beds in Delaware Bay by a disease caused by a spore-forming protozoan *Minchinia nelsoni,* commonly known as MSX. Resistant oyster strains did survive, however, and it appears that this resource may be gradually rebuilt. Oyster drills and starfish are ubiquitous and serious predators on oysters. Recent research has shown that these pests may be effectively controlled by coating sand or other materials with chlorinated hydrocarbons and/or other chemicals, and either broadcasting the treated sand over the oyster beds or surrounding the beds with a strip of the treated sand to keep the predators out.

Factors Affecting Quality of Frozen Oysters

Composition of Fresh Oyster Meats.—Studies on seasonal variation in composition of southern oyster meats show that solids content ranged from a low of about 9.0% in September to a high of 13.4% in March. Fat content follows a similar pattern ranging from 7.0% (dry-weight basis) in October to about 13.0% in April.

In addition to large variations in total solids, there is a great variation in the proportion of protein and carbohydrate in the solids. In May, these constituents each account for about 39% of the dry matter; but in August, protein has increased to about 62%, and carbohydrate has dropped to 12%. No definitive tests have yet been conducted to determine how these variations affect the frozen storage life of oyster meats.

Freshness of the Raw Material.—As with other food products, oysters approaching the limit of edibility have a very short frozen storage life— about 1 month of 0° F. (−18° C.).

A good indication of quality at the time of freezing may be obtained by measuring the pH. Pacific oysters when just harvested have a pH of about 6.4. During storage at 32° to 34° F. (0° to 1° C.), the pH decreases and at the point where samples were rated as being of fair quality, the pH was about 6.0. Below pH 6.0 the oysters were definitely stale.

There are similar results with Eastern oysters: Very fresh = pH 6.5 to 6.7; stale = pH 5.8 to 6.0. However, with *C. virginica* from the Gulf of Mexico, there are considerable seasonal variations in the pH of very fresh oysters—from 6.38 in February to 6.02 in June. The stale point (appearance of a sharp sour odor) also varies from 5.80 in the winter months to 5.52 to 5.69 during the summer. There does not appear to be any correlation between pH and the quality of the frozen-stored product, with little change during a 12-month storage period.

Free Liquor.—Some samples of frozen oysters will show well over 20% of "drip" upon thawing. The amount of free liquor depends on the conditions of blowing. As blowing time increases, more water is picked up by the oyster meats; the water is, however, very loosely held in the tissues and is released on thawing. Blowing in 0.75% salt water greatly decreases the amount of drip formed on thawing. A similar result was found with Pacific oysters, although the amount of drip (1.0 to 4.28%) was far less than was found in Eastern oysters.

Darkening.—Frozen oysters may darken somewhat during frozen storage, a condition worsened by slow freezing rates and by high storage temperatures. The addition of monosodium glutamate does not ameliorate the condition. Adding ascorbic acid to the oyster meats does not prevent or retard darkening. Similarly, adding antioxidant mixtures of (1) nordihydroguaiaretic acid, betahydroxyanisole, and ascorbic acid and (2) ascorbic acid plus citric acid have little or no effect on darkening.

Pink Yeast.—A phenomenon that, to date, has only been reported as occurring with frozen oysters is the growth of a pink yeast during frozen storage. Research has shown that this organism is capable of growth at temperatures as low as −35° F. (−37° C.). Apparently sanitation in the processing plant is the only weapon effective against this organism.

Packaging.—The package selected must, in addition to providing the usual eye-appeal, be sufficiently tight to prevent leakage of the contents before freezing and be sufficiently impervious to moisture vapor and oxygen transfer to prevent desiccation and rancidity during storage periods of about 12 months. Pacific oyster meats frozen individually, glazed with water, packed in polyethylene bags and held at 0° F. (−18° C.) will be of satisfactory quality for about 8 months.

However, Eastern oysters individually frozen and glazed with either 0.5 or 1.0% ascorbic acid, and packed in pint metal friction-top cans and stored at −40° F. (−40° C.) develop a slight "rancid fish" odor in 2 months.

Rate of Freezing and Storage Temperature.—Rapid freezing in brine at −10° F. (−23° C.) and storage at low temperatures −10° F. (−23° C.) greatly extends the storage life as determined by taste panel evaluation of simmered oysters in comparison with oysters slowly frozen in air at +10° F. (−12° C.) and stored at +10° F. (−12° C.). Lower storage temperatures appear to have slightly greater effect on extending storage life than does increasing the rate of freezing. The improvement in quality is masked, however, when the oysters are served in a stew.

Harvesting Oysters

There are basically three methods of harvesting the oysters—picking, tonging, and dredging.

Picking.—This method is confined to the West Coast, where some oyster beds are exposed at low tide. The operator selects an area to be harvested, boats are towed into place, and when the tide goes out the boats are filled with oysters picked by hand from the surrounding beds.

Tonging.—This method is used primarily on the East Coast. The tongs range from 12 to 20 ft. long and consist of two poles crossed like scissors and with toothed iron baskets about 3 ft. long at the end of each pole. The tongs are lowered to the bottom in the open position, and when closed by the operator, scoop up the oysters. They are then raised and opened to dump the catch into the boat. A day's catch will vary greatly with the density of the oysters on the beds, but a good tonger may get more than 25 bu. per day.

Dredging.—The conventional oyster dredge is made up of a steel frame with a toothed bar at the lower front edge and a bag of netting and chain at the rear to catch the oysters that are scooped up as the dredge is towed over the bottom. The size will vary from one capable of holding only 2 or 3 bu. up to one capable of holding 30 bushels or more.

There are two types of mechanized dredge in use; in the first, a conveyor belt carries the oysters up to the vessel, while in the second, water flowing through a large-diameter hose brings the oysters up from the bottom. The mechanized dredges are also very useful in controlling predators, because these are brought up with the oysters and can be culled from the catch and destroyed.

Preparation for Freezing

Various processing steps are necessary prior to freezing, as it has been found that adverse flavor changes occur very rapidly in oysters frozen in the shell. Details of these processing steps vary from one area to another and also from one processing plant to another within a given area. The following is, therefore, of necessity, a generalized description.

Unloading and Washing the Shell Stock.—Where large quantities are handled, the oysters are unloaded from the dredgers onto conveyors that carry the shell stock to a cylindrical washer; here the mud is removed by sprays of water. The oysters are then conveyed to the shucking benches.

Shucking the Oysters.—The shucker stands at a long bench. Immediately in front of him is a small anvil-like iron on which he rests the oyster while he breaks a small piece from the edge of the shell with a hammer. A special knife is then inserted between the shells, the adductor muscle is cut free from its attachment to the shell, the shell halves are separated, and the meat is removed. In many plants, the shucker sorts the oyster meats by size into 2 or 3 containers. In others, the sorting is done mechanically.

Table 8.2 shows the count per gallon for the various size categories of fresh and frozen Eastern and Pacific oysters. The Olympia oysters are uniformly small, averaging about 1,600 meats per gallon, and are not ordinarily sorted by size.

Washing, Culling, and Blowing.—The shucked oyster meats may be dumped on a washing table and given a preliminary rinse with a hand-held spray while the operator culls out large pieces of shell and torn or discolored oysters. The oysters then usually go to a blowing tank where they are violently agitated by compressed air from a perforated pipe in

TABLE 8.2

CLASSES AND SIZES OF FRESH AND FROZEN OYSTERS

| Class | Size (Counts) | Type I—Fresh | | | Type II—Frozen |
| | | Count per Gal. | Count per Quart | | Count per 6 Lb. |
			Largest[1]	Smallest[2]	
I—Eastern or Gulf (*Crassostrea virginica*)	Extra large	160[3]	—	44	Not more than 113
	Large (extra selects)	161–210	36	58	114–148
	Medium (selects)	211–300	46	83	149–212
	Small (standards)	301–500	68	138	213–352
	Very small	Over 500	112	—	353 and over
II—Pacific (*Crassostrea gigas*)[3]	Large	Not more than 64	—	—	Not more than 45
	Medium	65–96	—	—	46–68
	Small	97–144	—	—	69–101
	Extra small	More than 144	—	—	102 and over

Source: Federal specification—Oysters, Fresh (Chilled) and Frozen: Shucked *PP-O-956e* June 29, 1966.
[1] Least count.
[2] Maximum count.
[3] Largest oyster shall be not more than twice the weight of the smallest oyster within each size category.

the bottom of the tank. The blowing process serves to remove sand, silt, and shell fragments. It may also, unless controlled carefully, serve to reduce quality, as the oyster meats will absorb water readily. The water added by "floating" the meats is, however, lost readily when the frozen meats are thawed, resulting in excessive drip and loss of valuable nutrients. The Federal Specification for Oysters, Fresh (Chilled) and Frozen, *PP-O-956e* prepared by the U.S. Bur. of Com. Fisheries, Technological Laboratory in Gloucester, Mass., requires that:

"The total time that the oysters are in contact with water shall not exceed 30 min. The time of blowing the oysters shall not exceed 10 min. which shall count as 20 min. in computing the total time of 30 min."

Following blowing, the oysters are drained briefly and then packed in suitable containers.

Freezing and Storage

The freezing method should conform to accepted commercial practice for the package type used, such as compression plate for meats packed in waxed cartons and overwrapped, and blast tunnels for meats packed in cans or for those frozen individually. Freezing rates should be as fast as practicable, and storage temperatures should be as low as possible. In no case should they exceed 0° F. (−18° C.). Although frozen oyster meats may remain in good condition for 9 months or even longer if prepared from freshly harvested and shucked shell stock and stored at −20° F.

(−29° C.), in oxygen and moisture-vapor proof containers, it would be unwise for the processor to plan on a shelf-life in excess of 6 months, as raw material quality and storage conditions will vary greatly.

SCALLOPS

Historically, New Bedford, Mass., has been the leading sea scallop (*Placopecten magellanicus*) port of the nation. From New Bedford the 60 to 100-ft. vessels have ventured forth summer and winter to the Georges Bank area, 100 to 200 miles distant, to haul their dredges over the highly productive bottom.

A recent drop in the return from Georges Bank has, however, impelled the New Bedford, and other vessels to turn their attention to the Middle Atlantic scallop beds. The potential production of this area is largely unknown, so the future of the sea scallop fisher as a whole is also unknown. Hopefully, the Middle Atlantic beds will satisfy the market until the Georges Bank population has rebuilt itself.

The sea scallop is, however, not the only species of importance. The tiny bay scallop (*Pecten irradians*) is considered by many to more than make up in gastronomic delight for its small size. In addition, a fishery is developing for the calico scallop (*Pecten gibbus*) in the Gulf and South Atlantic areas. The meat of this small animal cannot be distinguished from that of the bay scallop except by the the use of sophisticated electrophoretic techniques. The calico scallop resource is, however, capable of much greater production if satisfactory methods of mechanically shucking out the meats can be developed.

Scallop Predators and Disease

In their larval stage, scallops are preyed upon by the many plankton eaters; later, as bottom dwellers, they make up part of the diet of cod and other groundfish. In addition, boring sponges, snails, and starfish take their toll of the larger scallops. Little is known of the impact of diseases on scallop populations, but they may well play an important role in the observed fluctuations in catch.

Harvesting Methods

Scallops are caught by dredges. In the harvesting of the sea scallop, the dredges are towed by vessels ranging from 60 to 100 ft. in length. Usually 2 dredges are towed simultaneously, 1 from each side of the vessel, and are simultaneously hauled back and emptied on the deck of the vessel. The dredge consists of a heavy steel frame, most commonly 12 ft. wide, to which is attached a bag made up of steel rings on the bottom and on part of the top; the rest of the top is rope netting. The steel rings are fastened together by steel links, and the rope netting is not ordinarily knotted but is held together with clips.

Gear used in catching calico and bay scallops is similar, except for size, to that used in the sea scallop fishery.

Factors Affecting Quality

Composition of Fresh Scallop Meats.—The protein content of scallop meats varies among samples from 14.8 to 17.5%; fat varies from 0.1 to 1.0%; and carbohydrate content is about 3.4%. Information on seasonal variation of scallop composition is not available in the literature.

Washing.—The washing process may range from a very brief rinse in sea water to a six-hour soaking. All too often, neither procedure is sufficient to rid the scallop meat of objectionable sand. A short wash in rapidly flowing sea water will almost entirely remove the sand. If, however, sand is left on the meat, subsequent washing at the shore plant tends only to drive the sand in between the muscle fibers, which have begun to separate during the several days of iced storage; consequently, an inferior-quality product results.

Iced Storage.—The practice of packing the scallop meats in 35-lb. capacity cotton bags undoubtedly contributes to quality loss, although fishery technologists have yet to come up with a better method of shipboard stowage. The geometry of the package is such that cooling rates, even though the bag is surrounded by ice, are very slow. Thus, scallop meats that may be at 60° F. (16° C.) (after soaking in sea water during the summer months) when packed in the cotton bags will require at least 48 hr. to cool to ice temperature. In one series of tests, it was found that the quality of scallop meats packed in 5-lb. cartons and slowly frozen aboard the vessel (approximately 24 hr. to reach 5° F.; −15° C.) was superior, during storage period of 12 months at 0° F. (−18° C.), to the quality of scallop meats held on ice in the usual cotton bags for 2 days, then packaged and plate frozen ashore.

Frozen Storage.—At 0° F. (−18° C.) scallop meats have a frozen storage life of 7 to 12 months. Information is lacking on permissible storage life at other temperatures.

CLAMS

Statistics are not readily available on the proportion of the 65 million pounds of clam meats produced in the United States that are frozen; however, the relative importance of the dozen or so species that make up the commercial catch is as follows: Surf clams (*Spisula solidissima*), which are harvested almost exclusively off the coast of New Jersey, account for 59%; the hard clams or quahogs (*Venus mercenaria*) from the New England, Middle Atlantic, and Chesapeake Bay areas account for 23%; the soft clams *(Mya arenaria)* from New England and Chesapeake Bay account for 17%; and the ocean quahog, razor clams, etc., account for the remaining 1%.

Predators and Disease

Clams in general are subject to attack by the same types of predators and diseases as are other shellfish.

In the larval stage, clams are relished by the many plankton eaters, and as adults are subject to attack by starfish, drills, bottom fish, and for soft clams in particular, the green crab—although its seriousness as a predator depends directly on water temperature; high temperatures promote predation, low temperatures greatly reduce predation.

Factors Affecting Quality

Composition.—There are only minor species differences in composition, with protein ranging from 10.1 to 15.1%, fat 0.9 to 2.8%, and carbohydrate averaging about 3.4%. The intraspecies differences are at least as great as the interspecies differences.

Freezing and Cold Storage.—Although only a very small part of the commercial harvest of clams may be frozen, only the best quality raw material should be selected for this process. Frozen storage life is limited to only 4 to 6 months at 0° F. (−18° C.), rancidity and toughening of the flesh being the major limiting factors.

Harvesting Methods

Dredges (such as those described in the section on oysters) took 79% of the clam catch, tongs (again similar to those used with oysters) took 9%, while rakes, hoes, and forks accounted for most of the remaining 12%.

Preparation for Freezing

Practices employed in preparing clams for freezing are largely dependent on the species being frozen.

Soft Clams.—The shell stock is washed, then the clams are opened—a relatively easy job, as the shells are not as tightly closed as in many other bivalves. The meats are washed, drained, and packed in containers ranging from 1-lb. cartons to 1-gal. cans.

Hard Clams.—The shells of this mollusk are tightly closed, so in some areas of the country, to facilitate insertion of the shucking knife, the edge of the clam is drawn over a coarse rasp, which removes some of the shell and thus makes it easier to insert the shucking knife and cut the adductor muscles. In other areas of the country, the knife is forced in between the shells near the hinge, the adductors cut, and the clam opened. Prior to being frozen, the meats are thoroughly washed and packaged.

Surf Clams.—The shells of the surf clams do not close as tightly as the quahog and so are somewhat easier to open. Treatment after opening is, however, radically different from that given soft and hard clams. Surf clams are eviscerated by squeezing the meats, which removes the stomach and other soft tissues. The eviscerated clams are then washed and, depending on intended use, may be chopped, sliced into strips, or left whole. The meats are then packed in containers of various sizes for freezing.

Freezing and Storage

The freezing method varies with the type of container used—that is, compression plate for rectangular 1- to 5-lb packages, and blast or shelf coil for large cans.

ABALONE

The abalone are gastropods, or snails, of the Haliotis species. The commercial fishery for this shellfish in the United States is confined almost entirely to the central and southern coasts of California and the Channel Islands where *Haliotis rufescens,* the red abalone, and *H. corrugata,* the pink abalone, account for almost the entire catch, which amounts to 817,000 lb. of meats.

Abalone meat contains about 17% protein and 1% fat. No information is available on seasonal or species variations in composition, nor on the frozen storage life that may be expected.

Harvesting Methods

Abalone are found on rocky shores from the intertidal zone out to 250 ft. or more, with the majority of the catch being taken in waters 20 to 80 ft. deep. Abalone are harvested by divers using either the traditional hard hat or the modern lightweight gear supplied through hoses by a surface vessel. The divers pry the abalone loose from its rock and place it in a bag or basket, which, when full, is hauled to the surface.

Preparation for Freezing

At the processing plant the abalone is cut from the shell and the foot is separated from the viscera. The tough outer surface is then removed and the muscle is sliced across the grain into steaks ½-in. thick. The most critical part of the processing is the tenderizing step. The steak slices are placed on a solid table and allowed to relax. The slices are then hit, just once, a smart blow with a wooden mallet.

About ¾ of the catch is frozen. Freezing and packaging are conventional, with the finished product being shipped primarily to restaurants in the state. California regulations do not permit shipment of abalone meats beyond the state boundaries.

ADDITIONAL READING

AM. FROZEN FOOD INST. 1971. Good commercial guidelines of sanitation for frozen soft filled bakery products. Tech. Serv. Bull. *74*. American Frozen Food Institute, Washington, D.C.

ARBUCKLE, W. S. 1972. Ice Cream, 2nd Edition. AVI Publishing Co., Westport, Conn.

ASHRAE. 1974. Refrigeration Applications. American Society of Heating, Refrigeration and Air-Conditioning Engineers, New York.

BAUMAN, H. E. 1974. The HACCP concept and microbiological hazard categories. Food Technol. *28*, No. 9, 30–34, 70.

DeFIGUEIREDO, M. P., and SPLITTSTOESSER, D. F. 1976. Food Microbiology: Public Health and Spoilage Aspects. AVI Publishing Co., Westport, Conn.

GUTHRIE, R. K. 1972. Food Sanitation. AVI Publishing Co., Westport, Conn.

HARPER, W. J., and HALL, C. W. 1976. Dairy Technology and Engineering. AVI Publishing Co., Westport, Conn.

HARRIS, R. S., and KARMAS, E. 1975. Nutritional Evaluation of Food Processing, 2nd Edition. AVI Publishing Co., Westport, Conn.

JOHNSON, A. H., and PETERSON, M. S. 1974. Encyclopedia of Food Technology. AVI Publishing Co., Westport, Conn.

KRAMER, A., and TWIGG, B. A. 1970, 1973. Quality Control for the Food Industry, 3rd Edition. Vol. 1. Fundamentals. Vol. 2. Applications. AVI Publishing Co., Westport, Conn.

MOUNTNEY, G. J. 1976. Poultry Products Technology, 2nd Edition. AVI Publishing Co., Westport, Conn.

PETERSON, M. S., and JOHNSON, A. H. 1977. Encyclopedia of Food Science. AVI Publishing Co., Westport, Conn.

RAPHAEL, H. J., and OLSSON, D. L. 1976. Package Production Management, 2nd Edition. AVI Publishing Co., Westport, Conn.

SACHAROW, S. 1976. Handbook of Package Materials. AVI Publishing Co., Westport, Conn.

TRESSLER, D. K., VAN ARSDEL, W. B., and COPLEY, M. J. 1968. The Freezing Preservation of Foods, 4th Edition. Vol. 2. Factors Affecting Quality in Frozen Foods. Vol. 3. Commercial Freezing Operations—Fresh Foods. Vol. 4. Freezing of Precooked and Prepared Foods. AVI Publishing Co., Westport, Conn.

WEISER, H. H., MOUNTNEY, G. J., and GOULD, W. A. 1971. Practical Food Microbiology and Technology, 2nd Edition. AVI Publishing Co., Westport, Conn.

WILLIAMS, E. W. 1976A. Frozen foods in America. Quick Frozen Foods *17*, No. 5, 16–40.

WILLIAMS, E. W. 1976B. European frozen food growth. Quick Frozen Foods *17*, No. 5, 73–105.

WOOLRICH, W. R., and HALLOWELL, E. R. 1970. Cold and Freezer Storage Manual. AVI Publishing Co., Westport, Conn.

9

Freezing of Dairy Products

Byron H. Webb
and Wendell S. Arbuckle

D airy products other than ice cream and other frozen desserts require only simple preparation for freezing. Ice cream, ice milk, and sherbets, discussed later, are complicated mixtures of ingredients and the freezing process plays a very special part in their manufacture. Ice cream is eaten in a frozen state, but other dairy products are frozen only to preserve them for future use. They are thawed before consumption. Freezing of dairy products is a process for maintaining them in a fresh state during necessary periods of storage.

THE FROZEN DAIRY PRODUCTS INDUSTRY

The annual production of milk in the United State is over 100 billion pounds. Somewhat less than half of this is consumed as fluid milk. Part of the remainder is manufactured into products which are not considered perishable but which benefit from cool storage. These include the sterilized and dried milk products. Another portion of the milk is manufactured into cheese and butter which are usually stored at cold temperatures but not necessarily at temperatures at which the water they contain is frozen. The freezing points of some dairy products are given in Table 9.1.

The freezing points should not be considered as the temperature that must be reached to preserve the products. In some cases preservation can be satisfactorily achieved by holding at temperatures above those at which ice is formed.

Cream, plastic cream, and butter are listed in the USDA cold storage reports as commercial products that are held at freezing temperatures. The cream and plastic cream are prepared and frozen from the spring and summer surplus. They are not consumer items but are held for food manufacture during fall and winter when prices of these commodities have advanced. Butter is the common form of milk product used by industry and government to store milkfat. Government purchases of

TABLE 9.1

FREEZING POINTS AND MOISTURE CONTENT OF SOME DAIRY PRODUCTS

Product	Freezing Point °F.	Freezing Point °C.	Moisture Content, %
Milk[1]	31.03	−0.54	87.5
Evaporated milk	29.5	−1.38	74.0
Concentrated milks			
Whole (10% fat, 23% SNF)	28.4	−2.0	67.0
Skim (36% TS)	26.4	−3.13	64.0
Cheese			
Cottage	29.8	−1.2	78.7
Cheddar (processed)	19.6	−6.9	38.8
Swiss	14.0	−10.0	34.4
Cheddar	8.8	−12.9	33.8
Roquefort	2.7	−16.3	39.2
Sweetened condensed milk	5.	−15.	27.0
Butter (water phase)			
Unsalted	32.	0.	15.8
Salted((2.0%)	15.8	−9.0	15.8
(3.5%)	−3.6	−19.8	15.8

[1] Cream, skimmilk, whey, and starter cultures have approximately the same freezing point as milk unless products are chemically altered in processing.

surplus milk products for price support have generally been in the form of butter and nonfat dry milk. Butter held at −10° F. (−23° C.) and nonfat dry milk below 40° F. (4° C.) remain acceptable for 1 to 2 years.

Fluid milk is sometimes frozen to preserve it for short periods. Its high water content (87%) makes freezing an expensive method of keeping milk. Nevertheless, some milk is frozen in cartons for use by the armed forces, on ships and where supplies of fresh milk are difficult to obtain. The freezing of a 3:1 milk concentrate has been practiced commercially by one company for special market tests. Very recently the freezing of starter cultures has been done on a commercial scale. Freezing is not an entirely satisfactory way to preserve cheese, but cold storage is desirable. Freezing of canned evaporated milk may cause rupturing of the can and subsequent spoilage. Sweetened condensed milk is preserved by sugar, which also lowers its freezing point to about 5° F. (−15° C.). There is no advantage in holding it below this temperature.

Storage Life of Refrigerated Dairy Products

The estimated storage life of dairy products when held under common storage conditions is shown in Table 9.2. Many of the products given in Table 9.2 are not held in frozen storage, but the life of all of them is prolonged by cold storage.

While freezing damages the body of some products, it prolongs the life of others. Static freezing, without agitation, is used for cream, plastic

TABLE 9.2

STORAGE LIFE OF DAIRY PRODUCTS

Product (Commercial Pack)	Approximate Storage Life at Specific Temperatures			Critical or Dangerous Storage Conditions
	Months	Temperature °F.	°C.	
Butter (in bulk)	1	40	4	Above 50°F. (10°C.) or damp or
	12	−10	−23	wet storage
Butteroil (sealed, full tins;	3	70	21	Above 75°F. (24°C.)
maximum moisture	6	50	10	
0.3%)	9	32	0	
Ghee (sealed, full tins)	6	90	32	Above 90°F. (32°C.)
	9	70	21	
	18	40	4	
Cream (50% fat)	12	−10	−23	Above 20°F. (−7°C.)
Plastic cream (80% fat)	12	−10	−23	Above 20°F. (−7°C.)
Frozen milk	3	−10	−23	Above 10°F. (−12°C.)
Frozen concentrated milk	6	−10	−23	Above 10°F. (−12°C.)
Frozen cultures	6	−10	−23	Above 10°F. (−12°C.)
Nonfat dry milk, Extra	6	90	32	Above 110°F. (43°C.)
Grade (in moisture-	16	70	21	
proof pack)	24	40	4	
Dry whole milk, Extra	3	90	32	Above 100°F. (38°C.)
Grade (gas pack; max-	9	70	21	
imum oxygen 2%)	18	40	4	
Sweetened condensed milk	3	90	32	Above 100°F. (38°C.) or below
	9	70	21	20°F. (−7°C.), or dampness suf-
	15	40	4	ficient to cause can rusting
Grated cheese (in	3	70	21	Above 70°F. (21°C.) or above 17%
moisture-proof pack)	12	40	4	moisture in the product
Cheddar cheese	6	40	4	Above 60°F. (16°C.) or below
	18	34	1	30°F. (−1°C.)
Processed cheese	3	70	21	Above 90°F. (32°C.) or below
	12	40	4	30°F. (−1°C.)
Sterilized whole milk	4	70	21	Above 90°F. (32°C.) or below
	12	40	4	30°F. (−1°C.)
Evaporated milk	1	90	32	Above 90°F. (32°C.) or below
	12	70	21	30°F. (−1°C.) or dampness suf-
	24	40	4	ficient to cause can rusting

cream, butter, bacteriological cultures, milk, and concentrated milk. Fast freezing in the package sometimes affords greater protection to physical properties but often the differences are not important. Unlike ice cream, these products are not eaten in the frozen state so that large ice crystals and a coarse body may not be objectionable.

Use of Additives

Usually only slight or no modifications are made in the composition or the processing of fluid milk, cream, butter, or starter cultures to prepare them for freezing. But milk concentrated to the 2:1 or 3:1 level requires special processing if it is to survive frozen storage in acceptable condition.

The industry has tried to avoid the use of additives in the preparation of dairy products for freezing preservation. When additives are used they must conform to state and federal requirements. Products that enter interstate commerce must be labeled in accordance with the FDA's definition and standard of identity. Protection against destructive physical effects of freezing (i.e., gelation in frozen concentrated milk associated with lactose crystallization and improved by lactase additive) is best sought by improvement or modification of processing techniques or by changes in the normal composition of the product rather than through the use of additives.

THE FREEZING OF CREAM

Cream is frozen to preserve it for use in food manufacture, usually in ice cream. Two fat concentrations are common—50% frozen cream and 80% plastic cream. Frozen cream is not prepared in small retail packages but is frozen in bulk for food preparation as a means of taking care of surpluses and shortages. A bibliography on frozen cream has classified the reasons for the freezing of cream into five categories. (1) Cream is frozen in summer to be churned into butter in the winter in Germany and Holland. This is linked with a favorable price structure and it may not be economical elsewhere. For example, cream frozen and held in Germany for 5 to 6 months made better quality butter than was possible with storage of the butter. This would not necessarily be so if the butter were high in quality and the storage was at a constant −20° F. (−29° C.) temperature. (2) In the United States cream is frozen for later use, largely in ice cream manufacture. In this case, the problems of texture and body are not of prime importance because the cream is processed into ice cream mix after thawing. (3) Frozen whipped cream has been considered as a retail product, but it has never attained significant volume. (4) Frozen table cream may be prepared for export or winter use. This is convenient for storing, but it would seem to have little advantage over a high-fat frozen cream. (5) Freezing of cream in any form to penetrate tariff barriers. These may specifically exclude butter and other high-fat dairy products but not cream.

Growth of bacteria in cream is stopped during frozen storage. Counts of most organisms, yeasts, and molds decrease substantially as the period of frozen storage advances.

Frozen Cream

Frozen cream contains 50% fat, in contrast to plastic cream of 80% fat. Pasteurized cream is usually frozen for food manufacture but a method was devised for farm freezing of raw cream that kept well for 7 days at 9° to 0° F. (−13° to −18° C.). It was essential to pasteurize the cream immediately on defrosting.

Flavor Changes in Frozen Cream.—Oxidation of the fat during frozen storage is the principal flavor defect of frozen cream. This must be prevented in order to produce an acceptable product. When good quality cream of low acidity is frozen, its flavor usually remains acceptable during storage. Copper-free cream, pasteurized either at 165° F. (74° C.) for 15 min. or at 185° F. (85° C.) for 5 min., does not develop an oxidized flavor. However, when 1 p.p.m. of copper was added to the cream after pasteurization, oxidized flavor invariably developed during frozen storage. Homogenization of such cream was found to have only a very slight inhibitory effect on the copper-induced oxidized flavor. It was concluded that when high-quality cream was produced and handled free of copper contamination, and was adequately pasteurized, homogenization was not necessary to retard the development of oxidized flavor. Homogenization had a slightly beneficial effect, however, in retarding the development of copper-induced oxidation. Creams which were susceptible to the development of off-flavor usually showed such a defect during the first three months of frozen storage.

Only high-quality cream of low bacterial count and low acidity should be prepared for frozen storage. Several workers have observed that creams of high acidity (0.15% or higher) become unacceptable after 2 or 3 months, developing oxidized and other off-flavors. Sugar added to the extent of about 10% of the weight of the cream helps to maintain a satisfactory flavor.

Studies have been made on the effects of adding recognized antioxidants to delay or prevent development of oxidized flavor in stored frozen cream. In the absence of copper, several antioxidants delayed development of oxidized flavor for at least six months. In the presence of added copper, only ethyl caffeate retarded off-flavor development beyond five months and only in summer cream; it was ineffective in winter cream. These results were obtained when the frozen cream was stored in standard glass milk bottles. When storage was in metal cans, the antioxidants were much less effective; ascorbic acid was not only ineffective in preventing the development of oxidized flavor but actually accelerated it. When ascorbic acid was combined with ethyl hydrocaffeate, with or without the presence of copper, the cream did not develop an oxidized flavor during twelve months' storage.

Body Changes in Frozen Cream.—When cream is frozen there is a tendency to disrupt the fat emulsion and to destabilize the milk protein. The physical equilibria of both components is changed, depending upon the severity of freezing conditions.

Half or more of the fat in 50% cream can be destabilized and will "oil off" on thawing the cream to temperatures above the melting point of the fat. Rapid freezing tends to lessen the amount of fat de-emulsified as the water in the cream freezes. The fat destabilization will not be noticeable or objectionable unless the cream is thoroughly melted before it is added to ice cream mix or other food. Added sugar lowers the freezing point of cream and this protects the fat emulsion to the extent that ice formation in the cream is lessened. The fat of frozen cream can be completely reemulsified by homogenization either as cream or as the complete food product in which the cream is used.

Cream of 50% fat contains only 1.7% milk protein in contrast to 3.3% in fluid milk and 10% in a 3:1 milk concentrate. Plastic cream of 80% fat contains only 0.7% protein. The seriousness of the protein destabilization problem in frozen dairy products decreases with decrease in protein. If the protein in frozen cream becomes difficult to disperse on thawing because of prolonged storage or fluctuating storage temperatures, mild heat and stirring will usually disperse it.

Preparation of Frozen Cream.—Cream of 50 to 55% fat may be frozen and held in storage for future use by the following procedure (Heinemann 1967). Good quality cream (free of copper contamination) is pasteurized, cooled to 40° F. (4° C.) or lower and run into round plastic lined 2½- or 5-gal. containers for freezing. The containers are put into a cold room, such as an ice cream hardening room, where they are stacked to freeze and hold until they are to be used. When the containers are round they can be stacked close together, permitting air circulation between them. Freezing can also be done in plastic pouches which can be placed between refrigerated plates. More rapid freezing can be attained by slush freezing (without air incorporation) in a scraped-surface freezer (see preparation of plastic cream).

The National Research Development Corp. has been granted a British Patent covering a method of freezing cream. Unhomogenized cream is filled into an oxygen-impermeable container of good thermal conductivity. Six to ten per cent of the internal volume of the container is left unfilled. The containers are sealed hermetically. The cream is frozen by immersion in a liquid at approximately −38° F. (−39° C.) or by exposure to a blast of cold air or gas at −50° F. (−46° C.). The containers are stored at approximately 5° F. (−15° C.). The frozen cream resulting has been stored for 12 months without deterioration, retaining the characteristics and whipping properties of the original cream when thawed.

At the time of use, frozen cream is removed from the freezer and held overnight to soften the surface in contact with the container. It is then easily removed. Where large quantities must be thawed quickly, the frozen cakes are passed through a suitable machine for breaking up and melting.

Containers for Frozen Cream.—During the early years of freezing cream in the United States, 30- or 50-lb. metal cans were used as containers. Wire fastened to the two handles held the lids secure. The 2½- or 5-gal. single-service, round plastic-lined fiber ice cream container has replaced the tin can. The round shape permits air circulation even though the containers are stacked close together in the freezing room. The polyethylene liner is fastened to the cardboard stock of the container so that when the cream is partially thawed it drops out free of the liner. When a flexible or loose polyethylene bag is placed within a rigid container, it is difficult to remove the cream completely—because part of the cream becomes enmeshed in the folds of the polyethylene liner, and the residual product cannot be easily or completely stripped out. Plastic coated rectangular boxes without other liners are also used as containers for cream. The filled containers occupy less space in storage than round cartons, but there must be provision for air circulation during freezing, or freezing must be done between plates.

Plastic bags (100 × 70 × 5 cm.) are used as containers for frozen cream in Europe. The bags permit the cream to be frozen in slabs between plates, and the slabs may be stacked on pallets for transportation and storage.

Thawing Frozen Cream.—Small quantities of frozen cream may be thawed by removing the bags, cartons, or cans from storage 1 or 2 days before use. The melted cream on the surface of the cake frees it from the container so that it can be dumped to mix with other liquid food. When quantities are large some mechanical aid must be employed for thawing. One U.S. company made an ice breaker into a frozen cream crusher by tinning the drum and making the bushings sanitary.

A German method for thawing frozen cream has been described. Frozen blocks of cream are crushed and defrosted in a vat provided with a central vertical stirrer and a system of pipes coupled to a pasteurizer. The pieces of frozen cream are mixed with whole or skimmilk at a temperature not exceeding 77° F. (25° C.) and the mixture is warmed by passing milk at 77° to 86° F. (25° to 30° C.) through the pipe system.

A cream thawing unit has been developed at the dairy experiment station at Kiel under direction of Dr. Ing. G. Walzholz. It is illustrated in Fig. 9.1 and 9.2. The cream to be defrosted should be frozen in slabs. The slabs are fed into the thawing unit (Fig. 9.1) which cuts them into thin slices (Fig. 9.2). The cutting is done by jets of warm milk as the slabs slide by gravity at a controlled rate down the chute. The slices of cream drop into a vessel in which the final defrosting takes place. The temperature of the nozzles is approximately 158° F. (70° C.); the height of the cream layer cannot exceed 30 cm.

Plastic Cream

Based on a specific quantity of fat, plastic cream of 80% fat occupies only ⅝ of the storage space of cream, requires fewer packages, and contains correspondingly less water to freeze. Cream of 50% fat contains about 45% water, whereas cream of 80% fat contains only about 18% water. Some ice cream manufacturers consider that plastic cream produces better flavor and body in ice cream than does frozen cream. In spite of the advantages of plastic cream as a vehicle for storage of milkfat, about 5 times more 50% cream is held in frozen storage than plastic cream. Two centrifugal separations are required to manufacture plastic cream, whereas one suffices for 50% cream. A further advantage of the lighter cream is its free-flowing property before freezing and after thawing. No special handling equipment is necessary.

Preparation and Storage of Plastic Cream.—Plastic cream is prepared by reseparation of 40% pasteurized cream to yield cream of 80% fat. The heavy product is chilled in a scraped-surface freezing unit or in a con-

From Gronau

FIG. 9.1. PILOT PLANT FROZEN CREAM THAWING UNIT, GRONAU/ROGGE SYSTEM

From Gronau

FIG. 9.2. FROZEN CREAM THAWING UNIT IN OPERATION, GRONAU/ROGGE SYSTEM

Other machines for melting frozen cream and butter have been devised using, for example, internally heated rotating discs.

verted ice cream freezer, and is drawn directly into suitable containers such as are used for frozen 50% cream. The cream emerges from the chiller at about 40° F. (4° C.) or lower. Suitable slush freezing units are pictured in Fig. 9.3 and 9.4. Figure 9.5 is a pump suitable for use with the slush freezer shown in Fig. 9.4. The scraped surface continuous coolers of Fig. 9.3 and Fig. 9.5 are versatile units for cooling, slush freezing, or plasticizing cream and high-fat, high-viscosity dairy products.

The containers of chilled plastic cream are placed in a hardening room at 5° F. (−15° C.). Cream packaged in this way and held for two years has been reconstituted to a fluid milk of satisfactory flavor. Oxidized flavor was never a problem during a decade of production.

Plastic cream has been used for the storage of milkfat by a Dutch butter factory: 40% cream is received, neutralized, pasteurized, separated to 80% fat, frozen in molds at −40° F. (−40° C.), and the slabs are wrapped in plastic and stored at 10° F. (−12° C.). Frozen summer cream blended with fresh winter cream makes excellent butter.

Plastic cream represents less of a thawing problem than does normal cream since it contains less ice. It may be run through an ice breaker on removal from storage or it may be allowed to remain out of refrigeration for about two days to soften, at which time it can be removed from the container and put through a crushing machine.

Frozen Whipping Cream

Neither frozen whipping cream nor frozen whipped cream is an article of commerce. Whipping cream should contain 30 to 35% fat. It should

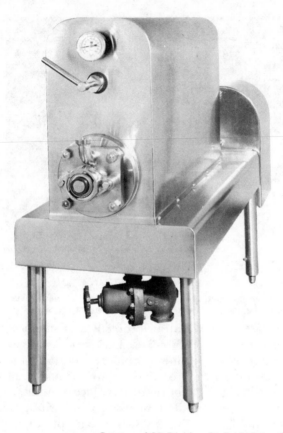

Courtesy of CP Division, St. Regis Paper Co.

FIG. 9.3. SWEPT-SURFACE HEAT EXCHANGER

This scraped-surface type of heat exchanger is used to chill and slush-
freeze milk products. It uses direct-expansion ammonia (or other refrigerant) as thechilling medium. The refrigerant accumulator, pressure reg-
ulator, and other controls are self-contained.

not be homogenized at a substantial pressure, because this would greatly
reduce whipping properties. Freezing damages cream body, but addition
of sugar before freezing, by decreasing ice formation, helps to produce a
smooth bodied product in thawed cream. Any of the frozen creams can be
reemulsified in prepared foods to overcome defects in body which develop during freezing.

A procedure for preparing frozen whipping cream of 32% fat was
developed using pasteurized cream filled into flat plastic containers or
lined cartons and frozen rapidly in 30 min. at −4° to −22° F. (−20° to

Courtesy of CP Division, St. Regis Paper Co.

FIG. 9.4. STAINLESS SANITARY ROTARY PUMP WITH VENTED COVER

This rotary pump has 5-lobe rotors made of a resilient rubber-like material. The model shown is equipped with a vented cover which permits recirculation or relief at a pre-set discharge pressure. Pumps of this type used in conjunction with slush freezers prevent damage to downstream equipment from excessive pressure due to freeze-up or other line stoppage.

−30° C.). Fast freezing produced fine ice crystals and tended to avoid destabilization. The product was satisfactory in smell, taste, and serum separation, but volume increase on whipping was impaired and flocculation was noticeable.

Frozen whipped cream can be prepared. Cream of 35% fat is pasteurized, and stabilizer, sugar, and vanilla flavoring are added. The cream is whipped, packaged in plastic bags, and frozen at −20° F. (−29° C.). After freezing and storage under various conditions, examination showed there is body and texture deterioration after three months. During the early storage period consumer acceptance was good.

STORAGE OF BUTTER

Butter is the most popular form of product for storage of milkfat. It keeps well at low temperatures for periods of a year or more and it is less sensitive to temperature changes than many other dairy products. For

Courtesy of Votator Division, Chemetron Corp.

FIG. 9.5. SCRAPED-SURFACE CHILLER-FREEZER-CRYSTALLIZER JACKETED FOR DIRECT-EXPANSION AMMONIA OR FREON

Unit consists of a rotating bladed shaft within a jacketed, insulated tube. Viscous products can be cooled from any temperature to slush-freeze if desired.

many years after World War II, the U.S. Government in its price support activities held large quantities of butter in storage. When the quality of the original butter was good and the storage temperature low and constant, the results were quite satisfactory.

Butter must be stored in the cold, but usually it is not considered as a product preserved by freezing. Butter will freeze at low storage temperatures but freezing has no effect on its characteristics that are noticeable after it has been thawed. Low-temperature storage of butter such as 0° F. (−18° C.) is advocated to prevent flavor deterioration and otherwise to keep it in a fresh, attractive condition.

Freezing Characteristics of Butter

Unsalted butter freezes at 32° F. (0° C.), but most butter is salted. If it contains 2% salt, the water freezes at 15.8° F. (−9° C.); at 3.5% salt, its water is frozen at −3.6° F. (−19.8° C.). The cryohydric point of sodium chloride is −4.2° F. (−21.2° C.). Below this temperature some salt crystallizes during frozen storage, but it readily dissolves when the butter is warmed. As butter is cooled there is a gradual formation of ice. The salt remains in the unfrozen portion until its concentration is such that it crystallizes. When butter is thawed there is a steady rise in temperature as the ice slowly melts and crystallized salt dissolves in it. The gradual freezing and thawing as temperature changes, which occurs in salted butter, avoid the destructive effects noted in products not provided with such a system buffered against freezing damage.

Both cooling and warming curves for salted butter are smooth and regular. They are affected to some extent by the fact that the specific heat of water is 1.0 while that of ice is only 0.5. Heat conductivity of butterfat is lower than that of water and air bubbles in butter contribute further to lowering its heat conductivity.

The characteristics of butter under various freezing and thawing conditions may be important in relation to its commercial handling. Water expands on freezing while butterfat shrinks, but these changes in volume apparently have not been correlated with changes in the physical properties of butter. When only a fraction of the water in salted butter is frozen, less refrigeration is required to lower it to a given temperature than would be required to freeze the water in unsalted butter at the same temperature. Repeated changes in the storage temperature of butter should be avoided, since the surface of the butter reaches air temperature more rapidly than the center of the block and frequent changes in temperature may therefore cause the surface butter to be inferior to that at the center.

Storage Temperatures.—Cold storage temperatures should be as low as possible, certainly no higher than −4° F. (−20° C.) if the holding time is to be several months. A preferred temperature is −20° F. (−29° C.) for long-term storage (1 year or more). If the butter is to be held for only 2 or 3 weeks, 40° F. (4° C.) may suffice. The temperature of storage has no significant effect on the grade of butter after storage when storage temperatures were between −10° F. (−23° C.) and 12° F. (−11° C.). The grade loss in score was between 0.9 and 1.7 points after 6 months of such storage. Cracking in the stored butter resulted from excessive handling when butter was frozen.

Contact freezing of butter was found to produce excellent results. The wrapped butter was passed through a multiple plate freezer, where it was frozen in single layers and where the time to reduce the temperature of the butter to −22° F. (−30° C.) was 57 min., compared to 71 hr. in a tunnel freezer and longer than 120 hr. in a cold storage room with the butter in tubs. The plate-frozen butter kept significantly better than the butter frozen by traditional methods. The freezer could be made part of the production line.

Preparation of Butter.—The prestorage treatment of butter has an important effect upon its subsequent keeping quality. Butter that is placed in a freezer within 2 or 3 hr. of its manufacture will show markedly better keeping quality than that which may have been exposed to 40° F. (4° C.) for several days before freezing.

Butter for storage may be manufactured by either the churn or the continuous process. The cream is always pasteurized so that it is free of proteolytic types of organisms and contains largely only starter cultures for flavor production.

The butter should be wrapped in gas-tight foil or film impervious to light, placed in cartons or boxes and transported immediately to cold storage. A study of several combinations of wrapping materials showed that those with an aluminum foil base held butter in the best condition during storage. Storage temperatures should not fluctuate, and the boxes of butter should be stacked so that air can circulate around them.

Most experts recommend that butter be packaged directly from the churn into retail units which after storage at 0° F. (−18° C.) may be supplied to the retail shop while still in a frozen state. Tests on 50-g. packages of aluminum foil-wrapped butter between refrigerated plates at −22° F. (−30° C.) indicate that they were frozen in 31 min., many hours sooner than large commercial blocks of butter. The small blocks of rapidly frozen butter kept better than that frozen in large blocks. The specific refrigeration requirement was 33.2 kcal. per kilogram butter, which is 27% less than for block butter.

Storage Changes in Butter

Butter is subject to deterioration in flavor and body during storage. Microbiological activity all but ceases at the lower temperatures and often actual numbers of organisms decrease, but not without leaving residual effects on flavor.

Flavor in Storage Butter.—Butter is held in cold storage in a frozen condition to preserve its flavor, but butter is notorious for the rapidity with which it will absorb flavor from its environment. It is therefore important to keep storage rooms free of foreign odors and to wrap the butter in foil or other wrapping material which will not permit the penetration of air and off-odors.

If the butterfat has been unduly exposed to oxygen in the presence of copper, oxidized flavor will develop. Addition of 0.015 to 0.02% ascorbic acid has been found useful in slowing the rate of this copper-induced oxidation. Tests reveal that lipid oxidation in cold-stored butter is in relation to its copper content and the pH of the serum. In butters with serum reactions between pH 6.6 and 8.0 the more alkaline butters have higher peroxide values and lower thiobarbituric acid values after storage.

The development of oxidized flavor in butter was found not to be significantly retarded by antioxidants during storage at either $-18°$ F. ($-28°$ C.) or 38° F. (3° C.).

Antioxidants and synergists applied to the paper wrapper were claimed to afford substantial protection to butter, and its keeping quality was doubled as compared with controls. But there was rapid quality deterioration and spoilage (probably enzyme action) after 20 to 26 months of storage at 5° F. ($-15°$ C.).

Surface taints on 65-lb. cubes of butter are controlled by use of impervious cardboard laminates and by coating the inner liner of the carton with a moisture- and gas-tight material.

Bacteriological Changes in Butter in Storage.—Butter, under normal storage conditions, is a poor medium for bacterial growth. In general, the numbers of bacteria decrease during storage until there may be only 10% of the original population viable after 8 to 12 months' storage at 0° F. ($-18°$ C.). However, butter containing large numbers of bacteria, especially nonlactic acid producers, has inferior keeping quality. In one test, control butter kept well for 6 months at $-0.4°$ F.($-18°$ C.), while test samples inoculated with coliform bacteria were in poor condition after 3 months. There are changes in total counts of proteolytic, lipolytic, and coliform bacteria, yeast and molds during storage of butter at 21° F. ($-6°$ C.). The counts decrease with time and decrease more rapidly with high initial contamination. While all of the organisms show a general decline in storage, the proteolytic bacteria are considerably inhibited by the presence of lactic acid but are not very sensitive to storage temperature.

Treatment of Butter After Storage

Butter that is removed from freezer storage should be placed in consumption channels at once. Bacteriological counts may have been lowered by storage treatment, but quality may have deteriorated. Undesirable physical changes sometimes occur. If a rapid return to higher temperatures is required, one may defrost large blocks by contact with electrically heated copper bands.

The body of butter that has become coarse or uneven in distribution of moisture or fat can be reworked to produce a uniform, fine crystallized structure and improved spreadability. The plasticizing section of a continuous butter plant or the Votator chiller-freezer-crystallizer shown in Fig. 9.5 may be used to do this.

Printing Butter After Storage.—Butter should not be frozen before molding. The equipment used in printing butter (cutting and packaging) may cause the butter to show the presence of free moisture on its surface after the bulk-stored product has been tempered and subject to the manipulation that takes place in the printing operation. When butter is removed from cold storage it is usually tempered in the range of 48° to 52° F. (9° to 11° C.) before printing. In examining the effect of four types of butter printers on the free moisture produced by the printing operation, differences were found in each printer. Printers employing extruders or rotating augers to force the butter into a mold caused moisture droplets to increase from less than 20μ to a range of greater than 20 to 100μ. Printers designed to handle soft butter out of the churn by forcing it with rotating polygonal rolls into the molding section did not change moisture distribution. The problem of free moisture and leaky butter resulting from the printing of cold storage butter could be eliminated by use of a homogenizing mechanism that would exert a high shearing stress.

FREEZING PRESERVATION OF MILK

Fresh fluid milk and milk concentrated to a ratio of 2:1, 3:1, or 4:1 have been frozen experimentally for preservation. Freezing of fluid milk and of a 3:1 concentrate have had limited commercial application.

Freezing of milk is satisfactorily accomplished by placing it in suitable retail containers in a cold storage room. However, attempts have been made to freeze milk rapidly by spraying it into a cold air stream at about −20° F. (−29° C.). The expense of such a procedure has thus far outweighed its advantages. A recent study examined the properties of milks quick-frozen at various temperatures.

Freezing Fluid Milk

Fresh milk may be frozen to preserve it for shipment to inaccessible places or to hold it for future consumption. During World War II frozen homogenized milk was used to supply fresh milk to patients on hospital ships. This successful use aroused considerable interest in the product. The milk was acceptable when it was thawed after storage in a frozen condition for three months. After this period the quality of the product varied, although some reports indicated acceptability for as long as 6 to 9 months of storage. During the longer periods of storage two major problems developed—flavor deterioration and separation of milk solids upon thawing. Development of these defects can now be delayed by proper processing.

Processing Milk to Prevent Oxidized Flavor.—The mild, delicate flavor of fluid milk products can be retained in frozen storage better than by any other method of preservation. The stale flavor that inevitably develops in concentrated and dried milks is rare in frozen milks. But frozen milks are susceptible to the development of a typical oxidized flavor.

There are several approaches toward prevention of oxidized flavor in frozen milks, including the use of ascorbic acid as an antioxidant. In general, findings indicate that ascorbic acid, while helpful, does not provide complete protection. Similarly, conventional antioxidants have a mild repressing effect on appearance of oxidized flavor. However, sodium gentisate is superior to the others. Milk containing 0.01% of this chemical could be held 6 months at $-10°$ F. ($-23°$ C.) without development of oxidized flavor, but if copper was added the flavor appeared in 4 months.

Heating above pasteurization temperature (161° F., 71.7° C.) will retard development of oxidized flavor, the condition for producing cooked flavor. Heating and homogenizing fluid milk is a simple and effective means of protecting it against oxidation. Oxidized flavor can then be retarded by a combination of measures: small additions of ascorbic acid; pasteurization at relatively high temperatures; homogenization; and finally, by being certain that the cows producing the milk receive adequate quantities of alpha-tocopherol (1g.) in their daily rations.

The volatile sufhydryl compounds developed in the milk by heat, which combat oxidized flavor, are dissipated in frozen storage after 1 to 2 weeks. When they are present in the milk the cooked flavor is pronounced, but this gradually subsides so that there is an actual improvement in flavor during the early storage period.

Processing Milk for Physical Stability.—Separation of milk solids during thawing of frozen milk can be delayed by addition of a small quantity of citric acid after pasteurization and by homogenization, procedures which also retard development of oxidized flavor. The sodium polyphosphates prevent casein insolubility by their peptizing action but they are usually not needed to stabilize fluid milk.

Homogenization alone is effective in protecting the fat emulsion of milk so that free fat will not oil-off on the surface of the thawed product.

Preparation of Milk for Freezing.—Fresh Grade A milk of low bacterial count and with a fat content of about 3.8% should be pasteurized, preferably by holding at 155° F. (68.3° C.) for 30 min., and packaged in half-pint, pint, or quart paper containers. The common polyethylene-lined paper milk containers provide excellent packages for frozen milk. The milk should be homogenized at about 135° F. (57° C.), as it is raised to pasteurization temperature. The homogenization pressure should be at least 1,500 p.s.i. Use of the following additives after pasteurization will essentially double the storage life of the frozen milk. For each liter (quart) of milk add 2 g. of sodium citrate and $1/10$ g. of ascorbic acid. Chemicals that stabilize both flavor and sedimentation of the milk are the most effective. After the stabilizers have been added, the milk should be packaged in paper cartons using conventional equipment.

Freezing and Storage of Milk.—Babcock froze and held his samples of milk in wax-lined cartons at 0° F. (−18° C.). Higher freezing temperatures are undesirable but lower temperatures may improve quality. Separation is not noticeable in the thawed milk until after it has been stored 150 days, and flavor is usually normal up to about 120 days, after which a slight oxidized flavor may develop.

Recent reports indicate that pasteurized milk (4% fat) frozen and stored at 14° F. (−10° C.) became destabilized in 40 days; at −4° F. (−20° C.) there was impaired stability in 6 to 11 months; at −22° F. (−30° C.) stability was retained for more than 11 months. Homogenization before freezing had a slight adverse effect on protein stability, but it prevented the formation of a surface layer of melted fat on the thawed milk.

It can be concluded that pasteurized homogenized milk frozen and held at 0° F. (−18° C.) can be shipped to almost any point on earth and should arrive in acceptable condition. Higher storage temperatures may be used, but temperatures above 10° F. (−12° C.) are not recommended because the storage life of the milk will be appreciably shorter.

Frozen Concentrated Milk

Milk can be concentrated in a vacuum without impairing its flavor, but preservation of the concentrate during long periods of storage has been a challenging problem. Reduction of the sterilized or cooked flavor of evaporated milk by substitution of high-temperature, short-time sterilization for the long-hold process has been effective in improving its quality. But its flavor is still more cooked than is acceptable for most beverage purposes, and room-temperature storage brings staling changes. Frozen concentrated milk more nearly meets the rigid quality requirements of the American public, but extensive use of this product has been delayed pending solution of technical and economic problems.

The same defects occur during the storage of frozen concentrated milk that occur in frozen pasteurized milk, but fat separation is less in the concentrate because the higher content of milk solids protects the fat emulsion and because the milk has been homogenized. Similarly, oxidized flavor in the concentrate can usually be avoided by heating the milk to a relatively high pasteurization temperature and homogenizing it before concentration. Thickening increases in magnitude as the concentration of the milk to be frozen is increased.

Preparation and Freezing of Concentrated Milk.—An early process patent describes a very simple procedure for preserving concentrated milk by freezing. Fresh milk was pasteurized, condensed under vacuum to ⅓ its original volume, cooled, sealed in cans, and frozen at 10° F. (−12° C.). The product was satisfactory after storage at this temperature for 2 or 3 weeks, but after that time the thawed reconstituted product showed fat separation and protein coagulation which made it unsuitable for commercial production. Research has shown how the undesirable change that occurs in frozen concentrated milk can be greatly delayed.

One of the advantages of freezing concentrated milk rather than fluid milk is that there is an attractive saving in container and shipping costs. Evaporated milk at a 2:1 concentration has always been popular, but it is very difficult to manufacture a sterilized milk at 35% solids because at this concentration the milk protein lacks the heat stability necessary if the product is to be sterilized. When the concentrate is frozen so that no sterilization is necessary, a 3:1 concentration becomes practical.

Concentration of Milk Under Vacuum.—Milk is concentrated under vacuum to reduce weight and volume, saving container, storage, and shipping costs. For food manufacture, as in production of bakery products, ice cream, and confections, either a concentrate or a powder is needed.

The concentration of milk as practiced in a large plant by means of a triple-effect falling-film evaporator is a very efficient operation. A diagrammatic outline of the process is shown in Fig. 9.6. Milk enters the first stage tube chest, descends as a boiling film, and enters the flash chamber where milk and vapor are separated. The vapor is used to heat the tube chest of the next stage, while the milk is further concentrated in that chest. This is repeated in the third stage at a still lower temperature. Vacuum concentration of milk removes off-flavors, and since the temperatures used are no higher than pasteurization, no objectionable cooked flavor is added.

Freeze-Concentration of Milk.—Processes have been devised whereby milk, whey, and other liquids are concentrated by freezing part of their water content and removing the ice crystals. Concentration by freezing should be an efficient operation because the latent heat of fusion of water is 143 B.t.u. per lb., whereas its latent heat of vaporization is 971 B.t.u.

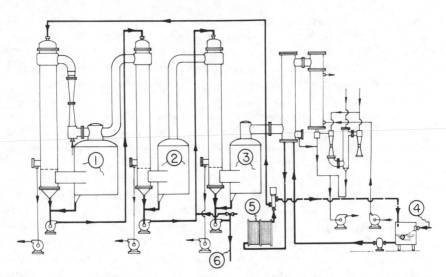

Information furnished by A. W. Baumann, Arthur Harris and Co., Chicago, Ill.

FIG. 9.6. TRIPLE-EFFECT FALLING-FILM EVAPORATOR WITH VAPOR COMPRESSION AND HEAT-
ERS FOR CONCENTRATION OF MILK

(1) First stage with operating temperature of 155° F. (68° C.); (2) second stage, with operating
temperature of 140°F. (60°C.); (3) third stage, with operating temperature of 115°F. (46°C.); (4) flow
of cold raw milk enters system, to be forewarmed by outgoing vapors from stage 3; (5) plate heater
for pasteurizing milk; and (6) flow of condensed pasteurized milk leaving system.

However, the cost of refrigeration and the mechanical difficulties attendant upon removing frozen ice crystals from milk or other dairy products
have thus far retarded extensive commercial adaptation of systems of
concentration by freezing. Energy costs might be greatly reduced in the
concentration of such liquids as skimmilk or whey if they could be exposed to subfreezing temperatures during suitable weather, and then
concentrated by removal of ice crystals. This would provide small milk
and cheese factories with a cheap means of removing water from their
liquid products, but uncertain weather would jeopardize the usefulness of
any process for concentration by natural freezing.

A number of freeze-concentration procedures have been developed.
One method provides for the controlled growth of ice crystals and their
removal in such a way as to minimize contamination of the crystals by
solids of the fluid being concentrated. In another procedure, whey is
concentrated to 30% solids, with a loss of 10% in the ice-crystal sludge.
One inventor advocates the use of a large stainless steel cylinder, surrounded by brine, within which is mounted a screw conveyor. Scraping
blades on the conveyor remove ice from the cylinder walls and carry it to a
centrifuge that separates the ice from any milk that may have been carried

over from the concentrator. In another process milk is frozen in the form of a soft block which fits into the basket of a centrifuge. The concentrate is spun off, the ice is removed from the centrifuge, and another block of frozen milk is placed in the centrifuge.

A process has been developed by Russian workers for the concentration of cheese whey by freezing. Whey containing 5.2% solids is frozen at 25° F. (−4° C.), filtered, and pressed to separate the liquid phase from the ice crystals; these are subsequently washed to obtain a liquid with 11.8% total solids. The freezing operation is repeated several times in order to obtain 25 to 30% whey solids. Another process for simultaneously concentrating and freezing milk has been described. The milk is pasteurized, homogenized, and atomized so that the droplets fall through an ascending cold air stream at −22° F. (−30° C.), while they are subjected to a vibratory motion. Obviously, both these concentration procedures would be cumbersome compared to modern methods of vacuum evaporation.

There is a further process in which milk is subjected to a nitrogen-stripping operation and then concentrated by freezing in absence of oxygen. The keeping quality of the N_2-processed concentrate is superior to that of air-processed milk.

Lactose Crystallization and Gel Formation in Frozen Concentrated Milk.—Coagulation is retarded by removal of some of the lactose or by suppressing its crystallization.

It is known that crystallization of the lactose in frozen concentrated milk appears to initiate protein flocculation. Only circumstantial evidence links the crystallization of lactose to the subsequent coagulation of milk protein during frozen storage. The relationship between the two observed changes is not well understood. Protein destabilization can usually be delayed for as long as lactose can be prevented from crystallizing in the concentrate. Seeding the concentrate brings rapid destabilization, but heating the concentrate or taking other precautions to avoid nuclei formation delays protein coagulation. The addition of freezing point depressants delays destabilization. Salt and sugar are effective in this respect.

Frozen Concentrated Milk Processes.—Six processes for manufacturing frozen concentrated milk will be described, but the last two are not commercially practical under present operating conditions. The processes described avoid three defects characteristic of early frozen concentrated milk products. These are coagulation of the milk protein during freezing, separation of fat, and development of an oxidized flavor.

Fat separation is retarded by increasing the solids content of the milk over that of normal milk and by homogenization. Oxidized flavor is retarded by homogenization and may be further delayed by use of antioxidants. The copper content of milk, when high, catalyzes the development of oxidized flavor and it is desirable to handle the milk under conditions such that copper contamination will not be possible. Stainless steel equipment should be used throughout for processing the milk. Milks of high natural copper and low tocopherol contents show an increased tendency toward development of oxidized flavor.

Enzymatic Hydrolysis of Lactose by Lactase.—To carry out the process, a supply of the enzyme lactase is necessary. This may be obtained commercially or it can be produced as a by-product during the growth of *Saccharomyces fragilis*. Much work has been done on the production and purification of a flavor-free lactase of high activity. To greatly prolong the storage life of frozen concentrated milk, it is necessary to hydrolyze only 10 to 15% of the lactose in the product. This is most easily done by separating 15 to 20% of the milk and hydrolyzing the lactose in the skimmilk fraction. The amount of enzyme preparation to use depends upon its potency and the quantity of lactose to be hydrolyzed. In general, 1.5 to 3% of enzyme by weight of lactose will hydrolyze 80 to 95% of the lactose. The enzyme is added to the skimmilk at 130° F. (54° C.) and the milk is held until about 90% of the lactose is hydrolyzed. Sufficient enzyme is used so that the hydrolysis is completed in about four hours. The enzyme is inactivated by pasteurization of the hydrolyzed skimmilk, which is then added to the raw whole milk. The cream previously removed from the hydrolyzed skimmilk fraction is also added back to the batch. The fluid whole milk, which now contains less than 90% of its original lactose, is pasteurized, preferably at 162° F. (72.2° C.) for 15 sec., homogenized at 1,500 p.s.i. or more pressure, concentrated under vacuum to 35% solids, cooled, packaged, preferably in metal cans, and frozen as rapidly as possible, usually in a cold room at about −20° F. (−29° C.). The enzymatic hydrolysis procedure more than doubles the storage life of the milk, which now reconstitutes to a smooth liquid product. Homogenization prevents fat separation, and together with concentration it usually retards the development of oxidized flavor. The concentrate is more resistant to oxidation than is the fluid milk.

Stabilization of Frozen Concentrated Milk by Polyphosphates.— Polyphosphates added to concentrated milk before freezing have been shown to stabilize it during frozen storage. The most suitable form of polyphosphate is a tetra-polyphosphate glass in the form of a straight-chain product having an average number of phosphorus atoms of about 4.8 per chain. Products with longer chain lengths have been used and are also acceptable. The cyclic polyphosphates should be avoided. The tetra-polyphosphate glass is available commercially and is made by fusion of mixtures of mono-basic and di-basic sodium phosphates. The milk should be pasteurized at 162° F. (72.2° C.) for 15 sec., and concentrated to 35+%

solids. By slightly overcondensing, the polyphosphate may be added, dissolved in water, when the milk is standardized to the desired solids content. The amount of polyphosphate to use may range from 0.5 to 1.5 lb. per 100 lb. of milk solids, depending upon the degree of stability desired. The storage life of a 35%-solids milk concentrate can be greatly extended by adding about 1.0 lb. of stabilizer per 100 lb. of milk solids. For example, milk held at 10° F. (−12° C.) begins to show sediment after three days' storage and this increases progressively during a storage period of 90 days, whereas test samples containing 0.8 lb. of polyphosphate per 100 lb. milk solids show no sediment during the 90-day experimental period. Polyphosphates increase storage life of frozen milk even when the milk is seeded with lactose before freezing.

Use of Soluble Additives to Suppress Lactose Crystallization.—Sugar at a 5 to 10% level (based on the weight of the milk) may be used to retard lactose crystallization. Sweetening of the milk to this extent is unacceptable for many beverage uses. Of various other solutes of low molecular weight available, salts of alkali and alkaline earth metals are particularly effective. Sodium choloride is used in the following procedure. Milk is pasteurized and concentrated under vacuum to 35% solids, of which the lactose content is approximately 13.3%. The milk is treated by dissolving sodium chloride in it at the rate of 0.25% by weight of the concentrated milk. After the salt has been dissolved, the milk is homogenized and again pasteurized and cooled. The treated milk is packaged in cartons or other suitable containers and placed in storage at 15° F. (−9° C.) or lower. After 20 weeks the concentrated milk, when thawed, should be of high quality and there should be no visible gel formation. A comparable batch of milk having the same solids content, but without added salt, and stored under the same conditions will show protein coagulation after six weeks of storage.

At this level of additive (0.25% NaCl), the flavor imparted to the milk on reconstitution to normal strength is scarcely noticeable. The effect of salt on flavor can be entirely overcome by adding flavoring materials such as chocolate or fruit flavors.

Processing to Retard Lactose Crystallization.—Another process has been commercially successful in limited production and distribution tests. With minor variations the process is essentially as follows: the raw milk is pasteurized at a temperature not exceeding 165° F. (74° C.) for 16 sec., homogenized at 2,500 p.s.i., condensed to 36% total solids at a temperature below 140° F. (60° C.), then packaged in cans. The canned product is heated at 155° F. (68° C.) for 25 min., cooled without agitation, and frozen under quiescent conditions at 0° F. (−18° C.). The most important step in the process is the postcondensing heat treatment of the concentrate. This dissolves the lactose nuclei which may have formed in the milk during condensing or canning. Any agitation after cooling below 100° F. (38° C.) or during the freezing of the concentrate reduces the beneficial effects produced by the postcondensing heat treatment. Frozen concentrated milk produced under these conditions remains acceptable in flavor and body for about four months' storage at 10° F. (−12° C.). Success of the method is dependent both upon dissolving lactose nuclei which may have formed and upon using extreme care that nuclei do not reform during handling before the milk is finally frozen. Fluctuations in storage temperature can be expected to trigger lactose crystallization and shorten the life of the milk.

Dialysis of Milk to Remove Lactose and Calcium.—About 50% of the lactose is removed by dialysis against a simulated milk ultrafiltrate of average composition, except that it is devoid of lactose. An alternate procedure is to partially remove the soluble calcium by dialysis against an ultrafiltrate devoid of calcium. The reader is referred to the original work for details of the dialysis procedure. After dialysis, the milk is forewarmed to 150° F. (66° C.) and vacuum-condensed to a concentration of 3:1. The concentrate is cooled quiescently to about 45° F. (7° C.) and then packed in plastic bags or other suitable types of containers. The milk is frozen at −15° F. (−26° C.) and held at this or at higher temperatures. When held at 15° F. (−9° C.) the milk should be stable for 30 weeks.

Crystallization of the Lactose and Its Removal from the Concentrate.— There is no present means to carry out this process on a commercial scale because of the high viscosity developed in the concentrate during lactose crystallization and the difficulty of removing the lactose crystals. Lactose can be removed from concentrated milk if sucrose is added to the milk before condensing. The sucrose has a diluting effect so that the concentrate remains thin during the period of several hours necessary to obtain lactose crystallization. In the absence of sucrose, 3:1 whole milk thickens in a matter of hours, during the time required to crystallize the lactose. The most practical way to remove the lactose is by centrifuging the concentrate, but sometimes the milk reaches a gel-like consistency before a substantial quantity of lactose has crystallized.

Thawing Frozen Milk

Milk and concentrated milk are usually frozen in retail cartons not exceeding the 2-qt. size. On removal from frozen storage it may be thawed by immersing the container in warm water. If more thawing time is available holding overnight in a refrigerator is a satisfactory method.

If flocculated casein, lumps, or a gel is apparent on thawing, the protein can often be redispersed by warming and stirring the milk. Gelation is reversible to a certain point, but eventually, as storage is prolonged, irreversible coagulation occurs.

PRESERVATION OF CHEESE BY FREEZING

The freezing of cheese is usually avoided because of a tendency toward physical breakdown in body and structural characteristics caused by ice crystal formation. The freezing points and moisture content of several varieties of cheese are shown in Table 9.1. Salt added to cheese during making, and the soluble constituents developed during ripening, lower the freezing points of most cheese. Unripened, high-moisture cheese such as cottage cheese, which is very perishable, has a freezing point of 29.8° F. (−1.2° C.). Frozen storage is useful in preserving cottage curd, but deleterious physical changes must be avoided or overcome.

Freezing Ripened Cheese

There is no sign that the changes in this process, now developing from an ancient art to a fast, mechanized operation, will overcome the deleterious changes caused by freezing.

Ripened cheese (Cheddar, Swiss) keeps well in cold storage above its freezing point, where microbiological changes slowly break down the cheese protein and fat to produce the mellow body desired in a well-ripened cheese. Excessive breakdown may occur if the ripening process is continued for too long a period of time. Ripening can thus be retarded if not altogether stopped by freezing the cheese. When this is done the smooth texture of the well-ripened cheese may became rough, mealy, and crumbly after freezing and thawing. There may be some recovery of body if aging and protein breakdown is allowed to continue. The flavor of ripened cheese is not affected significantly by freezing and thawing. A reducing atmosphere is developed during ripening which protects the fat against oxidation.

Several years ago, the University of Minnesota received numerous requests for information from locker patrons and others concerned with the problem of what to do with excess cheese. To answer such questions, researchers studied the freezing of ripened cheese and stated that one can freeze and store cheese if the temperature of the locker is 0° F. (−18° C.) or lower, and if the cheese is tightly wrapped in foil to prevent moisture loss. But the body and texture of ripened cheese sometimes become undesirable as a result of freezing and thawing. Water tends to separate from the protein; this produces a crumbly and mealy body and a general roughening of texture. In some cases, excess whey gives a wet appearance to the curd particles.

Ripened cheeses that are to be frozen should be cut and wrapped in moisture-tight foil or film so that the packages do not contain more than about one pound. Freezing should be done as rapidly as possible, preferably at a temperature of −10° F. (−23° C.) or lower. The cheese should be thawed slowly in a refrigerator set at a temperature above its freezing point. Indeed, some cheese can be frozen and stored for at least six months. The varieties they mention are Cheddar, brick, Port du Salut, Swiss, Provolone, Mozzarella, Liederkranz, Camembert, Parmesan, and Romano.

Chemical Changes in Frozen Ripened Cheese.—Chemical changes have been found to occur in ripened cheeses during storage at −10° F. (−23° C.). While there were only negligible changes in the free amino and free fatty acid contents of Romano, Provolone, Swiss, and Cheddar cheeses, there were appreciable losses of certain acidic carbonyl compounds. The alpha-acetolactic acid had completely disappeared from 2 of 3 Cheddar cheeses stored at −10° F. (−23° C.) for 1 year, and only a trace remained in the third. Oxalacetic acid had disappeared from all three cheeses, but the other keto acid constituents were unchanged. There was no change in the concentration of the keto acids in Romano cheese stored for 1 month, but after 2 months oxalacetic acid was missing and neither oxalacetic, acetoacetic, or alpha-acetolactic acid could be detected after 3 months. As a result of this study the authors recommend caution in the interpretation of results of analyses of cheese placed in cold storage. Other workers have found increases in soluble nitrogen during storage of cheese at low temperatures, but in many cases the temperatures used, while being below the freezing point of water, were not below the freezing point of the aqueous phase of the cheese.

Freezing Fresh Curd Cheese

Fresh curd cheese, particularly the cottage cheese of the United States and the "quarg" of Europe, are high in water content and do not freeze well in their usual merchandizable form. Yet, production of cottage cheese curd in the United States is substantial as is production of creamed cottage cheese.

The storage life of cottage cheese in normal refrigerators is about 15 days, but production tends to be seasonal with milk production. A simple method of freezing fresh curd to yield a satisfactory product is still not available. If the high-moisture cheeses are to be preserved by freezing, their composition and method of handling during manufacture may be altered to reduce freezing damage.

Oxidized flavor may develop in frozen high-moisture cheeses containing considerable quantities of fat. When cottage cheese is prepared for frozen storage, usually only the curd without fat is frozen. Thus, the development of oxidized flavor in frozen cottage cheese curd may be avoided.

In a review of methods of storing the curd, freezing was one of the best methods to preserve it. To overcome the weak and mealy condition of the thawed curd, salt was used to lower the freezing point of the water in the curd. The curd was salted lightly and stored in air-tight 50-lb. cans at 0° F. (−18° C.) or lower. Thawing was critical and should be done slowly over a period of 2 to 4 days. Cottage cheese stored for 9 months at 0° to −10° F. (−18° to −23° C.) was only slightly less acceptable than freshly-made cheese. A number of investigators have immersed cottage cheese curd in brine; this prevented the cheese from freezing, but stopped bacterial deterioration. Yet, workers have found that curd can be held in 3% brine for 14 days, in 4% brine for 1 month at 45° F. (7° C.), or 2 months at 35° F. (2° C.), or in 6% brine at 40° F. (4° C.) for 5 months or more. Curd preserved in brine should be packaged in plastic-lined containers, preferably in the absence of air.

Quarg is the fresh curd product made in Central and Eastern Europe; it resembles uncreamed cottage cheese curd. Russian workers have frozen quarg, packaged in polyethylene film and other materials, and preserved it successfully by freezing it rapidly at a temperature below −18° F. (−28° C.), with subsequent storage below 0° F. (−18° C.). In Czechoslovakia, quarg was stored for 6 months without significant deterioration of quality if it was held at 0° F. (−18° C.). In all cases storage caused increases in total solids, depending upon the packaging material and the storage temperature. There was further loss of moisture during defrosting. No change was found in titratable acidity, but soluble nitrogen increased particularly when the storage temperature was as high as 32° F. (0° C.). All quarg samples decreased in bacteria count during storage.

Preparation of Fresh Curd for Freezing.—Cottage cheese or the East European quarg may be prepared for freezing preservation. Cottage cheese should be made from clean-flavored skimmilk of low bacterial content which has been pasteurized at 161° F. (71.7° C.) for 15 sec., or at 145° F. (62.8° C.) for 30 min., or by some equivalent pasteurization process. A clean, active *Strep. lactis* starter culture should be used. At the time of cutting, the curd should have a titratable acidity of 0.48 to 0.50%. Either the "long-setting" or the "short-setting" process is satisfactory. Dry uncreamed curd of not more than 80% moisture is best for freezing. Yeasts, molds, and coliforms should be less than ten per gram. When prepared according to the above specifications, the flavor of the cheese should be stable for 6 months if held at 0° F. (−18° C.). The curd should be salted lightly, placed in suitable moisture-tight metal or fiber containers, and fast frozen in a freezer or between plates, preferably at −20° F.

(−29° C.) or lower. After the curd is completely frozen the storage temperature may be raised, but it should remain between 0° F. (−18° C.) and −10° F. (−23° C.).

FREEZING STARTER CULTURES

The use of frozen cultures of microorganisms for manufacture of cheese and cultured dairy products is a recent industrial development. It eliminates some of the uncertainties and expense involved in daily transfers, and provides a ready supply of active starters for dairy processing activities. Survival of bacteria is excellent when cultures are frozen and held in a deep frozen state until they are thawed for use. Survival of up to 75 to 100% of the cells may be expected under optimum conditions.

Preparation of Cultures for Freezing

Frozen cultures are prepared by selection and growth of the organisms on suitable media to get them quickly into the late logarithmic growth phase. The growth media must be compounded not only to meet the growth requirement of the organism, but also to permit it to have a full complement of required enzyme systems for subsequent activities.

Many investigators have been concerned with the composition of the growth media. Glucose tryptone broth, "deionized" dried whey, casein digest, and yeast extracts are among media components. The most active cultures were obtained when they were grown in skimmilk fortified with 2% added milk solids. Others have inoculated single-strain isolates into "freshly pasteurized antibiotic-free skimmilk" and frozen them in sterile polyethylene bags at 8° F. (−13° C.). Activity of cultures was "not substantially diminished" by frozen storage for periods up to 12 months.

Others have frozen suspensions of *L. bulgaricus* and *S. lactis* in 10% solids reconstituted skimmilk at −29° F. (−32.6° C.) for 6 months and found acid production of the thawed and incubated cultures to be 35 and 68% of subcultured unfrozen cultures, and that L-glutamic acid added to the medium before freezing stimulated growth after freezing and thawing.

Culture injury and death due to freezing is more pronounced when cells are frozen in water than in skimmilk, and different starters vary in their reaction to frozen storage. In many cases there is marked improvement in survival of cells if the starter is neutralized to 0.16% acidity before freezing.

Temperature and Time of Freezing

Cultures have usually been frozen and held at −4° F. (−20° C.). However, of 16 single strain *S. cremoris, S. lactis,* and *S. diacetilactis* cultures studied for their ability to withstand freezing and storage at 0° F. (−18° C.) and −10° F. (−23° C.) the survival rate was greater at the lower temperature, and numbers of surviving organisms decreased with storage time. The relative proportion of strains in mixed cultures might not remain constant under frozen storage.

Studies on frozen cultures of *L. acidophilus* at 14° F. (−10° C.), −8° F. (−22° C.), and −76° F. (−60° C.) for 8 weeks indicate survivals of 80, 82, and 88%, respectively. Recent work on freezing in liquid nitrogen indicates that this lower temperature, −320° F. (−196° C.), keeps cultures viable for longer periods. Data in Table 9.3 show the advantage of using liquid nitrogen.

Commercial liquid nitrogen-frozen cultures are available for cheese-making. The cultures are grown in special bacteriophage inhibitory media supplemented with pancreas extract to provide optimum activity, packaged in vials and frozen and stored in liquid nitrogen. The cheesemaker receives ten or more strains of cultures selected for cheesemaking characteristics and grows the seed, mother, intermediate, and bulk starter cultures from the stock of frozen culture. This is replenished every 30 days. The cultures have retained high viability at −320° F. (−196° C.) for up to three years. Figure 9.7 shows the plastic vial containing the cultures and the thermal container for storage and shipment of the cultures in liquid nitrogen. This program appears to offer an excellent plan for the control of phage and for maintenance of peak activity in starters.

Freezing Concentrated Cultures

After growth to maximum numbers the culture may be centrifuged from the growth medium, resuspended in a protective menstruum such as milk, and placed in frozen storage. The cells should be harvested for freezing when they near the peak of their logarithmic growth phase. The bacteria are readily removed from the medium by centrifugal separation.

TABLE 9.3

AVERAGE BIOLOGICAL ACTIVITIES OF FOUR SINGLE-STRAIN LACTIC STREPTOCOCCI
STORED AT −4° F. (−20° C.) AND −320° F. (−196° C.)

Storage Time (Days)	Cell Viability (% Survival)		Developed Acidity (%)		Proteinase Active (%)	
	−4°F. −20°C.	−320°F. −196°C.	−4°F. −20°C.	−320°F. −196°C.	−4°F. −20°C.	−320°F. −196°C.
0	100	100	100	100	100	100
3	54	93	69	98	30	82
30	42	89	31	91	24	82
60	5	81	19	83	24	77

Courtesy of Marshall Dairy Laboratories, Inc., Madison, Wis.

FIG. 9.7. LIQUID-NITROGEN FROZEN CULTURE IN 1-ML. PLASTIC VIAL IS REMOVED FROM THERMAL CONTAINER CONTAINING LIQUID NITROGEN

Containers and cultures are distributed to cheesemakers.

There has been commercial development in the production of frozen concentrated cultures. One contains the proper balance of organisms for cheese or cultured milk. Its use is said to eliminate the need for mother cultures, intermediate starter, and bulk starter for buttermilk, sour cream, and even for cottage cheese made by the overnight method. However, large quantities of bulk starter are required for manufacture of cottage cheese, so the cost of using concentrated preparations in lieu of bulk starters could become prohibitive. Another concentrated culture preparation which is frozen and maintained in liquid nitrogen is ready to be added direct to the milk for bulk starter. Some 10 cultures for cheese and 12 for buttermilk and sour cream are available commercially.

ICE CREAM

The general classification of frozen dairy foods includes ice cream, frozen custard or French ice cream, ice milk, sherbet, and water ices.

These may include approximately 30 products depending on composition, processing methods, ingredients, flavoring, size or shape, and condition of the product when sold.

A classification based on regulatory requirements may include the following: (1) ice cream; (2) frozen custard including French ice cream; (3) ice milk; (4) sherbet; (5) water or fruit ice; (6) quiescently frozen dairy confections; (7) quiescently frozen confections; (8) artificially sweetened frozen dairy foods; and (9) imitation ice cream.

The legal definition of ice cream and related products is set forth in the Federal Standards of Identity for Frozen Desserts and includes a number of requirements. These requirements usually include those shown in Table 9.4.

Packaged ice cream usually refers to ice cream in containers of the kind and size in which it reaches the consumer. Bulk ice cream pertains to ice cream that is to be repackaged or dispensed in portions for the consumer.

In fruit and nut ice cream a reduction in the fat and milk solids resulting from the addition of flavoring material in fruit, nut, and chocolate flavoring material is usually allowed. This usually amounts to at least 2% fat and 4% milk solids.

Basic Ingredients

When commercial ice cream was being introduced in this country, the ingredients were cream, fluid milk, sugar, and stabilizer. Later condensed milk, nonfat dry milk, and butter became popular ice cream ingredients. Technological developments and changes in marketing and economic

TABLE 9.4
COMPOSITION STANDARDS FOR FROZEN DESSERTS

	Ice Cream	Bulky Flavor Ice Cream	Frozen Custard or French Ice Cream		Ice Milk	Fruit Sherbet	Water Ices
			Plain	Bulky Flavor			
Minimum fat, %	10	8	10	8	2	1	—
Maximum fat, %	—	—	—	—	7	2	—
Minimum tms[1], %	20	16	20	16	11	2	—
Maximum tms[1], %	—	—	—	—	—	5	—
Minimum wt./gal., lb.	4.5	4.5	4.5	4.5	4.5	6	6
Minimum tfs[2] wt./gal., lb.	1.6	1.6	1.6	1.6	1.3	—	—
Maximum stabilizer, %	0.5	0.5	0.5	0.5	0.5	0.5	0.5
Maximum emulsifier, %	0.2	0.2	0.2	0.2	0.2	—	—
Maximum acidity, %	—	—	—	—	—	0.35	0.35
Total wt. egg yolk solids, not less than, %	—	—	1.4	1.12	—	—	—

[1] TMS—total milk solids.
[2] TFS—total food solids.

conditions have since encouraged the development and use of many other products.

A wide range of choice of ingredients for ice cream is now available from various sources. These ingredients may be grouped as (a) dairy products and (b) nondairy products. The dairy products group is most important as they furnish the basic ingredients of milkfat and milk-solids-not-fat (MSNF) which have essential roles in good ice cream. Some dairy products provide fat, others MSNF, others supply both fat and MSNF and still others supply bulk to the mix.

The nondairy product group includes sweetener solids, stabilizers and emulsifiers, egg products, flavors, special products, and water.

The basic ingredients in frozen dairy foods are milkfat, MSNF, sweetener solids, stabilizers, emulsifiers, and flavoring.

Manufacture of Ice Cream

Ice cream means the pure, clean, frozen products made from a combination of milk products, sugar, dextrose, corn syrup in dry or liquid form, water, with or without egg or egg products, with harmless flavoring and with or without harmless coloring, and with or without added stabilizer or emulsifier composed of wholesome edible material. It shall contain not more than 0.5% by weight of stabilizer and not more than 0.2% by weight of emulsifier, not less than 10% by weight of milkfat and not less than 20% by weight of total milk solids; except when fruit, nuts, cocoa, chocolate, maple syrup, cakes, or confection are used for the purpose of flavoring, then such reduction in milkfat and in total milk solids as is due to the addition of such flavors shall be permitted, but in no such case shall it contain less than 8% by weight of milkfat, nor less than 16% by weight of total milk solids. In no case shall any ice cream weigh less than 4.5 lb. per gal. or contain less than 1.6 lb. of total food solids per gal.

The mix consists of all ingredients with the exception of flavors, fruits, and nuts. The amount of ingredients needed is accurately calculated and is carefully compounded to give the proper composition and balance of fat, solids-not-fat, sugar, and stabilizer. Only the highest quality products should be used. The use of inferior products will result in an inferior ice cream and reduced sales.

The properties of the formulated mix should be such that it has the proper viscosity, stability, and handling properties and such that the finished ice cream will meet the conditions which prevail in the plant where it is to be produced.

A typical mix formula for 200 gal. of ice cream mix for a good average composition mix of 12% butterfat, 11% MSNF, 15% sugar, 0.3% stabilizer and 38.3% total solids might be as follows:

	Lb.	Fat	MSNF	Sugar	Stabilizer	Total Solids
Cream 40%	300.0	120.0	16.2			136.2
Condensed skim milk (27%)	247.6		66.9			66.9
Skim milk	299.4		26.9			26.9
Sugar	150.0			150.0		150.0
Stabilizer	3.0				3.0	3.0
Total	1000.0	120.0	110.0	150.0	3.0	383.0

The basic steps of production in manufacturing ice cream are composing and blending the mix, pasteurization, homogenization, cooking, aging, flavoring, freezing, packaging, hardening, and storage.

The first step of processing is composing the mix. This procedure may range in scope from the small batch operation where each ingredient is weighed or measured and added, to the large pushbutton operation where many of the ingredients are metered into the batch. The common procedure is to: (a) add liquid materials (cream, milk, or other liquid milk products) to mix vat or pasteurizer; (b) apply heat (optional) and then add dry solids such as egg yolk, gelatin, etc. (Mixing dry products with three parts of sugar and adding to the mix will aid in their dispersion); (c) add sugar when the mix reaches approximately 120° F. (49° C.); (d) use caution to insure that all materials are dissolved before pasteurization temperature is reached.

The pasteurization process consists of heating products for approved temperature and time. This may be:

Holding method	160° F. (71° C.) for 30 min.
High temperature-short time	175° F. (79° C.) for 25 sec.
Vacreation	193° F. (90° C.) for instant-about 3 sec.
Ultra high temperature	240° F. (115° C.) for instant

Pasteurization (1) renders the mix free of harmful bacteria, (2) brings into solution and aids in blending the ingredients of the mix, (3) improves flavor, (4) improves keeping quality, and (5) produces a more uniform product. There is a trend toward the higher temperature processes.

When the batch pasteurization method is used the mix is pasteurized by heating to 155° or 160° F. (68° or 71° C.) and held at that temperature for 30 min., thus killing all pathogenic types of bacteria and all or nearly all other objectionable organisms. The required time and temperature of pasteurization varies in different localities depending on state and city laws and ordinances. For that reason, temperature of pasteurization is carefully controlled and recorded.

When the continuous method of pasteurization is used the high-temperature, short-time (HTST) treatment is used. Mix processing arrangements vary greatly with the HTST treatment and in some arrangements the hot mix is homogenized before pasteurization is accomplished. Immediately following pasteurization, the hot ice cream mix is passed through the high-pressure pump of the homogenizer at pressures that range from 1500 to 3000 p.s.i.; this machine breaks the particles of butter fat into very small globules (Fig. 9.8).

The purpose of homogenization is to produce a homogeneous mix. The hot mix is pumped from the pasteurizer through the homogenizer. The optimum homogenization temperature may range as high as 180° F. (82° C.). This process reduces the size of fat globules.

The advantages of homogenization are as follows: (1) thoroughly blends the ingredients of the mix; (2) breaks up and disperses the fat globules, thus preventing churning of the fat during freezing; (3) improves the texture and palatability of the ice cream; (4) makes possible the use of different ingredients; (5) reduces aging and aids in obtaining overrun; (6) produces a more uniform product.

Homogenization reduces the size of fat globules to less than two microns (one micron = $1/25,000$ in.). The homogenization pressures generally

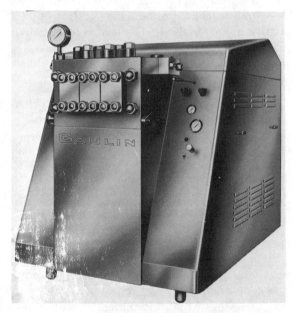

Courtesy of Manton-Gaulin Mfg. Co.

FIG. 9.8. MANTON-GAULIN HOMOGENIZER

used are 1,500 to 3,500 lb. for the single stage machine. For the two stage homogenizer, pressures of 2,000 to 3,000 lb. are used on the first stage and 500 to 1,000 on the second stage. The correct amount of pressure to apply for a given mix is influenced by the following: type of homogenizer, temperature of mix (low temperature-lower pressure), acidity of mix (high acid-lower pressure), composition of mix (high fat, stabilizer, and solids require low pressure to prevent excessive viscosity).

The smooth mix then flows to a cooler (Fig. 9.9) where the product is cooled as rapidly as possible in order to prevent bacterial growth. The

Courtesy of National Dairy Council

FIG. 9.9. COOLING THE ICE CREAM MIX

Sanitary pipelines carry the hot, pasteurized, and homogenized mix to a cooler. Here the temperature of the mix is reduced rapidly to about 40°F. (4°C.) as the mix flows down over refrigerated tubes, then out to the holding tanks.

cooler chills the mix to a temperature of 40° F. (4° C.) or colder. After chilling, the mix may go directly to the freezers, or it may go to small flavor tanks where liquid flavorings like vanilla or chocolate are added, or it may go to so-called aging tanks. Flavoring and aging tanks are insulated to maintain the temperature of the mix at 40° F. (4° C.) or lower. If the mix is aged, it is held from 4 to 24 hr. The aging step is believed to be necessary if gelatin is used as the stabilizer. By aging the mix, the gelatin has time to set and better accomplish the purpose for which it was added. Most authorities agree that a four-hour aging is ample, but many plants prepare a mix one day and hold it overnight for freezing the next day. Prolonged aging beyond seven days may result in abnormal product properties. Most vegetable stabilizers set up immediately upon being cooled and, if one of them is used, there appears to be less advantage in aging.

Changes that take place during aging include: (a) combination of stabilizer with water of the mix; (b) the fat solidifies; (c) the proteins may change slightly; (d) increase in viscosity; and (e) mix ingredients may become more stable.

Freezing the mix is one of the most important steps in the making of ice cream. The freezing process should be accomplished as rapidly as possible to insure small ice crystals and a smooth texture in either the batch freezer or the continuous freezer. The function of the freezing process is: (1) to freeze a portion of the water of the mix; and (2) to incorporate air into the mix. There are four phases of the freezing process: (1) lowering the temperature from the aging temperature (usually about 40° F.; 4° C.) to the freezing point of the mix; (2) freezing a portion of the water of the mix; (3) incorporating air into the mix; (4) hardening the ice cream after it is drawn from the freezer.

Freezing involves refrigerating the mix in a freezer cylinder which is surrounded by sub-zero ammonia or brine, as today most plants use the continuous freezer (Fig. 9.10). The cylindrical freezer is provided with blades which scrape the freezing mix from the refrigerated metal walls of the machine. During the whipping, air is forced into the mix increasing the volume of the frozen ice cream. Without this overrun, ice cream would be an almost inedible hard frozen mass.

The temperature at which the mix starts to freeze varies with the per cent total solids, but for the average formula that temperature is approximately 27° F. (−2.8° C.). When the ice cream is drawn from the freezer, its temperature will usually range from 25° to 20° F. (−3.7° to −6.7° C.).

The freezing time and temperature is affected by the type of freezer used. When the batch freezer is used, the freezing time to 90% overrun approximate is about seven minutes and the drawing temperature is about 24° to 26° F. (−4° to −3° C.); continuous freezer, the freezing time

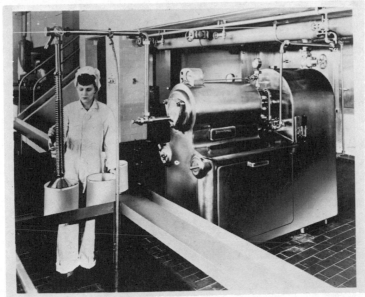

Courtesy of National Dairy Council

FIG. 9.10. CONTINUOUS ICE CREAM FREEZER

The soft ice cream which emerges from the freezer is run through another machine called a flavor feeder where fruits are added prior to packaging.

to 90% overrun approximate is about 24 sec. and the drawing temperature is about 21° to 22° F. (−6° to −5.6° C.); counter freezer, the freezing time to 90% overrun approximate is about 10 min. and the drawing temperature is about 26° F. (−3° C.); and for the soft serve freezer, the freezing time to 90% overrun approximate is about 3 min. and the drawing temperature is about 18° to 20° F. (−8° to −7° C.). About half the water of the mix is frozen in the freezer and most of the remaining water is frozen in the hardening room.

If fruit is to be added to the mix, the soft ice cream coming out of the freezer is run through a flavor feeder machine which adds the fruit. The soft ice cream then goes to a packaging machine where it may be packed in bulk containers or small packages (Fig. 9.11) for retail sale. From the packaging machine, the product goes to the hardening room, where the freezing process is completed. In the hardening room (Fig. 9.12), the packaged soft ice cream becomes firm within 24 hr. Hardening rooms are maintained at a temperature of −10° to −30° F. (−23° to −34° C.) either with or without forced air circulation. The ice cream is now ready for storage or delivery.

Courtesy of National Dairy Council

FIG. 9.11. PACKAGING ICE CREAM

A packaging machine opens flat paraffined packages, fills them with ice cream and closes each package. This machine packages 4,000 pints per hour.

Courtesy of Delvale Dairies

FIG. 9.12. ICE CREAM HARDENING ROOM

In the hardening room, the soft ice cream from the freezers becomes firm within 24 hr. Unit air cooler and ducts are shown which maintain this room at a temperature of −20° F. (−29° C.).

SOFT SERVE PRODUCTS

Soft serve products are those sold as drawn from the freezer without hardening. The product is drawn from the freezer at a temperature of about 18° to 20° F. (−8° to −7° C.). The mix composition of soft serve is usually lower in sweetener solids and total solids, in order to meet desired dryness and stiffness characteristics. The overrun on soft serve products usually ranges from 35 to 45%.

There is a marked demand for soft serve products and a high percentage of these products meet the standards for ice milk. As the soft serve products became commercially important, freezing equipment was developed which would dispense, at all times, a soft freshly frozen product. Batch and semi-continuous type freezers are available. Most retail sales establishments selling soft serve products do not attempt to produce their own mix, but buy the mix and freeze it in the soft serve freezers. Refrigerated storage space for the mix is provided.

ADDITIONAL READING

ASHRAE. 1974. Refrigeraton Applications. American Society of Heating, Refrigeration and Air-conditioning Engineers, New York.

ASHRAE. 1975. Refrigeration Equipment. American Society of Heating, Refrigeration and Air-conditioning Engineers, New York.

ASHRAE. 1976. Refrigeration Systems. American Society of Heating, Refrigeration and Air-conditioning Engineers, New York.

ARBUCKLE, W. S. 1972. Ice Cream, 2nd Edition. AVI Publishing Co., Westport, Conn.

ARBUCKLE, W. S. 1976. Ice Cream Service Handbook. AVI Publishing Co., Westport, Conn.

HALL, C. W., FARRALL, A. W., and RIPPEN, A. L. 1971. Encyclopedia of Food Engineering. AVI Publishing Co., Westport, Conn.

HARPER, W. J., and HALL, C. W. 1976. Dairy Technology and Engineering. AVI Publishing Co., Westport, Conn.

HELDMAN, D. R. 1975. Food Process Engineering. AVI Publishing Co., Westport, Conn.

HENDERSON, S. M., and PERRY, R. L. 1976. Agricultural Process Engineering, 3rd Edition. AVI Publishing Co., Westport, Conn.

JOHNSON, A. H., and PETERSON, M. S. 1974. Encyclopedia of Food Technology. AVI Publishing Co., Westport, Conn.

PENTZER, W. T. 1973. Progress in Refrigeration Science and Technology, Vol. 1–4. AVI Publishing Co., Westport, Conn.

PETERSON, M. S., and JOHNSON, A. H. 1977. Encyclopedia of Food Science. AVI Publishing Co., Westport, Conn.

TRESSLER, D. K., and SULTAN, W. J. 1975. Food Products Formulary. Vol. 2. Cereal, Baked Goods, Dairy and Egg Products. AVI Publishing Co., Westport, Conn.

TRESSLER, D. K., VAN ARSDEL, W. B., and COPLEY, M. J. 1968. The Freezing Preservation of Foods, 4th Edition. Vol. 1. Refrigeration and Refrigeration Equipment. Vol. 3. Commercial Freezing Operations—Fresh Foods. Vol. 4. Freezing of Precooked and Prepared Foods. AVI Publishing Co., Westport, Conn.

UMLAUF, L. D. 1973. The frozen food industry in the United States—Its origin, development and future. American Frozen Food Institute, Washington, D.C.

WILLIAMS, E. W. 1976. Frozen foods in America. Quick Frozen Foods 17, No. 5, 16–40.

WOOLRICH, W. R. 1965. Handbook of Refrigerating Engineering, 4th Edition. Vol. 1. Fundamentals. Vol. 2. Applications. AVI Publishing Co., Westport, Conn.

WOOLRICH, W. R., and HALLOWELL, E. R. 1970. Cold and Freezer Storage Manual. AVI Publishing Co., Westport, Conn.

10

Freezing of Egg Products

James M. Gorman, Orme J. Kahlenberg
and William D. Powrie

Frozen egg products marketed in the U.S. include whole eggs, yolks (plain, sugared, and salted), and albumen. Frozen whole egg is a mixture of whites and yolks in naturally occurring proportions with about 25 to 26.5% solids. According to the Standards of Identity under the Federal Food, Drug, and Cosmetic Act, frozen egg yolks must contain not less than 43% solids. Commercial yolk generally contains about 15 to 20% albumen. To prevent an alteration in the yolk viscosity due to freezing and thawing (gelation), 10% sucrose or 10% NaCl may be added. The solids content of frozen albumen is generally around 12%. The yolk content of albumen must not be over 0.09% to insure satisfactory foamability. In compliance with FDA standards, all egg products must be pasteurized or otherwise treated prior to freezing to destroy all viable Salmonella microorganisms.

Preservation of egg products by freezing has been on the upswing for the last 30 years, and the proportion of frozen egg products to combined cold storage and frozen eggs rose from about 37 to 89%.

Frozen egg products are used for the manufacture of a multitude of food products. Frozen whole eggs and albumen are utilized mainly for the manufacture of bakery products.

Mayonnaise and salad dressing manufacturers use frozen salted yolk as a source of emulsifying agents. On the other hand, frozen sugared yolk is preferred for bakery products and ice cream. Frozen plain yolk without sugar and salt is included in formulations for noodles and baby foods.

CHARACTERISTICS OF ALBUMEN AND YOLK

Composition

Shell eggs are made up of 8 to 11% shell, 56 to 61% albumen, and 27 to 31% yolk. On an edible portion basis, eggs consist of about 65% albumen and 35% yolk.

Albumen of the egg is made up of outer thin white, thick white, and inner thin white. The proportion of thick white in the albumen may be as high as 77%. The gelatinous nature of the thick white may be attributed to ovomucin-containing fibers. During the storage of shell eggs, carbon dioxide in the albumen diffuses through the shell and, as a consequence, the pH of albumen rises from about 7.6 (at laying time) to over 9. The pH of albumen in eggs stored at 36° F. (2° C.) rises to approximately 9.3 in 20 days. With a pH increase, the gel structure of the thick white is weakened, probably by the disruption of the ovomucin-containing fibers.

The solids content of albumen ranges from about 11 to 13%. Average albumen solids of 11.8 and 11.3% were recorded for two flocks over a one-year production period.

Albumen from eggs laid by hens in 20 flocks over the 1963–64 season had an average solids of 12.11%. The solids content of albumen is dependent on strain and age of hens.

With a decrease in the solids of albumen, the volume of angel food cakes decreased. Most of the albumen solids consists of proteins in the following composition: solids, 12.1%, protein 10.6%; carbohydrate, 0.9%; ash, 0.6%; and fat, 0.03%.

Yolk is made up of concentrically-oriented yellow and white layers caused presumably by diurnal rhythm of the hen. Yolk solids content, in the vicinity of 50%, is influenced by age of the layer, and age of shell egg. Researchers have calculated the average solids of yolk from the fresh eggs of five hen strains to be 52.7%, while others have found an average solids of 52.29% for fresh eggs from three strains. During the storage of shell eggs, water migrates from the albumen into the yolk through the vitelline membrane and thus the solids content of yolk decreases.

When eggs are held at 34° to 39° F. (1° to 4° C.) for 64 days, the solids content of yolk decreases from about 52.8 to 50%; the yolk solids drop from 53.5 to 49% when eggs are stored at 75° F. (24° C.) for 16 days, with an average yolk solids content of 50.09% for eggs stored at 39° F. (4° C.) for one week. The pH value of liquid yolk from freshly-laid eggs is in the region of 6. Studies indicate that yolk from both fresh and stored eggs (16 days at 55° F. (13° C.)) had a pH value of 6.0. When unwashed shell eggs were stored for periods up to 16 days at 75° F. (24° C.), the pH of yolk fluctuated between 6.0 and 6.2. The pH of yolk rises from 6.0 to 6.3 to 6.4 upon egg storage at 34° to 39° F. (1° to 4° C.).

The lipid content of yolk generally falls into the range between 32 and 36%. Variability of the yolk lipid content can be attributed to the strain of the layers. Studies show that average lipid values of yolk from these strains of hens are 35.50, 32.67, and 31.95%. Yolks from eggs collected over an 11-month period from White Leghorn hens of the same strain and age had lipid contents between 32 and 33.5%. The yolk lipid composition was 65.5% triglyceride, 28.3% phospholipid, and 5.2% cholesterol. Palmitic, oleic, and linoleic acids are the major fatty acids of lipids in yolk from hens on a corn-soya diet. When vegetable oil was added to a normal layers diet, the fatty acid composition of yolk lipid was altered dramatically.

The protein content of egg yolk has been reported as 15.7% and 16.6%. Using the data, the protein content of yolk with 50% solids would be 16.15%.

Chemical and Physical Properties

Egg White Proteins.—Fresh egg white (albumen) may be regarded as a protein system consisting of ovomucin fibers in an aqueous solution of numerous proteins. Several excellent reviews on egg white proteins have been written. The protein compositions of the thin and thick layers of albumen differ only in the ovomucin content. The principal protein fractions in albumen can be prepared by the stepwise addition of ammonium sulfate. The protein composition of egg white is presented in Table 10.3. Many of these protein fractions are heterogeneous, as shown by starch gel electrophoresis. Ovalbumin, for example, has been separated into three bands on a gel electrophoretogram. Ovalbumin, the major protein fraction of egg white, has an approximate molecular weight of 45,000 and an isoelectric point of 4.6. Upon heating a solution of ovalbumin, the protein is converted to S-ovalbumin, and also is denatured and coagulated. S-albumin, being more heat-stable than ovalbumin, may not be a desirable component of egg white for bakery products. In contrast to ovalbumin which is completely precipitated during shaking, the conalbumin fraction does not undergo surface denaturation. However, conalbumin is more heat sensitive than ovalbumin. At pH values in the region of 6 to 7, the conalbumin has minimum heat stability.

The conalbumin, with a molecular weight of about 76,000 and an isoelectric point of about 6.6, has the ability to bind metal ions such as ferric, cupric, and aluminum above pH 6. The ferric-conalbumin complex is colored red, while the aluminum-conalbumin complex is colorless.

When conalbumin is complexed with metal ions, the protein is resistant

TABLE 10.1

THE PROTEIN COMPOSITION OF EGG WHITE

Protein Fraction	% of Total Protein
Ovalbumin	64.3
Conalbumin	13.6
Ovomucoid	9.1
Lysozyme	3.4
Globulins	8.6
Ovomucin	1.1

to thermal denaturation and coagulation. Ovomucoid, a glycoprotein with 20 to 22% carbohydrate, has antitryptic activity. Although the heating of ovomucoid solutions with pH values between 6 and 9 does not cause coagulation, antitryptic activity is reduced. Another egg white glycoprotein of functional importance is ovomucin. Ovomucin, with a carbohydrate content of about 10%, is the major protein in the fibers of thick white. Upon dilution of egg white with water, ovomucin is precipitated. Ovomucin is soluble only in alkaline solutions.

The functional properties of egg white are dependent on specific types of proteins. The foaming power of egg white can be attributed to the globulin fraction (including lysozyme), whereas ovomucin, aggregating during the whipping period, stabilizes the foam. Heat-coagulable proteins, such as ovalbumin, are essential for the semisolid cell walls of baked angel cake.

Pasteurization of liquid egg white at 140° to 144° F. (60° to 62° C.) for about 3 min. causes protein coagulation. The heat stability of albumen proteins (particularly conalbumin) can be increased by lowering the pH to about 7 and adding a small amount of an aluminum salt. Egg white treated in this way can be pasteurized commercially prior to freezing.

Microstructure of Egg Yolk

Egg yolk is a complex system composed of particulate spheres, lipid droplets, and low-density lipoproteins in a continuous aqueous phase (live-tin solution). Spherical masses with diameters between 25 and 150 μ are observed in yolk. Using a microscope, two types of spheres can be seen: those in the white yolk with only a few droplets throughout, and those in the yellow yolk with numerous droplets. The white and yellow spheres have diameter ranges of 4 to 75 μ and 25 to 150 μ, respectively. With the aid of electron microscopy three types of sphere surfaces are found: the lamellated capsule, the unit membrane-like structure, and naked surface. The majority of yolk spheres have naked surfaces. Free-floating lipid droplets in yolk have diameters in the vicinity of 2μ, whereas the diameters of round profiles (low-density lipoprotein) are around 250 Å. The low-density lipoprotein diameters estimated by analytical ultracentrifugation are between 117 and 480 Å.

When yolk is subjected to high-speed centrifugation, the granules sediment. The clear supernatant is called the plasma. The granules, which constitute about 11.5% of the liquid yolk, can be broken down by the addition of 0.3 M NaCl solution. Thus, frozen salted yolk (10% NaCl) does not contain granules. The major portion of iron and calcium ions reside in the granule. Granules are made up of 70% α and β-lipovitellin (high-density lipoproteins), 16% phosvitin, and 12% low-density lipoprotein.

Plasma, with a moisture content of about 49%, is made up of α-, β- and γ-livetins, and low-density lipoproteins. The low-density lipoproteins occur as spheres, each with a triglyceride core upon which phospholipids and proteins are layered. So far, two low-density lipoprotein (LDL) fractions have been separated from yolk plasma. The lipid contents of the two LDL fractions isolated were 86 and 89%. Since LDL fractions of plasma are probably involved in the gelation of yolk, the surface characteristics of these micelles should be known. Protein moieties of LDL must be on the surfaces of the micelles, since papain can hydrolyze the protein molecules. About two-thirds of the surface of an LDL micelle can be covered by protein if the thickness of protein molecules is assumed to be 8 Å. When one adds phospholipase D to an LDL dispersion, 95% of the phospholipids are hydrolyzed. Apparently, the phosphate groups of the phospholipids are near the LDL surface. Using equilibrium dialysis with methyl orange, at least 100 cationic sites are located on the surface of each LDL micelle.

FREEZING OF ALBUMEN AND YOLK

During the removal of heat from fluid albumen or yolk, the temperature may drop several degrees below the freezing point, without ice crystal formation, as a supercooled mass. An egg product can be maintained in the supercooled state under suitable conditions for a few seconds up to several months. The duration of supercooling is dependent on the type of egg product, rate of cooling, storage temperature, and the presence of suitable nuclei. When albumen was cooled in a brine bath at 18.7° F. (−7.4° C.), this egg product was supercooled to 21.6° F. (−5.8° C.), but immediately thereafter, the temperature rose within half-a-minute to the freezing point.

The freezing point of egg white has been reported as 31.20° F. (−0.45° C.). No supercooling of egg white was noted when the freezing temperatures of the brine were at 13.1° F. (−10.5° C.) or below. Similar results on supercooling were obtained with yolk. After supercooled yolk reached a temperature of 21.6° F. (−5.8° C.), (18.7° F., −7.4° C. brine bath), 4 min. were required for ice crystallization. The freezing point of yolk was estimated to be 30.2° F. (−0.65° C.) and 30.96° F. (−0.58° C.). Judging from the results available, egg white and yolk in the shell can be supercooled at temperatures above 21° F. (−6° C.) for periods up to 3 months, and at 12° F. (−11° C.) for 7 days. Apparently, shock and temperature fluctuation promoted ice crystallization of supercooled shell eggs.

The amount of ice formed in albumen and yolk has been calculated as per cent of total initial moisture (Table 10.2). At 30.2° F. (−1° C.) frozen egg white contained 48% ice, and as the temperature was lowered to

TABLE 10.2

PERCENTAGE OF ICE IN ALBUMEN AND YOLK AT VARIOUS TEMPERATURES

Temperature		Ice Content, %	
°F.	°C.	Egg White, 86.5% H₂O	Yolk, 50% H₂O
30.2	−1	48	42
28.4	−2	75	67
24.8	−4	86	77
23.0	−5	87	79
14.0	−10	92	84
−4.0	−20	93	87
−22.0	−30	94	89

−22° F. (−30° C.), the ice content increased to 94%. The low salt and sugar concentration in albumen (0.6% ash and 0.9% carbohydrate) may account, in part, for the extensive ice formation at relatively high freezing temperatures. The ice content of yolk (50% initial moisture) was somewhat lower than that for albumen at the same temperature. A considerable amount of water is undoubtedly bound strongly to the yolk solids and would not be expected to be transformed into ice crystals.

Alteration of Physical and Functional Properties of Frozen Albumen and Yolk

Albumen.—When shell eggs were frozen at temperatures between 26.6° and −13° F. (−3° and −25° C.) and then stored at 26.6° F. (−3° C.) for 24 hr., the thawed albumen had a greater percentage of thin white than unfrozen controls. The cryothinning of albumen was dependent on the temperature reached during the freezing period, rather than on the freezing rate. Apparently, greater amounts of ice crystals in the frozen white caused more extensive damage to the ovomucin-containing fibers. With prefreezing disruption of the thick white by mixing or homogenization, the viscosity of the unfrozen and thawed (previously stored at −9° F., −23° C.) egg white had similar viscosity values. The foaming volume of egg white was unchanged when this egg product was previously stored at −9° F. (−23° C.) for periods up to 2 months, but the 3-month frozen stored albumen had a much higher foamability than the product stored for a shorter freezer time. After 4 months of storage at 26.6° F. (−3° C.), distinct white fibers were apparent in the thick white. Such particulate matter would undoubtedly contribute to the stability of foam lamellae.

Yolk.—Unfrozen yolk is a viscous, nonNewtonian fluid. Upon plotting shear stress (dynes/cm.^{2}) against shear rate (sec.$^{-1}$) for yolk at 77° F. (25° C.), the curve deviated slightly from a straight line. The apparent viscosity, expressed in poises or centipoises, of yolk is dependent on the temperature and moisture content of yolk as well as rate of shear. Using a Brookfield viscometer, researchers reported that yolk with solids contents of 52.5 and 50.8% had viscosities of 1,452 and 782 centipoises, respectively, at 77° F. (25° C.).

Gelation.—When yolk is frozen and stored below 21° F. (−6° C.), the viscosity of the thawed product is much higher than that of native yolk. This irreversible change of yolk fluidity has been termed gelation. The following factors have an influence on the rate and extent of gelation: (1) rate of freezing; (2) temperature and duration of frozen storage; and (3) thawing rate.

Rate of Freezing.—Researchers froze liter quantities of yolk at freezing rates between 0 and 39 hr. to evaluate the influence of freezing rate on the degree of yolk gelation. Freezing rate was defined as the time for yolk to decrease in temperature from 45° to 20° F. (7.2° to −6.7° C.) at the center of a yolk sample. The viscosity of thawed yolk previously stored at 0° F. (−18° C.) for periods up to 8 months was greater as the freezing rate rose from 0 to 39 hr. Similar results were obtained with mixed whole egg although yolk was more viscous than mixed whole eggs under the same freezing-thawing conditions. When they froze yolk rapidly in liquid nitrogen (B.P. −321° F., −196° C.) and stored it at 0° F. (−18° C.) for 7 days, the yolk after rapid thawing had a low degree of gelation.

Temperature and Time of Frozen Storage.—Thawed shell eggs, previously stored in the frozen state at temperatures between about 30° and 21° F. (−0.7° and −6° C.) for several months, possessed ungelled yolks. On the other hand, this researcher found that the viscosity of thawed yolk from eggs stored at 12° F. (−11° C.) increased with storage time up to about 24 hr. Yolk frozen at 19° F. (−7° C.) for periods up to 750 min. had about the same fluidity upon thawing as unfrozen yolk. However, with storage temperatures of 14° F. and 7° F. (−10° and −14° C.), yolk gelatin increased markedly over a few hours of storage. The rate of gelation is increased as the temperature is lowered to −58° F. (−50° C.). Experimenters showed that degree of gelation of whole egg magma at 0° to −6° F. (−18° to −21° C.) reached a maximum between 60 and 120 days of storage.

Thawing Rate.—Rapid heat penetration into a frozen yolk mass during the thawing period can prevent extensive gelation especially when the yolk was frozen rapidly to inhibit freezing damage. Upon rapid thawing at 86° F. (30° C.), yolk previously frozen in liquid air had the same fluidity as unfrozen yolk, whereas the yolk thawed slowly became pasty.

The functional properties of yolk are altered somewhat during the freezing-storage-thawing cycle. The blending of thawed gelled yolk with other ingredients has been reported to be difficult for the preparation of cake batters. When pasty yolk was used for the preparation of sponge cakes, the cake volume was considerably lower than that for cakes prepared with unfrozen yolk. Custards prepared with gelled yolk were soft and their crust contained hard, yellow lumps.

Prevention of Yolk Gelation.—Gelation of yolk can be minimized by prefreezing treatments such as the addition to yolk of chemical protective agents and proteolytic enzymes, or the passage of yolk through an homogenizer or colloid mill. Sugars such as sucrose, arabinose, galactose, and glucose at the 10% level act as cryoprotective agents for preventing yolk gelation.

Using whole egg magma, a concentration of sucrose as low as 1% can inhibit yolk gelation. The effectiveness of sucrose as a cryoprotective agent in yolk increased as the concentration increased from 0.34 to 3.79% (g. per 100 g. of yolk). Glycerol at the 5% level can inhibit gelation. None of the above-mentioned polyhydroxy compounds alter appreciably the properties of native yolk. On the other hand, the addition of NaCl to yolk as a gelation preventative brought about an increase in translucency and viscosity. Presumably granule disruption by sodium and/or chloride ions is the cause of these changes. Concentrations of NaCl above 8% are sufficient for preventing the gelation of yolk.

The use of enzymes for the prevention of yolk gelation is possible by incubating yolk with added papain or trypsin for a short period of time prior to freezing. The most effective enzymatic prefreezing treatment for the prevention of gelation was the addition of 0.05% commercial papain to yolk and the incubation of the yolk-enzyme mixture for 15 to 20 min. at 75° F. (24° C.). Enzyme-treated yolk does not have as high an emulsifying action as native yolk.

Homogenization and colloid milling can inhibit but not prevent yolk gelation. When yolk is passed through a colloid mill three times with a 0.003-inch clearance, the degree of yolk gelation is low. Colloid milling as a prefreezing treatment prevents completely the gelaion of whole egg magma.

Mechanism of Yolk Gelation.—In order to bring about the gelation of yolk, a specific amount of water (80%) must be frozen and the frozen mass must be stored for at least a few hours. Supercooling of yolk, without ice crystal formation, will not initiate any alterations of the yolk components and thus will not cause gelation. Undoubtedly the particulate matter such as granules and low-density lipoproteins (LDL) is involved in the gelation mechanism. Plasma having LDL as naturally-occurring components become somewhat pasty when frozen at −4° to −13° F. (−20° to −25° C.) for 24 hr. prior to thawing. Although the livetins of plasma are not altered during the freezing process, LDL aggregates to such an extent that only 15% of the total LDL in thawed plasma is soluble in 10% NaCl. A considerable amount of LDL in gelled yolk is unable to migrate electrophoretically on paper. Judging from the rheological data, both granules and LDL interact during frozen storage to form highly-hydrated cmplexes, with the consequence of a pasty yolk mass.

COMMERCIAL FREEZING OPERATIONS

Preparation

Hand Breaking.—In early breaking operations, it was necessary to separate the yolk from the white by flipping the yolk back and forth between halves of the shell until the whites drained off (Fig. 10.1). This process was very slow and inefficient. In 1912, Harry A. Perry invented the hand separator, which contributed to the development of large-scale breaking operations. The "hand separator" essentially consists of a sliding hinged separator which is attached to a breaking knife. The separator itself consists of two parts—a receptacle about the size of the egg yolk and a sharp-edged ring just large enough to fit over the yolk. This ring is

Courtesy of Tranin Egg Co.

FIG. 10.1. HAND-OPERATED BREAKING ROOM WITH THREE COMPLETE LINES OF BREAKING CONVEYORS, FORMERLY USED BY TRANIN EGG CO.

hinged so that it can be raised or lowered to cut the white from the yolk in the receptacle. The whites fall into a small cup, and the receptacle containing the yolk is tipped to one side, causing the yolk to drop into a second cup. The cups, when filled with three eggs, are first smelled by the operator to detect any off-odors and are then emptied into a larger container. An efficient operator can break and separate 2 to 3 cases (60 to 90 doz. eggs) per hour. The use of the egg separator greatly improved the speed and efficiency of egg breaking and contributed to large-scale breaking operations.

Automatic Egg Breakers.—The first commercial egg breaking machine, invented by L. Sigler in 1943, was manufactured in 1949–1950 by the Barker Poultry Equipment Co. A second commercial automatic egg breaking machine was developed by C. H. Willsey and further improved by The Seymour Foods Co. (Fig. 10.2).

An operator-inspector can break and separate 20 cases (600 doz. eggs) per hour with the Willsey automatic machine. Automatic egg breakers lower production costs through saving of labor, and require less floor space than conventional hand-breaking operations. In addition, the quality of the product is improved because the egg meats are individually inspected after breaking. The modern automatic egg washer and break-

Courtesy of Seymour Foods Co., Division of Norris Grain Co.

FIG. 10.2. A BATTERY OF NINE SEYMOUR BREAKING MACHINES

ing machine can wash, sanitize the shell, break, and separate the yolks from the whites, and this results in better yields and much lower bacterial counts of the liquid raw material than the tedious hand breaking methods. In modern plants all the eggs are automatically broken and separated; if the desired product is whole egg, the separate streams of white and yolk from the machines are combined.

Pasteurization.—The first commercial pasteurization of liquid eggs was done by Henningsen Bros. in 1938. The process was used during World War II when the Armed Services required that all whole eggs prepared for them be pasteurized before the liquid egg was dried. Considerable research has been done since that time on the pasteurization of liquid eggs under commercial conditions. The USDA, under its voluntary inspection program, requires now that all egg products be pasteurized, regardless of whether they are to be distributed in liquid, frozen, or dried form. This action was found necessary because pasteurization and heat treatment are the only effective ways known today to eliminate the food poisoning organism Salmonella from egg products. Liquid whole eggs, under these regulations, must be flash-heated at a minimum temperature of 140° F. (60° C.) for at least 3½ min., and other egg products, such as yolks, blends, sugared, and salted yolk products, be pasteurized at such temperatures and held for such time periods as will give equivalent assurance of *Salmonella*-free products.

When, however, liquid whites are subjected to this temperature for this length of time, they coagulate. In order to pasteurize whites and at the same time prevent coagulation, the USDA Western Regional Research Laboratory (W.R.R.L.) developed a process for adding lactic acid and aluminum salts to liquid egg whites before pasteurization. The lactic acid is used to adjust the pH of the whites to pH 7.0 because at this pH all of the egg white proteins, except for a small amount of conalbumin, can withstand temperatures of 141° to 143° F. (60.7° to 61.7° C.) for a period of 3½ to 4 min. The conalbumin is stabilized by the addition of iron or aluminum salts through the formation of a complex that is more heat-stable than the free protein. The finished W.R.R.L.-processed liquid whites have a pH of 7.0, distinctly different from that of normal liquid whites (pH 8.0 to 9.0). Since there is some denaturation of the protein, resulting in functional impairment, triethyl citrate or triacetin is added to diminish the damage.

The pasteurization of liquid whites without coagulation can also be accomplished by the use of temperatures of 125° to 127° F. (51.7° to 52.8° C.) and hydrogen peroxide. In this method, the liquid whites are heated to 125° to 127° F. (51.7° to 52.8° C.) for 1½ min. Hydrogen perioxide at 0.075 to 0.10% is metered into the egg white and held at the same temperature for 2 more minutes. The hydrogen peroxide is then eliminated by the addition of the enzyme catalase. Because of the low temperature and short time period required for this "Armour" process, there is no need to use additives to protect against denaturation of the egg white proteins. The pH of the finished pasteurized Armour-processed egg white is normal.

In the pasteurization of liquid yolk, higher temperatures are required than for liquid whole eggs. It is generally agreed that liquid yolks should be flash-heated to not less than 144° F. (62.2° C.) and held at this temperature for not less than 3½ min.

The British method of pasteurization requires a temperature of 148° F. (64.4° C.) for 2½ min. It has been shown also that the sensitivity of alpha amylase could be used as an indicator that liquid whole eggs had been pasteurized at 148° F. (64.4° C.) for 2½ min.

Since May 19, 1966, the U.S. FDA has required that processed egg products in interstate commerce be pasteurized or otherwise treated to destroy all viable Salmonella microorganisms. Mayonnaise and acidified salad dressing manufacturers are exempt from this law in that they are permitted to use nonpasteurized egg products. The processing of mayonnaise and other acidified dressings results in an environment detrimental to Salmonella and therefore these products are considered not to be dangerous to public health.

Freezing Methods

The washing, breaking, and canning operations for frozen eggs are carried out in many plants as a continuous process. Buckets of broken-out whole eggs, yolks, or whites are mixed separately and filtered to remove all shell particles, membranes, and other foreign materials. When egg yolks are frozen separately, it is necessary to add certain percentages of salt or sugar to prevent gelation. During the process of freezing yolk, water separates from the yolk solids, and this produces small lumps of yolk which become harder as further separation of water takes place. When the frozen yolk is defrosted, it does not return to its original smooth consistency but becomes a lumpy, gummy mass—a state of gelation—which makes it difficult to mix with other ingredients.

Ten percent of salt has long been used as an additive to yolk for the mayonnaise and salad dressing manufacturers, whereas the baking, confectionery, and ice cream industries require the addition of 10% sugar to yolk. In the presence of these additives the colloidal nature of the yolk is not destroyed during freezing. Since gelation does not take place, moisture is reabsorbed during thawing and the consistency of the yolk is restored.

Courtesy of Container Corp. of America

FIG. 10.3. FILLING A PLASTIC BAG WITH LIQUID EGGS FOR FREEZING

There is no chemical breakdown of egg whites as a result of freezing; when they are thawed, however, the whites are more liquid than they were originally.

In commercial practice, eggs are generally frozen at temperatures ranging from −10° to −40° F. (−23° to −40° C) and then stored in a holding room at temperatures ranging from 0° to −10° F. (−18° to −23° C.). Present commercial practice involves initial sharp freezing in cans at −20° F. (−29° C.) or lower, in order to assure that the contents are frozen hard or are reduced to a center temperature of 10° F. (−12° C.) or lower within 60 hr. from the time of draw-off. Egg products are usually frozen in the type of container the product is shipped in and sold. The common standard 30-lb. metal can has recently been replaced in some plants with either a 30-lb. plastic container, or with master shipping containers holding four 10-lb. plastic bags, five 8-lb. plastic bags (Fig. 10.3), four 1-gal. plastic jugs, or four 8-lb. plastic-coated milk cartons (Fig. 10.4). After 60 hr. in the freezer, the frozen product is next transferred to a holding room at 0° F. to −10° F. (−18° to −23° C.) for storage purposes. To prevent gelation, salted yolk is usually frozen at 0° F. to −10° F. (−18° to −23° C.) and then held at this temperature until ready for shipment.

Defrosting Operations

In order to maintain the original high quality of the egg product, it is necessary that proper defrosting operations and sanitary handling be practiced. Details of the defrosting operations required of plants operating under USDA supervision are described in USDA Regulations governing the grading and inspection of egg products. Frozen whole eggs, combinations of whites and yolks, and yolks must be brought to a liquid state in a sanitary manner as quickly as possible after defrosting has begun. Frozen whites which are later used in the production of dried

Courtesy of Anheuser-Busch, Inc.

FIG. 10.4. EIGHT-POUND PLASTIC-COATED MILK CARTON USED FOR FROZEN EGGS

TABLE 10.3

PLAIN AND MIXED WHOLE EGGS, ALBUMEN, AND YOLK PRODUCTS[1]

Estimated Percentages of Each Used by Food Industries and Institutions, Purchased Under Government Programs, and Exported.[2]

Egg Product	Principal Users	Proportion of Each Type Product (%)
Plain and Mixed Whole Eggs		
Frozen[3]	bakeries, institutions[4]	63.0
Dried	bakeries, cake mix manufacturers, institutions	11.0
	U.S. Dept. Agr.	26.0
Total		100.0
Albumen		
Frozen[3]	candy makers	—
	bakeries	58.0
Dried	cake mix and meringue powder manufacturers	21.3
	export trade	17.2
	candy makers	3.5
Total		100.0
Yolks		
Frozen (salted)[3]	mayonnaise and salad dressing manufacturers	32.7
Frozen (sugared)[3]	bakeries, baby food processors, ice cream manufacturers	26.0
Frozen (plain)[3]	noodle makers, baby food processors	17.2
Frozen (other)	export trade	2.7
	miscellaneous	0.2
Dried (plain)	doughnut and cake mix manufacturers	19.5
	export trade	1.7
Total		100.0
Total, All Egg Products		

[1] Liquid weight equivalents.
[2] Total quantities (liquid, frozen, and dried) were equivalent to annual production.
[3] Includes liquid eggs used for immediate consumption.
[4] Hospitals, hotels, restaurants, U.S. Military establishments, etc.

albumen may be defrosted at room temperature. Frozen whole eggs, combinations of whites and yolks, and yolks can be tempered, or partially defrosted, within 48 hr. if the room temperature is below 40° F. (4° C.), and within 24 hr. if the room temperature is above 40° F. (4° C.). If frozen eggs are packed in metal or plastic containers, they can be placed in running cold water without submersion to speed defrosting.

PATTERN OF USES OF FROZEN EGGS

In a survey of the U.S. food manufacturing industry, it was determined that on a liquid-equivalent basis, bakeries, the largest users of eggs, purchased about 58% of their egg requirements in frozen form and 35%

in dried form. Food manufacturers in the miscellaneous group, which included firms making baby food, meat and fish products, noodles, macaroni, ravioli, mayonnaise, salad dressing, and a variety of specialty products, used mostly frozen eggs and relatively large quantities of shell eggs. Of the confectioners included in the survey, 15% used frozen whites for certain products, 91% used dried albumen in other kinds of goods; only 3% used liquid whites, and 47% used egg substitutes.

In an extensive study, it was observed that bakeries were the principal users of egg products (Table 10.3). Institutions, on the other hand, utilized mostly whole eggs, both frozen and dried. Purchases of whole egg solids by the USDA amount to more than 10% of the total production.

The largest consumers of frozen salted yolk are the mayonnaise and salad dressing manufacturers. Frozen sugar yolks are utilized by bakeries, baby food processors and ice cream manufacturers. The principal users of frozen plain yolk are the noodle companies and baby food manufacturers, while the cake mix and doughnut manufacturers utilize plain yolk solids (Table 10.3).

ADDITIONAL READING

AM. FROZEN FOOD INST. 1973. Good commercial guidelines of sanitation for frozen, soft filled bakery product. Tech. Serv. Bull. 74. American Frozen Food Institute, Washington, D.C.

ASHRAE. 1974. Refrigeration Applications. American Society of Heating, Refrigeration and Air-Conditioning Engineers, New York.

BAUMAN, H. E. 1974. The HACCP concept and microbiological hazard categories. Food Technol. 28, No. 9, 30–34, 70.

DeFIGEUIREDO, M. P., and SPLITTSTOESSER, D. F. 1976. Food Microbiology: Public Health and Spoilage Aspects. AVI Publishing Co., Westport, Conn.

GUTHRIE, R. K. 1972. Food Sanitation. AVI Publishing Co., Westport, Conn.

HARRIS, R. S., and KARMAS, E. 1975. Nutritional Evaluation of Food Processing, 2nd Edition. AVI Publishing Co., Westport, Conn.

JOHNSON, A. H., and PETERSON, M. S. 1974. Encyclopedia of Food Technology. AVI Publishing Co., Westport, Conn.

KOMARIK, S. L., TRESSLER, D. K., and LONG, L. 1974. Food Products Formulary. Vol. 1. Meats, Poultry, Fish and Shellfish. AVI Publishing Co., Westport, Conn.

KRAMER, A., and TWIGG, B. A., 1970, 1973. Quality Control for the Food Industry, 3rd Edition. Vol. 1. Fundamentals. Vol. 2. Applications. AVI Publishing Co., Westport, Conn.

PETERSON, M. S., and JOHNSON, A. H. 1977. Encyclopedia of Food Science. AVI Publishing Co., Westport, Conn.

RAPHAEL, H. J., and OLSSON, D. L. 1976. Package Production Management, 2nd Edition. AVI Publishing Co., Westport, Conn.

SACHAROW, S. 1976. Handbook of Package Materials. AVI Publishing Co., Westport, Conn.

STADELMAN, W. J., and COTTERILL, O. J. 1973. Egg Science and Technology. AVI Publishing Co., Westport, Conn.

TRESSLER, D. K., and SULTAN, W. J. 1975. Food Products Formulary. Vol. 2. Cereal, Baked Goods, Dairy and Egg Products. AVI Publishing Co., Westport, Conn.

TRESSLER, D. K., VAN ARSDEL, W. B., and COPLEY, M. J. 1968. The Freezing Preservation of Foods, 4th Edition. Vol. 2. Factors Affecting Quality in Frozen Foods. Vol. 3.

Commercial Freezing Operations—Fresh Foods. Vol. 4. Freezing of Precooked and Prepared Foods. AVI Publishing Co., Westport, Conn.

WEISER, H. H., MOUNTNEY, G. J., and GOULD, W. A. 1971. Practical Food Microbiology and Technology, 2nd Edition. AVI Publishing Co., Westport, Conn.

WILLIAMS, E. W. 1976A. Frozen foods in America. Quick Frozen Foods *17*, No. 5, 16–40.

WILLIAMS, E. W. 1976B. European frozen food growth. Quick Frozen Foods *17*, No. 5, 73–105.

WOOLRICH, W. R., and HALLOWELL, E. R. 1970. Cold and Freezer Storage Manual. AVI Publishing Co., Westport, Conn.

11

Precooked Frozen Foods

Samuel Martin
and Thomas J. Schoch

P repared frozen foods—products which have been either cooked or converted into a processed or convenience item—are the giant products of the industry. So vast has this frozen food segment become, that, for purposes of statistical grouping, French-fried potatoes and the family of kindred products, potato puffs, rissole potatoes, hash browns, and other variants (which quite obviously are prepared and precooked foods) are usually included with regular vegetables. Products such as breaded shrimp, fish sticks and fish portions have been classed as seafoods.

This placement may be necessary statistically, but by definition they must properly be regarded as prepared and precooked frozen foods. In this context they represent the largest single grouping of the frozen food industry both in poundage and in dollar value. The product categories broadly included in the reckoning of frozen prepared foods include, in addition to processed frozen potatoes, breaded shrimp, fish sticks, and fish portions, other products such as dinners, bakery foods, nationality foods, prepared vegetables, entrées, meat pies, fruit pies, cream pies, seafood specialties, soups, breaded and precooked poultry, as well as a broad miscellaneous category involving hors d'oeuvres, snacks, candy, hush puppies, frozen raw fruit preserves, sauces, crepes suzette, vegetable creams, synthetic meats, and a considerable variety of other items.

The feasibility of frozen prepared foods was evaluated seriously right from the beginning. The "father of frozen foods," Clarence Birdseye, in his Gloucester, Mass., laboratories, experimented on a wide range of prepared products including bread, rolls, cakes, pies, and other bakery foods; fried and broiled poultry as well as more exotic poultry dishes; various pre-cooked meats and meat dishes; precooked fish, shellfish as well as many fish and shellfish soups, chowders, stews; many precooked vegetables and fruits; in fact, the entire gamut and range of prepared foods. This research was directed by Donald K. Tressler. Because of the early state of development of the freezing industry, the primitive state of distribution, the scarcity of suitable holding facilities at that time, and the economic severity of the depression, all this basic work resulted in patents, technical papers, and stacks of laboratory notes.

As time went by, the Western Utilization Research and Development Division of the USDA, at Albany, Calif., did experimental freezing of chicken à la king, cream sauce, cheese fondue, cream of celery soup, Cape Cod clam chowder, and lamb stew (with and without vegetables). Various universities independently conducted research projects on the formulation and frezing of a variety of prepared items, including many unusual items.

One of the first prepared products of any consequence was chicken á la king, marketed by Birdseye Frosted Foods and later emulated by several other firms. Why only this product should have found favor is difficult to say. Among factors in its favor was that it retained its character extremely well under freezing, had a long storage life, could be produced to sell at a reasonable price, could be packaged compactly, and would have been difficult for the consumer to prepare without special effort at home.

In view of the long list of precooked foods on the market, one might assume that practically all foods result in satisfactory products if carefully cooked, rapidly frozen, and then held at a low temperature until used. This is a long way from the truth, since many precooked foods are greatly changed by freezing, subsequent storage, and reheating for use.

Cooked food may be classified into four categories:

(1) Those which may be frozen, stored, and thawed without marked change; for example, applesauce, winter squash, various pies, bread, rolls, cookies, most cakes, and clear soups.

(2) Those which are greatly changed by freezing, storage, and reheating, but which by certain changes either in the method of cooking or in the recipe, may be so modified as to be well suited for freezing. Creamed chicken and turkey, poultry pies, most sauces and gravies, and cream soups fall in this category.

(3) There are also many precooked products which, when freshly prepared, are excellent but which deteriorate relatively rapidly at ordinary storage temperatures, and consequently have a short storage life and so must be held at unusually low temperatures [e.g., $-20°$ F. ($-29°$ C.) or lower] if they are to be stored for long periods. Turkey dishes, fatty fish, and shellfish are examples of such products.

(4) Those which are greatly changed by freezing and reheating and which are difficult or impossible to improve. Custards, cooked egg whites, and vegetable salads belong to this class of products.

In the listing of problems which are encountered in freezing, storing, and handling precooked frozen foods, mention should be made of bacteriological and food spoilage problems which may give trouble unless pre-cooked foods are chilled immediately after preparation, frozen rapidly, held at temperatures well below freezing, and then rapidly reheated without permitting slow thawing. Detailed consideration of these bacteriological problems will be presented in Chapter 13.

PROBLEMS ENCOUNTERED

Some of the changes occurring when prepared foods are frozen definitely may be attributed to physical phenomena (e.g., separation of emulsions), others partly to physical and partly to chemical changes (e.g., wilting of lettuce, celery, and other unheated vegetables), and still others may be caused by chemical actions.

Physical

When French dressing is frozen, the oil and aqueous phases separate. Freezing causes the water to crystallize as ice, and when the dressing thaws it does not again emulsify the oil. The "breaking" of many other dressings, sauces, and emulsions affected by freezing and thawing is wholly or largely caused by a similar physical phenomenon.

When ice cream, sherbets, and ices are stored for several months they usually become grainy because the ice crystals increase in size. Fluctuating temperatures and relatively high storage temperatures accelerate crystal growth.

Sucrose hydrates often crystallize in cold processed frozen fruit spreads during storage. These crystalline deposits cause deterioration in the appearance and the texture of the spreads. They first appear as white, mold-like, spherulitic formations at the surface of the product, slowly increase in size during storage, and eventually involve the entire mass of the product. Some of the sucrose in these products can be replaced by corn syrup or by invert sugar, but if too much dextrose is used dextrose hydrate appears as white, bead-like deposits throughout the mass during storage. Samples packed in paper cups were found to be affected by sucrose hydrate crystallization, whereas samples packed in hermetically sealed jars or cans were not, if they had not been seeded with crystals of the hydrate. The appearance of sucrose hydrate crystals is slowest at $-30°$ F. ($-34°$ C.), successively faster at $+10°$ F. ($-12°$C.), $0°$ F. ($-18°$ C.), and fastest at $-10°$ F. ($-23°$ C.) storage. Fruit spreads with 30% of the sucrose requirements replaced by invert sugar showed only a very slight growth of sucrose hydrate crystals.

Physical and Chemical

Most changes occurring when cooked foods are frozen result because of a combination of chemical and physical actions. The coagulation, or curdling, of custards is in part caused by the crystallization of water as ice and partly by the continuing denaturation of the egg proteins.

When frozen and thawed, many gravies and sauces (e.g., white sauce) curdle. This occurs for much the same reason as the coagulation of many types of custards. The crystallization of water causes retrogradation of the starch solution. The wider the temperature fluctuations and the higher the storage temperature, the greater the liquid separation from frozen sauces, other conditions being the same. This indicates clearly an important advantage of uniformly low temperatures for storage of frozen sauces and gravies.

Freezing and thawing wilts most vegetables and fruit tissues. Since greens must be crisp to be satisfactory for use in tossed and other vegetable salads, these salads are not satisfactorily preserved by freezing. Freezing wilts vegetables because of the crystallization of water which reduces the turgidity of the cells.

Wheat flour is not a wholly satisfactory stabilizer for gravies and white sauces. If, however, it is used in combination with gelatin or some colloid, the coagulation may be retarded for considerable storage periods, and perhaps prevented altogether. White sauces and gravies in which the wheat flour is largely or wholly replaced by waxy maize or waxy rice flour are relatively stable. There has been a comprehensive study of the development of a curdled appearance and liquid separation in white sauces and gravies subjected to freezing and frozen storage. The use of amylopectin starches and flour minimizes these defects but since amylopectin starches, in contrast with some waxy cereal flours, have a long paste character, only certain flours appear suitable for use in sauces and gravies. Based on tests with the samples of waxy cereal flour available, waxy rice flour appears to be superior to waxy corn flour and waxy sorghum flour. But, if the appearance and liquid separation are important only in the heated sauces and gravies, then the waxy corn and rice flours are essentially interchangeable. (Starches will be discussed in detail later in this chapter.)

The loss of gas from batters during freezing and thawing is in part caused by the separation of ice and the resultant concentration of the carbon dioxide in the remaining liquid phase, which eventually becomes so high that it will not stay in solution; and partly on account of the reaction of the carbonate of the soda and the acid in the more concentrated aqueous solution produced because of the separation of water as nearly pure ice. Further, it is probable that some denaturation of proteins occurs with resultant coagulation, thus permitting escape of carbon dioxide.

Yeast doughs also deteriorate in quality if stored for long. The yeast cells gradually lose their viability; consequently, the longer the doughs are stored the more slowly the dough will rise after it is thawed and warmed. Further, during storage some carbon dioxide is lost because of the separation and growth of ice crystals. This causes loss of gas for the same reasons given in the preceding paragraph for loss of carbon dioxide during freezing and storage of batters (except for the statement concerning reaction of the baking powder ingredients).

Changes in proteins, with resultant toughening or coagulation, occur in many foods. These changes are particularly noticeable in foods composed principally of proteins. Raw egg white is not markedly affected by freezing and thawing. However, freezing makes the cooked product tough and rubbery. One can conclude that the damage to cooked egg white by freezing is caused by the mechanical effects of the ice crystals formed.

During freezing, the water in the elastic gel of the cooked egg whites (denatured protein) migrates to increase the size of crystals wherever nuclei are present. As the crystals grow, they penetrate the gel and separate the structure, thus releasing a part of the elastic tension. The migration of the water from within the gel structure plus the force exerted by the growth of ice crystals and the release of elastic tension by mechanical cleavage, cause the gel structure to contract. That this contraction is largely irreversible is demonstrated by the liquid-filled spaces remaining after thawing. The structure remaining is naturally tougher, since it contains a considerably higher proportion of protein than the original gel. Thus, if a gel before freezing contains 12% solids and if a 55% liquid separation occurs during freezing, as occurred in many experiments, then the remaining gel would contain upward to 27% solids. The actual remaining solids will be somewhat lower than 27% because the liquid will contain part of the soluble solids.

Chemical

Many chemical actions occur during freezing and storage of cooked foods; few of these are well understood. Lobster, crab, and shrimp gradually toughen during long-continued storage, probably because of continuing denaturation of proteins. The higher the temperature at which the frozen shellfish are held, the more rapidly these products toughen. Lobster meat also often changes from a red color to a yellow one. The change in color is believed to be due to oxidation.

Frozen cooked crab meat which had been steamed for 40 min. (including 15 min. at 15 lb. steam pressure) contained active l-malic dehydrogenase, an enzyme which is capable of producing oxalacetic acid, which nonenzymatically decarboxylates, causing a loss of carbon dioxide. This conclusion is surprising in view of the long cooking given the crabs. A slight increase in pH was also noted. These changes are believed responsible for the toughening which usually occurs during long storage of precooked crab meat.

Oxidation and resultant rancidity are chemical changes which are likely to occur during the freezing and subsequent storage of certain fatty foods. Turkey fat is particularly subject to oxidative rancidity. This has been a factor limiting the use of turkey in precooked frozen foods. In the case of creamed turkey, rancidity may be detected immediately after the product has been prepared, and increases during frozen storage.

During preparation and storage of frozen creamed turkey, rancidity development may be greatly retarded when small amounts of edible antioxidant are present during cooking of the turkey. Workers have shown the importance of adding the antioxidant during the cooking rather than after the cooking has been completed.

Many frozen food research workers have assumed that the fats of cooked meat, in particular pork, turn rancid more rapidly than the fat of the uncooked product. Such is not the case; the rancidification of raw and cooked pork is explained as follows:

(1) The rate of rancidification of raw ground pork increased rapidly with decreasing pH of meat within the pH range of 6.5 to 4.8. The pH had no effect on rancidification of precooked ground pork. Except at the upper limits of normal pH range of fresh pork, the precooked kept better than the raw.

(2) Certain salts (sodium chloride, sodium nitrate, sodium acetate, magnesium chloride, and potassium nitrate) had a marked accelerating effect on rancidity development in raw ground pork in freezing storage, but not on precooked ground pork. Other salts (potassium chloride and magnesium chloride) had no effect.

(3) The effects of acids and salts on rancidity development in raw pork are believed to be due to the activity of a fat peroxidizing enzyme, possibly hemoglobin. Decomposition of the hemoglobin, with a resulting discoloration of the meat, accompanies rancidification.

Cooking, of course, causes inactivation of the peroxidizing enzyme or enzymes, and consequently rancidification occurs more slowly in the precooked product. Some results are summarized in Tables 11.1 and 11.2.

During frozen storage, cooked ham, cured shoulder (picnic ham), Canadian bacon, and wieners usually lose their red color, and turn brown and then gray or dull green, and change in flavor.

TABLE 11.1

DEGREE OF RANCIDIFICATION OF FROZEN RAW AND COOKED PORK AFTER STORAGE FOR 4.5
MONTHS AT 0° TO + 5° F. (−18° TO − 15° C.)

Lactic Acid Added %	pH of Raw Meat	Peroxide Number After Storage	
		Raw	Cooked
None	6.5	2.0	3.3
0.031	6.4	1.6	3.7
0.103	6.1	5.9	2.9
0.206	5.6	16.9	3.6
0.617	4.8	25.2	4.7

These changes occur relatively quickly in the case of sliced meats and slowly in larger pieces. They may be greatly retarded by incorporating sodium ascorbate in the product, or by using the phosphate cure instead of the usual cure containing potassium nitrate and nitrite.

The fading and discoloration of cured meats is due to the oxidation of nitric oxide myochromogen (which is formed by the heat denaturation of nitric oxide myoglobin). The oxidation products consist of (1) the ferric pigment, metmyoglobin, which is brown in color, and if the reaction proceeds far enough, (2) the green or faded decomposition products of the porphyrin ring.

Ascorbic acid is the only antioxidant that has shown any great promise in the protection of meat color. In the presence of nitrite, ascorbic acid accelerates methemoglobin reduction at all temperatures. It protects cured meat surfaces from fading when exposed to air.

EFFECTS OF FREEZING ON STARCHES

Uncooked or ungelatinized starches are not affected in any discernible way by freezing in the presence of water or by prolonged cold storage. While the cell structure of uncooked starchy vegetables (e.g., fresh legumes, corn-on-the-cob, potatoes) may undergo considerable mechanical damage by improper freezing, the starch granules themselves remain intact and unchanged. However, if the granules have been gelatinized by cooking in the presence of water, the resulting pasted starch becomes susceptible to various changes in physical character on freezing. Typical instances where the starch is in a pasted state include bread and other baked goods, cooked starchy cereals and vegetables, and the wide variety of prepared foods where flour or refined starch is added as a thickening or stabilizing agent (e.g., cream soups, white sauces, toppings, pie fillings, baby foods).

TABLE 11.2

EFFECT OF VARIOUS SALTS ON RANCIDITY DEVELOPMENT IN FROZEN RAW AND COOKED GROUND PORK

Salt Added	pH of Raw Meat[1]	Peroxide Number After 10 Months' Storage	
		Raw	Cooked
None	6.3	5.5	2.6
1.5% sodium chloride	6.3	15.2	2.4
1.0% sodium nitrate	6.3	13.7	2.3
1.25% potassium nitrate	6.3	12.3	—

[1] pH of cooked samples 0.1 to 0.2 higher than raw.

The undesirable physical changes which occur during the freezing of pasted starch systems include the following: (1) increase in opacity of the product; (2) development of a coarse grainy structure and pulpy "mouth-feel"; (3) excessive increase in consistency and even congelation; and (4) the loss of water-holding capacity. As examples, the crumb structure of bread becomes hard and coarse (staling), starch thickened sauces may set up to rigid gels, and frozen pie-fillings may show the syneresis of a watery phase on thawing.

Theory of Starch Instability

The above physical changes are all attributable to associative hydrogen-bonding between starch molecules, whereby the pasted starch progressively becomes more desolvated or insoluble, and simultaneously loses its ability to hold water. In order to understand the effects of cold storage and freezing on starchy systems, it is necessary to consider certain pertinent aspects of the physicochemical behavior of starch molecules. Most common starches (e.g., corn, sorghum, wheat, rice, potato, tapioca) contain two types of polysaccharides. The minor component (termed the linear fraction or "amylose") is an extended chain molecule of some 500 to 2000 glucose units in length. The major component (the branched fraction or "amylopectin") is a highly branched or tree-like molecule with some hundreds of linear branches, each of which is 20 to 30 glucose units in length.

The linear-chain molecules show a strong tendency to associate with one another through hydrogen-bonding between hydroxyl groups; this phenomenon has been termed "retrogradation" and becomes apparent in either of two ways (Fig. 11.1).

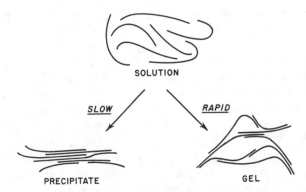

FIG. 11.1. MECHANISMS OF RETROGRADATION OF THE LINEAR
STARCH FRACTION

Dilute solutions slowly deposit an insoluble precipitate (left).
Concentrated solutions rapidly set up to rigid gels (right).

(1) In dilute starch systems, the linear molecules slowly align themselves in parallel fashion to give insoluble bundles or "micelles," which cause opacity and eventual precipitation.

(2) In more concentrated systems (e.g., a cooked 5% cornstarch paste), the linear molecules rapidly associate in random fashion to give the reticulated network of a gel. Thus cooked starchy foods develop firmness or gel characteristics on cooling and particularly on refrigeration, as exemplified by boiled potatoes, cooked oatmeal, and the old-fashioned type of molded cornstarch pudding.

Since the branched starch molecules are relatively globular in shape, and since they contain only short linear branches, they do not undergo the pronounced retrogradation of the linear molecules. However, these short linear branches are still capable of some degree of inter-and intra-molecular association through hydrogen-bonding. This probably involves a folding-up of the extended branches, and the slow progressive development of associative bonding both within and between branched molecules. Actually, the mechanism of association of branched molecules is identical in principle with retrogradation of linear molecules; the only difference lies in the strength of the bonding. Thus a strongly retrograded gel or insoluble precipitate of linear molecules can be liquefied or redissolved only by heating to super-temperatures of 284° to 302° F. (140° to 150° C.). In contrast, an associated system of branched molecules can be readily dissociated merely by warming to 122° to 140° F. (50° to 60° C.). Both types of association occur in bread. The elastic gel structure of normal fresh bread (i.e., not "softened" with monoglyceride) is attributed to an associated network of linear molecules of the wheat starch developed during baking and cooling of the loaf. The subsequent hardening or staling of the crumb structure is due to the gradual association of branched molecules; this may be readily reversed simply by reheating or toasting the bread.

In contrast with starch molecules, the polysaccharide glycogen has extremely high physical stability. This substance is present in certain shellfish (oysters, scallops), in animal liver, and in golden sweet corn. It is a highly branched glucose polymer with an average branch length of 9–11 glucose units. Hence its structure is tight and "bushy," as compared with the loose tree-like molecules of the branched starch fraction. Because its branches are too short to associate, glycogen shows no evidence of retrogradation or insolubilization. Solutions of relatively high concentration may be maintained indefinitely in cold storage, or may be subjected to repeated freezing and thawing, without any evidence of opacity or precipitation.

COMPLETE MEALS

Frozen complete meals on a plate or platter were introduced in 1945. In that year, the Maxson Food Systems, Inc. produced 18 different "Strato-Plates," designed primarily for the feeding of airplane passengers.

Most of these meals were quite satisfactory, although the texture of the omelet was not good, the color of the Canadian bacon soon faded in storage, and its flavor soon became undesirable. As a consequence, the French toast, glazed apples, and potato puff was the only satisfactory breakfast menu.

The value of frozen dinners packed annually in the United States is far greater than that of any other precooked specialty. According to Quick Frozen Foods it surpassed by more than $100 million that of frozen baked goods.

As previously indicated, meals on a platter were originally used to feed airplane passengers and crew. During the last few years, the chicken, turkey, seafood, and some of the other dinners have become very popular with the public. Recently, there has been a trend toward frozen entrées (the main course of dinners). When these are used, one has the opportunity of providing one's own favorite vegetables.

Leading restaurants that have gone into the frozen food business have had outstanding success in merchandizing frozen entrées.

A debate is now in progress as to whether entrées or platters possess the greater sales potential. The leading packers of meals on a plate or platter feel that these dinners offer the ultimate in convenience and that the complete tray dinner, which normally carries potatoes and a vegetable, along with the entrée, has greater attraction than the entrée alone. Recently, many packers of the frozen tray dinners have put out a line of popular entrées. They believe that the meals on a platter will always sell well, but that there is also a big market for the frozen entrées for the following reasons:

(1) Any person who dislikes mashed potatoes or peas is disqualified as a customer for most dinners.

(2) Fifteen to 40 min. of hot oven isn't very convenient—especially in the summer.

(3) Many chains are putting in vegetables under their own labels and we (the packers of frozen precooked foods) want to place ourselves in the market as an adjunct to the chains' own efforts.

(4) Vegetables take only eight minutes to cook, and an entrée that doesn't take much longer has in some ways greater convenience than the platter.

(5) Several packers of entrées lay great stress on the 'fact' that platters are not a 'company' dish. One of them puts it this way: 'Would you think of inviting company to the house and plunking a dinner in a compartmentalized tray, similar to those used by roadside diners, in front of him?"

Problems

One of the most difficult problems in producing meals on a platter is to work out a system of preparing, cooling, freezing, packaging, and reheat-

Courtesy of Quick Frozen Foods

FIG. 11.2. AN ATTRACTIVE TURKEY DINNER. A COMPARTMENTALIZED TRAY OF SLICED TURKEY WITH GRAVY, GREEN PEAS AND MASHED POTATOES READY FOR SERVING

ing the products so that the meal will taste freshly cooked and not like warmed-up leftovers. Certain products, such as ordinary ham and Canadian bacon, when sliced, cooked, and frozen on a platter, quickly lose color and flavor. Scrambled eggs and most omelets markedly change in texture and often also in flavor. Cooked sausage can be frozen without noticeable change in texture or flavor, but the frozen product may become rancid and develop other off-flavors during even a short storage period. Since ham, Canadian bacon, sausage, and eggs are the principal items on many breakfast menus, the number of entirely satisfactory frozen breakfast menus now available is not large.

Potatoes, especially whipped or mashed potatoes, are likely to give trouble, as they often change in texture, becoming rather soggy, and take on a flavor resembling the warmed-over product. French-fried, au gratin potatoes, and potato puffs are less changed by freezing and reheating. However, it is difficult to reheat French-fried potatoes on a platter and obtain a product which tastes like a freshly fried potato.

It is not easy to produce frozen precooked green vegetables (e.g., peas, green beans, and spinach) of the same bright color and fresh flavor as those prepared in the home by cooking frozen blanched vegetables. One of the reasons for this is the difficulty of cooking, cooling, and freezing the vegetables rapidly enough to prevent marked loss of green color and fresh flavor. If vegetables are cooked in boiling water in lots larger than a few pounds, special equipment will be required to effect rapid cooking followed by fast cooling. However, rapid cooking can be effected by the use of steam under pressure, e.g., in a speed cooker, but this equipment does not handle more than a few pounds of product. Vegetables can be quickly cooled if they are spread out in a shallow layer in an aluminum pan, but this exposes the product to air which rapidly oxidizes the warm product with loss of color and vitamin C. Some of the more difficult problems encountered by researchers in their study of frozen meals carried out for the U.S. Navy follow.

First, it was necessary to choose single food items that had been successfully reheated for serving after freezer storage. Then it was important to combine them in such a way that the food was not only satisfactory in nutritive value, color, shape, texture, and flavor, but was uniformly hot. Otherwise, one food on the plate was overcooked and another still had ice in the center. The rate of heat penetration depends mainly on the nature of the food, the amount, and the shape. In some plates, it was found that by the time the meat was hot enough for serving, the vegetable was overdone. Heat penetrates protein foods relatively slowly. It was found also that the fairly solid mass of a mashed vegetable required a longer time for reheating than did a 'loose' vegetable such as broccoli. Therefore, broccoli was only scalded before freezing, and its cooking was finished when the plate was reheated (15 to 25 min.).

Meats and poultry tended to dry out unless covered with a gravy or sauce of some kind . . .

Rapid chilling of the cooked foods before freezing is necessary to keep bacterial growth to a minimum. The gravy was made very thick, and then chilled quickly by adding ice cubes in an amount equal to the omitted water . . .

During storage, some items on each plate lost quality sooner than did others. Hence it is important to use the meal before the items that have the shortest storage life begin to lose quality, or to use foods that have about the same storage life.

In reviewing the development of the Maxson "Strato-Plates," the importance of using sauces and gravies on cooked foods which are to be frozen on a plate can be stated as follows:

> Advantages result from the use of sauces in cooked frozen foods. In the first place, they provide products with an ideal protective coating. Dehydration and oxidation are minimized. Secondly, they facilitate molding, and the removal of the frozen products from the molds. Thirdly, they enhance flavor characteristics, since each sauce is prepared for a specific product. Sauces, however, are extremely difficult to freeze, since they may separate or gel. Separation or gelling was overcome after the sauce ingredients and methods of preparation were carefully investigated.

Since it is very important to protect as many as possible of the foods frozen on plates and platters by a covering of gravy or sauce, special consideration must be given to the starch, flour, or other thickener used, as otherwise the sauce or gravy may curdle or separate during thawing and reheating. Separation and curdling is of much more serious consequence in sauces and gravies used on meals on a plate or platter than it is in separately packaged products which are removed from the package prior to reheating. In the latter case, the sauce or gravy, or product containing it, is usually stirred occasionally during reheating, thus mixing the sauce or gravy and restoring its normal appearance.

Products Which Freeze and Store Well

It is of great importance to select only those items, to be included in plate dinner menus, which retain their flavor, texture, color, and appearance well during freezing, storage, and reheating. It may be well to point out which precooked foods have been found to be best for freezing on a plate or platter. Beef, lamb, veal, and poultry are considered to be wholly satisfactory. Pork tenderloin and fresh ham freeze perfectly but have a relatively short storage life. Ground and cured pork products are usually not considered satisfactory because of a very short storage life, even at 0° F. (−18° C.). Potato puffs and croquettes or patties, scalloped potatoes, potatoes au gratin, and French-fried potatoes are usually found to be best, although whipped and mashed potatoes are included in many of the menus. Many precooked vegetables freeze well (provided they are slightly undercooked before being placed on the plate or platter). The list of vegetables includes asparagus tips, broccoli, carrots, cauliflower, green lima beans, green snap beans, mixed vegetables, onions, spinach, and sweet corn. Most precooked lean fish and fish sticks freeze well. Salmon is satisfactory provided the meals are not to be held in storage for very long. As a rule, crab, lobster, and shrimp dishes are quite satisfactory, but they may have relatively short storage life. Waffles and pancakes freeze well but special care should be taken in reheating them.

Varieties of Menus Packed

The various kinds of frozen meals are usually classified according to the entrée provided: thus there are chicken, turkey, fish, fish stick, ham, lamb, poultry, seafood, sparerib, Swiss steak, stuffed pepper, and veal meals. In addition, they are grouped by the method of cooking employed, e.g., Chinese, Mexican, Spanish, Hungarian, Italian, and Kosher.

A great variety of menus is frozen for sale to the public. It includes poultry, beef and veal, pork, ham, and shellfish meals. Numerous foreign dinners are offered.

Who Eats Frozen Dinners

Surveys conducted with thousands of consumers, in an effort to learn who is eating frozen dinners and what they thought of them, revealed the following: in a considerable number of families the dinners were eaten principally by children. About 60% of those purchasing dinners regularly preferred chicken, 40% turkey, 32.6% beef, and only 15% the fish dinners. Most of those serving dinners accompanied the dinner with one or more side dishes such as a salad and a beverage. Many complained that the size of the portions was too small; in fact, this was the most common comment concerning the meals.

Meals Served by Airlines

Airline catering has become big business. The Pan American airline freezes a large proportion of its meals. Pan Am freezes all its own products and is able to buy in season at the lowest prices in the best markets. Frozen foods make it possible to offer a wider choice, cut down on waste and maintain consistently high quality. Pan Am production centers are located in New York, San Francisco, and Paris. The New York plant supplies millions of meals a year to South America, Africa, the Caribbean, and part of Europe. San Francisco supplies a million meals to the Pacific area; Paris furnishes one-half million meals to Europe and the Middle East. The frozen meals are stored in flight kitchens strategically located in Mexico, Spain, Africa, etc., where the local staff prepares snacks, salads, and any other fresh additives. The frozen meals are reheated in specially designed, fast heating ovens.

Menus Offered by Airlines.—Many different meals are frozen for serving on board airplanes. The following list of meals is representative of the variety served on airplanes:

Lunch or Dinner

Braised chicken, mashed potatoes, mixed vegetables.
Breaded veal cutlets, peas, scalloped potatoes.
Braised sliced beef, French-cut green beans and scalloped potatoes.
Roast turkey, dressing, peas, and croquette potatoes.
Sliced roast veal, mixed vegetables, and scalloped potatoes.
Pork chops with apple sauce, scalloped potatoes, French-cut green beans.
Filet mignon, croquette potatoes, asparagus tips.
Veal scallopini, mashed potatoes, cauliflower au gratin.
Roast breast of chicken, croquette potatoes, peas.
Roast turkey and cranberry jelly, mashed potatoes, and French-cut green beans.
Roast lamb, mint jelly, scalloped potatoes and peas.
Mixed grill with potato balls and French-cut green beans.
Roast turkey, giblet gravy, dressing, mashed potatoes, peas.
Pot roast of beef, gravy, mashed potatoes, peas.
Chopped beef (Salisbury steak), gravy, mashed potatoes, mixed vegetables.
Swiss steak, gravy, mashed potatoes, peas.
Sirloin steak, au gratin potatoes, green peas.

Breakfasts

Waffle, pork sausage, glazed apples.
Omelet, Spanish sauce, sausage patty, sweet roll.

Use by U.S. Air Force

On many long flights, the U.S. Air Force uses precooked frozen meals which are reheated in a B-4 electric oven. The meal is put in an expendable aluminum tray or casserole and covered with a sheet of aluminum foil which is crimped under the protruding lip. The only preparation required is placing the trays in the B-4 oven which heats primarily by conduction, rather than by convection as do conventional ovens.

The five menus listed in Military Specification, MIL-M-13966C are described in Table 11.3.

Since the commercial methods employed in cooking foods preparatory to freezing have been presented in other sections of this book, they will not be considered in this chapter. It should be pointed out, however, that the preparation and cooking of a number of different foods for packing on trays or platters requires coordination of all of the preparation, cooking, cooling, packing, and freezing operations to an extraordinary degree. Each operation must be so timed and coordinated with every other operation that each of the products progresses steadily through the plant

TABLE 11.3

DESCRIPTION OF MENUS PURCHASED BY THE DEPARTMENT OF DEFENSE

Menus	Net Weight
Menu No. 1—Turkey with dressing and gravy	
Turkey	3 oz.
Dressing	2 oz.
Gravy	2 oz.
Mixed vegetables	2½ oz.
Butter	1 pat
Mashed sweet potatoes	3 oz.
Butter	1 pat
Total	12½ oz.
Menu No. 2—Swiss steak with gravy	
Swiss steak	4 oz.
Gravy	2 oz.
Peas	2½ oz.
Butter	1 pat
Au gratin potatoes	3½ oz.
Total	12 oz.
Menu No. 3—Beef Steak	
Beef Steak	4 oz.
Corn	2½ oz.
Butter	1 pat
Mashed potatoes	3 oz.
Butter	1 pat
Total	9½ oz.
Menu No. 4—Beef pot roast with gravy	
Beef pot roast	4 oz.
Gravy	2 oz.
Green beans	2½ oz.
Butter	1 pat
Mashed potatoes	3 oz.
Butter	1 pat
Total	11½ oz.
Menu No. 5—Waffles	Not less than 1½ oz.
Sausage links	2 oz.
Applesauce	3 oz.
Total	6½ oz.

with a minimum of delay anywhere along the line. Otherwise, both the flavor and sanitary qualities of the frozen meals will be poor.

It should not be necessary to indicate that the plant must be well-equipped and well laid out. The meat cutting department must be arranged so that it receives the meat directly from a receiving ramp. This department should have a walk-in cooler (32° to 34° F.; 0° to 1° C.) used solely for meat. If halves, quarters, or other wholesale cuts of meat are used, it should have an overhead track so that the meat can be hung and moved with a minimum of handling. If meals including a meat item are packed and shipped in interstate commerce, the plant must be operated under Federal Inspection; therefore, all facilities must meet Federal requirements for sanitation, etc.

The kitchens must have adequate cooking facilities and equipment, which will include a battery of kettles, ovens, steam cookers, grills, and deep-fat fryers. In addition, there must be a variety of specialized equipment such as meat and vegetable dicers, slicers, and comminuting machines, and also homogenizers, mixers, and blenders. Large walk-in coolers must be provided not only for fresh vegetables, fruits, milk, cream, butter, fish, shellfish, poultry, and other perishable items, but also to hold partially processed foods overnight and during short periods when certain processing equipment may be shut down for repairs, etc.

The assembly line where each of the items is placed on the tray or platter must be especially well laid out or the labor cost of assembly will be entirely too high.

As a rule, each of the compartmentalized aluminum foil trays holding a meal is covered with lightweight aluminum foil. In some plants, each tray

Courtesy of Quick Frozen Foods

FIG. 11.3. THE "EKCO TOPPER"

This aluminum tray of food has been reheated on top of a saucepan of boiling water, thus converting an ordinary pan into a double boiler, and eliminating the nuisance of putting food into a double boiler, transferring it to a serving dish, and also making unnecessary the washing of a pan and a dish.

is then slipped into a shallow carton. The packages containing the meals on a tray are then automatically overwrapped and placed on shelves on a wheeled rack for easy movement into an air-blast freezer. In other plants, the foil covered trays are first frozen on shelves on a wheeled rack in an air blast, and, when frozen, put into the cartons and overwrapped. Sufficient space should be left between the trays or cartons on the rack to permit air circulation around each carton during freezing.

PACKAGING PREPARED FROZEN FOODS

Although the basic principles of packaging of frozen foods are presented in Chapter 12, the problems encountered in packaging precooked and prepared foods are sufficiently different to necessitate a special chapter on the packaging of precooked foods in this volume. The following are a few of the special problems encountered in packaging precooked foods:

(1) Many of them are very fragile and cannot be sold if their appearance or texture has been damaged, e.g., cakes, puff pastry, and stuffed peppers. (2) Some of them are difficult to transfer to other utensils and therefore must be reheated in the container in which they were frozen, e.g., batters, doughs, pot pies, and pies. (3) In many cases, it is desirable to reheat and serve the product in the container in which it was frozen, e.g., pies, pot pies, shortcakes, complete meals, pizza pies, "thermidors," etc. (4) Many precooked foods are covered with a sauce or gravy and consequently the package must not only be watertight, water- and greaseproof, but in many instances must also be able to stand heating in a hot oven. Examples of such products are chicken and other pot pies, and meat stews. (5) Many precooked foods are usually filled hot; therefore, they must be packaged in containers which are waterproof even at high temperatures.

Rigid Aluminum Foil Containers

Of the requirements listed above, the rigid aluminum foil container probably comes closest to satisfying all of the demands which may be made on a package to be used for precooked frozen foods. As is evident from Fig. 11.4, this type of container is available in about every conceivable shape and size to meet the needs and requirements of the industry.

The thickness of aluminum used in rigid foil containers ranges from 0.0025 to 0.0125 in., the latter being in reality a light aluminum sheet. The rigidity of the finished foil container depends not only on the physical design of the container, but also on the gage of the foil used and whether it is hard or soft temper.

Courtesy of Ekco Containers, Inc.

FIG. 11.4. ASSORTMENT OF RIGID ALUMINUM FOIL CONTAINERS OF
THE TYPES USED IN PACKAGING FROZEN FOODS

The properties of rigid foil containers are such that these containers can withstand very high and very low temperatures, far higher than commercial oven temperatures and much lower than those employed in commercial freezing operations. Aluminum foil has no odor and it provides an effective barrier against transfer of odors from adjacent products. Assuming that the foil used is free from pinholes, neither moisture nor moisture-vapor can pass through. Of course, proper seals or closures must be used. Aluminum foil is nontoxic even when pitted by saline solutions. It is insect resistant and has excellent heat conductivity; the latter property is important both during food freezing and also during reheating. Aluminum foil containers can be fabricated in light weights, often an advantage both in freezing foods and in reheating them.

Although foil is highly resistant to the weak organic acids found in many food products, it is attacked by strong mineral acids. Foil is less resistant to weak alkaline products, hence with such products it is desirable to use foil coated with a protective film of a polyvinyl plastic. Therefore, a packer considering foil containers for a saline or slightly alkaline food should conduct actual storage and handling tests prior to commercial adoption of the package.

Thermoplastic coatings have been devised (approved by the FDA) that can be used for coating aluminum foil for protection against strong food acids and alkaline foods which may corrode the aluminum.

Food in plastic coated foil containers may be reheated in a microwave oven without the objectionable arcing which occurs when frozen food is thawed and reheated in plain aluminum foil containers in this type of oven. These coatings are not damaged by oven heat because of their thermosetting properties.

Aluminum pie plates have been perfected in which a pie can be baked with a nicely browned bottom crust free from sogginess. This was accomplished by perforating the bottom of the pie plate with a few pinholes and by applying a black coating to the bottom which absorbs instead of reflects the radiant heat.

At a small cost, vivid decorative colors can be applied to the outside of an aluminum container. Careful preparation and proper printing of aluminum foil give a depth and luminosity to inks and also a dramatic printing effect. Color coatings have been devised that are acceptable to the U.S. FDA for inside application to aluminum foil containers.

Methods of Manufacture.—Foil is fabricated into rigid containers by two principal methods. All having central ridging or other special shape modifications are formed by the single stroke of a die into the metal. These containers are referred to as seamless, having no folds or joints. They have the characteristic corrugation or wrinkled sidewalls, which gives them considerable strength. This corrugation is sometimes controlled to produce a fluted effect. These containers have a slight shoulder offset for additional rigidity and are finished at the rim in one of four ways. When a raw edge is grooved to add rigidity, it is referred to as trenched. That type rim and the semi-curl (which is a raw edge rolled downward to give a skirting effect) are today being superseded by the full-curl rim, which is tightly rolled and beaded, greatly increasing rigidity; and by the vertical flange rim, which is a combination flat flange and vertical projections designed to receive and crimp over a flat cover. Practically all round and many rectangular foil containers are made by this seamless method in a variety of sizes and shapes.

The second method of manufacturing rigid foil containers is by folding. The foil is handled like paperboard and folded into a rectangular shape. These containers have folded seams at the corners, a flat, hemmed rim; there are no corrugations in the sidewalls. Obviously, folded containers are square or rectangular only. They are available in a variety of sizes and have the advantage of being somewhat easier to handle on high speed overwrapping equipment. However, they lack the reinforcing effect of corrugations in the sidewalls and the rigidity that is added by the full curl rim of the seamless container. Like the seamless containers, the walls are tapered to permit nesting during shipment. Solid foil recessed covers are available for folded containers, as well as a variety of closure equipment.

The frozen food packer who plans to use rigid foil containers should choose between these two general types based on the specific packaging and marketing characteristics of his products.

Equipment for Closing Foil Containers.—The basic closure styles are: (1) aluminum foil hooding; (2) board or plastic covers in vertical flange containers; (3) snap-in lids of paper, foil, or plastic; and (4) plastic snap-on lids. In addition, shrinkable films, conventional overwraps, and bagging are used.

In recent years great strides have been made in the perfection of equipment for putting the container cover in place and crimping it securely in position. In the case of equipment for vertical flange containers to be fitted with a foil-board laminated cover, the basic principle is the use of a male and female two-part die. Automatic equipment (Fig. 11.5) will effect 120 closures a minute. Simple foot-press machines, which operate on the same principle, can make 10 or 12 closures per minute.

Hermetic closures are a recent development for rigid aluminum foil containers. Vinyl interior coatings are applied both to the container and a foil-board cover. Heat and pressure seal the two vinyl surfaces together, forming a completely airtight, dustproof, leakproof seal.

New Shapes.—Attractively shaped aluminum containers are now available made of heavy-gage aluminum sheeting, ranging up to 0.01 in. or even more in thickness. These sturdy packages may have snap-in heavy-gage aluminum lids; easy opening, tear-off closures, or snap-on plastic lids, or even double-seamed can-type lids that are airtight.

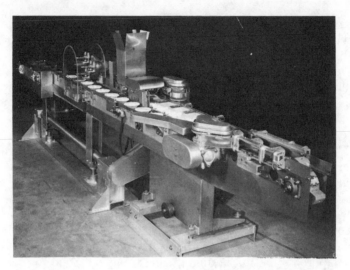

Courtesy of the Marathon Corporation

FIG. 11.5. POT-PIE PACKAGING MACHINE IS FULLY AUTOMATIC

Aluminum Foil Laminations

Containers made of aluminum foil, laminated to kraft or other paper, have a limited use in packaging certain frozen food products. A satisfactory container for frozen citrus and other juices and beverages is made by sandwiching several layers of kraft paper between a liner and an exterior of aluminum foil. The liner is coated with polyethylene as the barrier material, thus preventing the corrosion of the aluminum.

Wide Use of Aluminum Foil and Aluminum Foil
Laminated Containers

Aluminum foil containers are available in a wide assortment of shapes and sizes. They are used for packaging a large number of different precooked frozen food items. With the possible exception of the poly pouch, no other container is used in such great numbers. Some of nearly every precooked item on the market are packaged in foil. The manufacturers of foil packages are very ingenious; they have been successful in designing a package to fit almost every product commonly packed. Further, most of the packages are well suited for (1) freezing, and (2) reheating the product. Often the hot food is served in the same foil container in which it was packed.

Flexible Aluminum Foil

The properties of rigid aluminum foil have already been considered;

the chemical properties of flexible aluminum foil are the same as those of the rigid product. However, its use in packaging precooked and prepared frozen foods is quite limited. A sheet of it is sometimes used to cover the top of a rigid foil container. This is satisfactory provided the entire food container is packaged in a carton.

Flexible foil does not possess the strength to be used by itself as a packaging material. When laminated to paperboard, it can be used to make excellent covers for rigid foil containers. Because of its grease- and moisture-proofness laminated with paperboard, it is generally used for this purpose. Laminations of foil and paper make very good pouches and are used by some packers for products to be reheated by the "boil-in-the-bag" method.

Pouches and Preformed Bags

Unless packed in a protective carton, preformed bags have a rather limited use for packaging precooked foods. They are mostly used for baked goods which will withstand some rough handling. High density polyethylene preformed bags have found an important use in packaging precooked products (entrées, vegetables, etc.) for institutions, hotels, and restaurants when the "boil-in-the-bag" procedure is to be used for reconstitution (reheating). For satisfactory use, a vacuum must be pulled and the bag sealed while the product is still under vacuum.

Pouches.—A pouch is "a flexible package configuration which can be formed continuously from a roll of material, filled, and heat-seal closed on automatic machinery."

Large quantities of both partially cooked vegetables in butter and cream sauces, and precooked entrées are packed in evacuated pouches usually made of high density polyethylene. The quality of these products is retained very well indeed for two reasons: (1) they are vacuum packed in a pouch; this prevents desiccation and retards oxidation; and (2) there is little loss of quality during reheating.

The pouch in a carton package is not costly because the pouches are made in the food packing plant at the time the food is packed. The Bartelt machine used in filling, evacuating, and sealing the bags occupies relatively little floor space and is substantially automatic. The making and filling of the pouches by the Bartelt machine is as follows:

(1) The 'Scotchpak' film feeds out into a series of rollers leading to pouch formation, after a worker controls the level of incoming vegetables on an overhead conveyor and inspects for discoloration and damaged products. A single spring-loaded roller keeps the film tension as even as possible during and between indexing stops of the machine. A break in the film makes this roller drop into a switch which stops the line.

(2) The film passes around a plow which folds it to form the front and

back of the pouch. The machine first forms the bottom, then the side seal. A pressure crimper prevents the pouch from curling inward prior to forming the top seal. The feed roll stops the film as the continuous line of formed pouches passes a knife which cuts successively formed pouches apart. The static bond is broken by air releases, separating the front and back of each pouch and thus prepares it to receive the vegetables.

(3) As the machine is forming the pouches, the frozen vegetables drop from an overhead conveyor into the machine's large hopper. Vegetables from the hopper feed into measuring cups of the continuously rotating filler turret and then pass into each pouch through small tapered chutes.

(4) Filling is done at two stations. At the first, each pouch receives approximately one half of its full contents. Then the pouch moves along the line where a second filler turret discharges the final half of the contents.

(5) A timer interval valve operates a nozzle. This injects the butter sauce into each pouch. A vacuum draw snorkel then removes the air after which a flat seal bar puts on the top seal. The pouch then passes to a round heat-seal bar which the machine applies to the pouch to insure a good top seal.

(6) Each filled pouch drops from an indexing chain down a slide and goes into one of the individual cartoner conveyor pockets. A cam-operated arm pushes the pouch into the carton further along the line.

(7) The carton end is then closed. The closed carton then moves to heat-sealing equipment shown in Figs. 11.6 and 11.7 after which the cartons are conveyed into the freezer storage.

Courtesy of Green Giant Co.

FIG. 11.6. AFTER PASSING HOSES THAT FORCE HOT AIR AGAINST THEIR SEALING FLAPS, THE CARTONS OF BOIL-IN-THE-BAG VEGETA-BLES MOVE BETWEEN COOLING BARS THAT SET THE SEAL FIRMLY ELIMINATING DOG EARS AND PROVIDING GREAT STRENGTH

Courtesy of Green Giant Co.

FIG. 11.7. AFTER "HOT-MELT" SEALING CARTONS OF BOIL-IN-BAG
FROZEN VEGETABLES ARE CONVEYED TO A FREEZER FOR STORAGE

Paperboard Cartons and Overwraps

In the early days of the quick frozen food industry, the usual carton
was made from either ordinary or cold waxed paperboard. The frozen
food was protected from desiccation by a sheet of moistureproof cellophane or plastic liner and a heat-sealed lithographed waxed paper
overwrap. When precooked frozen foods were introduced, they were
usually packaged in this type of carton with or without the plastic sheet
liner. Many frozen food packers have now shifted over to the use of
polyethylene coated paperboard and have dispensed with both the overwrap and the liner, lithographing the carton itself.

Polyethylene coated paperboard is grease-resistant, and substantially
moisture-vapor proof. Foods do not stick to the inside of the carton nor
does the polyethylene flake off into the food. Since polyethylene is heat-
sealable, the cartons may either be heat-sealed or they can be closed by
conventional gluing equipment.

A number of packers of precooked frozen foods still use a plain paperboard (sulfite) carton overwrapped with lithographed paper, coated with
one of the following: polyethylene; ordinary paraffin; paraffin modified

either with 10% or less of a modifier (microcrystalline wax, butyl rubber, or polyethylene or other plastic, e.g., ethylene vinyl acetate copolymer), or a so-called hot-melt coating. A hot-melt coating is usually a formulation of paraffin containing more than 10% of modifiers. The modifiers used in hot-melt coatings may include those listed above and also ethyl cellulose, cyclized rubber, and butadiene copolymers. There are some promising "hot-melts" that contain no paraffin, but consist only of resins. Hot-melt coatings are characterized by extremely strong seals and fair blocking resistance.

The details of the contruction and modes of closing paperboard cartons is considered in Chapter 12. However, it may be pointed out that because of the very great number of different precooked foods packed, and the wide variation in their shape and size, a large number of shapes and sizes of cartons is required.

Cans

With a few exceptions, ordinary tin cans and composite cans (those with laminated bodies and metal ends) are not extensively used for packaging precooked and/or prepared foods. However, the company that packs more frozen soup than any other puts their product in tin cans. Further, nearly all fruit juice concentrates and fruit purées are packed and frozen in either tin cans or composite cans (e.g., Sefton cans). Melon balls in light syrup are also packed in Sefton cans.

One of the important innovations in composite cans is the rather general use of easy opening devices. One of these has a strip called a Mirastrip around one end of the can. When this is pulled, the entire end of the can comes off so that the contents can be emptied after only enough thawing to release the grip of the ice. Another is known as the T-Tab which, when pulled, pulls the end out of the can. The third is called Easy-O which also removes the can end.

Combination Packages

Thus far in this chapter certain basic packaging materials have been considered including coated papers and boards, foil sheets, and laminated paper and foil sheets, bags and pouches, plastic pouches and bags, both metal and composite cans, and molded plastic and molded pulp containers. Few frozen food packages consist of only one of these items.

Some of the frozen precooked or prepared foods that are often packed in a "package" consisting of only a single component are the following: melon balls in either enamel-lined tin cans or in Sefton cans (a composite can); various fruit and "punch" concentrates in enamel-lined cans; whip toppings in molded plastic containers; bagels in plastic bags; bread dough in bags. Bags made of kraft paper laminated with polyethylene are often used for institutional packs of French-fried potatoes.

On the other hand, the great majority of the prepared and/or pre-cooked foods are packed in a package having at least two components. Many foods are put first in a rigid foil container which in turn is placed in a paperboard container. In some instances the paperboard container is wrapped and sealed in a coated lithographed wrapper. The coating may be ordinary paraffin wax, a modified wax or polyethylene. Or, the carton may not be overwrapped but is made of coated lithographed sulfite paperboard.

As a rule the rigid aluminum foil container in which the food is placed for freezing is used by the housewife as the container in which the product is reheated, and also often as the "dish" in which the product is served. For example, a pot pie is put in a small aluminum foil pan, which in turn is placed in a lithographed carton. When the consumer prepares the pot pie for the table, it is removed from the carton, placed still in the aluminum pan, in a hot oven, and, when it is thoroughly reheated, it is served in the neat little foil pan in which it had been frozen.

Since the container in which the food is packaged is usually placed on the consumers' dining table, the rigid foil containers are neat and attractive. Indeed, some are so well constructed that they may be saved by the housewife and later used again as a cooking or serving utensil.

Pouches and bags containing food which is to be reheated by immersion in boiling water are almost invariably packaged in cartons; some of these are made of lithographed sulfite paperboard coated with polyethylene, and consequently need no additional barrier to prevent desiccation; others, which are not coated, require overwrapping with coated litho-graphed paper. Two of the reasons why the pouches and bags of this type are always packed in cartons are: (1) since they have been evacuated, they are not attractive in appearance because they are so wrinkled; and (2) if they are damaged in handling, they become leakers and the food cannot be constituted by the "boil-in-the-bag" procedure without loss.

A few frozen baked goods are attractively packaged in rigid aluminum foil pie or cake pans with a cover composed of a piece of coated litho-graphed paperboard held in place by crimping the foil container. Even if the foil pans are not placed in a carton, this type of packaging is satisfactory provided the aluminum foil is sufficiently heavy to withstand the rough treatment it may get during marketing.

Shipping Containers

Corrugated fiberboard boxes are standard shipping containers for all frozen foods, including fresh, precooked, and prepared. Of the numerous types of products that fall into the latter classes, there are many in which the initial container is not solidly packed; this makes the product more fragile during handling. Likewise some products, like frozen fruit pies, are easily damaged by shipping unless amply protected against breakage.

A special corrugated container for frozen pies, which is lightweight and expendable and offers savings in transportation and damage costs is shown in Fig. 11.8. This replaces the nonexpendable metal case with wood trays once used as shipping containers for frozen pies. A six-sided corrugated insert piece fits snugly around each of the pies, which are packed in six layers of two each. These inserts are high enough to give clearance for pies, and they support corrugated "shelf" inserts. As may be noted (see Fig. 11.8), printing on the box cautions to keep the top side up and to keep the contents frozen.

To guard against weakening of the box from condensation formed when warm air strikes its cold surfaces and from other exposure to moisture, the box and inserts are made of water-resistant V3C board, a type much used for military overseas shipments. For extra stability, the

Courtesy of Stone Container Corp.

FIG. 11.8. CORRUGATED CONTAINER FOR FROZEN PIES. FROZEN PIES, WHICH PRESENTED A SPECIAL CHALLENGE IN DAMAGE-FREE SHIPMENTS, ARE NOW BEING PACKAGED SUCCESSFULLY IN THIS SPECIAL CORRUGATED BOX

"shelf" inserts of the package have flanges that are folded down and wedged against the container walls. The package also has corrugated pads at top and bottom for added protection.

The reader is referred to Chapter 12 for more detailed information on the general subject of the packaging of frozen foods.

STORAGE OF PRECOOKED FROZEN FOODS

Of all the questions about production and handling of precooked frozen foods, the ones most frequently asked concern optimum storage conditions and permissible length of storage at various temperatures. This is not surprising since so little concerning the storage of precooked frozen foods has been published, and that to be found in the literature is often contradictory.

In order to understand the reason for these contradictions, it is necessary to know something about the more important factors affecting the storage life of these foods, which are the following.

Factors Affecting Storage Life

The Kind of Food Stored.—Some, such as doughs, batters, frankfurters, ham, most white sauces and gravies thickened with wheat flour, and certain kinds of sandwiches, do not remain long in excellent condition under ordinary frozen storage conditions. Others, such as most soups, bread, and many meat dishes, hold their original quality remarkably well if the products are properly packaged and stored at 0° F. (−18° C.) or below.

The Condition of the Product Entering Storage.—(1) If the product is to be held for many months, it must be in excellent condition at the time it is frozen.

(2) In the case of cooked fruits and vegetables, the variety of the fruit or vegetables is an important factor to be considered in estimating storage life.

(3) Another factor of major importance is the method of freezing employed. If products are packed in cases or large containers before freezing, they are frozen so slowly that they may not be of high quality when placed in storage; in other words, much or all of its commercial storage life may already have been exhausted.

The Method of Cooking Employed.—Stews will keep better than fried or broiled meats. Cream-style sweet corn will retain its flavor better than whole grain corn and far better than corn on the cob. Apple sauce can be stored longer than baked apples.

The Degree of Doneness.—In general, foods should be removed from the heat before they reach the well-done stage. Green vegetables, for example green beans, peas, spinach, Swiss chard, and broccoli, that are thoroughly cooked before freezing, gradually lose their bright green color during long storage. Berries should be given a minimum of heating, if they are to retain their bright colors during long storage.

Unstable Flavoring Ingredients Should Not Be Used.—Spice cake and gingerbread deteriorate far faster than cakes which do not contain spices. The flavor of onion gradually fades during storage.

The Kind of Pack, Method of Packaging, and Packaging Materials Used, and Completeness of Fill.—Whenever possible, the food should be solidly packed so that a large amount of surface is not exposed to the air. Packaging in an atmosphere of nitrogen retards oxidative deterioration, as is shown in Fig. 11.9. Further, the packages should not permit desiccation, because the glaze of ice over the surface of the product helps to retard oxidation.

Last and Most Import of All of These is the Maintenance of a Uniformly Low Temperature Throughout the Storage Period.—In gen-

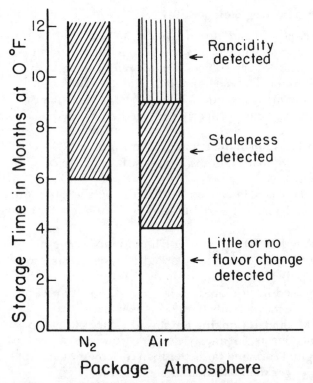

FIG. 11.9. EFFECT OF PACKAGE ATMOSPHERE (AIR VS. NITROGEN) ON OFF-FLAVOR DEVELOPMENT IN FRIED CHICKEN STORED AT 0° F. (−18° C.)

eral, it can be said that the speed of chemical reactions is increased two and one half times when the temperature is raised 18° F. (10° C.). In the case of precooked foods, assuming that the products are held solidly frozen, most of the deterioration and change is caused by these chemical reactions. Conversely, when the temperature is raised 18° F. (10° C.), the permissible storage period is more than cut in half. Thus, if this chemical rule is assumed to hold for frozen foods, if a product retains its fresh quality for only two and one-half years at −36° F. (−38° C.), it can be kept for one year at 0° F. (−18° C.), and for only two months at +18° F. (−8° C.). Actually, the rate of deterioration about triples for every 18° F. (10° C.) rise in temperature. It is safe to say that, if a food retains its fresh quality for three years at −18° F. (−28° C.) to −20° F. (−29° C.), it can be held in good condition for one year at 0° F. (−18° C.), but only for six months at +10° F. (−12° C.).

Such generalizations can be applied only to temperatures at which foods remain substantially completely frozen. If the food contains a considerable percentage of sugar or other water-soluble solids, when the temperature is raised to the point where all or a portion of the food is liquid or semi-liquid, deterioration occurs at even a greater rate. For example, fruit pies containing a high percentage of sugar soften at 10° to 15° F. (−12° to −10° C.), and deteriorate rapidly at or above this temperature. Strawberry shortcake also has a short storage life at any temperature above +10° F. (−12° C.), whereas it will remain in good condition at 0° F. (−18° C.), for six months.

Deterioration is Cumulative

In general, it can be said that the deterioration occurring during storage of frozen foods is cumulative. However, a food does not deteriorate *faster* at a higher temperature because it has previously been held at a lower temperature, and vice versa. Let us assume that food X has a storage life of 12 months at 0° F. (−18° C.), 6 months at 10° F. (−12° C.), and only 3 months at 15° F. (−10° C.). If this food is held 6 months at 0° F. (−18° C.), then 1 month at 10° F. (−12° C.), $\frac{2}{3}$ of its storage life would have been used up, i.e., $\frac{6}{12} + \frac{1}{6} = \frac{2}{3}$, and if the food were placed in a cabinet at 15° F. (−10° C.), it would become unacceptable in a single month.

Further, it makes no difference whether the storage at higher temperature precedes or follows holding at the lower temperature or temperatures. Thus, if food "X," the storage characteristics of which have been described in the preceding paragraph, is held first for 1 month at +10° F. (−12° C.), then for 6 months at 0° F. (−18° C.), $\frac{2}{3}$ of its storage life would have been used up, and if it were placed in a cabinet at 15° F. (−10° C.), it would become unacceptable in a single month.

Effect of Fluctuating Temperatures

When under fluctuating temperature, if the maximum temperature to which a frozen food is subjected is sufficiently low so that the product remains solidly frozen and no liquid separates, the rate of deterioration of many foods (excepting cream sauces and thickened gravies) is approximately the same as that which occurs when the product is held uniformly at the mean temperature. Thus, if the temperature of a food fluctuates between $-6°$ and $+5°$ F. ($-21°$ and $-15°$ C.), the rate of deterioration is approximately the same as that which would occur if the food were held at a constant temperature of $0°$ F. ($-18°$ C.).

On the other hand, if a chiffon pie of high sugar content were held at a temperature fluctuating between $+5°$ and $+15°$ F. ($-15°$ and $-9.4°$ C.), each time the temperature rose to $+15°$ F. ($-9.4°$ C.) it is probable that there would be some separation of liquid and that the amount of free liquid which appeared at $+15°$ F. ($-9.4°$ C.) would gradually increase, and the deterioration which occurred would be similar to that occurring at a constant $+15°$ F. ($-9.4°$ C.) temperature.

Fluctuating storage temperatures are considerably more detrimental to white sauces, thickened gravies, cornstarch, and custard puddings than a constant temperature which is the mean of the maximum and minimum temperatures (see Fig. 11.10). Fluctuating temperatures also accelerate the formation of frost and "cavity ice" inside packages.

Products Which Have Short Storage Life

Although the majority of precooked frozen foods and baked goods will remain in excellent condition for six months or longer, there are a few which have a very short storage life. In Table 11.4, the more important of these are listed.

Some of the reasons why these products deteriorate so rapidly have been discussed previously. Roll bread and other yeast doughs fermented for longer than 15 min. at $80°$ F. ($27°$ C.) deteriorate rapidly because of the gradual death of yeast cells and loss of viability of those which remain. Batters and other unbaked goods containing baking powder have a relatively short storage life for two reasons: (1) The acid and soda of the baking powder slowly react. (2) The carbon dioxide formed from this reaction gradually escapes from the batter, causing loss of leavening power. Spice cakes deteriorate rapidly because of loss and change of flavor of the spices. Sponge cakes made with egg yolks cannot be kept in storage for long periods, because of undesirable changes in flavor of the egg yolk. Strawberry shortcake, and other products in which the fruit is not immersed in syrup or other liquid, and, consequently, is exposed to the oxidative action of air, also have a short storage life even at $0°$ F.

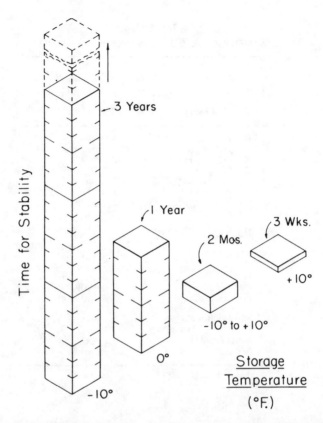

FIG. 11.10. EFFECT OF CONSTANT AND FLUCTUATING STORAGE TEMPERATURES ON STABILITY OF FROZEN WHITE SAUCE THICKENED WITH WAXY RICE FLOUR

(−18° C.). The whites of hard cooked eggs are undesirably toughened by freezing, even though the storage period is very short.

Cream sauces and gravies thickened with wheat flour or ordinary cornstarch coagulate in a short time when stored at temperatures much above 0° F. (−18° C.)

If these precooked foods and baked goods which have a short storage life at 0° F. (−18° C.) or higher are held at lower temperatures, storage life will be increased somewhat, but few of them will retain their original qualities for six months even though held at −20° F. (−29° C.).

Products Which Have Storage Life of Medium Length

In Table 11.5, a considerable number of foods are listed which ordinarily will reain in good condition for 6 to 8 months at 0° F. (−18° C.). Some

TABLE 11.4

PRECOOKED PRODUCTS WHICH HAVE SHORT STORAGE LIFE

Product	Maximum Storage Life at 0°F. (−18°C.)
Bacon, Canadian	2 weeks
Batter, gingerbread	3–4 months
Batter, muffin	2 weeks
Batter, spice	1–2 months
Biscuits, baking powder	1–2 months
Bologna, sliced	2 weeks
Cake, sponge, egg yolk	2 months
Cake, spice	2 months
Dough, roll	1–2 months
Frankfurters	2 weeks
Gravy[1]	2 weeks
Ham, sliced	2 weeks
Poultry giblets	2 months
Poultry livers	2 months
ham	
Sandwiches cheese	2 weeks
bologna	
Sauce, white[1]	2 weeks
Sausage	2 months

[1]Thickened with ordinary wheat flour or corn starch.

TABLE 11.5

PRECOOKED PRODUCTS WITH STORAGE LIFE OF MEDIUM LENGTH[1]

Batter, devil's-food[2]	Pies, chicken
Batter, white cake[2]	Pies, fruit, unbaked[3]
Cakes,[2] various kinds [3]	Pies, meat
	Potatoes, French-fried
Chicken, fried[2]	Sandwiches
Crab	Roast beef
Fish, fatty[1]	Various spreads
Fruit purées	turkey
Ham, baked, whole[2]	liverwurst
Lobster	Soups[3]
Meals on a tray[1]	Shrimp
Meat balls	Turkey[2]
Meat loaf	

[1]Will remain in good condition at 0° F. (−18° C.) for 6 to 8 months.
[2]Storage life 4 to 6 months.
[3]The storage life varies widely depending on the kind and also the flavoring or seasoning used. Many retain quality for a year or longer.

research workers will not agree that all of those listed should be included. However, the authors have evidence to indicate that if sufficient care is taken in the selection of the formulas used, the ingredients chosen, the method of cooking employed, and the packaging materials and method of packaging used, products of the kinds listed can be produced which will retain their fresh qualities at least as long as six months at 0° F. (−18° C.). However, it is dangerous to draw too sweeping generalizations concerning the probable storage life of any class of products. For instance, many fruit pies, e.g., prune, blueberry, and raisin, can be stored for a year or even longer without serious loss of quality.

On the other hand, peaches, fresh apricots, and apples must be specially prepared if the fresh quality is to be retained for as long as six months.

Some kinds of cakes retain their flavor, color, and texture during frozen storage better than others. This is shown in Table 11.6 in which the information available in the literature on cake storage is summarized.

Cooked lobster, crab, and shrimp meat toughen to a greater or lesser degree during freezing and storage, probably due to denaturation of

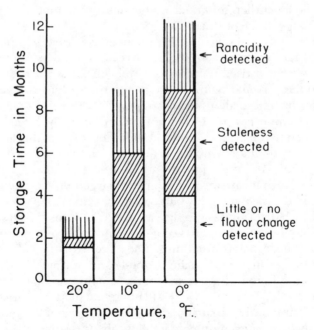

FIG. 11.11. EFFECT OF STORAGE TEMPERATURE ON OFF-FLAVOR DE-VELOPMENT IN FROZEN FRIED CHICKEN IN RETAIL PACKAGES STORED AT 0°, 10°, AND 20° F. (−18°, −12°, AND −6.7° C.)

TABLE 11.6

LENGTH OF TIME WHICH VARIOUS KINDS OF CAKE RETAIN THEIR ACCEPTABILITY

Kind of Cake	Temperature of Storage °F.	°C.	Acceptability Retained Months
Angel food	−10° to 0°	−23° to −18°	4
Angel Food	−10° to 0°	−23° to −18°	6+
Cheese cake, French	−15°	−26°	9+
"Fathers Day"	−15°	−26°	14
Fruit cake	0°	−18°	12+
Plain, yellow	0° to −10°	−18° to −23°	9+
	−10°	−23°	4+
	−10°	−23°	8+
	0°	−18°	6+
Plain, white	0°	−18°	7+
Spice	+12°	−11°	4
Sponge (egg yolk)	−10° to 0°	−23° to −18°	2
Sponge (whole egg)	−10° to 0°	−23° to −18°	6+

proteins, as noted earlier, and their highly unsaturated fats that oxidize easily and change in flavor during long storage.

Although the fats and oils in which potatoes are fried are composed principally of fats which are not highly unsaturated, still, since much of the fat on the French-fried potatoes is largely on the surface, and, consequently, in direct contact with the oxygen of the air, the storage life of the product is limited to about ten months at 0° F. (−18° C.). This same statement holds for other products fried in deep fat. If the fat used is not nearly saturated (from a chemical standpoint), or if it remains hot in the frybath for more than a few hours, foods fried therein may not remain free from rancidity for even eight months at 0° F. (−18° C.).

The storage life of meals on a plate depends largely on the components of the meal. If the meal includes sausage, ham, Canadian bacon, or any other food which deteriorates rapidly (see Table 11.4), in frozen storage, the meal cannot be considered to be suitable for holding in zero storage (−18° C.) for longer than four months. The placement of meals on a tray in the list of foods which will remain in good condition at 0° F. (−18° C.) for 6 to 8 months is, therefore, based on the assumption that no component listed in Table 11.4 (those which will not retain their fresh quality for longer than four months), is included. Since packaging of the foods on a tray is not such as to exclude contact with air, the storage life of the components is not as great as it would be if each were individually packaged in air-tight containers with little or no headspace.

The More Stable Precooked Foods

In Table 11.7 are listed various food products which, if properly prepared and packaged, should remain in good condition at 0° F. (−18° C.) for at least 12 months. The list includes stews and sauces which can be solidly packed and thus eliminate air, and also the possibility of serious desiccation.

TABLE 11.7

PRODUCTS WHICH SHOULD REMAIN IN GOOD CONDITION at 0° F. (−18° C.)
FOR ONE YEAR OR LONGER

Applesauce	Cookies
Apples, baked	Doughs, cookie
Bread	Fish, lean
	Peanuts
Bread (rolls)	Pecans
Blackberries	Plums
Blueberries	Stew, beef
Cake, fruit	Stew, veal
Candies[1]	Waffles
Cherries	
Chicken, creamed[2]	
Chicken à la king[2]	

[1]Some candies do not freeze well.
[2]If thickened with waxy rice flour.

The list also includes bread, rolls, and waffles, cereal products containing much air, which, however, are not particularly subject to oxidative deterioration.

This list cannot be considered as complete. Undoubtedly there are other cooked foods which are equally stable.

ADDITIONAL READING

AM. FROZEN FOOD INST. 1971A. Good commercial guidelines of sanitation for the frozen vegetable industry. Tech. Serv. Bull. *71*. American Frozen Food Institute, Washington, D.C.

AM. FROZEN FOOD INST. 1971B. Good commercial guidelines of sanitation for frozen soft filled bakery products. Tech. Serv. Bull. *74*. American Frozen Food Institute, Washington, D.C.

AM. FROZEN FOOD INST. 1971C. Good commercial guidelines of sanitation for the potato product industry. Tech. Serv. Bull. *75*. American Frozen Food Institute, Washington, D.C.

AM. FROZEN FOOD INST. 1973. Good commercial guidelines of sanitation for frozen prepared fish and shellfish products. Tech. Serv. Bull. *80*. American Frozen Food Institute, Washington, D.C.

ASHRAE. 1974. Refrigeration Applications. American Society of Heating, Refrigeration and Air-Conditioning Engineers, New York.

BAUMAN, H. E. 1974. The HACCP concept and microbiological hazard categories. Food Technol. *28*, No. 9, 30–34, 70.

DEFIGUEIREDO, M. P., and SPLITTSTOESSER, D. F. 1976. Food Microbiology: Public Health and Spoilage Aspects. AVI Publishing Co., Westport, Conn.

GUTHRIE, R. K. 1972. Food Sanitation. AVI Publishing Co., Westport, Conn.

HARRIS, R. S., and KARMAS, E. 1975. Nutritional Evaluation of Food Processing, 2nd Edition. AVI Publishing Co., Westport, Conn.

JOHNSON, A. H., and PETERSON, M. S. 1974. Encyclopedia of Food Technology. AVI Publishing Co., Westport, Conn.

KRAMER, A., and TWIGG, B. A. 1970, 1973. Quality Control for the Food Industry, 3rd Edition. Vol. 1. Fundamentals. Vol. 2. Applications. AVI Publishing Co., Westport, Conn.

PETERSON, M. S., and JOHNSON, A. H. 1977. Encyclopedia of Food Science. AVI Publishing Co., Westport, Conn.

RAPHAEL, H. J., and OLSSON, D. L. 1976. Package Production Management, 2nd Edition. AVI Publishing Co., Westport, Conn.

SACHAROW, S. 1976. Handbook of Package Materials. AVI Publishing Co., Westport, Conn.

STADELMAN, W. J., and COTTERILL, O. J. 1973. Egg Science and Technology. AVI Publishing Co., Westport, Conn.

TRESSLER, D. K., and SULTAN, W. J. 1975. Food Products Formulary. Vol. 2. Cereal, Baked Goods, Dairy and Egg Products. AVI Publishing Co., Westport, Conn.

TRESSLER, D. K., VAN ARSDEL, W. B., and COPLEY, M. J. 1968. The Freezing Preservation of Foods, 4th Edition. Vol. 2. Factors Affecting Quality in Frozen Foods. Vol. 3. Commercial Freezing Operations—Fresh Foods. Vol. 4. Freezing of Precooked and Prepared Foods. AVI Publishing Co., Westport, Conn.

WEISER, H. H., MOUNTNEY, G. J., and GOULD, W. A. 1971. Practical Food Microbiology and Technology, 2nd Edition. AVI Publishing Co., Westport, Conn.

WILLIAMS, E. W. 1976A. Frozen foods in America. Quick Frozen Foods 17, No. 5, 16–40.

WILLIAMS, E. W. 1976B. European frozen food growth. Quick Frozen Foods 17, No. 5, 73–105.

WOOLRICH, W. R., and HALLOWELL, E. R. 1970. Cold and Freezer Storage Manual. AVI Publishing Co., Westport, Conn.

12

Packaging of Frozen Foods

Bernard Feinberg
and Robert P. Hartzell

A successful package is one that protects a product or content from an environment for a period of time and at reasonable cost. Some of the factors which must be taken into account in selecting a package for frozen foods are protection of the contents from: (1) atmospheric oxygen, (2) loss of moisture, (3) flavor contamination, (4) entry of microorganisms, (5) mechanical damage, and (6) exposure to light. In addition to the protective functions listed above, it is desirable that materials used for packaging foods to be frozen in the package have a high heat transfer rate to facilitate rapid freezing.

Because of the importance of impulse appeal in today's self-service markets, a frozen food package serves not only to protect but also has a selling function. A well designed package is sometimes called "the silent salesman." The package should not only be attractive and informative, but easy to open, and, when the contents are only partially used, easy to close.

The modern frozen foods processor is fortunate that he can select from a wide range of protective materials in designing his package. Today, even the consumer who wishes to wrap food for home feezing has a wide variety of papers, films, and foils from which to select. These can easily be purchased at the corner supermarket or hardware store where the selection would probably include heavy aluminum foil, moisture-proof heat-sealable cellophane, and various thermoplastic films such as Dow "Saran," DuPont's "Mylar," Goodyear's "Pliofilm" and Cordite plastic wrap. Both plastic films and metal foil are available as rolls in handy dispensers with cutting edge, or as bags in a wide variety of sizes ranging from ½-pint to 2-gal. capacity. The films are transparent, lightweight, resistant to puncture, and flexible at low temperatures. They are free from off-odors and off-flavor, are nontoxic, frequently are heat-sealable, and have a low moisture-vapor transmission rate. Various kraft papers coated with wax or plastic are also available. A favorite wrap for home freezers is laminated paper, available in many combinations where two or more materials are permanently bonded to each other by adhesive or by heat and pressure, combining the desirable properties of each of the component materials in one sheet. Semirigid plastic containers in many sizes, with easily fitted friction caps, can be found in most hardware stores.

The pioneers of commercial freezing would have been envious indeed at this treasure house of packaging materials. Clarence Birdseye related some of the trials and tribulations in packaging frozen fish in the late 1920's:

> Quick freezing called for transparent moisture-vapor-proof wet-strength tasteless and odorless wrapping materials but no such sheets had ever been produced. My company first tried ordinary waxed paper, but that gave practically no protection against moisture-vapor loss during storage and disintegrated when the fish thawed and became wet. Then we used plain vegetable parchment, which had wet-strength but no moisture resistance, and which frequently stuck to the thawed product. Next we tried paraffining the parchment paper, and found that, though it did not stick to our product, it did not prevent drying out during storage. In the meanwhile, we had experimented with uncoated Cellophane, imported from France, but it was both pervious to moisture and went to pieces on contact with the wet fish. Finally, in desperation, we persuaded duPont to give some Cellophane a moisture-vapor-proof coating, and that product was so promising that it was put into production especially for our company, which for a number of months, was the sole purchaser. Then followed its use in wrapping cigars and cigarettes and a mushrooming demand which fathered a revolution in plastic sheeting.

Packaging Responsibility

Depending on the size of the organization, the selection of the packaging materials and construction may be left to the purchasing department, the research or quality control department, or, if the plant is large enough, a packaging coordinator. There are many factors involved in the packaging function in a food packing organization.

TYPES OF CONTAINERS

There are essentially two classes of containers in the frozen food industry: (1) the unit or primary package, and (2) the outer package or shipping container. The unit or primary package is the one in direct contact with the contents and usually comprises the retail unit of sale.

Primary Packages

Drums and Barrels.—Wooden barrels and wooden boxes which were widely used in the earlier years for frozen packs of small fruits have virtually disappeared. They have been largely replaced by the 55-gal., 18-gage, steel drum with a polyethylene liner. The disadvantage of large size containers such as drums or barrels is their slow rate of freezing and thawing; however, they are well adapted for slush-frozen fruit purées or IQF (individually quick frozen) berries.

Large quantities of frozen fruits to be remanufactured into preserves are packed in drums. Frozen fruit juices are sometimes packed and frozen in 55-gal. drums for use in blending with juices of a different acid-Brix ratio later in the season or even with other types of fruit juices. For example, one packer in Michigan packs single-strength filtered, clarified cherry juice in 55-gal. steel drums with a polyethylene lining. The juice is frozen and stored at 0° F. (−18° C.) for blending later in the season with apple juice to make an apple-cherry blend.

Fiber drums with a 2–3 ml. polyethylene liner are sometimes used in lieu of metal drums. Fiberboard drums consist of straight-sided kraft cylinders with tops and bottoms of fiberboard or metal. Polyethylene may be coated directly on the interior of the drums in lieu of liners. There are usually special provisions for locking rims.

Boxes.—Wooden boxes are used in shipping frozen poultry. The boxes are lined with parchment or waxed paper and the covers are wired down. Wooden boxes are slowly being replaced with strong wax-impregnated or coated corrugated cartons.

Both wood bins and strong corrugated bins holding approximately 1,000 lb. have been used for bulk storage. They are almost always lined with a 2- to 3-ml. polyethylene liner.

Cartons.—The folding paperboard carton is still the most important package in the retail marketing of frozen foods. Despite the flood of new plastic films, laminates, metallic foils, etc., paperboard remains an economical, versatile, strong, easily converted packaging material. Some of the criteria which must be looked for in carton packaging are resistance to grease, strength for stacking in the freezer, and a seal which is resistant both to water vapor loss and tampering by the curious consumer. Figure 12.1 illustrates the most common cartons used for frozen foods. Despite the many advances in frozen food packaging, the cartons illustrated are still those most commonly used in the frozen food industry. All of these cartons require a wax or plastic coating plus an exterior overwrap for protection of the contents from loss of moisture.

The grade of boxboard used for making folding cartons for frozen food is a solid bleached sulfate board from 0.012 to 0.026 in. caliper. It is used because it is strong, white in color, and has excellent bending properties which make it adaptable to high-speed automatic filling machines.

The Fibreboard Corp. has recently introduced a new package for retail sales consisting of a board coated on both sides with polyethylene. This "Barriermatic" carton is heat-sealed, polyethylene to polyethylene, and

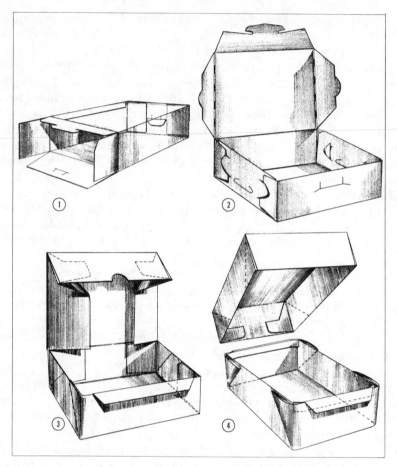

Courtesy of Marathon Corporation

FIG. 12.1. LINE DRAWINGS OF TYPICAL CARTONS USED FOR FROZEN FOODS

(1) Marapak end opening—designed for high-speed automatic filling—used for free flowing vegetables and also corn-on-the-cob. (2) Kliklok—Charlotte style with hinged cover and front arrow lock—designed for FMC filling machine and also hand packing—used very extensively for frozen vegetables. (3) No. 5 style top opening Klik Top Cover Lock (minimum height 1 in.)—designed for fast hand setup or automatic setup on the FMC No. 15 line—used for frozen vegetables and 1-lb. fish fillets. (4) Two-piece telescopic cover-designed for plate freezing poultry.

reportedly has a moisture-vapor transmission loss only half as great as an overwrap carton. It can be erected, filled, and heat-sealed at speeds up to 300 a minute. The cartons can be either printed with full color gravure, then coated, or lithographed directly over the coating. The high-gloss polyethylene coating over the print gives an excellent surface appearance while protecting the surface from soiling and scuffing. The carton has an ingenious perforated thumb-hole tab on one side which enables the consumer to quickly open the carton by pressing in on the tab and pulling up on the carton top. The carton is "jet sealed" by blowing hot air under the end flap and compressing. For retail packaging of frozen fruits a new Fibrematic carton from the same company, consisting of a hermetically sealed polyethylene liner within a sealed-end carton, has been successfully used (Fig. 12.2). This carton also has a polyethylene-to-polyethylene heat seal.

Although the one-piece polyethylene-coated container has several advantages, it requires a relatively large operation, packaging many cartons on a single run, to be economical. The many small custom packers prefer a plain carton with a waxed overwrap. It is much easier to carry a stock of one or two sizes of paperboard cartons and numerous rolls of waxed

Courtesy of Fibreboard Paper Products Corp.

FIG. 12.2. THE FIBREMATIC CARTON

A hermetically sealed polyethylene-coated liner within a sealed-end carton for packaging of frozen fruits.

paper imprinted with various trade names, items, and grades than to carry a similar stock of preprinted cartons. Polyethylene-coated cartons, whether single-piece or overwrapped, reportedly retain their rigidity better than waxed cartons when a breakdown in the filling or packaging lines requires the holding of cartons of wet fruits or vegetables for 5 to 15 min.

Retail sizes of paperboard cartons are either end-opening and end-filled, or top-opening and top-filled. The end-opening cartons are adaptable to easy-to-fill fresh or IQF vegetables or fruits, and have the advantage that less paperboard is used in making the carton and a much smaller area requires sealing. For such items as asparagus and broccoli, and for such large food pieces as fish sticks, poultry, shrimp, etc., it is necessary to use a top-opening carton. Frozen French fries are packaged in an institutional pack carton holding five pounds, and are sealed with hot-melt adhesive. One advantage of packaging in such a container is the ease of pouring material out of the carton.

Package Sizes.—In the early years of the frozen food industry a 10-oz. net weight was standard for frozen vegetables. The consumer, however, can now find frozen vegetables packed in cartons with net weights of 5½, 6, 7, 8, 9, 10, or 12 oz. Ten-ounce cartons, however, are still the most common. Because of the various densities and shapes of different vegetables, the dimensions of cartons in the retail cabinet show considerable variation; by contrast, most freezers who pack for institutional or bulk users use one or two standard packages and vary the weights according to the properties of the vegetable being frozen. The usual depth of a retail size carton is 1¾ in. so that cartons are sometimes identified as "1¾-in. cartons."

Weight Loss Through Package.—Although frozen foods are free from some of the storage problems encountered in foods stored at room temperature, such as protection from rodents, insects, and mold, they do encounter special problems of their own. The most important of these is desiccation, as previously noted.

Peas packaged in plain paperboard cartons, with no paraffin coating or overwrap, lost 25% of their in-going weight after 12 months of storage at −5° F. (−21° C.). By contrast, peas packaged in paraffin-coated cartons with carefully sealed ends lost almost no weight under the same conditions. High moisture products, if they are not to lose appreciable weight in frozen storage over a considerable period of time, must not only be packaged, but must be very thoroughly packaged in material which provides a good water-vapor barrier and this material must be applied thoroughly and completely, without providing any opportunity for water-vapor escape because of improper sealing.

Several researchers have demonstrated the need for an overwrap around 10-oz. cartons in a storage test of frozen peas. After 6 months of storage at 0° F. (−18° C.) peas packaged in a paperboard carton without overwrap showed a 0.05% weight loss per day, while the same carton with a waxed paper overwrap had a loss of only 0.003% per day. There is a direct correlation between the weight of the waxed paper and the rate of moisture loss, the heavier waxed paper overwrap reducing the rate further. Waxed paper overwrap with a weight of 45 lb. per ream gave appreciably more protection from water-vapor loss than an overwrap of 32 lb. per ream. No relationship was found between the weight of the carton board and the rate of moisture loss for the vegetables examined.

Bags.—Polyethylene-coated kraft paper bags are used by some freezers to pack free-flowing vegetables such as peas, French-fried potatoes, and diced carrots in quantities from 20 to 50 lb. for use in remanufacturing such items as canned soups or stews, or for institutional use. Large kraft paper bags have been also used as shipping containers for such items as ice cream. One firm which formerly used two such kraft bags to bundle four half-gallons of ice cream for shipping to retail stores has replaced them with a polyethylene wrap. The film is folded at the sides of the bundle and heat-sealed. The polyethylene wrap has the advantage that the various flavors of ice cream are visible through the transparent film, thus eliminating the necessity for other means of identification, such as the use of color tapes.

Cans.—Metal cans made of tin-plate or aluminum have been used to package strawberries and small fruits. Their use for this purpose has declined and cans are now used almost exclusively for the packaging of fruit juice concentrates and frozen soups. Their use had earlier been a matter of some controversy, since the consumer could conceivably place cans of frozen food on the same shelf in the kitchen with heat-sterilized canned foods. Although the packaging industry has long felt that frozen products intended for retail sale should not be packed in metal cans, there has been no reported difficulty with foods so packed. Metal cans fill many of the packaging requirements for frozen foods listed earlier in this chapter. They have excellent WVTR (water-vapor-transmission rate) properties, may be hermetically sealed, are strong, and are easily adapted to high-speed filling and handling. Food packed in cans may also be frozen by immersion in a refrigerated liquid, a procedure which might be undesirable if the refrigerant made direct contact with the food.

Hermetically sealed lithographed metal cans have long been used for packaging frozen juice concentrate. Special enamel linings to protect the contents have been developed for such containers. Number 12 cans holding one full gallon of frozen fruits are frequently used for institutional packaging. Hermetically sealed cans of various sizes are also used for such items as crab meat and peeled shrimp.

For various production reasons spiral-wound construction is the most common in composite cans. In another type of construction, "convolute winding," a web of paper the width of a can body height is fed onto a winding mandrel which turns a specified number of times, wrapping the paper around itself. The can body thus formed is sheared from the parent roll of stock. The Container Corporation makes a convolute can for frozen concentrates which is lined with a "plastic-wax" coating instead of aluminum foil. Convolute cans are available not only as cylinders, but also as the oblong cans, commonly known as "Sefton" cans, and long used for packaging frozen fruits in syrup.

A variety of devices, including tear-strips and pull-off tops, are presently used for easy-opening cans of frozen juice concentrates. The American Can Co.'s Mira Strip easy-open cans for frozen concentrates features a narrow strip of plastic wrapped around one end of a fiber-foil composite body. The end of the strip forms a handy tab which easily opens the carton. The entire top of the can is released and the contents can be removed without splatter or spilling. The Continental Can Co. can features an end unit which includes a ring and an aluminum end that is scored completely around the circumference. The ring and lid are removed in one piece simply by lifting.

Interior and exterior enameled cans approximately 10¾ in. in diameter and 13 in. in height, commonly known as 30-lb. tins, have long been used for freezing and storage of eggs, purée, whole fruits and berries, and fruit juices. The cans are closed with a friction top lid. One disadvantage of the 30-lb. tin is that it takes up just as much storage space empty, as it does full. A recently developed tapered can, wherein emptys can be stacked one within another, has become increasingly popular with freezers. Because of their design, tapered 30-lb. tins have more space between them when stored close together. This increased space permits greater circulation of cold air for faster freezing or of warm air for faster thawing (Fig. 12.3).

Fruits packed in 30-lb. tins are protected from oxidation by immersion in a heavy sugar syrup or by covering with dry sugar which soon forms a syrup. During the freezing process, however, the fruit will expand and rise above the level of the protecting syrup. To prevent this, special inserts are placed in the can which will hold the fruit beneath the syrup while allowing the syrup to rise and expand. Such head-space depressors may be made of waxed board or may consist of a perforated metal disk plus spring.

Courtesy of Rheem Manufacturing Co.

FIG. 12.3. TAPERED CANS

Nine tapered 30-lb. frozen food tins, on the right, take up the same space as
three conventional cans. Resulting saving in storage and transportation
costs are large.

Slab Packaging

Freezing in rectangular molds and sawing the resulting frozen slab into
small blocks offers opportunities for mechanized packaging. For example, crab meat has been frozen into 15-lb. blocks which are sawed into
package-size units and automatically wrapped in Saran-coated cellophane
which provides a tough vapor barrier. Since this film gives little mechanical protection, the film-wrapped block is packaged in a paperboard carton
which is sealed automatically with hot-melt adhesive. Crab meat packed in
this manner comes in several sizes, including a 6-oz., twin 8-oz. packages,
16-oz., and twin 1½-lb. units for use in restaurants. The use of twin
individually wrapped blocks enables the consumer to use half the contents
and store the other half in the freezer. Spinach has also been commercially frozen and packaged using the slab-freezing technique.

Plastic Films

Plastic packaging in the form of flexible films, semirigid packages, and rigid packages has expanded so rapidly, both in diversity of materials and in packaging shapes, that it is difficult to squeeze even an outline of their types and properties into the limits of this book. It is estimated that more than 600 flexible film combinations are commercially available today.

Terms frequently used in describing flexible film packages are:

Coatings.—A plastic or wax applied in solution or as a liquid on some substrate such as paper, metallic foil, or film. The thickness of the coating may vary according to the requirements of the coated material.

Laminations.—Two or more substrates, such as paper, plastic film, or metallic foil, cemented together by adhesive or heat, and pressure.

Flexible Films.—A bewildering variety of flexible films is used in frozen food packaging as bags, pouches, overwraps, rolls of films to be converted to pouches, and in laminates. Almost all of these films are made from organic polymers, a term given to large molecules made of long chains of smaller molecules, called monomers, chemically hooked together in a repetitive pattern. Starch, cellulose, and rubber are examples of natural polymers which have been converted into films, some of which have proved useful in food packaging.

Many of the names of the new plastics begin with "poly." Polyethylene, for example, is a long-chain polymer of ethylene, a gas derived from natural gas or petroleum. Polyethylene is a useful film in frozen food packaging; it is low in cost, has excellent clarity and gloss, is resistant to tearing, has relatively low WVTR, is flexible at low temperatures, and is heat-sealable. Because of this last property polyethylene is frequently used as a coating or as a layer in a laminate with nonheat-sealable films. Important properties such as heat sealability, moisture resistance, and flexibility of a polyethylene film depend in large measure on the density of the polyethylene resin from which it is made. Commercially polyethylene is classified into three density ranges, expressed as gm. per cc.: low—0.910 to 0.925, medium—0.926 to 0.940, and high—0.941 to 0.965. Most polyethylene films used in frozen food packaging are in the low and medium density ranges.

Nylon is a name given to a group of long-chain polymers with recurring amide groups as a part of the chain. Nylon films have excellent properties of toughness, tear, and breaking strength. They are useful both as single films and in many laminates.

Pliofilm is the trade name for rubber hydrochloride film. It has historical significance in the packaging industry, since it was the first transparent film which could be heat-sealed. It was widely used in the early years of the frozen food industry, but for many of its former uses it has been replaced by newer films.

Polypropylene is a relatively new film in frozen food packaging, but its clarity, strength at low temperatures, and low WVTR make it useful in laminates, and in coating paper or aluminum foil. It is heat-sealable when it is treated with a variety of coatings.

Cellophanes.—As pointed out earlier in this chapter, cellophane was one of the earliest packaging materials for frozen foods. Cellophane is made from regenerated cellulose; when it is coated with other plastic resins, such as polyethylene or Saran, a wide variety of cellophanes result, each designed for a specific use. These make excellent packaging films. The various types are listed by various code designations by different producers, for example, DuPont MSAD cellophane is similar to cellophane products made by other manufacturers under different code names; it designates a strong, moisture-proof, heat-sealable film flexible at low temperatures.

Polyvinylidene Chloride.—Polymers of vinylidene chloride are better known as Saran and are available not only as films, but as water-based emulsions in which form they are frequently used as coating resins for paper substrates. Saran is superior to polyethylene as a water vapor barrier material and has a low oxygen permeability. The film is excellent for use as a wrap around odd shaped poultry or meat. Because of the difficulties of heat sealing such a package, the edges are frequently closed with a special metal clip. This closing operation is a rather slow one and thus fairly expensive (Fig. 12.4).

Polyesters.—This strong film is well known under the familiar trade name of Mylar. It is frequently used in laminations with other films, especially with polyethylene, to make pouches which can be filled and heat-sealed. These laminates are now used for boil-in-the-bag pouches. Polyester films are strong, durable, transparent, nontoxic, inert, and easily stretched. They also have the advantage of being flexible at very low temperatures ($-95°$ F., $-71°$ C.). They are sometimes known as nylon-type films and because of their toughness are useful in packaging sharp-pointed, irregular foods such as lobster.

Plastic Pouches

Although the per capita consumption of frozen vegetables has zoomed during the past ten years, the growth of frozen fruits in retail sizes has been relatively slow. Part of this slow growth of frozen fruits is because of lack of convenience. Composite cans of frozen fruit may require 1 to 2 hr. at room temperature to thaw before serving. One possible answer to this

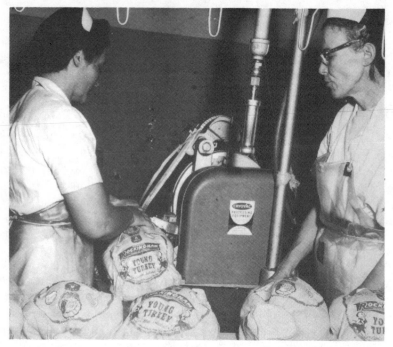

Courtesy U.S. Dept. Agr.

FIG. 12.4. THIS MACHINE ATTACHES A METAL CLIP TO THE BAG CONTAINING THE TURKEY AND CUTS OFF THE BAG'S LOOSE END

problem is the packaging of frozen fruit in plastic pouches; such a package requires less than 15 min. immersion in warm water to thaw to exactly the right consistency. One packer packs such a pouch holding 2 servings in a 10-oz. carton. The pouch used is made from a laminate of polyethylene, nylon, and Mylar films. Like boil-in-the-bag vegetables, the pouches used for thawing fruits must be vacuum packed; this eliminates the insulating effect of air and the floating of the package, and helps to preserve the quality of the fruit.

Another packer offers two 8-oz. pouches of frozen vegetables packed in butter sauce, with the 2 pouches inserted in a polyethylene bag. Both the bag and the two inner pouches incorporate easy-opening tear features which eliminate the need for the consumer to cut the package and possibly damage the inner pouches or contents. A ½-in. lip which surrounds the pouch makes it easier for the consumer to remove the pouch by hand from the boiling water. A perforated nylon film bag has been designed for products requiring direct contact with boiling water such as frozen macaroni or frozen quick-cooking lima beans.

IQF vegetables packed in polyethylene bags are liked because the product is visible to the purchaser, the bag is easy to open and reseal, the consumer can use only as much as he needs, and can see how much remains.

One of the weakest links in plastic bags is the seal, both from the standpoint of vapor leakage and that of bag breakage at the seal. A broken bag is particularly annoying to the supermarket manager, since he not only loses the contents but must use expensive labor to clean up the freezer case.

Special Packages

Frozen bread dough has been packaged in a special bag made from a nylon film. The loaf is proofed, shaped, and baked in the bag. Pies are usually packaged in a foil pan inserted into a waxed carton, although some pies are first packed in a special inner wrap. The rapidly growing specialty item of whipped topping made from soybean oil and various gums is packaged in a white polyethylene tub with a resealable plastic lid. Pancake batter may be packaged and frozen in a conventional milk-type carton. Such a package is particularly convenient because it enables the user to pour the thawed batter from the carton as needed.

One company packages fresh frozen eggs in a 2-gal. disposable plastic-coated carton similar to that used for holding liquid milk, holding 16 lb. Two cartons are packed to a case, with a net weight of 32 lb. A similar carton is available in 1-gal. size holding 8 lb. The cartons can be formed, filled, and sealed in one sanitary operation. They are liked by restaurants and bakers because the containers are easy to pour from and easy to use.

A new "boil-in-the-tray" package consisting of a polyethylene tray with a heat-sealed laminated film cover is adaptable to microwave heating, as well as to heating in boiling water. The cover is removed by grasping the overlapping edge and peeling it back across the top of the tray.

Occasionally it is necessary to ship frozen fruits and vegetables in quantities up to 100 lb. without refrigeration. Special fiberboard containers lined with polyurethane, or other insulation, and reinforced with aluminum strips are available for this purpose. When they are packed in dry ice and carefully sealed, such containers will hold the contents in the frozen state for 48 hr. or longer. Rigid aluminum containers, heavily insulated, have also been used for this purpose.

Miscellaneous Problems

Because price-marking is such an important operation in today's self-service markets, the surface of the package must quickly absorb and hold ink after stamping. However, the surface must also be resistant to staining. The frequency of smudged prices and soiled exteriors, evident in the freezer case of many markets, shows that these problems have not yet been completely solved. Many plastic packaging materials which are excellent in all other respects will not hold the price marking. Some foods, such as whole crab, for example, are irregular in shape; when they are wrapped, the package is difficult to stamp.

One frequent complaint of consumers is the difficulty of keeping the cooking instructions intact after the frozen food package is opened. When the wrapper is torn to get to the carton, the instructions are, for all intents and purposes, lost. Single-piece cartons without overwrap are free from this problem. It might be useful if instructions were inserted in a carton on a small printed sheet or even printed directly on the carton.

Clear polyethylene permits the passage of light. This can result in the bleaching of certain vegetables such as peas. A white opaque polyethylene has therefore been used for packaging 1½ lb. bags of vegetables. The material presents an excellent printing surface and a full-color illustration of the product can be printed on the opaque white film. The same effect may be obtained with laminates of clear polyethylene and white paper.

Shipping Containers

Individual packages must be further packed in a properly engineered container for shipping and storage. Corrugated containers are the most widely used form of container for shipping frozen foods. Veneer wooden crates and foam polystyrene shipping containers are also used but to a much lesser degree. Corrugated boxes may have optional interior packing such as dividers, trays, pads, etc. Retail packages of frozen food, such as the common 10-oz. paperboard containers are usually packed 24 cartons per case; institutional packs such as the common 3-lb. carton are packed 12 cartons per case.

Corrugated fiberboard is made of fluted sheet glued between two liners. The use of a corrugated sheet provides exceptional crush-resistance. Four types of corrugated board are commercially used—A, B, C, and E-flute. The difference lies in the number of flutes and board thickness: A-flute, 35 flutes per in., $^3/_{16}$ in. thick; B-flute, 50 flutes per in., ⅛ in. thick; C-flute, 41 flutes per in., $^5/_{32}$ in. thick; E-flute, 90 flutes per in., $^5/_{64}$ in. thick. Each type has some specific advantage over the others. For example, A-flute has a high capacity for absorbing shock; B-flute, because of its greater number of corrugations, has higher crush resistance; C-flute combines the best points of A and B, while E supplies high-strength corrugated board at minimum board thickness.

The corrugated containers used for shipping frozen foods in folding cartons are 200 to 275-lb. test, much lighter than those used for shipping canned goods. Specifying a corrugated box is not a job for amateurs. The best source of information is a reliable corrugated box manufacturer. He can show the shipper how to save money in many different ways: by reducing the overall weight of the package; by making the best use of automation to reduce in-plant labor costs; by creative and intelligent use of the various regulations; by careful consideration of auxiliary materials; and in many other ways.

A recent innovation in shipping cases is the wrap-around system, in which the packages are placed directly on a single sheet of board which is wrapped around the packages, forming a shipping container, and then sealed.

Although freezing a shipping container packed with unfrozen packaged foods is not recommended, it is, unfortunately, still practiced to some extent. Some freezers have experimented with shipping containers that have several holes ½ to 1 in. in diameter punched in the container to allow better air circulation. However, the freezing rate within the container is still undesirably low.

Various government and industry standards have been established to protect the corrugated container buyer and to guide him in the selection of the proper container. It is essential, therefore, to consult the various rules and regulations pertaining to corrugated shipping containers, for example: Uniform Consolidated Freight Classification, Rule 41; Motor Truck Classification, Rule 18; Postal laws and regulations; Official Air Freight Rules, Tariff 1; and Freight Container Tariff.

Corrugated containers used for shipping paperboard cartons, both consumer and institutional size, are not customarily treated for moisture protection. With normal care and handling it is usually not necessary to provide protection against humid atmospheres even though the shipping containers may be exposed to high-moisture conditions, even momentary light rain, during a part of their movement from the shipping point to the retail store. Common sense and/or understanding of the limitations of the untreated container are essential considerations when handling frozen foods in untreated corrugated containers.

In the shipping of frozen poultry and red meats for institutional sale, a plastic- or wax-coated container is generally used because the products are shipped unwrapped. Therefore, the shipping container must possess all the barrier characteristics required for proper handling of frozen foods. Impregnating or coating the container with wax, vinyl resin, or other comparable substances provides an excellent moisture-resistant board. Various coating methods have been used for many years; however, the most recent technique is called curtain coating, wherein the coating formulation is applied by a machine designed for that purpose. The continuous film coating acts as an effective barrier, not only against free water, but also against water vapor, gas, grease, oil and other foreign substances. It is necessary to use a coated shipping container if there is a possibility of the container being exposed to top icing, high humidities, or excessive moisture conditions during handling. It is also desirable to use coated containers if the frozen product has a high fat content which may stain untreated corrugated board.

The wooden veneer wire-bound shipping container is still used by some poultry packers. One of the obvious advantages in the use of wire-bound veneer containers is their water resistance where the product is either ice-packed or subjected to extremely wet conditions during handling and transit.

The foam polystyrene container for shipping fresh frozen foods is relatively new. It has the advantage of being water resistant and having excellent insulating properties after the frozen product is placed inside, the foam lid applied, and the container sealed. It has the disadvantages of being somewhat more costly and harder to handle than either the corrugated or the wire-bound shipping container. Since they cannot be taken to the packinghouse in other than a completed form, the empty containers are bulky to handle and store. However, because of its light weight and insulating properties, the polystyrene container is particularly adaptable to the air transit of frozen foods.

Institutional Pack

Portion control has become extremely important in institutional feeding, for example, in restaurants, hospitals, and school cafeterias. One large New York elementary school lunch room receives most of its main courses in frozen form. For example, chicken is supplied to the school packed one hundred 1½-oz. pieces of chicken to a case.

The boilable pouch offers interesting possibilities for portion control. The problem of designing a package which will serve a variety of customers, ranging from the single man or woman to the family with six children, has long been a difficult one. A special problem for the frozen food industry is the individual consumer; to serve this market, one large processor has marketed single-portion pouches of vegetables two-to-a-carton. Restaurants find that large boxes of vegetables kept on steam tables for several hours lose much of their flavor and texture. Individual boil-in-the-bag servings of frozen vegetables offer a logical answer to this problem.

One ingenious California freezer packs and freezes sliced freestone peaches, in heavy syrup, in single-serving individual high-impact extruded polystyrene containers similar to those used on airlines as disposable cocktail glasses. This item has had an excellent reception in hospital feeding and on some airlines.

Soups are packaged for institutional use in boil-in-the-bag pouches in one-quart and half-gallon size. The soups are prepared by heating directly in the bag. Special sauces, such as Newburg sauce, are also frozen and then prepared by the boil-in-the-bag method. Special entrées, such as beef stew with vegetables, macaroni and cheese, and chicken à la king, are also packaged in 4-lb. aluminum containers. The entrée is frozen, stored, heated, and served from the same container, and this is discarded after use.

The packaging requirements for various types of institutions are closely related to the number of people served at a single meal. For example, hospitals like a package of vegetables, or entrées, that will handle from 16 to 25 people, while college cafeterias, serving thousands at a single meal, prefer a package holding 100 servings. Boil-in-the-bag entrées are limited in size by the physical problems of handling a hot bag taken from the boiling water, cutting, and removing the contents. Items such as macaroni and cheese or beef stew are seldom packed more than five pounds per bag.

Aluminum foil pans have the disadvantage that they cannot be used in microwave ovens since aluminum reflects microwave energy. Some frozen foods are packaged for institutional feeding in polyethylene containers, reinforced with polyester-polyethylene ribs, and holding up to five pounds of product. The frozen food, usually a special entrée, may be reheated in boiling water, pressure steamer, or microwave oven. One large freezer uses a "slab pack." This is a relatively simple packaging method wherein the prepared foods are frozen in a mold of any shape, for example, a mold that would fit a steam table pan. After freezing, the solid block is removed from the mold and wrapped in polyethylene film for later use.

Disposable aluminum pans have now been developed for use in the standard steam table sizes used in restaurants and institutions. A variety of frozen foods is packed and frozen directly in the pan; the pans of food are cooked, then heated on the steam table, and served directly therefrom. The kitchen staff need only reheat the frozen prepared foods, serve, and throw away the empty pans. A wide variety of products has been packaged in these pans—frozen vegetables, spaghetti and meatballs, stuffed peppers, casseroles, etc. The products can be packed in a full-size pan measuring 12 × 20 in., or in half-size pans 12 × 10 in., or one-third size pans 12 × 7 in. Some advantages of this type of packaging are: fast freezing and rapid thawing and heating because of the excellent conductivity of aluminum foil; better control; the ability to store unused portions in the heating container; and improved sanitation.

Shrink Packaging

Shrink packaging was introduced by the Cryovac Company in 1948 as a protective package for frozen poultry. The Saran-type film which was first used for this kind of packaging offered excellent protection against freezer burn and loss of weight by desiccation, and many other types of film are also used today for this purpose. In the Cryovac process the whole bird, or a large piece of red meat, is inserted into a plastic bag, air is removed by vacuum so that the bag is drawn in tightly against the bird, the bag is sealed by twisting and clamping, and the film is shrunk into a skin-tight wrap by dipping the package into hot water (about 195° F., 90° C.) for a few moments.

At one time, there was an attempt to use wax for coating meat and poultry products for freezing. However, this process was never entirely acceptable and there are few wax-coated meat or poultry products frozen today.

Tests and Specifications for Packaging Materials and Containers

Packaging is a complex science and the requirements for packaging materials for frozen foods are diverse and sometimes difficult to measure. These requirements might include: tensile strength and elongation, tear, impact strength, stiffness, bursting strength, fold endurance, compression, grease resistance, water-vapor transmission, etc. The various tests have been carefully detailed. Copies may be obtained from: American Society for Testing and Materials, 1916 Race Street, Philadelphia, Pa. 19103, or the Technical Association of the Pulp and Paper Industry, 360 Lexington Avenue, New York, N.Y. 10017. Their publications contain detailed instructions for determining the rate of water-vapor transmission through papers and boards, and through packages at low temperatures.

The test for water-vapor transmission rate (WVTR), for example, requires test dishes, an analytical balance, desiccant, and a test chamber in which the temperature and relative humidity are closely controlled. Moisture in the chamber permeates the film and is picked up by the desiccant. After a measured period of time, the test dish is reweighed and the weight of water-vapor transmitted is calculated. WVTR is significant for selecting packaging material for a product which must be prevented either from drying out or from picking up moisture from the surrounding atmosphere. One must not confuse *water* proofness with *water-vapor* proofness. A packaging material may be water-proof without being water-vapor proof. A water-proof material will hold water and will not go to pieces in water. A water-vapor proof material prevents vapor from passing through it.

While the moisture-vapor barrier characteristics of a packaging material may be excellent, if the package is not properly sealed after it is filled the desirable properties possible with the material may be of little value. A test developed recently measures the total barrier characteristics of a sealed carton and takes only a few days instead of the weeks required by earlier techniques. The test reveals moisture-vapor losses through seams, scores, and unavoidable vents, and can thus lead to structural improvement of a package by identifying points of deficiency.

PACKAGING MACHINERY

The packaging operation can be divided into four major steps: (1) forming the package; (2) filling the package; (3) closing and sealing the package; and (4) placement in the shipping container.

Forming the Package

In case the product is packaged in a metal can, glass or other rigid material, the package is usually formed at some location other than the freezing plant. However, if folding cartons are to be used the carton is conveyed to the packaging area in a flattened form and by use of a forming machine is brought into the desired configuration. This process is accomplished by a wide variety of machines specifically designed for that purpose, all highly specialized as to carton design, size, and method of sealing.

Rigid plastic containers are becoming increasingly useful in the frozen food industry for such specialty items as whipped toppings, fruit-ices, and sliced fruit. These containers are ususally made by plastic converters in special molding machines. Commercial converters use rather expensive and complex machines with enough versatility to handle different plastics and make containers of different shapes and sizes. Such equipment requires skilled operators. Plastic molding equipment is also available specifically for in-plant use by food processors. This equipment is less expensive and is relatively simple to operate because it is designed to form only one particular container from only one type of plastic.

Filling the Package

The product, either blanched and cooled or in the frozen state, is filled into the package manually or by means of machines specifically designed for the product. Whether the product is filled manually, or by means of machines of various degrees of automation largely depends upon two factors: (1) the size, shape, and fragility of the product; and (2) the size of the container. Because of either size, shape, or fragility, blanched and cooled products such as asparagus spears, broccoli, cauliflower, and corn-on-the-cob are filled manually or by means of semiautomatic equipment. Products such as cut corn, cut green beans, Lima beans, peas, peach slices, and strawberries are readily adaptable to automatic filling equipment. Products such as cut corn, cut green beans, lima beans, peas, large containers (30-lb. tins or 55-gal. drums) the operation is usually manual, or at best, semiautomatic.

Product characteristics also dictate the type of feeder used on a filling machine. Important characteristics which must be considered are: (1) density, weight per unit volume; (2) flowability, natural angle of repose, compressibility, and the like; (3) particular size, minimum, and maximum; and (4) fragility of the product to be packaged.

The type of container determines the general category of filling machinery to be used. The size of container is largely dependent upon whether the product is for institutional or for consumer use, and the size and shape characteristics of the product to be packaged. It is interesting to note that when the packaging is being done for the consumer market, the size of container is normally adjusted to the size required to hold some constant weight, say 10 oz., of the product; whereas if the packing is for institutional use, the container is kept at a constant size and the weight is adjusted accordingly.

There are three basic categories of speed in the design of fillers: (1) semiautomatic machines that require an operator to accomplish some part of each filling operation. Machines of this nature are usually simple, inexpensive, and flexible, but are limited to the speed of the operator. Most semiautomatic machines have a maximum capacity of 10 to 20 units a minute. (2) Intermittent-motion machines that are fully automatic and usually fill a group of containers at one time to obtain speeds ranging up to 120 containers a minute. (3) Continuous-motion machines are the ultimate step in high-speed filling. Speed in excess of 400 containers per minute is obtained on various carton and rigid-container machines, depending on the type of measuring system and the number of filling stations provided.

When free-flowing frozen or unfrozen fruits or vegetables (such as peas, corn, or cut beans) are being filled, an end-opening folding carton is usually used. However, for filling corn-on-the-cob, French-cut string beans, broccoli, cauliflower, asparagus spears, and other bulky or fragile items, the top opening folding carton is used. To try to fill products such as broccoli through an end-open carton is not only difficult, but it is not adaptable to high-speed filling equipment.

Filling packages to a prescribed weight is an important packaging operation. If the packages are filled to more than the stated weight there will be economic losses. If the packages are under-filled, the packer may run into difficulty with state and federal regulatory agencies. Some of the causes responsibile for short weight in frozen food packages may be (1) defective filling equipment, (2) loss of moisture transmitted either through the package material or through defective seals, or (3) careless fill-control. The job of filling a package must be done rapidly and accurately. Unfortunately, speed does not correlate well with accuracy in the packaging operation. Various in-line check-weighing devices can be installed directly after the filling and packaging machines to remove mechanically under-filled or over-filled packages. Whether or not automatic check-weighers are used a progressive packer will maintain a quality control system which makes frequent checks of package weights.

In choosing filling equipment, it is important to determine whether the product will be frozen before packaging or after packaging. Therefore, the type of filling device used will depend upon the characteristics of the product to be filled, the size and type of container, and whether the product is to be frozen before or after packaging.

Closing and Sealing the Package

Closing and sealing the package is one of the most important operations in the packaging process. Since most of the frozen food packages for fresh products used today are the Kliklok type of folding carton, where the closed container provides little moisture-vapor barrier protection, it becomes not only desirable, but absolutely necessary that a good overwrap be used. Wax-coated or poly-coated paper is the commonly used material. However, it is important to keep in mind that even though the wrapping material may have excellent barrier characteristics, the seal which one obtains with a wrapping machine is of vital importance. Even the best material, subjected to a poor sealing job, gives, at best, poor barrier protection.

There are several methods for sealing frozen food cartons, depending on whether or not they are over-wrapped. The over-wrapping machine automatically cuts the wrapping material, wraps it around the carton, and heat-seals or glues seams and flaps. The sealing method generally used with the over-wrap is to pass the wrapped carton between two hot bars. Sealing methods used with the single-piece unwrapped cartons are: high temperature air under the flap; hot-melt adhesive; and a combination of low heat, applied directly to the mating surfaces of the carton flaps, with hot air applied as a lubricant to cushion the flap as it slides past the heating surface.

Both primary packages and shipping cartons are frequently sealed with adhesive. Liquid adhesives require a long compression line to give time for the solvent to evaporate and achieve a bond. Hot-melt applicators have become popular on high-speed packaging lines where they have replaced glue pots. There are two basic types of hot-melt applicators—the nozzle and the wheel. The nozzle-type applicator passes a cord-like adhesive from a reel directly to the applying nozzle. The wheel-type applicator picks up adhesive as it rotates and transfers it to the paper surface.

Cans up to the No. 12 size are usually sealed on automatic equipment which produces a double seam. Thirty-pound cans and 55-gal. drums are normally manually friction sealed. Boil-in-the-bag pouches are always heat-sealed.

Placement in Shipping Container

Ideally the product should not be placed in the shipping container until it is frozen. This may be either a manual or an automatic operation. Canned frozen foods up to the gallon size are usually shipped in high-test, corrugated containers, whereas 30-lb. cans and 55-gal. drums are palletized and shipped with no other containerization.

LABELING

Labels for frozen foods usually are of the wrap-around or over-wrap type and are normally printed in full color with a large vignette or picture of the contents on the primary panel. Like other food labels, the emphasis is frequently on appetite appeal.

A frozen food label is the sum total of information presented in written, printed, graphic, and pictorial fashion on the immediate package or overwrap that encloses all or part of the package. A good frozen food label will fulfill legal requirements by providing the consumer with correct information on four fundamentals. These are: (1) the true name of the frozen food; (2) the quantity or amount of frozen food in the package; (3) the name and business address of the packer or distributor; and (4) the ingredient statement for those frozen foods made from more than a single substance. Part of the label, namely the price, is usually stamped on at the retail store.

The front panel of the package, sometimes called the "shoppers' panel," is probably the most important part of the label. It should be clean, bright, and well designed to capture the consumer's attention. A good shoppers' panel will identify the product by its common name, will give information about the style, size, variety, and quality of the product. A legible declaration of the net quantity together with the name and address of the packer or distributor will complete a well designed shoppers' panel.

It is important to realize that, regardless of the brand on the package, or method of distribution, compliance with the provisions of federal, state, and municipal regulations on labeling and packaging is ordinarily the responsibility of the packer of the frozen food product. However, if the frozen food product is distributed by a party other than the company packing it, this fact is declared by an appropriate statement, such as "Packed for (or Distributed by) Doe Frozen Food Company."

The name of the frozen food is declared in an accurate manner by using the common or usual name the consumer normally associates with the product. Legal product names are established by the FDA in published mandatory standards of identity. Only six frozen foods are covered by such standards at this time. These are: frozen concentrated orange juice, frozen orange juice, orange juice for manufacturing, frozen concentrate for lemonade, frozen concentrate for pink lemonade, and frozen raw breaded shrimp. Additional standards for a variety of frozen foods are expected to be established in the near future.

There are federal, state, or municipal regulations, or a combination of these, controlling the labeling of commodities moving in commerce locally, nationally, or internationally. Although the legal aspects of packaging today are complex and constantly changing, the subject is far from inscrutable. The laws have been written for specific areas and usually with very specific product categories in mind. Except for frozen meat, meat products, poultry, and poultry products, the labeling provisions of the Federal Food, Drug, and Cosmetic Act in 1938, as amended, must be met by labels of all frozen food shipped in interstate commerce. In the case of meat products, mandatory label review and approval are required by the USDA for shipment in interstate commerce. For all other frozen foods the Food, Drug, and Cosmetic Act of 1938, the Food Additives Amendment of 1958, The Hazardous Substance Labeling Act of 1960, and The Fair Packaging and Labeling Act of 1966 may apply in part or totally.

The Fair Packaging and Labeling Act is a very important consideration in designing a new label. The Act directs the Secretary of Health, Education, and Welfare, in the case of food, drugs, devices, and cosmetics, to promulgate rules governing the regulation of all such commodities. These are referred to as mandatory regulations. The Act authorizes the promulgation of additional regulations, referred to as discretionary regulations, for particular commodities if the agency finds that such regulations are "necessary to prevent deception of consumers or to facilitate value comparisons as to any consumer commodity."

The regulations for the enforcement of the Federal Food, Drug, and Cosmetic Act under The Fair Packaging and Labeling Act were issued in the Federal Register, Vol. 32, No. 140, dated July 21, 1967. They became effective on July 1, 1968, for all packages introduced to interstate commerce except as to any provisions that may be stayed by the filing of proper objections.

State, county, and city regulations on labeling are generally based on model regulations developed by the National Conference on Weights and Measures. Section 12 of The Fair Packaging and Labeling Act specifically directs that "it is hereby declared that it is the express intent of Congress to supersede any and all laws of the state or political subdivisions thereof insofar as they hereafter provide for the labeling of the next quantity of contents of the packaging of any consumer commodity covered by this Act, which is *less stringent* or *require information different from* the requirements of Section 4 of this Act." In plain language this means that state, county, and city laws which are less stringent or require less information than the Federal Act do not apply. In many cases, local packaging laws have been superseded.

For more detailed information on labeling of frozen food packages, the following sources are suggested: *Program of Labeling Practices for Labeling in the Consumer Benefit,* National Association of Frozen Food Packers, 919 Eighteenth Street, N.W., Washington, D.C. 20026; and *The Guide to Packing Law,* by the editors of *Modern Packaging Encyclopedia,* McGraw-Hill, Inc., 330 West 42nd Street, New York, N.Y. 10036.

We might sum up this chapter very briefly by saying that today's excellent frozen foods deserve—and need—excellent protective packaging at all stages of their production and distribution, so that they will remain excellent right up to the moment of consumption.

PACKAGING REFERENCES

The field of packaging has become so complex and is changing so swiftly, that it is impossible to give a comprehensive survey of this field in a single chapter. There are several excellent reference books, some of which are revised every year. They contain not only reviews of recent developments in packaging, but a list of manufacturers of equipment and wrapping materials. One such reference is *Modern Packaging Encyclopedia,* issued once each year and available from Modern Packaging, McGraw-Hill, Inc. 330 W. 42nd Street, New York, N.Y. 10036; another is the book, *Packaging: A Guide to Information Sources,* available from Gale Research Co., 1400 Brook Tower, Detroit, Mich., 48226.

ADDITIONAL READING

AM. FROZEN FOOD INST. 1971A. Good commercial guidelines of sanitation for the frozen vegetable industry. Tech. Serv. Bull. *71.* American Frozen Food Institute, Washington, D.C.

AM. FROZEN FOOD INST. 1971B. Good commercial guidelines of sanitation for frozen soft filled bakery products. Tech. Serv. Bull. *74.* American Frozen Food Institute, Washington, D.C.

AM. FROZEN FOOD INST. 1971C. Good commercial guidelines of sanitation for the potato product industry. Tech. Serv. Bull. 75. American Frozen Food Institute, Washington, D.C.

AM. FROZEN FOOD INST. 1973. Good commercial guidelines of sanitation for frozen prepared fish and shellfish products. Tech. Serv. Bull. 80. American Frozen Food Institute, Washington, D.C.

ASHRAE. 1974. Refrigeration Applications. American Society of Heating, Refrigeration and Air-Conditioning Engineers, New York.

BAUMAN, H. E. 1974. The HACCP concept and microbiological hazard categories. Food Technol. 28, No. 9, 30–34, 70.

DeFIGUEIREDO, M. P., and SPLITTSTOESSER, D. F. 1976. Food Microbiology: Public Health and Spoilage Aspects. AVI Publishing Co., Westport, Conn.

GRIFFIN, R. C., and SACHAROW, S. 1972. Principles of Package Development. AVI Publishing Co., Westport, Conn.

GUTHRIE, R. K. 1972. Food Sanitation. AVI Publishing Co., Westport, Conn.

HARRIS, R. S., and KARMAS, E. 1975. Nutritional Evaluation of Food Processing, 2nd Edition. AVI Publishing Co., Westport, Conn.

JOHNSON, A. H., and PETERSON, M. S. 1974. Encyclopedia of Food Technology. AVI Publishing, Co. Westport, Conn.

KRAMER, A., and TWIGG, B. A. 1970, 1973. Quality Control for the Food Industry, 3rd Edition. Vol. 1. Fundamentals. Vol. 2. Applications. AVI Publishing Co., Westport, Conn.

MOUNTNEY, G. J. 1976. Poultry Products Technology, 2nd Edition. AVI Publishing Co., Westport, Conn.

PETERSON, M. S., and JOHNSON, A. H. 1977. Encyclopedia of Food Science. AVI Publishing Co., Westport, Conn.

RAPHAEL, H. J., and OLSSON, D. L. 1976. Package Production Management, 2nd Edition. AVI Publishing Co., Westport, Conn.

SACHAROW, S. 1976. Handbook of Package Materials. AVI Publishing Co., Westport, Conn.

SACHAROW, S., and GRIFFIN, R. C. 1970. Food Packaging. AVI Publishing Co., Westport, Conn.

STADELMAN, W. J., and COTTERILL, O. J. 1973. Egg Science and Technology. AVI Publishing Co., Westport, Conn.

TRESSLER, D. K., and SULTAN, W. J. 1975. Food Products Formulary. Vol. 2. Cereal, Baked Goods, Dairy and Egg Products. AVI Publishing Co., Westport, Conn.

TRESSLER, D. K., VAN ARSDEL, W. B., and COPLEY, M. J. 1968. The Freezing Preservation of Foods, 4th Edition. Vol. 2. Factors Affecting Quality in Frozen Foods. Vol 3. Commercial Freezing Operations—Fresh Foods. Vol. 4. Freezing of Precooked and Prepared Foods. AVI Publishing Co., Westport, Conn.

WEISER, H. H., MOUNTNEY, G. J., and GOULD, W. A. 1971. Practical Food Microbiology and Technology, 2nd Edition. AVI Publishing Co., Westport, Conn.

WILLIAMS, E. W. 1976A. Frozen Foods in America. Quick Frozen Foods 17, No. 5, 16–40

WILLIAMS, E. W. 1976B. European frozen food growth. Quick Frozen Foods 17, No. 5, 73–105.

WOOLRICH, W. R., and HALLOWELL, E. R. 1970. Cold and Freezer Storage Manual. AVI Publishing Co., Westport, Conn.

13

Microbiology of Frozen Foods

M. F. Gunderson
and Arthur C. Peterson

A ny discussion of the microbiology of frozen foods must begin with an understanding of the effect of temperature on microbial growth. Each microorganism has an optimum temperature for growth, a maximum temperature permitting growth and, fortunately for food preservation purposes, a minimum temperature below which microbial growth does not occur. Preservation of food by freezing is based on the retardation of microbial growth to the point where decomposition due to microbial action does not occur.

EFFECT OF TEMPERATURE ON MICROBES

In order to provide an effective storage life of many weeks, some foods need to be frozen and then stored substantially below the freezing point of water. These low temperatures are needed to preserve the flavor, odor, color, and texture of the food by retarding chemical changes, by retarding the action of food enzymes, and by eliminating the growth of microorganisms capable of growth near or below 32° F. (0° C.).

Minimum Growth Temperature

In general, as the temperature is lowered to freezing, fewer organisms are capable of growth and the rate of multiplication of these organisms becomes progressively slower. Psychrophilic microorganisms have been described as microorganisms capable of appreciable growth at refrigeration temperatures (32° F., 0° C.). This name implies that the microorganism grows best at cold temperatures, whereas actually most of the microorganisms so designated have high optimum growth temperatures in the range 68° to 99° F. (20° to 37° C.). Most psychrophilic microorganisms may be capable of significant or even rapid growth at refrigeration temperatures, but still not as rapid as at the higher optimum temperature. Most of the microorganisms growing at low temperatures are molds and yeasts, but some bacteria will grow below 32° F. (0° C.). True psychrophilic microorganisms are rare in nature and in foods.

It should be emphasized that the psychrophilic genera *do not include human pathogens,* particularly those characteristic of food-borne illnesses.

Optimum Growth Temperatures

Microorganisms reproduce most rapidly at their optimum growth temperature. For most organisms, this temperature is probably substantially above those encountered by frozen foods, even during unintentional thawing. The thawing temperatures of frozen foods are below the thawing temperature of water and may be as low as 15.2° F. (−9.3° C.). During thawing, the temperature of a frozen food passes from below the freezing point of water to above it. These low temperatures favor first the growth of microorganisms which have low optimum growth temperatures. Unless the food is deliberately thawed by heating, the rise in temperature is usually slow. Only when a frozen food is exposed to the extreme heat of summer for a prolonged time or when the refrigeration equipment malfunctions for a considerable length of time are temperatures above 80° F. (27° C.) ultimately attained.

Effect of Temperature on Microorganisms of Public Health and Sanitary Significance.—Most food-borne pathogens (Salmonella, Staphylococci, *Clostridium botulinum,* and *C. perfringens*) are not psychrophilic.

The temperature relationship for growth of food poisoning and psychrophilic food spoilage organisms is shown in Fig. 13.1, demonstrating that food poisoning organisms do not grow at the usual refrigeration temperatures. Neither growth nor toxin formation has been reported at temperatures below 50° F. (10° C.) for *Clostridium botulinum* Types A, B, C, and D. A development prompts a word of caution. It has been shown that *C. botulinum* Type E. grows and produces toxin at 38° F. (3° C.) after prolonged incubation. The minimum temperature for Staphylococcus growth was shown to be about 40° F. (4° C.).

Whether staphylococci can form toxin at this temperature is still not known. The minimum growth temperature of *Salmonella typhimurium* is 44° F. (7° C.). Th enterococci, particularly those involved in gastrointestinal disorders, have a minimum growth temperature above 50° F. (10° C.). Proper processing, storage, and handling techniques will prevent the multiplication of these food-borne pathogens. The food technologist and the public health sanitarian must be alert, however, to exclude whenever possible, or to minimize to the greatest degree, the presence of pathogens in the product prior to freezing.

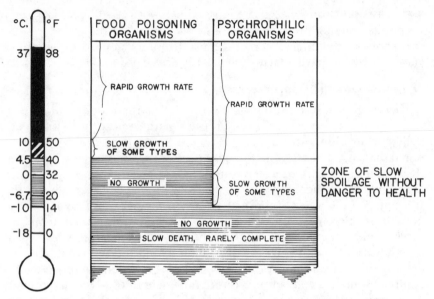

FIG. 13.1. THE TEMPERATURE RANGE OF GROWTH OF FOOD POISONING AND PSYCHROPHILIC
ORGANISMS

Coliforms, particularly *Escherichia coli*, and the enterococci, particularly the fecal streptococci, have been used as indicator organisms for assessing the sanitary quality of foods and their relative probability of freedom from pathogenic organisms from human fecal sources. Most of these indicator organisms can be considered as mesophiles with optimum growth temperatures around 77° to 99° F. (25° to 37° C.). With occasional exceptions among the coliforms, most of the indicator organisms do not grow below 41° F. (5° C.), although their growth has been reported at 32° F. (0° C.) in rare instances.

Because of the lack of specificity of derivation of the coliform organisms, food microbiologists have sought other microbial indices of specific fecal contamination, particularly from human sources. The enterococci were believed to meet this criterion. Unfortunately, they also are ubiquitous, having been widely found in soils, on plants, and in insects. They have been known to become common food processing plant contaminants, so that their presence in a frozen food does not imply direct fecal contamination. The enterococci offer a possible advantage over the coliform organisms as a microbial index of sanitary quality in that they are not subject to the rapid loss in viable numbers upon freezing and storage that the coliform organisms experience.

The frozen food industry has in the coliform count and the enterococcus count useful tools for determining and maintaining sanitary conditions of processing and handling. The results of these examinations, however, must be interpreted in the context of a specific situation. It should not be supposed that a low coliform content necessarily indicates freedom from the presence of pathogens. For example, even a low coliform count of frozen eggs does not indicate that the material is Salmonella-free.

Desirably, all foods should be free of pathogenic microorganisms. *This is not always possible.* Fortunately, man can usually ingest small numbers of some pathogens, for example the Staphylococci and Clostridium spores, without ill effect. The discerning food processor can use the staphylococcal count as one of the indices of the reliability of his processing and sanitation techniques. Frozen peas, beans, and corn should contain only very low numbers of coagulase-positive Staphylococci when the products are prepared under normal commercial practices. The hands of the employees are a major source of the Staphylococci found in these foods. A very high staphylococcal count would strongly indicate contamination, but a low staphylococcal count would not necessarily indicate that the food was free of the possibility of staphylococcal food poisoning.

Until recently, there has been little necessity for determining clostridial species other than *Clostridium botulinum*. It is likely, however, that *C. perfringens* was the cause of some unexplained incidents of food poisoning.

Recent developments in the use of gas-impervious films for food packaging prompt some caution. Frozen foods packaged in such plastic films should be kept frozen until used. It is likely, however, that ballooning and bursting of the pouch would occur in the event of serious thawing of this type of product. This action, due to the growth of putrefactive anaerobes, would occur before botulism toxin would be produced. Recently developed techniques allow quantitative determination of the number of clostridia by selective plating techniques.

Because of the increased occurrence of Salmonella infections in man, and the great significance of contaminated food materials in the transmission of the disease, the determination of viable Salmonella organisms in frozen foods is important. All materials particularly subject to this type of microbial contamination should be examined. These are particularly frozen eggs, egg products, low-acid dairy products, meats, and poultry. Processors (and purchasers where possible) should be assured that the food involved is Salmonella-free. Public health agencies will be increasingly vigilant in testing for food-borne Salmonella organisms.

SPECIFIC FOOD MICROBIOLOGY

Fruit Juices

The growth of fungi in appreciable numbers on the fruit leads to flavor defects in the juice. Molds are likely to appear in fruit juices, *Penicillium expansum,* and *P. digitatum* being the most common on citrus fruits. Monilia, Aspergillus, Alternaria, Clostridium and Fusarium also occur. Processing of juices in the factory, however, usually eliminates fungi from the juices. The low pH of most fruit juices (usually between 3 and 4) excludes most bacteria, especially putrefactive ones. However, development of Clostridia and butyric acid bacteria has been observed in low-acid fruits. Under modern processing conditions, where juices are handled in bulk in the absence of air, yeasts are more favored than other organisms. The thermoresistant molds, which are capable of withstanding the temperatures used in fruit juice pasteurization, are important. These include species of Byssochlamis, Monascus, Paecilomyces and Phialophora.

Bacteria found in noncitrus fruit juices are principally members of the acetic acid and lactic acid bacteria. Pathogenic bacteria found in these juices cannot grow in the medium and generally die off rapidly. A wide variety of yeasts, including those classified as osmophilic, is found in fruit juices. These can develop in sugar-rich juices and on the surfaces of concentrates, and produce a weak alcoholic fermentation. Molds are an important source of quality losses in noncitrus fruit juices. These are principally species of Mucorales, Aspergillaceae, Penicillium and various Fungi Imperfecti (mainly *Hypomycetes*).

The microbiology of the processing equipment becomes very different according to the nature of the fruit processed and is difficult to discuss except in generalities. All of the fruit handling equipment must be frequently cleaned. Water used to move fruit can become highly contaminated; it then presents an obvious hazard if it is recycled. The juice-extraction equipment is a major source of infection for the juice. Pasteurizers can themselves be the source of large numbers of organisms, as was shown for orange juice. Multiple-effect evaporators where the first stages operate continuously, where the temperature is the lowest and the concentration low, can contribute to high counts of Lactobacillus organisms. These have been observed to cause off-flavors in orange juice.

There is a considerable volume of literature on fruit juice and concentrates which demonstrates that the significance of coliform organisms and enterococci in these products is doubtful. In view of a demonstrated lack of health hazards of bacterial origin, the usual bacterial indices of sanitation are not applicable. The lactic acid bacteria constitute the best possible index of processing sanitation for high-quality frozen citrus products. On the other hand, there is evidence that pathogens like Salmonella can survive in the same products for considerable periods of time.

Fruits

Fruit usually carries a heavy microbial load on its surface. Aspergillus, Penicillium, Cladosporium, Fusarium, Alternaria, Rhizopus, Sterigmatocystis, Mucor, Saccharomyces and Torula are the most common genera. Bacteria are also present in significant numbers. These are usually *Bacillus termo, B. subtilis,* and *Staphylococcus aureus.* Systematic information about the nature and number of microorganisms on fruits is generally lacking, but it is known that these vary greatly with the season and climate. Fruits are more subject to fermentation by yeasts and lactobacilli than by other microorganisms because they contain considerable quantities of sugars and a relatively high percentage of organic acids. The fully ripe fruit tissues are soft and easily invaded by yeasts, molds, and bacteria. Heat accelerates the growth of yeasts, molds, and bacteria, and also softens nearly all fruits so that invasion by microorganisms is facilitated. For this reason, fruits should be kept cool and should be processed as rapidly as possible. Thorough washing of the fruit is important; bactericidal or fungicidal substances may be added to the wash water.

Fruits and vegetables should not be held after packing but before freezing for longer than 24 hr. at 40° F. (4.4° C.), 5 hr. at 50° F. (10.0° C.), or 2 hr. at 80° F. (26.7° C.) before freezing. The inhibitory effect of the acidity of fruits and berries helps to maintain good microbiological quality, but, on the other hand, the sugar present helps to protect microorganisms against the lethal effect of freezing.

The processing factors are similar to those for fruit juices. Equipment, such as bins for moving the fruit from fields to the plant, can be a major source of microbial contamination. The processing equipment must be frequently cleaned, because heavy slimes build up rapidly. Similarly, water used for conveying the fruit can be a microbial hazard if recycled.

If fruits, fruit purées, or fruit juices are to be frozen in large containers, e.g., 30- or 50-lb. cans or cartons, or barrels, it is especially important that the product be cooled down to about 50° F. (10° C.) before being placed in the container for freezing. In large containers, freezing is so slow that unless that fruit product has been precooled, microorganisms can multiply enough to cause spoilage even though the temperature of the freezer is considerably below 0° F. (−18° C.).

Vegetables

The inner tissue of unbruised, nondiseased fresh vegetables and fruits has long been considered as sterile or nearly so. However, microorganisms are found in the inner tissues of sound fruits and vegetables. During the preparation for freezing, the outer protective cover of the vegetable is usually removed and the tissues are bruised. The husking of corn and cutting of the kernels from the cob, the removal of peas and beans from the pod, and the snipping and cutting of beans are processes which cause contamination of the inner vegetable tissues with microorganisms of many kinds. This type of microbial contamination renders the original microflora of the vegetable relatively unimportant. The adhesiveness of corn juice to processing equipment causes special sanitation problems in the processing of that vegetable. In most processing plants, conveyer belts are a chief source of microbial contamination, and without continuous cleaning counts on these belts can reach into the millions of organisms per square inch. Vegetable tissues and juices are excellent media for the growth of microorganisms. Consequently the prepared vegetables should be blanched, cooled, and frozen immediately after preparation, or else be immediately refrigerated. During blanching, vegetables are usually heated to nearly 212° F. (100° C.) and nearly all of the vegetative microorganisms on them are killed. Relatively few spores are destroyed, however. Because blanching makes the vegetables more easily infected, blanched vegetables should be cooled promptly to below 60° F. (16° C.) to prevent rapid bacterial growth during the remainder of the preparation and packaging process.

The equipment for handling vegetables is difficult to keep free of microorganisms. Likewise, the water flow system can be an important source of microbial contamination. For this reason, the recycling of processing water in contact with vegetables needs strict control. Following blanching, the number of bacteria on the blanched vegetables increases considerably during the subsequent handling. The numbers of bacteria on peas during various stages of processing immediately following blanching (Table 13.1) suggest it should be possible to keep the numbers of bacteria on frozen vegetables below 100,000 per gram, provided that the processing plant is kept scrupulously clean. Lack of practical means of controlling microbial growth was the principal factor in the heavy contamination of these vegetables. Total counts in excess of 200,000 organisms per gram are common on frozen green beans and in excess of 100,000 organisms per gram on peas and corn. With large vegetables which have irregular shapes, such as broccoli, cauliflower, and asparagus, it is difficult to keep the bacterial count down, and this may easily exceed 100,000 organisms per gram.

The bacterial flora of frozen vegetables usually consists of members of the following genera: Bacillus, Aerobacter, Erwinia, Flavobacterium, Achromobacter, Alcaligenes, Cellulomonas, Chromobacterium, Strep-

TABLE 13.1

EFFECT OF PROCESSING ON BACTERIAL COUNT OF PEAS AT VARIOUS STAGES OF PROCESSING

Point of Sampling	Thousands of Bacteria per Gram of Peas
Platform	11,346
After washing	1,090
After blanching	10
End of flume	239
End of inspection belt	410
Entrance to freezer	736
After freezing	560

tococcus, Lactobacillus, Leuconostoc, Micrococcus, Neisseria, Pseudomonas, Mycobacterium, Phytomonas, Sarcina, Staphylococcus, Serratia and Vibrio. From frozen peas, beans, and corn, 40 to 70% of the isolates are species of Leuconostoc and Streptococcus. A characteristic microflora, which is dependent on the type of vegetable being handled, develops on processing equipment in several different processing plants.

Meats

The bodies of meat animals are not sterile. The microbial content in and on carcasses may be due to slaughtering practices, infection, and air-borne contamination. Because of the contamination which occurs during the removal of skin from the carcass, the subsequent handling, and the presence of air, bacteria and molds grow rapidly on the surface of the meat and soon make it unfit for consumption unless it is kept at 32° F. (0° C.) or below. The higher the humidity, the more rapid is the growth of microorganisms on the surface. After slaughter, beef is ordinarily rapidly chilled to about 35° F. (1.7° C.) to 38° F. (3.3° C.). Then the whole carcasses, half-carcasses, or quarters are aged for periods varying from 5 days to 6 weeks. The length of the aging period depends upon the grade of beef, the tenderness desired, and the practice of the packing or locker plant. If the meat is aged at too high a temperature, or humidity, or for longer than about three weeks at 35° F. (2° C.), the surface of the meat becomes moldy and sometimes slimy. It is then necessary to trim off the exterior before packaging and freezing the meat.

The principal genera of bacteria on meats held at low temperatures are Achromobacter and Pseudomonas. Bacillus, Lactobacillus, Serratia, Flavobacterium, Chromobacterium, and Staphylococcus are sometimes found in surface slimes. Molds appear on meats primarily as a result of long holding periods and high humidities maintained in the storage rooms. They may also appear as the result of poor sanitary practices. Aspergillus, Thammidium, Alternaria, Monilia, Rhizopus, Penicillium, and Mucor are genera of fungi associated with meats. Contamination of beef ranges from 100 to 100,000 organisms per gram while pork varies from 5,000 to 1,000,000 organisms per gram. When meat is held near 34° F. (2° C.), microbial growth is, of course, limited to organisms capable of growth at that temperature. An example of the importance of low-temperature organisms in the quality of fresh meats is seen in the characteristic flavor of aged beef, due to the growth of microorganisms on the cut surfaces. Some believe satisfactory organoleptic characteristics will not develop without microbial aid.

Glandular organs such as kidneys, livers, sweetbreads, etc., are more susceptible to spoilage by microorganisms than is muscle meat (e.g., steaks and roasts). These meats are never aged, but are either sold in a few days or are frozen.

Veal, pork, and lamb are chilled in much the same way as beef carcasses. Rapid chilling is essential so that microorganisms will not build up a appreciable numbers. Since these meats are relatively tender, there is no need for aging. Mutton is improved in flavor and tenderness by aging.

Ground meat (e.g., hamburger) often becomes badly contaminated during and subsequent to grinding, and cannot be kept for long at temperatures above freezing without danger of spoilage. Very high counts are common in ground meats and are attributable to the tremendous increase in surface of the meat. Sharp decreases have been observed in the microbial content of meats following freezing. Quick freezing of hamburger steak reduces the average bacterial count per gram. Studies have also shown, however, that some organisms survive even extended frozen storage.

Poultry

The microflora of eviscerated poultry is largely derived from the feathers, feet, feces, and skin. The procedures of killing and bleeding affect the quality of the carcass, but have little influence on its microbiology. The picking of feathers is mostly done by the semiscald procedure, in which the carcass is immersed in 125° to 135° F. (52° to 57° C.) water for 30 to 60 sec. The scald water is not an important source of additional microbial contamination provided that it is changed continuously. The greatest decrease in the various microbial groups comprising the load on the skin occurs as a result of scalding and picking. The picking process does affect the keeping quality of the bird. The microbial load on the skin is usually less than 250,000 organisms per square centimeter at this point, and the count in the flesh is much lower. Evisceration of the bird adds to the surface microbial load of the carcass. Much of the surface contamination, both microbial and debris, can be removed by efficient washing. The carcasses should be promptly and rapidly cooled to about 35° F. (1.7° C.)

following dressing and drawing. The number of bacteria on poultry meats is generally a good indication of the sanitary condition of the processing plant. The organisms most frequently found on fresh carcasses and as common contaminants in poultry processing plants are species of Pseudomonas, Achromobacter, Alcaligenes, Micrococcus, Flavobacterium, and coliforms.

Cutting up and boning contribute heavily to the contamination of the meat. Cooking the carcass before boning greatly reduces the microbial flora, but this always increases during subsequent handling. Freezing only partly eliminates the surface bacteria of eviscerated poultry, even on long storage. Species of Pseudomonas and Alcaligenes are always associated with off-odor and slimy chicken, and in tests these organisms multiplied appreciably at 32° F. (0° C.). Microbial populations reached values of approximately 10^8 per square centimeter before off-odor and sliming were observed.

Eggs and Egg Products

The microbial flora of frozen egg products is important because it affects the functional properties of the egg product, greatly affects the keeping qualities of the egg products, and affects the safety and wholesomeness of other foods into which the egg products may be incorporated. Most freshly laid eggs are bacteriologically sterile; however, the bacterial flora of the shell affects the bacterial count of the product. Much of the dirt and microbial load can be removed by proper washing and drying of eggs. Detergent sanitizers frequently are used in the wash water. In breaking, very close control is necessary because a single bad egg can contribute billions of bacteria and spoil the entire batch. The sanitation of the breaking areas is of great importance in determining the microbial load of liquid egg. The liquid egg should be refrigerated quickly because bacteria can grow rapidly when the natural protective mechanisms of the albumen are overcome by mixing with the yolk. The total count of commercial frozen whole egg is apt to be high, ranging usually from 500,000 to 10,000,000 organisms per gram. Pseudomonas, Alcaligenes, Flavobacterium, Proteus, Acrobacter, and Escherichia are the predominating bacterial genera found in unpasteurized liquid eggs.

Pasteurization is recommended for liquid egg and egg products. This procedure, when properly applied, will result in the elimination of 92 to 99% of the total viable count, including coliforms and pathogenic bacteria. This process is particularly important for the elimination of viable Salmonella organisms. Care must be exercised to avoid postpasteurization contamination, however.

The freezing and storage of liquid eggs and egg products results in a rapid reduction in the viable count, but coliform, enterococcus, and Salmonella organisms can survive for considerable periods of time. Frozen eggs should be defrosted rapidly to prevent bacterial multiplication.

Seafoods

Fish, shellfish, and other seafoods are the most susceptible of all of the flesh foods to microbial spoilage. These seafoods have a natural indigenous psychrophilic microflora. The conditions on the fishing boat are an important factor in determining the microbial load of the seafood. Wood in contact with seafoods becomes heavily impregnated with microorganisms, and these soon become impossible to eliminate even with chemical sanitizers. When the boat operates far from the processing plant, preservation measures must be applied on the boat. Even though fresh fish are packed in ice, bacteria grow on them. These bacteria are generally members of the Pseudomonas-Achromobacter genera and are usually the cause of spoilage. If fish are allowed to become warm spoilage occurs quickly. Much of the microbial contamination of seafoods is external and is removed by washing off slime and dirt in the preparative procedures. When applied at sea, the use of antibiotics as a primary preservative treatment for seafoods has been restricted by the tendency for antibiotic-resistant strains of pseudomonads to develop; these become the dominant microflora and cause spoilage despite antibiotic treatment.

Shellfish are even more subject to bacterial decomposition than are fish. Because fish and shellfish spoil so easily, it is very important that they be handled in such a way as to keep microbial contamination to a minimum. Prompt handling in the processing plant, with rapid freezing, is essential in preserving quality. All processing equipment which comes into direct contact with fish and shellfish should be sterilized frequently. Sterilization can be accomplished by means of a germicidal solution, such as a hypochlorite. However, the sanitizer must not impart an objectionable flavor to the product. Chlorinated water can be used to keep the bacterial count down, but it may detract from the keeping qualities of the frozen fish; a "salt-fishy" odor and flavor may develop during subsequent cold storage sooner than it would have if chlorine had not been used.

Fabrication procedures, such as filleting and breading, contribute significantly to increases in the total aerobic count and in coliform count, and require close microbiological control. Microorganisms do not cause spoilage difficulties at the temperatures commonly used for the storage of frozen foods. However, they can cause deterioration during freezing (especially if the foods have been packed in large containers without prechilling) and also after thawing. The freezing process reduces the number of viable organisms considerably, but significant numbers survive for considerable periods of time. Freezing and thawing have been observed to exert selective lethal effects on the microflora of seafoods. In review of the role of sanitation in seafood processing, except for shellfish, seafoods have not caused public health problems in western countries.

Dairy Products

Good sanitation and the maintenance of low temperatures are the primary tools for preserving the quality of dairy products. Total quality of these foods is so dependent on their microbiology that the effect of holding microbial populations in check can hardly be overemphasized. Modern milking practices and improved sanitation, coupled with bulk-tank handling on the farm using mechanical refrigeration, have routinely produced raw milk with a microbial population so low as to be previously attainable only by exceptional producers.

Processing-plant sanitation is equally important to the production of dairy products with low levels of microbial contamination. Pasteurization of milk, cream, and ice cream mixes provides effective control of microorganisms. Postpasteurization contamination is the usual cause of quality defects of microbial origin. The best ice cream and related products are made only from sweet unneutralized cream. The microbiological quality of the additives is extremely important. Some of those which are added after the mix is pasteurized, such as nuts and fruits, can be important sources of microbial contamination.

Members of the genera Pseudomonas, Proteus, Alcaligenes, Aerobacter, Achromobacter, and certain coliforms are the psychrophilic organisms associated with defects in dairy products. In frozen milk, the normal microflora die off rapidly. Similarly, the microbial content of cream declines during frozen storage. While the bacterial content of butter decreases during frozen storage, a considerable number of microorganisms remains because of the protective action of the fat.

Parasites.—The interaction of time of exposure and freezing temperature on the elimination of *Trichinella spiralis* larvae from pork tissues is particularly important. Although when pork is frozen and refrigerated to 0° F. (−18° C.) all of the *Trichinella* organisms are destroyed, the USDA Meat Inspection Program recommends a holding period as follows:

At any stage of preparation and after preparatory chilling to a temperature of not above 40° F. (4.4° C.) or preparatory freezing, all parts of the muscle tissue of pork or product containing such tissue shall be subjected continuously to a temperature not higher than one of those specified in Table 13.2, the duration of such refrigeration at the specified temperature being dependent on the thickness of the meat and the inside dimensions of the container.

TABLE 13.2

REQUIRED PERIOD OF FREEZING AT TEMPERATURE INDICATED TO CAUSE DEATH OF *SPIRALIS TRICHINELLA*

| Temperature | | Group 1 | Group 2 |
°F.	°C.	Days	Days
5	−15	20	30
−10	−23	10	20
−20	−29	6	12

Source: USDA, Meat Inspection Program.

Group 1 comprises meat or product in separate pieces not exceeding 6 in. in thickness, or arranged on separate racks with the layers not exceeding 6 in. in depth, or stored in crates or open boxes not exceeding 6 in. in depth, or stored as solidly frozen blocks not exceeding 6 in. in thickness.

Group 2 comprises meat or product in pieces, layers, or within containers, the thickness of which exceeds 6 in. but not 27 in. Such containers include tierces, barrels, kegs, and cartons having an inside diameter not exceeding 27 in.

The meat or product undergoing such refrigeration or the containers thereof shall be so spaced while in the freezer as to insure a free circulation of air between the pieces of meat, layers, blocks, boxes, barrels, and tierces in order that the temperature of the meat throughout will be promptly reduced to not higher than 5° F. (−15° C.), −10° F. (−23° C.), or −20° F. (−29° C.), as the case may be.

MICROBIOLOGY OF PRECOOKED FOODS

Like all other frozen foods, precooked frozen foods are preserved by freezing storage at temperatures below the minimum for microbial growth or action. Precooked frozen foods differ from other frozen foods in several important respects. All of the precooked frozen foods have received heat treatments in their processing which have materially and significantly modified the microbial flora of the ingredients and hence of the final product itself. But, generally there is no terminal sterilization. More significantly, these foods may be eaten without further heat treatment or may receive heat treatments which range from warming to pasteurizing effects.

For the purposes of this discussion, all precooked and prefabricated (prepared) frozen foods which are not simply a commodity type of food will be considered. These foods range from raw or partially cooked doughs, rolls, breads, breaded seafoods, boil-in-the-bag uncooked vegetables with fabricated sauces, to prepared dinners, entrées and some boil-in-the-bag entrée items which only require warming, to some cakes and cream pies which need only to be defrosted. Some precooked foods are completely cooked during processing, for example, some frozen soups. Other foods are mixtures of raw fabricated components and completely cooked ones like the meat pot pies and frozen dinners which may require more than warming or pasteurizing heat treatment by the consumer to cook the raw portion.

In all of these precooked and prefabricated foods, the potential lurks, even if only very vaguely, for food-borne infections or intoxications. This threat remains vaguely potential, however, only with the continuous, conscientious, knowledgeable efforts by the processor. The efforts by the processor must begin with ingredients of excellent quality, especially microbiologically, carefully processed under conditions of excellent sanitation and temperature control. They proceed through proper packaging and expeditious freezing to 0° F. (−18° C.) or below and storage under the same condition. The story does not end there, however, because these foods should be warehoused, transported, distributed, retailed, and even stored by the consumer without having been thawed and having been preferentially maintained at 0° F. (−18° C.) or lower. Such temperatures are necessary to maintain the original quality of the product as produced by the processor.

Frozen food processors have long realized that the freezing process cannot sterilize, pasteurize, or improve the microbiological quality of the product. They have likewise recognized that adverse temperature effects on product quality due to thawing cannot be reversed by refreezing. These facts need to be emphasized to all concerned with frozen foods, and especially to the consumer.

Precooking a food before freezing does not necessarily render it free from pathogenic microorganisms. In those foods which do not receive terminal sterilizing heat treatment, there is, as mentioned, a recognized potential hazard of contamination with microbial pathogens and their growth or toxin production in the food (a fortunately not materialized hazard). This serious health hazard is least likely to develop during processing. It could occur as a result of thawing at some point in the distribution chain but would most likely occur as a result of mishandling by the consumer. The frozen food industry's educational program, concerning the care and protection which frozen foods require, has been directed through the distribution chain, but processors should recognize their responsibility to disseminate further and amplify this information at the retailing and consumer levels.

A subcommittee on Milk and Food Protection of the Committee on Environmental Health Problems (PHS) observed that "technological changes are occurring so rapidly in the food field that their public health implications are not receiving proper attention by either government or industry." Similar views were expressed by an international committee which recommended microbiological standards for frozen foods, particularly for those precooked. The concern for microbiological health problems associated with precooked frozen foods has been well summarized in a report of The Food Protection Subcommittee of the National Academy of Sciences as follows: "because of the great amount of food produced in a short time in such processing operations and because of the rapid and widespread distribution of food products, a very large population is at risk in the event of a malfunction or error in the food process." This latter fact, that large populations may be endangered, is the crux of the entire matter of the public health aspects of precooked frozen foods.

For these reasons, considerable efforts have been made by various agencies, especially those charged with responsibilities for public health, to establish legal microbiological standards for precooked frozen foods. Such standards would be less stringent only than those for market milk, with which a frequent but fallacious parallel is drawn. The problem of microbiological standards for precooked frozen foods, however, is even more complex than the problem for fresh frozen foods which was discussed previously. Without again delving deeply into the problem, the implications of such standards are most important and extend far beyond the selection of any particular numerical microbial index. These include especially the actual or tacit assumption of responsibility by the processor for his product from the time of manufacture to actual consumption by the purchaser. This would be true even when the processor no longer has actual control of his product during its passage through many hands in the distribution chain, and even when the processor usually does not even have legal ownership of the food. Even the presence on the package of a reliable thawing indicator which effectively integrates the effect of exposure to various times at thawing temperatures as a true picture of quality changes in the product cannot resolve the larger issue.

Microbiological Examination of Precooked Foods

The problems of microbiological analysis of precooked frozen foods are many. Perhaps most serious of these are the nonhomogeneity of samples from a given production run and even of parts of the same sample. The problem of obtaining replication of counts within the same laboratory as well as between different laboratories is much more severe than for milk.

Many have studied the applicability of various microbiological assay techniques to the examination of precooked frozen foods. The most useful and authoritative compilation of procedures is given in the second edition of *Recommended Methods for the Microbiological Examination of Foods* (APHA).

Total Aerobic Count.—Most prominent in discussions of the microbiology of frozen foods, especially precooked ones, is the value of the total aerobic count. Microbiologists are generally agreed, however, that most foods involved in food poisoning and food-borne infections were those with large microbial populations and that the total count was not a reflection of either the acceptability or the safety of frozen foods. The total count is a good measure of sanitation, however. These workers noted that the total count was a summation of many kinds of contamination and the time and temperature at which the ingredients were held (Table 13.3).

Frozen pot pies have been extensively investigated. For example, 75% of 95 samples of frozen pot pies produced in 5 different plants had total counts less than 50,000 organisms per gram, while 16% had counts in excess of 100,000 bacteria per gram. Similar data show the total aerobic count of 60 frozen beef pie samples was between 1,000 and 13,000 bacteria per gram, with an average of 3,000 bacteria per gram. The bacterial counts of frozen precooked turkey meat pies, shown in Table 13.4, were part of the Delaware Valley Survey carried out by the National Association of Frozen Food Packers. Thirty-seven percent of all brands of turkey pies had total aerobic counts less than 1,000 organisms per gram. Of the samples examined, 64% were under 5,000; 87% were under 50,000 and 90% were under 100,000 bacteria per gram.

The total aerobic count of frozen prepared dinners in good commercial practice should be low as those shown in Table 13.5. An examination of precooked frozen meals prepared for the military found that 86% of the samples had total counts less than 50,000 organisms per gram. They concluded that a total count standard of less than 100,000 bacteria per gram should be easily met by any producer. These authors stressed the importance of sanitary supervision in the quality control program and the importance of starting with good quality raw materials to insure a finished product with a low bacterial count.

Macaroni and cheese frozen prepared dinners had counts ranging from 880 to 63,000 per gram. Line survey data for the same dinners illustrated that the cheddar cheese topping was the principal source of the total aerobic count, coliform, and Staphylococcus organisms found in the finished product. It was found that 76% of 195 samples of precooked frozen foods had total aerobic counts of less than 100,000 bacteria per gram; generally poultry products appeared to be of poorer microbiological quality than other precooked foods.

An extensive bacteriological survey of the frozen precooked foods

TABLE 13.3

TOTAL PLATE COUNTS ON FROZEN CHICKEN PIES

Production Line (Week of January 6 to January 13)

Time	Jan. 6	Jan. 9	Jan. 10	Jan. 11	Jan. 12	Jan. 13
7:00 A.M.	1,500	7,000	23,000	1,400	14,000	2,300
8:00 A.M.	5,500	4,000	16,000	3,800	4,000	11,000
9:00 A.M.	3,400	6,500	21,000	9,500	9,200	6,900
10:00 A.M.	13,000	7,200	6,200	12,000	16,000	5,100
11:00 A.M.	21,000	12,000	20,000	4,600	5,200	1,300
12:00 P.M.	4,100	11,000	11,000	8,800	3,500	2,600
1:00 P.M.	18,000	9,000	17,000	13,000	3,900	2,200
2:00 P.M.	15,000	3,200	11,000	9,700	7,500	29,000
3:00 P.M.	19,000	25,000	2,900	2,100	120,000[1]	19,000
4:00 P.M.	12,000	27,000	13,000	1,400	4,500	19,000
5:00 P.M.	3,600	22,000	5,500	11,000	25,000	3,500
6:00 P.M.	3,000	16,000	200,000[1]	11,000	54,000	6,600
7:00 P.M.	2,200	33,000	17,000	3,200	65,000	4,200
8:00 P.M.	2,900	8,600	11,000	24,000	13,000	18,000
9:00 P.M.	2,100	6,900	7,200	20,000	63,000	6,600
10:00 P.M.	2,700	6,300	8,400	41,000	23,000	30,000
11:00 P.M.	5,700	3,200	5,800	11,000	1,200	16,000
Daily Average	7,900	12,000	23,000	11,000	25,000	11,000

[1] Such counts are subject to rechecks and complete bacteriological studies.

TABLE 13.4

PERCENTAGE OF THE TOTAL NUMBER OF ORGANISMS, COLIFORMS AND STAPHYLOCOCCI PER GRAM OF RETAIL SAMPLES OF TURKEY PIES BELOW SPECIFIC LEVELS FOR EACH BRAND AND FOR ALL BRANDS

Brand	Percentage of Samples with Total Numbers of Organisms per Gram of Sample below						
	1,000	5,000	10,000	50,000	100,000	500,000	1,000,000
A	4.17	12.50	29.17	54.17	70.83	83.33	100.00
B	29.17	79.17	87.50	100.00	100.00	100.00	100.00
C	58.33	91.67	100.00	100.00	100.00	100.00	100.00
D	100.00	100.00	100.00	100.00	100.00	100.00	100.00
E	25.00	75.00	91.67	100.00	100.00	100.00	100.00
F	0.00	12.50	45.83	91.67	95.83	95.83	95.83
G	8.33	16.67	16.67	16.67	25.00	37.50	41.67
H	12.50	62.50	79.17	100.00	100.00	100.00	100.00
J	25.00	75.00	95.83	100.00	100.00	100.00	100.00
K	0.00	58.33	79.17	95.83	100.00	100.00	100.00
L	91.67	95.83	95.83	100.00	100.00	100.00	100.00
M	0.00	37.50	45.83	66.67	79.17	83.33	87.50
N	95.83	100.00	100.00	100.00	100.00	100.00	100.00
All Brands	34.62	62.82	74.36	86.54	90.06	92.31	94.23

	Percentage of Samples with Coliforms per Gram of Sample below					
	10	50	100	500	1,000	5,000
A	54.17	66.67	70.83	83.33	87.50	87.50
B	100.00	100.00	100.00	100.00	100.00	100.00
C	95.83	95.83	100.00	100.00	100.00	100.00
D	95.83	95.83	100.00	100.00	100.00	100.00
E	100.00	100.00	100.00	100.00	100.00	100.00
F	75.00	79.17	83.33	95.83	95.83	95.83
G	45.83	62.50	66.67	83.33	83.33	100.00
H	100.00	100.00	100.00	100.00	100.00	100.00
J	95.83	100.00	100.00	100.00	100.00	100.00
K	91.67	100.00	100.00	100.00	100.00	100.00
L	100.00	100.00	100.00	100.00	100.00	100.00
M	79.17	87.50	87.50	100.00	100.00	100.00
N	100.00	100.00	100.00	100.00	100.00	100.00
All Brands	87.18	91.35	92.95	97.12	97.44	98.72

	Percentage of Samples with Staphylococci per Gram of Sample below[1]						
	10	50	100	500	1,000	5,000	10,000
A	4.17	4.17	8.33	33.33	45.83	62.50	79.17
B	37.50	62.50	66.67	91.67	100.00	100.00	100.00
C	41.67	87.50	91.67	100.00	100.00	100.00	100.00
D	95.83	100.00	100.00	100.00	100.00	100.00	100.00
E	50.00	75.00	91.67	100.00	100.00	100.00	100.00
F	12.50	16.67	16.67	29.17	37.50	87.50	91.67
G	8.33	12.50	12.50	20.83	29.17	45.83	45.83
H	54.17	79.17	83.33	100.00	100.00	100.00	100.00
J	66.67	95.83	95.83	100.00	100.00	100.00	100.00
K	4.17	16.67	29.17	70.83	83.33	100.00	100.00
L	87.50	91.67	91.67	100.00	100.00	100.00	100.00
M	20.83	45.83	50.00	91.67	91.67	100.00	100.00
N	66.67	100.00	100.00	100.00	100.00	100.00	100.00
All Brands	42.63	60.58	64.42	79.81	83.65	91.87	93.59

[1] Each brand represents seven different sample series, of three pies each, obtained from seven different retail outlets for a total of 21 samples per brand.

TABLE 13.5

BACTERIAL COUNTS—FROZEN TURKEY DINNERS

		Production Line (Week of January 5 to January 12)					
Time	Component	Jan. 5	Jan. 6	Jan. 9	Jan. 10	Jan. 11	Jan. 12
7:00 A.M.	Meat	1,000	2,100	1,000	1,500	1,600	15,000
	Gravy	1,400	1,000	1,000	1,000	4,200	1,200
	Dressing	1,000	1,000	3,200	6,700	2,500	32,000
	Peas	4,200	55,000	12,000	15,000	2,700	23,000
	Potatoes	2,200	6,000	4,900	6,100	2,800	12,900
9:00 A.M.	Meat	1,000	5,200	8,700	3,100	1,000	1,800
	Gravy	1,000	1,000	1,000	1,000	1,000	1,000
	Dressing	1,000	3,300	1,400	7,100	1,900	4,600
	Peas	1,000	3,800	26,000	42,000	6,000	7,200
	Potatoes	2,100	2,700	15,000	3,300	2,300	3,400
1:00 P.M.	Meat	10,000	1,300	2,800	1,900	1,100	2,200
	Gravy	1,700	1,000	1,200	1,300	1,000	1,000
	Dressing	2,500	4,400	6,900	9,100	1,400	24,000
	Peas	2,600	15,000	4,900	8,400	3,400	22,000
	Potatoes	3,700	11,000	3,400	8,300	2,600	33,000
3:00 P.M.	Meat	1,000	1,500	3,600	3,300	1,200	26,000
	Gravy	1,000	1,300	1,000	1,000	1,000	36,000
	Dressing	2,400	4,300	4,800	4,400	3,200	27,200
	Peas	1,000	2,500	11,000	16,000	1,000	22,000
	Potatoes	1,000	7,100	5,200	8,400	3,500	27,000

industry was carried out by the FDA. They studied 81 products with over 3,000 samples, but did not include any frozen dinners. The products were classifed into four groups based on the degree of cooking which the product might get in the home and on the amount of cooking received by the product during manufacture. Group I, comprised of finished products which do not receive additional heat treatment by the consumer, was bakery products. Of these, 84% had total aerobic counts of less than 100,000 bacteria per gram. Group II items were primarily main course items like meat knishes, and macaroni and cheese which are cooked early in their manufacture and only warmed by the consumer. Seventy-five percent of Group II products had less than 100,000 bacteria per gram, but 11% contained more than 1,000,000 bacteria per gram. Group III products were cooked late in the processing, but again only required warming by the consumer and included such items as chop suey, creamed chicken, fish cakes, and oyster stew. Of these, 83% had total aerobic plate counts less than 100,000 bacteria per gram while 8% had total counts in excess of 1,000,000 bacteria per gram. Group IV products were those which required cooking by the consumer and included such items as pot pies, pizza, raw breaded shrimp, and spinach loaf. Of the products in this group, 58% contained less than 100,000 bacteria per gram. The authors concluded that the total aerobic plate count and the coliform count varied with the product and production processes. It was noted that these counts served as rough guides to plant sanitation.

About 70% of 192 assorted samples of frozen precooked pot pies, dinners, seafoods, and entrée items had total aerobic counts less than 100,000 bacteria per gram. In examining precooked frozen foods such as pizza pie, high total counts should be expected because of the cheese topping which incorporates starter culture microorganisms in the manufacture of the cheese. Such foods cannot be included in the group to which restrictive total count bacterial standards might apply. The frozen food industry undertook a comprehensive microbiological survey of its products, particularly of frozen pot pies and precooked frozen dinners. From those restults, the National Association of Frozen Food Packers concluded that such products should have total aerobic bacterial counts of less than 100,000 bacteria per gram.

It can be said that for many prepared and precooked frozen foods, the processor should be able to provide products with low total microbial counts. Each product type must be considered, however, with consideration given to the microbiological quality of the available ingredients and the extent of the heat treatment which the product receives during processing. High total counts are to be expected in some precooked foods which incorporate microbially inoculated foods like cheese in their manufacture. The total aerobic plate count can be useful in conjunction with other microbial indices as a measure of sanitation and care of handling of ingredients. Processors of precooked frozen foods must be aware of the fact that present concepts of the public health microbiology of these foods dictate not only that such foods be safe, but that they have been prepared under esthetically satisfactory conditions.

An important "built in" aspect of the public health safety of frozen prepared foods is the action of the natural saprophytic, psychrotrophic microflora of these foods in destroying the organoleptic acceptability of these foods before the development of hazardous pathogenic microbial populations.

It should be observed that this mechanism cannot protect the consumer against the presence of infectious pathogenic microorganisms which might be present in products which have never been thawed. Even legal standards using the usual microbial index counts will not guarantee freedom from this danger, however.

Sources of Microbial Contamination

Facilities.—The production of precooked or prepared frozen foods is a maximum sanitation operation. It cannot be effectively implemented in the immediate proximity of a dirty operation like the killing and eviscerating of poultry or of the peeling of vegetables. Ventilation must be properly arranged so that air from highly contaminated areas is not circulated to maximum sanitation areas. Even aerosols set up by improperly functioning floor drains have been shown to be an important source of coliform bacteria. Separate locker rooms and washup facilities should be provided for personnel preparing precooked frozen foods. These areas should be apart from those of other workers.

People.—It is desirable to train all plant personnel in the concepts of sanitation and personal hygiene in order to ensure sanitary conditions in the prepared foods areas. They must also be instilled with a sense of responsibility for the quality of the product. Clean uniforms, head covers, gloves, and the necessary essential hand tools provided by the processor are an excellent practical demonstration to the workers that the processor is interested in the personal appearance and hygiene of the workers. Workers who prepare precooked frozen foods need constant supervision and reminding of the necessity for washing the hands with germicidal soap before beginning work and before returning to work after every break. They need to be taught to keep their hands out of the ingredients and product unless handling is absolutely required. The hands of workers can be an important source of microbial contamination in the slicing, weighing, and placing of meats. Workers on these jobs need to wash their hands several times during the production shift. Workers must avoid resting pots, pans, tools, and other equipment on top of foods. Mechanics must wash their hands before working on equipment which contacts food. They must avoid resting the parts of such equipment on foods. Equipment contaminated by handling during repair or modification should be sanitized and rinsed before use. Workers must be made aware that stirring oars, ladles, thermometers, gage sticks, sanitary fittings, and gaskets cannot be allowed to rest on the floor or in their pockets between uses. They must be taught the importance of their part in keeping the foods uncontaminated.

The processor must realize that optimum results will be obtained with a stable, well-trained work force accustomed to performing the same tasks in an approved manner and who understand the principles of sanitation and personal hygiene. For this reason, it is very desirable to avoid transferring workers from other less fastidious operations to precooked frozen food production on a day to day basis, as production requirements vary, and labor force requirements might seemingly dictate.

Equipment.—There is a need for and an excellent opportunity for significant improvements in the microbiological aspects of the development of processing equipment. It should be designed with sanitary standards of the food and ease of cleaning in mind. Too often, equipment is adapted for use to a product or process for which it is neither suitable, sanitizable, or really economical. A meeting of minds between equipment design engineers and food sanitarians similar to that which has occurred in the dairy industry is needed for the entire food industry.

Particular attention must be directed to comminuters, colloid mills, homogenizers, plate coolers, pumps, slicers, dicers, and filling machines in order to be sure that these are regularly disassembled, washed (brushed), sanitized, and rinsed. Many of these pieces are difficult and laborious to disassemble, and production and cleanup workers will attempt to clean them by water flushing them only. These pieces cannot be satisfactorily cleaned in this manner. Plants and equipment for precooked frozen food production must be better than just "eyeball clean." Quality control personnel must periodically establish that equipment is satisfactorily clean on the basis of microbiological examination based on swab samples and contact plates. If serious microbiological problems are encountered, a line survey involving the ingredients and product before and after each step of processing and of every piece of equipment contacting the food may be needed. Because of the viscous, adhering nature of many precooked foods, in-place cleaning has not been widely successful. Where possible, equipment and lines should be washed (brushed) and sanitized at the start of the production shift, at the lunch break, and at the end of the shift. It is important that all ingredients and containers should be removed during this process so that splashings cannot contaminate them. Dead end piping and thermometer wells need particular attention. These should be eliminated where possible and kept scrupulously clean where their presence is absolutely required. Equipment used in the precooked frozen food operation should *not* be shared with other operations, particularly raw food preparation. Equipment used in the precooked frozen food operations should be washed and sanitized in a separate location from that used for other washing operations.

Ingredients.—One of the most serious problems facing precooked frozen food processors is the obtaining of high-quality ingredients, especially those with low microbial content. As has been previously stated, frozen foods must be good from the beginning. It is impossible to prepare low bacterial count precooked and prepared frozen foods from high-count ingredients. Quality frozen precooked frozen foods cannot be prepared from high-count ingredients which have been "laundered" to reduce their microbial content or to conceal incipient spoilage. Particular attention should be directed to spices as a source of microbial contamination. All ingredients used should have microbiological limits and qualifications included in their production or purchasing specifications. Meat slicing and dicing operations and equipment can be an important source of microbial contamination. Meats, in particular, should be refrigerated during the lunch break. Unused food materials should be properly dated, tagged, and refrigerated at the end of each shift and most of those to be held more than 12 hr. should be quick-frozen. It should be pointed out that meats, gravies, sauces, and other cooked foods in bulk returned to refrigeration for holding after preparation, need prechilling with agitation in the case of liquids.

Gravies and sauces in large containers can require many hours to reach effective refrigeration temperatures even in a low temperature environment. Cooked ingredients should be refrigerated separately from raw ones and those rejected for high bacterial counts. Satisfactory gravies cannot be made from puréed, soured, high-count meats.

The time during which product remains in the room temperature range during processing and freezing should be as short as possible; therefore lengthy holding of ingredients and products on the production line during line breakdowns and long freezing times are to be avoided.

In a study of the cooking and boning of poultry meats, it was shown that the bacterial content of the meat was drastically reduced during cooking. Boning of cooked meats involves contact with the hands of workers and much equipment. It also involves prolonged holding at room temperatures which results in rapid increases in bacterial numbers in the boned meats. The workers' hands may not only be an important source of bacterial contamination but may inoculate the meat with pathogens. The equipment used, especially pans, knives, slicing and dicing machines, may become heavily contaminated early in the production shift. Unless the equipment is cleaned and sanitized periodically during the production day, the equipment will serve as a source of bacterial inoculation for other clean, low-count meats with which it comes in contact. One of the original investigations of frozen food ingredient quality demonstrated that the bacterial counts of boned diced poultry meats should be low, as shown in Table 13.6. Comparison of bacterial counts of cooked diced chicken meats shows that the original concept of low microbial populations in these meats was being maintained (Table 13.7).

Effect of Processing.—Fabricating procedures are the most serious source of microbial contamination for precooked and prepared frozen foods. Seafoods provide an excellent example of this fact. Ninety percent of the bacterial load on fish fillets was observed to be picked up from the fillet cutting board. Similarly, utensils were reported as a major source of contamination of shellfish, while the breading operation was a major source of bacteria on frozen fishery products.

The finding of coagulase positive Staphylococci in precooked frozen foods is usually related to extensive handling by the workers. Inves-

TABLE 13.6

TOTAL BACTERIAL COUNTS BONED DICED CHICKEN MEAT

Weekly average—two producers

Week Ending	Plant A	Plant G
7- 9	2,400	13,000
7-16	35,000	50,000
7-23	19,000	32,000
7-30	10,000	42,000
8- 5	57,000	50,000
8-12	50,000	57,000
8-19	38,000	51,000
8-26	52,000	26,000
9- 3	15,000	23,000
9-10	12,000	15,000
9-17	26,000	16,000
9-24	48,000	16,000
10- 1	7,300	12,000
10- 8	52,000	14,000
10-15	52,000	20,000
10-22	17,000	25,000
11- 5	14,000	9,000
11-12	14,000	27,000
11-19	9,000	11,000
11-26	20,000	14,000
12- 3	6,700	16,000
12-10	21,000	9,000
12-17	4,900	13,000

tigators have noted that a number of poor plant practices are reflected in high bacterial counts in the finished product.

The initial cutting operation can cause a tenfold increase in the counts of seafoods. The battering and breading operations further increased the contamination of the product with coliforms, enterococci, Staphylococci, hemolytic Streptococci, and anaerobes. The precooking process reduces the heat sensitive portion of the bacterial flora. The total count, coliform count, and streptococcal counts are reduced; the enterococci, Staphylococci, and anaerobes are only slightly affected.

Blanching and precooking procedures are usually responsible for sharp reductions in the microbial content of prepared and precooked foods. Sterility is usually not achieved, however. Subsequent handling contributes to an increase in the microbial counts of these foods.

The microbial counts of hot-filled and cold-filled chicken á la king reveal that the cold-filled product has a somewhat higher bacterial count than the hot-filled product. This is ascribed to post-heating contamination during handling and filling prior to freezing. Hot-filling followed by quick-freezing at 0° F. (−18° C.) produces a nearly sterile product. This was not the situation when cold-filling was employed although the counts

TABLE 13.7

A SUMMARY OF BACTERIAL COUNTS ON WEEKLY BASIS ON COOKED, DICED FROZEN CHICKEN MEAT

Week No.	Total Count (SPC)/Gram			Coliforms Count/Gram			No. of Samples
	High	Low	Average	High	Low	Average	
1	110,000	18,000	57,000	—	—	—	6
2	43,000	4,600	19,000	50	0[1]	5	15
3	28,000	6,700	17,000	30	0	10	4
4	30,000	4,700	13,000	20	—	2	8
5	44,000	7,200	16,000	10	0	2	8
6	17,000	2,700	11,000	60	0	24	7
7	41,000	4,600	16,000	600	0	120	12
8	52,000	8,800	29,000	80	50	65	4
9	94,000	9,900	35,000	400	0	100	20
10	59,000	9,200	26,000	90	0	17	16
11	76,000	6,900	31,000	160	0	25	15
12	47,000	1,000	16,000	100	0	18	15
13	100,000	2,900	17,000	20	0	4	37
14	89,000	3,000	12,000	100	0	8	56
15	120,000	3,700	31,000	90	0	9	24
16	17,000	11,000	13,000	0	0	0	4
17	11,000	2,300	7,800	60	0	15	14
18	61,000	1,000	13,000	40	0	2	23
19	53,000	1,200	16,000	900	0	73	20
20	28,000	500	11,000	230	0	39	24
21	18,000	600	4,800	30	0	10	7
Overall average			20,000				30

[1] (0) indicates no growth at 0.1 dilution.

were low. This indicates the great importance of reducing the time to a minimum in which a food is allowed to remain in the bacterial incubation danger zone of 50° F. (10° C.) to 130° F. (54° C.) regardless of whether a product is cooled before or after packaging. It also clearly demonstrates the significance of proper temperature control during processing and the necessity for rapid freezing without delay as soon as processing and packaging have been completed (Table 13.8). Results show how rapidly bacteria can multiply in the favorable environment of a precooked food at room temperatures. Working with creamed chicken and creamed turkey artificially inoculated with S. aureus, it was concluded that precooked foods should not be held more than two hours before freezing. The development of even small populations of saprophytic bacteria in precooked foods can lead to organoleptic impairment of the product. The growth of Salmonellae and Staphylococci in foods illustrates that the danger zone for multiplication of these pathogens is between 40° F. (4° C.) and 120° F. (49° C.).

It can be seen that management has a most important role in recognizing the scope of the problem, in providing the necessary tools and super-

TABLE 13.8

EFFECT OF DELAY IN FREEZING, TIME OF FROZEN STORAGE, AND ELAPSED TIME DURING DEFROSTING ON INCREASE IN AEROBIC PLATE COUNTS OF PRECOOKED FROZEN CREAMED CHICKEN INOCULATED WITH *STAPHYLOCOCCUS AUREUS*[1]

Time of Frozen Storage at —30°F. (—34°C.)	Ratio of Final to Initial Aerobic Plate Counts for Samples Defrosted and Held at 77°F. (25°C.)			
	Control[2]	6 Hr.	11 Hr.	18 Hr.
	Samples Placed in Freezer Immediately After Preparation			
2 days	1.2	1.3	—	66.0
14 days	1.1	1.3	—	19.0
28 days	1.2	1.0	—	1.0
3 months	1.2	1.3	1.2	—
6 months	1.4	1.1	1.7	—
12 months	1.2	1.1	24.0[3]	—
	Samples Held at 77°F. (25°C.) for 2 Hr. After Preparation Before Placing in Freezer			
2 days	4.1	16	—	66
14 days	2.5	4	—	420
28 days	4.9	3	—	12
3 months	3.6	10	13	—
6 months	3.2	3	15	—
12 months	2.8	3	175[3]	—
	Samples Held at 77°F. (25°C.) for 5 Hr. After Preparation Before Placing in Freezer			
2 days	32	66	—	66
14 days	29	52	—	607
28 days	35	42	—	231
3 months	56	59	110	—
6 months	26	48	202	—
12 months	31	28	1035[3]	—

[1] Initial count 370,000 per ml. based on the average of two containers.
[2] Defrosted immediately in warm running water.
[3] Room temperture went up to 86°F. (30°C.).

vision and in providing the necessary checks to be sure that the products are being produced under conditions of care and good sanitation. Likewise, conscientious precooked frozen food processors will be fully aware of the microbiological condition of their products.

Freezing and Thawing.—The detailed effects of the freezing process and of the thawing process on microbial survival and of extended storage in the frozen state on microbial survival were considered previously.

Blast freezing was found to reduce the total numbers of microorganisms in chicken pot pies. The fecal Streptococci, Staphylococci, coliforms, and yeasts and molds were sharply affected, but not the anaerobes.

There are reports on the effect of storing frozen chicken à la king for a 15-day period in which the temperature to which the food was exposed was cycled from 20° F. (−7° C.) to 30° F. (−1° C.) and back to 20° F. (−7° C.) in a five-day cycle. After three cycles, the product was returned to 0° F. (−18° C.) for five days and then examined. Temperature cycling was not found to have any appreciable effect on the acceptability of the product. Product which had been cold filled had a significant decrease in bacterial count as a result of this treatment. Hot-filled product showed essentially no change in microbial content because its count was already very low. Thawing of packages of chicken à la king at 40° F. (4° C.) for three days produced a slight increase in bacterial content which was reduced on refreezing. Alternate freezing and thawing did not increase the total microflora of chicken pies unless growth was initiated in the thawed state. They observed no growth at 36° F. (2° C.), 45° F. (7° C.) and at 65° F. (18° C.) in 48 hr. Growth of bacteria was observed after 10 hr. at 70° F. (21° C.) and 90° F. (32° C.) Cyclic freezing and thawing was observed to cause a decrease in the numbers of bacteria of public health significance in seafoods. Fluctuating temperatures have been shown to be responsible for moisture migration in frozen foods to localized areas in the food. This migration resulted in sufficient available water for mold growth even during short periods of exposure to high temperatures in which the product did not thaw completely. Precooked frozen foods can be thawed and refrozen with safety. This process cannot be recommended to consumers, however. Thawing, particularly with protracted holding in the thawed state, invariably leads to quality losses which are not reversed by refreezing.

Reconstitution—Heating

The effect of thawing frozen chicken à la king, beef stew, and creamed seafood in a household refrigerator at 43° F. (6° C.) and then heating to 181° F. (83° C.) followed by refrigeration for 48 hr. at 43° F. (6° C.) and reheating to 185° F. (85° C.) reveals the multiplication of bacteria. Cooking was observed to reduce the numbers of all kinds of microorganisms, but it failed to completely eliminate any type originally present (Table 13.9). Studies on the effect of reheating on the bacterial populations of various meat dishes including creamed chicken and rice, chicken paprika and gravy, spaghetti and meatballs, and ham patties indicate that the initial counts were less than 2,000 bacteria per gram and were very low after heating to 185° F. (85° C.). There are reports that frozen broccoli with an initial total aerobic count of 55,000 bacteria per gram and frozen green beans with 40,000 organisms per gram had 5 and 10 bacteria per gram respectively after heating to 185° F. (85° C.).

In the early phases of roasting frozen stuffed poultry, bacterial multiplication occurs. The center of the stuffing must reach 165° F. (73.9° C.)

TABLE 13.9

AVERAGE VALUES FOR BACTERIA PER GRAM IN FROZEN PRECOOKED FOODS SUBSEQUENT TO KITCHEN HANDLING

Condition of Product	Total Bacterial Count (Tryptone Glucose Extract Agar) Count/Gm.[2]	Micrococci Count (7.5 % NaCl Phenol Red Mannitol Agar) Count/Gm.[2]
Chicken à la king[1]		
Thawing in refrigerator at 43°F. (6°C.)	2,310,100	213,100
Heating at 185°F. (85°C.)	341,000	3,100
Household refrigeration for 48 hr. following heating	815,000	10,000
Reheating to 185°F. (85°C.) after household refrigeration for 48 hr., following first heating	497,000	1,100
Beef stew[3]		
Thawing in refrigerator at 43°F. (6°C.)	4,400	3,700
Heating to 185°F. (85°C.)	100	57
Household refrigeration for 48 hr. following heating	250	191
Reheating to 185°F. (85°C.) after household refrigeration for 48 hr., following first heating	40	5
Creamed sea food[3]		
Thawing in refrigerator at 43°F. (6°C.)	3,400	1,400
Heating to 185°F. (85°C.)	720	200
Household refrigeration for 48 hr. following heating	430	160
Reheating to 185°F. (85°C.) after household refrigeration for 48 hr., following first heating	330	70

[1] Samples taken from eleven packages.
[2] Each figure represents, on an average, counts of 73 plates.
[3] Samples taken from 14 packages.

in order to kill food-borne microbial pathogens and provide a margin of safety. Time required to reach that temperature varied with the initial temperature of the bird, stuffing and size of the bird, and oven temperature. Nonspore forming bacteria were completely killed or reduced to less than one percent of the original numbers when meat pies were baked according to the manufacturer's directions. Spore forming bacteria survived, however. There is a reduction in population of 57.5 to 100% of the total aerobic count on baking of precooked frozen foods. Similarly, the microorganisms present in frozen pot pies are essentially eliminated by baking as directed. Sterility was not achieved, however.

OVERALL PERSPECTIVES

Frozen foods have become an important adjunct of modern life. They provide qualities not obtainable by any other method of food preservation, with economic advantages comparable to those of home preparation. It must be emphasized that cleanliness and sanitation of equipment and personnel are extremely important in producing clean and wholesome frozen foods. Because a single processing plant can produce an enormous quantity of frozen foods which can be widely distributed over a multistate area in a very short time, the microbiological quality of these foods is highly important. Enlightened management of frozen food processing plants should be well aware of the relations between methods of handling the raw ingredients and procedures of processing and packaging, and the microbiological character of the product. It should be recalled that most frozen foods are not sterile, and that they need the same care and consideration that other foods do once they are removed from their protective environment. Most spoilage of frozen foods occurs from only a very few causes: these include mishandling before freezing (which rarely happens), accidental thawing during distribution, and, most often, mishandling by the consumer after the food is thawed.

Precooked and prepared frozen foods represent a special category of convenience foods which are protected from microbial damage by storage at freezing temperatures. In general, they are prepared from ingredients which are superior in microbiological quality to those available to the consumer. The precooked and prepared foods are also generally superior in microbiological quality to those which can be prepared by the consumer. Nevertheless, precooked and prepared frozen foods are perishable and need the same consideration as other foods once they are removed from their protective environment.

ADDITIONAL READING

AM. FROZEN FOOD INST. 1971A. Good commercial guidelines of sanitation for the frozen vegetable industry. Tech. Serv. Bull. *71*. American Frozen Food Institute, Washington, D.C.
AM. FROZEN FOOD INST. 1971B. Good commercial guidelines of sanitation for frozen soft filled bakery products. Tech. Serv. Bull. *74*. American Frozen Food Institute, Washington, D.C.
AM. FROZEN FOOD INST. 1971C. Good commercial guidelines of sanitation for the potato product industry. Tech. Serv. Bull. *75*. American Frozen Food Institute, Washington, D.C.
AM. FROZEN FOOD INST. 1973. Good commercial guidelines of sanitation for frozen prepared fish and shellfish products. Tech. Serv. Bull. *80*. American Frozen Food Institute, Washington, D.C.
ASHRAE. 1974. Refrigeration Applications. American Society of Heating, Refrigeration and Air-Conditioning Engineers, New York.
BAUMAN, H. E. 1974. The HACCP concept and microbiological hazard categories. Food Technol. *28*, No. 9, 30–34, 70.
BYRAN, F. L. 1974. Microbiological food hazards today—based on epidemiological information. Food Technol. *28*, No. 9, 52–66, 84.
BYRNE, C. H. 1976. Temperature indicators—state of the art. Food Technol. *30*, No. 6, 66–68.

CORLETT, D. A., JR. 1973. Freeze processing: Prepared foods, seafood, onion and potato products. Presented to FDA Training Course on Hazard Analysis in a Critical Control Point System for Inspection of Food Processors, Chicago, July and August.

DeFIGUEIREDO, M. P., and SPLITTSTOESSER, D. F. 1976. Food Microbiology: Public Health and Spoilage Aspects. AVI Publishing Co., Westport, Conn.

GRIFFIN, R. C., and SACHAROW, S. 1972. Principles of Package Development. AVI Publishing Co., Westport, Conn.

GUTHRIE, R. K. 1972. Food Sanitation. AVI Publishing Co., Westport, Conn.

HARRIS, R. S., and KARMAS, E. 1975. Nutritional Evaluation of Food Processing, 2nd Edition. AVI Publishing Co., Westport, Conn.

JOHNSON, A. H., and PETERSON, M. S. 1974. Encyclopedia of Food Technology. AVI Publishing Co., Westport, Conn.

KAUFFMAN, F. L. 1974. How FDA uses HACCP. Food Technol. 28, No. 9, 51, 84.

KRAMER, A., and FARQUHAR, J. W. 1976. Testing of time-temperature indicating and defrost devices. Food Technol. 30, No. 2, 50–53, 56.

KRAMER, A., and TWIGG, B. A. 1970, 1973. Quality Control for the Food Industry, 3rd Edition. Vol. 1. Fundamentals. Vol. 2. Applications. AVI Publishing Co., Westport, Conn.

MOUNTNEY, G. J. 1976. Poultry Products Technology, 2nd Edition. AVI Publishing Co., Westport, Conn.

NAT. ACAD. SCI. 1969. Classification of food products according to risk. An evaluation of the Salmonella problem. NAS-NRC Publ. 1683. Natl. Acad. Sci.—Natl. Res. Council, Washington, D.C.

PETERSON, A. C., and GUNNERSON, R. E. 1975. Microbiological critical control points in frozen foods. Food Technol. 28, No. 9, 37–44.

PETERSON, M. S., and JOHNSON, A. H. 1977. Encyclopedia of Food Science. AVI Publishing Co., Westport, Conn.

SMITH, C. A., JR., and SMITH, J. D. 1975. Quality assurance system meets FDA regulations. Food Technol. 29, No. 11, 64–68.

STADELMAN, W. J., and COTTERILL, O. J. 1973. Egg Science and Technology. AVI Publishing Co., Westport, Conn.

TRESSLER, D. K., VAN ARSDEL, W. B., and COPLEY, M. J. 1968. The Freezing Preservation of Foods, 4th Edition. Vol. 2. Factors Affecting Quality in Frozen Foods. Vol. 3. Commercial Freezing Operations—Fresh Foods. Vol. 4. Freezing of Precooked and Prepared Foods. AVI Publishing Co., Westport, Conn.

USDA. 1967. Market Diseases of Fruits and Vegetables. Agriculture Handbook 66. U.S. Govt. Printing Office, Washington, D.C.

WEISER, H. H., MOUNTNEY, G. J., and GOULD, W. A. 1971. Practical Food Microbiology and Technology, 2nd Edition. AVI Publishing Co., Westport, Conn.

14

The Nutritive Value
of Frozen Foods

Bernice K. Watt

I n the mid-1940's, Miss Miriam Birdseye, then Nutrition Specialist for the Federal Extension Service of the USDA, and sister of Clarence Birdseye created a sensation among her professional friends by serving a full course dinner for which every item on the menu had been obtained in frozen form from the Birdseye Laboratories. The great interest in this innovation and uncertainty about its future seem incredible today to those who did not experience the excitement of the early days in this new development in processing of foods.

Frozen foods not only were readily accepted but greatly increased variety in the American diet. The market span for countless perishable products has been extended from an "in season" period of a few days or weeks to the 12 months of the year. Of the 6,000 to 10,000 food items regularly stocked in the larger food stores, there is an ever-increasing number in the frozen food section from major production categories of fruits and vegetables; meat, poultry, fish; dairy products; grain products; and other foods, mainly fats and sweets. The directory of the National Association of Frozen Food Packers has about 1,000 frozen food items.

The nutritive values of frozen foods are of concern both to the consumer and the manufacturer. Advances in technology of freezing foods have been rapid and nutritional research teams have not been able to keep pace. On the whole the nutritive values of foods preserved by freezing are well retained. However, losses of nutrients occur in one or more steps between the time of production and the ultimate use by the consumer. The significance of the losses depends in part upon the proportion of the nutrient lost and in part upon the value of the food item as a source of the nutrient affected. These "significances" will be discussed later as they apply to specific foods.

A general knowledge of the nutritive values of our food supplies is basic to understanding the changes that occur when the products undergo various forms of treatment. This review is particularly concerned with nutritional changes in foods before, during, and after freezing, and in the frozen state.

MAJOR SOURCES OF NUTRIENTS

The percentages of energy value and of several important nutrients provided by major foods in the national food supply for the civilian population are shown in Table 14.1. Of the more than 50 nutrients known to be required by man, only the long-recognized key nutrients for which dietary requirements have been well established are included. The contribution of protein, calcium, iron, five vitamins, and energy value (calories) from five major food groups to the food supply is shown in this table.

Protein

Protein is present in nearly all foods, exceptions being a few highly refined manufactured products. The principal sources of protein in diets of this country are foods of animal origin, cereal grains, legumes, and nuts.

Protein was one of the first nutrients recognized as essential for life and from the beginning of its recognition was given a place of prominence. The name "protein," derived from the Greek, means first place. It is an indispensable component of every cell in the body. Proteins are usually made up of 22 or more amino acids. Proteins differ in the structural arrangements, combinations, and proportions of the amino acids they contain. These differences account for the great number of proteins that occur, the wide range in properties they exhibit, and their many functions.

In the course of metabolism the body can make most of the amino acids it needs in building and maintaining its protein tissues, but at least eight amino acids must be supplied preformed. Proteins in foods of animal origin such as eggs, milk, meat, fish, game, and poultry are considered of excellent quality or of high biological value because they provide each of the amino acids required by the body.

Grain products, legumes, and nuts furnish about a fourth of the total protein provided by the national food supply. If appropriately selected, these foods of plant origin can provide a mixture of proteins that also have high biological value.

Calcium

Calcium is supplied chiefly by milk and the dairy products having the nonfat solid portion of milk. These include the dry, fluid, and evaporated

TABLE 14.1
DISTRIBUTION (IN PERCENT) OF NUTRIENTS IN THE FOOD SUPPLY

	Food Energy (Calories)	Protein	Calcium	Iron	Vitamin A value	B-Vitamins Thiamine	B-Vitamins Riboflavin	B-Vitamins Niacin	Ascorbic acid
Fruits and vegetables	9	7	9	20	51	19	9	17	94
Fruits, including melons	(3)	(1)	(2)	(5)	(8)	(4)	(2)	(3)	(33)
Vegetables	(6)	(6)	(7)	(15)	(43)	(15)	(7)	(14)	(61)
Meat, poultry, fish, eggs, dry legumes, and nuts	24	49	9	42	28	35	31	51	1
Meat, poultry, fish, eggs	(21)	(44)	(6)	(35)	(28)	(29)	(29)	(<45)	(1)
Dry legumes, nuts	(3)	(5)	(3)	(7)	(<1)	(6)	(2)	(<7)	(<1)
Dairy products	12	24	77	2	12	10	44	2	5
Grain products	21	19	3	28	<1	35	15	24	0
Other foods, mainly fats and sweets	34	1	2	8	9	<1	1	6	0
All foods	100	100	100	100	100	100	100	100	100

forms of whole and skim milk; fluid and dry buttermilk and fermented milks; cheeses and cheese foods; ice cream and other frozen dairy desserts; whey solids; and the many prepared foods containing these products. Primarily because of the quantities used, milk with its products is the outstanding source of calcium in the diets of this country. Diets that do not include milk or its products in fairly generous servings are likely to be deficient in this important mineral. However, a number of foods are more concentrated sources of calcium than milk or its products.

Several of the dark green leafy vegetables have more calcium than the same weight of milk would provide. Turnip, mustard, and dandelion greens, collards, and kale are especially rich sources. Other foods that are notable sources of calcium include egg yolk, nuts, molasses, and dry legumes, especially soybeans and their products. While muscle meat and other flesh foods have a relatively low content of calcium, fish can be a rich source if the bones are eaten, as they frequently are, especially in some small fish and in processed items as canned salmon and canned sardines.

Cereal grains when harvested contain relatively little calcium. However, some flours and meals are the means of conveying calcium since it is in mineral mixtures usually added in the manufacture of many prepared mixes and self-rising flours and meals. Most, but not all, baking powders contain considerable calcium.

Iron

Iron is supplied to some extent by all of the major food groups. A fourth of the total amount in the food supply is from meat alone, with small additions from fish and poultry. Fruits and vegetables together supply another fourth. Somewhat over a fourth is from whole grain and enriched or restored cereals. Much of the remainder is from eggs and dry legumes, which are concentrated sources. Cocoa and molasses are rich sources and make an appreciable contribution. Milk contains very little iron, but in the quantities produced it contributes a small amount to the food supply.

Vitamin A

Vitamin A values of foods may be from the preformed vitamin, from one or more of its precursors which the body can convert to the vitamin, or from a mixture of the preformed vitamin and one or more of its precursors. The value is ordinarily expressed in international units since this is a convenient common denominator for the several compounds each showing some vitamin A activity but differing among each other in biological potency.

The most important sources are vegetables that are dark-green or deep-yellow in color, including the highly pigmented sweet potatoes. Their vitamin A value is from several precursors of the vitamin, but largely from beta-carotene. A number of the fruits contribute some vitamin A value, especially apricots, yellow peaches, and some of the melons. Butter and the butterfat in other dairy products, liver of all kinds, and egg yolk are concentrated sources. Fatty tissues of beef also contain some vitamin A. Most of the vitamin A value of these foods of animal origin is present as the preformed vitamin. The most important other single source of vitamin A value is margarine since it is customary to fortify all margarine that is produced for sale in this country with 15,000 international units per pound, about the year-round average value found for butter.

Thiamin

Thiamin (vitamin B-1) is provided to some extent by each of the major groups of food but notably by whole grains, enriched or restored grain products, and by meat, especially pork. Eggs, dry legumes, and nuts, which are frequently used in meals as alternates for meat, are notable sources; also, potatoes, dairy products containing the nonfat portion, and peas and beans.

Riboflavin

Riboflavin occurs in a wide variety of foods, but nearly ¾ is provided by foods of animal origin. Milk, cheese, and other dairy products containing the nonfat portion provide over 40% of the total amount in the food supply. Meats, poultry, eggs, and enriched cereals are the principal additional sources.

Niacin

Niacin is contributed chiefly by meat, poultry, and fish; this group furnishes about 45% of the total in the food supply. Next in importance are the grain products, particularly the whole grain, enriched, and restored products. Peanuts and peanut butter are outstanding as concentrated sources. Potatoes have only moderate amounts, but in the quantities produced furnish about 7% of the total in the food supply.

Ascorbic Acid

Ascorbic acid (vitamin C) is provided almost exclusively by vegetables and fruits. Nearly ⅔ of the total amount in the food supply is in vegetables and close to ⅓ in fruits. Wide variation in content of this vitamin occurs among different kinds of fruits and vegetables. Citrus fruits, strawberries, dark-green leafy vegetables, and cauliflower are some of the especially rich sources. Potatoes, because of the quantity produced, are an especially significant source, furnishing 20% of the total amount of ascorbic acid in the food supply.

Other Important Nutrients

Vitamin E has not generally been included in tables of food composition because not enough information has been available to provide data on individual foods. Also, its importance as a dietary essential has only rather recently come to be recognized. Vitamin E, like the many other nutrients required but not listed in Table 14.1, is believed to be present in sufficient quantity in a varied diet that provides the key nutrients already discussed. It is concentrated in the fat of foods and is especially abundant in the germ of the cereal grains and in vegetable oils. Vitamin E may not be well retained in fried foods that are held in storage. A study was made of the content of alpha-tocopherol (the tocopherol with the highest vitamin E activity) in vegetable oils extracted from fried foods, and found little loss during deep fat frying but large losses after storage at either room temperature or in the frozen state.

Fats are necessary in the diet and are furnished in some degree by each of the major food groups. Fats contain fatty acids including certain polyunsaturated fatty acids regarded as necessary in nutrition. The fats in cereals, as corn and wheat, and the fat in the seeds of cotton, safflower, and soybean plants are particularly concentrated sources of linoleic acid, an essential unsaturated fatty acid. Smaller amounts of linoleic acid are present in the fat of many other foods of plant or animal origin. Protection against oxidation is especially important for unsaturated fats. Antioxidants are sometimes used; also conditions of low temperature and low humidity.

EFFECTS OF FREEZING AND FREEZER STORAGE ON NUTRIENTS IN FOODS OF THE DIFFERENT GROUPS

Numerous studies bearing on some aspect of the nutritive value of frozen foods are continuing. Knowledge about changes that occur in the content of the nutrients during the preparation of freezing and frozen storage is growing steadily. Results reported by different investigators sometimes appear in conflict, but this is not surprising in view of the many variables operating at different steps in the production of frozen products. Findings from some of the studies on nutrients in frozen foods are summarized in this chapter by food groups.

More research has been conducted on fruits and vegetables than on most other kinds of foods. In the late 1930's and early 1940's workers at the N.Y. State Agricultural Experiment Station carried out a comprehensive study of the effect of freezing and low temperature storage on the vitamin content of vegetables and fruits.

In the mid-1950's the National Association of Frozen Food Packers sponsored a study of the nutrients in frozen fruits, fruit juices, and vegetables packed for commercial distribution. The study was based on extensive sampling of fruit and vegetable items and made an outstanding contribution to information on the nutritive value of these frozen foods.

The data from this study and others have been summarized in the USDA Agriculture Handbook No. 8, *Composition of Foods . . . raw, processed, prepared.* Data are presented for about 100 frozen food items from the various food groups in terms of food energy (calories), protein, fat, carbohydrate, calcium, phosphorus, iron, sodium, potassium, vitamin A, thiamin, riboflavin, niacin, and ascorbic acid.

Fruits and Vegetables

Freezing is an excellent means of preserving fruits and vegetables; it stops or retards physiological processes in the harvested produce which if permitted to continue would reduce the high eating qualities and some vitamin values.

The ascorbic acid lost is frequently taken as an indication of the extent of possible losses in other nutrients; ascorbic acid is more easily lost than most other important nutrients. Measures that protect ascorbic acid are considered to be protective of other nutrients as well.

The vitamin C value of a food is customarily measured in terms of the amount of reduced ascorbic acid (ASA) present. Dehydroascorbic acid (DHA), an oxidized form, also has vitamin C activity and occurs in small but significant proportions in a few freshly harvested fruits and vegetables, including cabbage, cantaloupes, squashes, and strawberries. Another oxidation product of ascorbic acid, diketogulonic acid (DKA), one which the body does not use as vitamin C, has been reported in freshly harvested grapefruit, cantaloupe, and strawberries. Data on strawberries are shown in Table 14.2.

Some of the reduced ascorbic acid in freshly harvested fruits and vegetables becomes oxidized to the dehydro form after harvest, especially if the tissues are cut or bruised. For nutritional purposes, the total vitamin C value is the sum of the reduced and dehydro forms, and the value would be somewhat underestimated if only the reduced form were determined, as is usually done. Inasmuch as dehydroascorbic acid is particularly unstable and seldom determined, reliance for vitamin C values has usually been based on ascorbic acid, which unless otherwise designated referred to the reduced form.

TABLE 14.2

COMPARISON OF ASA, DHA, AND DKA IN FRESHLY HARVESTED STRAWBERRIES

Sample	As Ascorbic Acid, Mg./100 Gm.				% of Total as ASA
	ASA	DHA	DKA	Total	
1	82	5	3	90	91
2	69	4	3	76	91
3	65	5	2	72	90
4	62	4	1	67	92
5	59	5	2	66	89
6	56	3	4	64	88
7	53	6	3	62	85
8	53	5	1	59	90
9	52	2	4	58	90
10	49	4	2	55	89

Attention is called here to the difference in meaning between total ascorbic acid as used in food composition tables for the purpose of indicating vitamin C value (ASA + DHA) and total ascorbic acid (ASA + DHA + DKA) as used for studying some quality changes occurring in fruits and vegetables that are frozen. The latter should not be considered a true measure of vitamin C.

If packed in crushed ice the leafy, dark-green vegetables and broccoli keep practically all of their ascorbic acid for several days, but under refrigeration at 40° to 50° F. (4° to 10° C.) they retain only half of it after 5 days. These vegetables have such high initial values that even after this substantial loss they are excellent sources. Nevertheless, if vegetables for freezing cannot be prepared and frozen within a few hours after harvest, they should be chilled below 40° F. (4° C.) immediately and held at that temperature until they can be processed.

Many fruits held at room temperature do not lose vitamin C as rapidly as do vegetables. Citrus fruits when whole keep well several days without refrigeration. Freshly squeezed orange juice retains its ascorbic acid for several days in the refrigerator. In fact, no serious loss occurs if the juice remains outside the refrigerator for a few hours. By comparison, berries are highly perishable. They should be kept whole and dry and refrigerated if they are not to be used for a few days.

It is equally important to use cold, preferably iced, water for the washing of fruits, especially berries. Iced water firms fruit and makes it easier to handle without becoming mushy, while warm water softens berries and causes "bleeding" or loss of juice to the wash water. After washing, the fruit must be handled without delay, especially if it is to be frozen in large containers. If warm fruit, fruit purée, or fruit juice is placed in a barrel or other large container for freezing, it will not only lose flavor and vitamins, but fermentation may occur before freezing begins.

The importance of chilling in retaining reduced ascorbic acid in fruit prepared for freezing is clear. Oxidative changes in the ascorbic acid content of whole stemmed strawberries held at 35° F. (2° C.) occur only about $^1/_5$ as rapidly as in the berries at 70° F. (21° C.). The curves in Fig. 14.1 illustrate the rapid losses in reduced ascorbic acid with the corresponding increase in the oxidized products for whole strawberries held at room temperature 70° F. (21° C.) up to 48 hr. and the much slower rates at 35° F. (2° C.). Total ascorbic acid in strawberries remains constant over a wide range of time and temperature conditions. For strawberries the change in the proportions of the three forms of ascorbic acid is useful in measuring the cumulative effect of adverse factors and conditions to which the berries have been subjected.

Freshly harvested berries have approximately one-tenth of the total ascorbic acid present in the oxidized forms. Slicing causes an increase in content of ascorbic acid oxidation products from 12% before slicing to 22% immediately after slicing and sugaring. The loss of reduced ascorbic acid and the corresponding increase in oxidation products in relation to the total ascorbic acid found when sliced, sugared strawberries were held as long as 24 hr. at 70° F. (21° C.) is shown in Fig. 14.2. This particular lot of strawberries had a total ascorbic acid content of about 56 mg. per 100 g. or about 50 mg. of reduced ascorbic acid if it is assumed that 90% of the total was present in the reduced form when the berries were harvested. The content of reduced ascorbic acid in the sliced sugared berries held 24

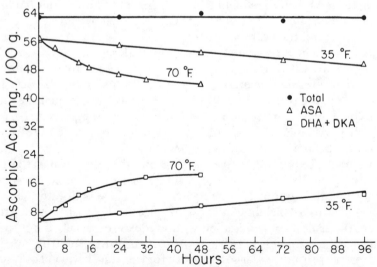

FIG. 14.1. EFFECT OF TIME AND TEMPERATURE ON ASCORBIC ACID CHANGES IN FRESH, WHOLE, STEMMED STRAWBERRIES

hr. at room temperature was about 24 mg. per 100 g. of berry or only about half the original level.

The temperature and duration of freezer storage are important factors affecting nutritive values. Many frozen fruits and vegetables lose some ascorbic acid unless they are held at temperatures well below their freezing point—at 0° F. (−18° C.) and for some even lower. In actual practice frozen foods experience temperatures above 0° F. (−18° C.) during wholesale and retail storage and distribution and storage in the home. Information on consumer practices in the handling and storage of frozen foods in the home reveals that adverse temperature histories, reflected in changes in ascorbic acid content, occur.

As there is no significant growth of food microorganisms at 15° F. (−9° C.) or below, public health is not directly menaced by frozen foods if their temperature rises at times above 0° F. (−18° C.) into the range up to 15° F. (−9° C.). It is important, however, to consider the effect on nutrients of storage that may be encountered at different temperatures above 0° F. (−18° C.). Also the type of container and packing have been found to be important in affecting nutritive value.

Fruits.—Results of some of the more recent studies on the effects of preservation by freezing in a few kinds of fruits are noted below.

Citrus Fruits.—Oranges, grapefruit, lemons, limes, and tangerines are the best known among the citrus fruits that are used in frozen products. All of the citrus fruits are notable as sources of ascorbic acid, but on the average oranges and lemons have a higher content than the other kinds of citrus. In freezing and during reasonable storage periods afterward,

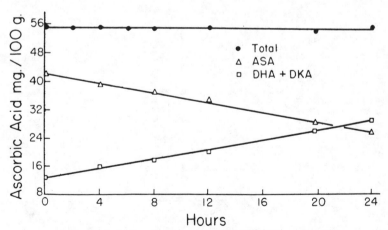

FIG. 14.2. EFFECT OF DELAY AT ROOM TEMPERATURE (70° F.; 21° C.) ON ASCORBIC ACID CHANGES IN FRESH SLICED SUGARED STRAWBERRIES (4 + 1)

citrus products retain nearly all of their high initial content of the vitamin.

Numerous workers have studied the effect of freezing and subsequent storage at different temperatures on the content of ascorbic acid, particularly in orange juice and orange juice concentrates. They have found that little loss of the vitamin occurs even during long storage although the temperature of the frozen product may at times rise well above 0° F. ($-18°$ C.).

Strawberries.—A comprehensive study of commercially frozen strawberries found that the type of container was a highly important variable affecting ascorbic acid retention during a 72-day storage period at 20° F. ($-7°$ C.). It should be noted that 20° F. ($-7°$ C.) is a much higher temperature than the usual 0° F. ($-18°$ C.) temperature at which fruit is stored commercially. Figure 14.3 shows the effect of storage at 20° F. ($-7°$ C.) on the content of reduced ascorbic acid in 2 lots of strawberries when packed in hermetically sealed enameled cans and in composite containers. Most retail frozen strawberries are packed in composite containers, that is, in containers with metal ends and paperboard sides. The curves show rapid loss of ascorbic acid in the composite container until most of the ascorbic acid is oxidized, whereas the loss in the metal containers is more gradual and levels off while there is still a high proportion of reduced ascorbic acid present.

It appears that the ascorbic acid oxidized in the hermetic containers is limited to the amount of oxygen originally sealed into the container. When this amount of oxygen is used up, little or no additional oxidation

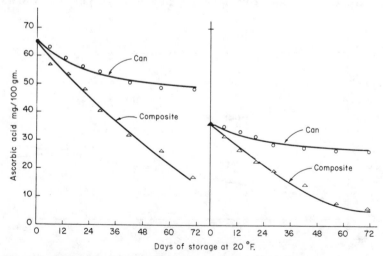

FIG. 14.3. EFFECT OF CONTAINER ON ASCORBIC ACID RETENTION IN FROZEN STRAWBERRIES ON VARYING ORIGINAL ASCORBIC ACID CONTENT

takes place. Since the composite carton is not hermetic, the fruit is in constant contact with oxygen supplied by the breathing action of the container, and consequently oxygen does not become a limiting factor.

The total vitamin C value of strawberries would include any dehydro-ascorbic acid present in the berries if they could be eaten before it was further oxidized. A third lot of frozen strawberries provided some information on changes in the proportion of the two oxidized products during storage. The berries had an initial total ascorbic acid content of not quite 80 mg. per 100 g. of berries, and at the beginning of the observation period had approximately 7 mg. in the physiologically inactive form of diketogulonic acid. During 120 days of storage at 20° F. (−7° C.), the content of diketogulonic acid increased steadily to approximately 45 mg. per 100 g. The proportions of the different forms are shown in Fig. 14.4. Dehydroascorbic acid would materially increase the vitamin C value if included with reduced ascorbic acid, but in view of the instability of this oxidized form it is questionable whether it would be present when the frozen berries are thawed for serving.

The effect of type of container, hermetically sealed can versus composite packaged, exerts progressively less effect on the loss of reduced ascorbic acid as the temperature is decreased to 0° F. (−18° C.). This is

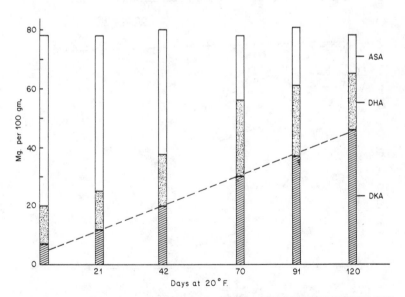

FIG. 14.4. ACCUMULATION OF VARIOUS OXIDATION PRODUCTS OF ASCORBIC ACID
DURING STORAGE AT ELEVATED TEMPERATURES

ASA = ascorbic acid. DHA = dehydroascorbic acid. DKA = diketogulonic acid.

shown in Fig. 14.5. After storage at 0° F. (−18° C.) for 120 days, the berries in both types of containers had nearly as much reduced ascorbic acid as when packed, but at each of the higher temperatures of storage 10°, 15°, 20°, and 25° F. (−12°, −9°, −7°, −4° C.), the hermetically sealed can gave the better protection; oxidation had leveled off before a fourth of the reduced ascorbic acid had been lost.

Strawberries are packed extensively in bulk for subsequent manufacture in preserves, ice cream, and other products. Researchers have determined the amounts of total and reduced ascorbic acid in the top and in the subsurface layers of 30-lb. containers of frozen sugar-packed strawberries. As would be expected, the rate of loss of reduced ascorbic acid was higher in the top layer with its greater exposure to air than in subsurface

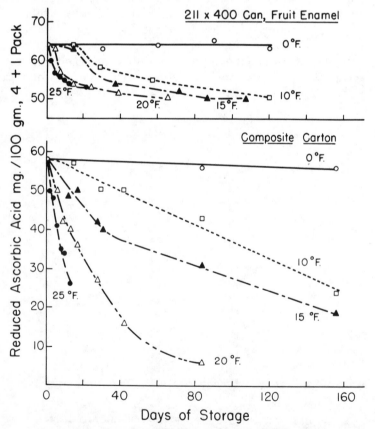

FIG. 14.5. EFFECT OF TIME AND TEMPERATURE ON ASCORBIC ACID CHANGES IN FROZEN STRAWBERRIES PACKED IN DIFFERENT CONTAINERS

layers. Since the top layer constitutes only a small fraction of the bulk, the total percentage of ascorbic acid lost is relatively small. These frozen strawberries in 30-lb. bulk pack containers retained about 72 to 76% of their reduced ascorbic acid content during the period which, according to various flavor and color standards, is designated as the high-quality life of the frozen strawberry at 0° to 30° F. (−18° to −1° C.).

Peaches.—Peaches usually have ascorbic acid added as an antioxidant to delay the onset of browning. The added ascorbic acid may be considerably more than the amount naturally present in the fruit. Retention of the total amount of reduced ascorbic acid in commercially packed frozen peaches was found to be determined by temperature and length of storage period. Fig. 14.6 shows that at 0° F. (−18° C.) only a small loss in reduced ascorbic acid has occurred after 60 days of storage, and after a year less than 20% has been lost. However, only a negligible amount remained in the peaches packed in metal end cartons after storage for two weeks at 25° F. (−4° C.). Fig. 14.6 also shows that in the range of temperatures somewhere between 15° and 20° F. (−9° and −7° C.), the rate of oxidation proceeds much more rapidly than at 15° F. (−9° C.) or lower. The increase in rate of ascorbic acid oxidation from 0° to 15° F. (−18° to (−9° C.) is approximately 20% of the increase in rate observed at 15° to 20° F. (−9° to −7° C.).

The loss of ascorbic acid in frozen peaches packed in hermetically sealed cans is not extensive even at 25° F. (−4° C.). This is illustrated by the data in Fig. 14.7. It has been observed that losses of ascorbic acid have

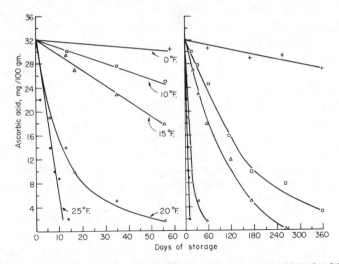

FIG. 14.6. EFFECT OF STORAGE TEMPERATURE ON THE ASCORBIC ACID CONTENT OF COMMERCIALLY PACKED FROZEN PEACHES

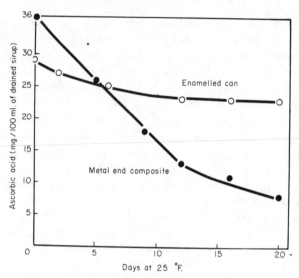

FIG. 14.7. EFFECT OF CONTAINER ON ASCORBIC ACID RETENTION IN
FROZEN SLICED PEACHES

been decreased by decreasing headspace and amount of air entrapped with the frozen foods.

Raspberries.—Raspberries are considered more stable than many other kinds of frozen fruits. They have much less ascorbic acid than most varieties of strawberries that are frozen and than most frozen peaches, since the latter often have ascorbic acid added when they are frozen. Time-temperature tolerance studies on the ascorbic acid content of frozen raspberries are of less practical interest for nutritional purposes than for information on the general behavior of fruits. For raspberries, as for peaches and strawberries, a storage temperature of about 0° F. (−18° C.) is needed to minimize oxidation of ascorbic acid. Losses in this vitamin when frozen raspberries were stored at 0° F. (−18° C.) and at several higher temperatures are shown in Fig. 14.8. No significant loss in ascorbic acid occurs during the freezing process or storage for 2 months at −10° F. (−23° C.) in raspberries without sugar or with concentrations of sugar syrup in the range of 40 to 68%. Sugar and flash-heating each exert a protective effect on retention of the vitamin in raspberry purée.

The addition of sugar before freezing has some protective action on ascorbic acid in frozen raspberries. However, less ascorbic acid remains in sweetened purée (sugar-free basis) than in purée to which no sugar has been added; oxidation from air incorporated while stirring in the sugar causes the loss.

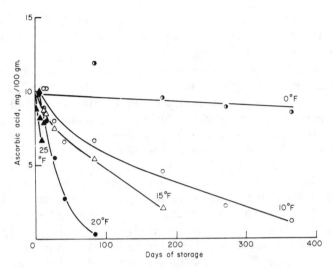

FIG. 14.8. EFFECT OF TIME AND TEMPERATURE ON ASCORBIC ACID
CONTENT OF RASPBERRIES AND SYRUP

Separate analyses of drained raspberries and drained syrup show that at 10° F. (−12° C.) and higher storage temperatures, ascorbic acid migrates from the fruit to the syrup. Fortunately, it is customary for consumers to use the drained syrups along with the fruit.

Raspberries frozen in 30-lb. containers and stored at 0° to 30° F. (−18° to −1° C.) for reuse in other foods are found to retain 70 to 80% of their original level of 16 mg. per 100 g. of reduced ascorbic acid throughout their high-quality life.

Blackberries.—Blackberries also are frozen in bulk for subsequent use in food manufacturing. From 60 to 70% of their original level of 7 mg. per 100 g. of reduced ascorbic acid was found to be retained throughout their high-quality life when stored at 0° to 30° F. (−18° to −1° C.).

Vegetables.—As a group, vegetables are an important source of vitamins and minerals, but individual kinds differ in their content of the various nutrients. Also, when vegetables are preserved by freezing, they differ in the extent to which they retain their nutrients. Many more studies have been made of the retention of ascorbic acid than of other nutrients. Like fruits, vegetables may lose ascorbic acid through oxidation; in addition, many vegetables lose some of this vitamin through solution when they are cooked. As other nutrients in vegetables are considered to be retained at least as well if not better than ascorbic acid, conditions that protect ascorbic acid during freezing procedures are also considered protective for other nutrients. Findings from some of the many studies on the nutritive value of frozen vegetables and on factors affecting retention of these values are summarized for several important vegetables in the following pages.

Most vegetables are blanched prior to freezing. This heat treatment partially precooks vegetables and makes them porous; the water soluble nutrients become more vulnerable to leaching after blanching than before. Some losses can be expected during water-cooling, brine-flotation, and other processes in which the vegetables come in contact with water. Blanching is beneficial, however, as it is effective in inactivating enzymes which otherwise cause the development of off-flavors, undesirable texture, and the rapid loss of carotene and vitamin C during storage. For example, studies on the effects of blanching on carotene in spinach, asparagus, peas, snap beans, and lima beans stored at $-4°$ F. ($-20°$ C.) for 6 or more months, reveal improved retentions during low-temperature storage over those not blanched before freezing. Also, during 12 months' storage at $-4°$ F. ($-20°$ C.) unblanched green beans lost about 90% of their vitamin C, 74% of their thiamin, and 39% of their riboflavin, whereas properly blanched beans lost 47, 22, and 3%, respectively.

Asparagus.—Asparagus is a fairly good source of many nutrients, and the frozen form has been studied for its content of several of them. Relatively little loss in ascorbic acid is reported in asparagus. The market sample used contained only about 18 mg. per 100 g. and after it had been blanched 3 min. in boiling water, it retained about 90% of its original amount. When frozen and stored at $0°$ F. ($-18°$ C.) for 6 months, the content was not significantly different.

Earlier studies reveal a small loss of vitamin C in frozen asparagus stored at $0°$ F. ($-18°$ C.) for 6 months, as well as at different temperatures from $-40°$ to $+16°$ F. ($-40°$ to $-9°$ C.). At the beginning of the period the samples had 38 mg. of ascorbic acid per 100 g., which is within the range, but above the average value usually found in frozen asparagus. After 10 months of storage at $-40°$ F. ($-40°$ C.) no loss of ascorbic acid was found, but at temperatures of storage above $0°$ F. ($-18°$ C.) significant losses had occurred by the sixth month. There are also reported losses of 24% for ascorbic acid, 28% for thiamin, 42% for riboflavin, and 24% for carotene in asparagus during commerical freezing procedures.

Snap Beans.—Green snap beans are considered among the least stable of the popular frozen vegetables, as changes in their color and texture develop readily. Low values for ascorbic acid, however, are not necessarily an indication of poor handling practices because there is wide range in the original content of this vitamin in beans used for freezing. Much higher ascorbic acid values are found in snap beans graded size 6, which are more mature beans, than in those graded size 4, and this calls attention to tests showing seeds to contain 2 to 3 times as much ascorbic acid as the pod material. Studies on the effect of blanching conditions on color stability of frozen snap beans, which included determinations of ascorbic acid, reveal better retention of ascorbic acid in beans blanched 1 to 5 min. at 200° to 212° F. (93° to 100° C.) than when blanched for a longer time at a lower temperature.

Water-blanched (2 min.) green snap beans (Bountiful variety) in E-Z Seal all-glass jars lose 33% of their original vitamin C content in preparation and freezing at −4° F. (−20° C.). An additional 14% loss is noted after a year's storage at this temperature. Unblanched beans of the same variety contain only 10% of their original vitamin C content after freezing and storage for a year at −4° F. (−20° C.).

Beans prepared according to a usual procedure suggested for home freezing (2-min. water-blanch) can lose up to 13% vitamin C during blanching and an additional 6% loss during chilling in ice water for about 2 min. Freshly processed, uncooked frozen snap beans that receive a 3-min. blanch in boiling water and cooled 2 min. in ice water have 83% as much ascorbic acid as the fresh raw beans, and a decrease of about 17% in the frozen beans during storage for 10 months at −40° F. (−40° C.) and a decrease of about 30% when storage is at 0° F. (−18° C.).

Measurements of ascorbic acid changes in frozen green beans held over a wide range of temperatures indicate that they lose little ascorbic acid in 1 year at −20° F. (−29° C.). Small but appreciable losses occur at 0° F. (−18° C.). Above 0° F. (−18° C.), the rate of deterioration increases rapidly, approximately doubling with each 5° F. (2.8° C.) increase in storage temperature from 0° to 25° F. (−18° to −4° C.).

Freshly harvested green beans of the Top Crop variety grown at the University of Minnesota were analyzed before and after blanching (3½ min. in boiling water), after chilling and freezing, and after storage at 0° F. (−18° C.) for 6 months. The raw beans contained 17 mg. of ascorbic acid per 100 g. and when blanched retained 60% of the original amount. After freezing they dropped to 54%, and after storage at 0° F. (−18° C.) for 6 months they retained 47% of the original amount in the fresh raw beans. For the frozen blanched beans this was a drop of about 15% during the 6-month storage period.

On the average there is a range of 3 to 20 mg. of ascorbic acid per 100 g. in retail packages of frozen beans. This variation is not surprising in view of the range of ascorbic acid in fresh green beans used for freezing and extent of loss indicated when frozen beans are subjected to different storage temperatures.

Thiamin appears to be fairly well retained by snap beans during preparation for freezing and good conditions of frozen storage. A loss of about

17% of the thiamin in the home preparation of beans for freezing occurs. Freshly processed uncooked snap beans contain about 86% of the amount of thiamin found for the fresh beans; the data show no loss of thiamin during storage at 0° F. (−18° C.) or below for 10 months.

Riboflavin also appears to be well retained by snap beans. Riboflavin appears to be stable to freezing, canning, storage, and dehydration, with a total loss of riboflavin of 18% during preparation for home freezing. Approximately ¾ of this loss occurs during water blanching, the remainder during chilling of the blanched beans. No appreciable loss of riboflavin is noted during partial thawing (three hours) of the frozen beans.

Lima Beans.—Lima beans have been popular as a frozen vegetable since the beginning of the frozen vegetable industry and were selected for analysis in some of the earliest research on the nutritive value of frozen foods. Early and more recent studies have indicated losses of ¹/₅ to ²/₅ of the ascorbic acid during blanching and cooling of lima beans. In general, the longer the blanching time the greater the loss. Exploratory studies indicated that during storage for 6 months, losses of ascorbic acid in lima beans were relatively small at 10° F. (−12° C.) or below, while at 16° F. (−9° C.) 80% was lost.

The vitamin content of lima beans and peas prepared in a local cannery as frozen or canned products analyzed before and after blanching, after freezing, and after storage at −10° F. (−23° C.) for intervals up to 1 year, reveal that about ²/₃ of their ascorbic acid and thiamin were lost in the blanching (5 min. in hot water), and after they were frozen were found to have 46% of their ascorbic acid and 62% of their thiamin. Retentions of riboflavin and niacin in the frozen lima beans were better and were about 75%. During the first 6 months of storage not much additional change occurred in the contents of any of these 4 vitamins, but after 12 months the frozen lima beans retained only 36, 45, 42, and 57%, respectively, of the amounts of ascorbic acid, thiamin, riboflavin, and niacin present in the beans before they were blanched.

A comprehensive study of the composition of beans of the baby lima and the Fordhook types before and after freezing (steam-blanched for 3 min., cooled in tap water spray, frozen, and stored at −20° F. (−29° C.) for 3 to 4 months) evidenced little change in content of total solids, alcohol insoluble solids, starch, protein, fat, calcium, and iron. In most, but not all instances, losses in vitamin A value, thiamin, and ascorbic acid were observed, with losses of ascorbic acid exceeding 50% in some cases. Values for riboflavin, niacin, phosphorus, and total sugars were all lower in the frozen than in the fresh lima beans. However, neither frozen nor fresh lima beans are eaten raw; they are cooked. Since frozen lima beans require less cooking than fresh, vitamin loss may also be less.

In baby green lima beans the ascorbic acid content after blanching and freezing was found to be about 60% of that in raw lima beans, and from 10 to 24 mg. per 100 g. for the Fordhook type obtained in retail stores. Their average value, 17 mg. of reduced ascorbic acid per 100 g., was close to the value of 19 mg. per 100 g. for samples obtained directly from a frozen food processor and to the average value of 20.4 mg. per 100 g. reported.

Broccoli.—Broccoli has become one of the most important of the commercially frozen vegetables. It is an excellent source of a number of nutrients but the parts—flower bud, stalk, and leafy portion—differ in their composition. This presents practical problems in preparing comparable samples for measuring the effect of different treatments. Lack of agreement in findings of investigators studying the nutritive values of broccoli before and after freezing may in part reflect different proportions of the three parts: i.e., 38 to 93 mg. of reduced ascorbic acid per 100 g. of frozen broccoli spears for a number of packages obtained in retail channels; 59 to 102 mg. of reduced ascorbic acid per 100 g. for frozen broccoli prior to shipment to retail outlets; and wide variations for many of the other nutrients, from about 1,200 to 5,400 I.U. per 100 g. for vitamin A measured as beta carotene, from 0.04 to 0.11 mg. per 100 g. for thiamin, and from 0.09 to 0.20 mg. per 100 g. for riboflavin. Studies of ascorbic acid in commercial samples of frozen broccoli held at 0° F. (−18° C.) showed that after 2 weeks the frozen broccoli had about 94 mg. of reduced ascorbic acid per 100 g. and 4.7 mg. of dehydroascorbic acid. At 25 weeks the samples of heads plus stems had 78 mg. of reduced ascorbic acid per 100 g. and 13 mg. of dehydroascorbic acid. After 36 weeks of storage, the heads analyzed had 81 and 19 mg. of reduced and dehydroascorbic acid respectively per 100 g., and the stems 94 and 11 mg. Carotene in frozen broccoli held at 0° F. (−18° C.) for 17 weeks and 61 weeks shows no decrease in this period.

Using raw broccoli having an average content of 87 mg. of ascorbic acid per 100 g. purchased in retail markets, about ⅔ of the ascorbic acid remains after a blanch of 4 min. in boiling water. There is no further loss after the broccoli has been chilled in ice water and frozen, or after 6 months' storage at 0° F. (−18° C.). However, a loss of 10% of the ascorbic acid can be expected in frozen broccoli stored 1 year at −20° F. (−29° C.).

Cauliflower.—Cauliflower has been studied less than most frozen vegetables. Like broccoli, kale, Brussels sprouts, and other members of the cabbage family, fresh cauliflower is an especially good source of vitamin C. Often values above 70 mg. per 100 g. are reported for market samples of cauliflower. Although appreciable loss of vitamin C occurs in blanching and continues during frozen storage even when temperatures near 0° F. (−18° C.) are maintained, cauliflower may still be a good source.

Studies on the effects of blanching, freezing, frozen storage, and cooking on ascorbic acid in cauliflower purchased through regular market channels indicate an average of 78 mg. of ascorbic acid per 100 g. when blanched (4 min. in boiling water), cooled in ice water, frozen, and stored at 0° F. (−18° C.) for 6 months. Immediately after it had been frozen, the cauliflower retained about 80% of its original ascorbic acid and 60% after 6 months of frozen storage.

In a study of quality changes, data on ascorbic acid in ten lots of frozen cauliflower representing different varieties grown in different years and places, packaged in retail-sized cardboard cartons, and commercially processed (blanching was with steam in all but one lot), shows that below 0° F. (−18° C.) some loss occurs but the rate is lower than at higher temperatures. A summation is shown in Fig. 14.9 with comparable data for peas and green beans.

Peas.—Peas contain appreciable amounts of vitamins and have been studied by numerous workers at various stages before and after freezing. Overall losses during the preparation for freezing and the freezing of peas have indicated that up to a third of the ascorbic acid may be lost. Early studies on the effects of varying the length of the blanching period showed that when peas were blanched in boiling water, the longer the blanching period, the greater the loss of ascorbic acid. However, when the

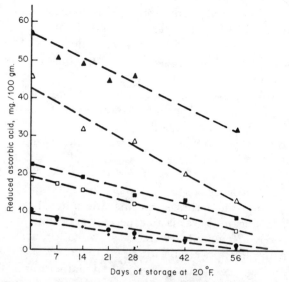

FIG. 14.9. LOSS OF REDUCED ASCORBIC ACID (ASA) IN CAULIFLOWER, PEAS, AND GREEN BEANS AT 20° F. (−7° C.)

peas were blanched in steam, the rate of loss became negligible after the first minute.

Work of many investigators indicates that frozen peas retain their nutrients very well when stored at 0° F. (−18° C.) or below; when stored at higher temperatures oxidation of ascorbic aacid occurs at increasingly rapid rates. Several studies were done on the retention of ascorbic acid, thiamin, riboflavin, niacin, and carotene in peas before and after they had been blanched in hot water (3 min.), after they had been frozen, and after they had been stored at −10° F. (−23° C.). Blanching caused a loss of 33% of the ascorbic acid, about 20% of the riboflavin, 10% of the niacin, 5% of the thiamin, and no loss of carotene. Any further losses that had occurred by the time the peas had been frozen were negligible, except for ascorbic acid and niacin for which the losses had increased to 45% and 25%, respectively. During the first six months of storage, vitamin losses were moderate or negligible. By the end of 12 months, retention was about 70% or more for the various B-vitamins, but only 33% for ascorbic acid. Studies with 3 varieties of peas, which had been blanched (60 sec. in water), water-cooled, and quality-separated, show they retained about 17 to 30% as much thiamin per 100 g. at the beginning of the frozen storage period as was in 100 g. of the fresh vegetable. No appreciable loss of thiamin was found in the frozen peas stored for 1 year at−7° to −10° F. (−22° to −23° C.). The changes in the content of reduced ascorbic acid and its oxidation products in frozen peas after a year at 0° F. (−18° C.) reveal that the loss was negligible, but at 10° F. (−12° C.) very little reduced ascorbic acid remained at the end of the year. The data in Fig. 14.10 show the retentions of ascorbic acid found in peas stored at four temperatures, 0°, 10°, 20°, and 30° F. (−18°, −12°, −7°, and −1° C.). Other studies show that deterioration in ascorbic acid more than doubled for each 5° F. (2.8° C.) increase in temperature between 0° F. (−18° C.) and 25° F. (−4° C.). That lower temperatures close to 0° F. (−18° C.) exert greater protective effect on ascorbic acid may be observed.

Spinach.—Spinach, like all the dark leafy greens, is a rich source of vitamin A value and vitamin C, and a moderately good source of riboflavin and other B vitamins. Leafy portions and stems differ in their content of nutrients and complicate studies of nutritive value since both stem and leafy portions are used and proportions can vary widely.

Studies made on the effects of preserving spinach by freezing have indicated that ¾ and often more of the carotene is retained if the spinach is properly blanched. No significant change with time was found at any given temperature in the content of carotene in frozen spinach that had been adequately blanched and stored for 2 years at −20° F. (−29° C.) or at 0° F. (−18° C.), or for 1 year at 20° F. (−7° C.), or for 7 days at 40° F. (4° C.).

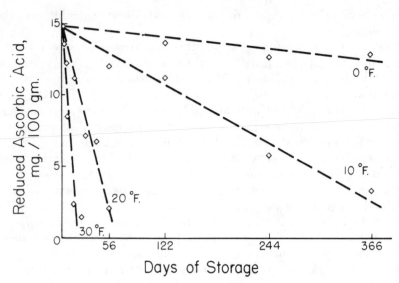

FIG. 14.10. LOSS OF REDUCED ASCORBIC ACID (ASA) IN LAXTON PEAS AT 0°, 10°, 20°,
AND 30° F. (−18°, −12°, −7°, AND −1° C.)

Ascorbic acid in spinach is not as well retained as carotene during the preparation for freezing and frozen storage. The effects of temperature of storage on reduced ascorbic acid in a number of fruits and vegetables, including leaf spinach, show that at −20° F. (−29° C.) the loss of ascorbic acid after 1 year was 10% of the amount present at the beginning of the storage period. At 0° F. (−18° C.) the losses by the end of the year were much greater, 55 and 90% respectively. There was little or no loss of ascorbic acid in frozen leaf spinach packed in 10-oz. wax-overwrapped cartons and stored at −20° F. (−29° C.) for a year. A very slight loss was found for the spinach stored at −10° F. (−23° C.) and a small but appreciable loss occurred when stored at 0° F. (−18° C.). However, after a year at 10° F. (−12° C.), the losses were 50 to 75%. At higher temperatures deterioration in ascorbic acid progressed at a much more rapid rate.

Other Vegetables.—Many other vegetables and mixtures of vegetables are available and are finding good consumer acceptance—Brussels sprouts, corn, mixed vegetables, and squash, to name a few. Studies of their nutritive value are appearing in the scientific and technical literature. As methods of preparing and freezing of vegetables continue to improve, even higher levels of retention of nutrients can be expected.

Meats, Fish, Poultry, Eggs, Dry Legumes, Nuts

The main dish of a luncheon, dinner, or supper is likely to consist of a meat, poultry, or fishery product, or to be made from legumes such as baked beans, or occasionally from nuts. These foods are often grouped together because they are particularly good sources of protein. In addition, each is a good source of one or more of the various other dietary essentials.

Very little research has been done on the effects of freezing and storage on the nutritive values of these foods; the few studies conducted have dealt mainly with meat and fishery products. It is generally believed that nutrients in meat, fish, poultry, dry legumes, and nuts will be well retained provided the techniques for freezing and storage allow for maintaining high quality in such other respects as flavor, color, and texture.

Meat.—Meat in good condition can be frozen and stored easily, but should be packaged to exclude as much air as possible and stored at 0° F. (−18° C.) or below. Frozen meat does not keep indefinitely, however, Recommended maximum storage periods for home-frozen meats and fish have been stated in a publication issued jointly by the USDA and the U.S. Dept of the Interior. At 0° F. (−18° C.) the recommended maximum periods vary from less than 1 month for bacon to about 8 months for beef or lamb roasts or beef steaks.

Different kinds of meat (beef, lamb, pork, and veal) are good sources of the B vitamins. Pork is a particularly good source of thiamin. Studies of the effects of freezer storage and thawing of meats have indicated some variation in retention of the different members of the B complex. Although results from different studies are somewhat conflicting, it appears that extent of ripening prior to freezing, length of freezer storage period, and procedures in thawing are the more important factors influencing the retention of each of these vitamins.

The thiamin content of pork roasts is not much changed when the frozen roasts are stored for periods of 4, 8, and 12 months at 0° F. (−18° C.). Even at the higher temperatures of 10° F. (−12° C.) and at a regularly fluctuating temperature pattern from 0° to 20° F. (−18° to −7° C.) which permits the roasts to reach 20° F. (−7° C.) for a total of 36 hr. per 6-day period, the retention of thiamin is good. Although the thiamin is stable under these circumstances, the fat of all the samples stored above 0° F. (−18° C.) has incipient rancidity by 4 months, and after 8 and 12 months the pork roasts have noticeably rancid odors and high peroxide values.

Studies on retention of thiamin, riboflavin, and niacin in pork loin chops frozen within 24 hr. of slaughter and held in freezer storage at 0° F. (−18° C.) and at −15° F. (−26° C.) indicate that raw pork chops after 3 months of freezer storage have undergone a loss of about 1/5 of the original thiamin, 1/3 of the riboflavin, and about 1/10 of the niacin. After six months in frozen storage the loss in thiamin extends to about a fourth of the original amount. No significant differences are observed between the vitamin losses in the pork chops stored at 0° F. (−18° C.) and at −15° F. (−26° C.).

Small differences are reported for the percentages of thiamin, riboflavin, pantothenic acid, and niacin retained in pork loins that have been aged 1, 3, and 7 days at 30° F. (−1° C.) and for 7 days at 40° F. (4° C.) before the loins were frozen and stored at 0° F. (−18° C.). The retention of thiamin after 8 weeks of storage ranges from 81 to 96%, and after 16 weeks from 74 to 87%. Retentions of riboflavin and pantothenic acid are within the same range after eight weeks of storage. Retention of niacin is 95% or more, except in samples aged only one day at 30° F. (−1° C.); for these samples retentions are 91 and 84%, respectively, after 8 and 16 weeks of storage.

In ripened beef, compared to beef muscles frozen without ripening and held at 0° F. (−18° C.) for 3 years, there are increases of about 30 to 35% for thiamin, calculated on the fat-free, dry basis. However, samples that are ripened 21 days at 34° F. (1° C.) prior to freezing have only slightly more thiamin than the unripened samples and show slight losses in content of thiamin during freezer storage. The content of riboflavin after the 21-day ripening period is practically the same as on the day of slaughter, but after 3 years of storage at 0° F. (−18° C.) shows a small increase of approximately 10% in ripened and nonripened samples. The content of niacin decreases approximately a third during the ripening period and shows little or no further loss during the three-year frozen storage period. The content of niacin in the samples of beef muscles frozen without ripening shows a loss of about 20%.

Meat may be cooked without thawing, or it may be thawed by any one of a number of procedures before it is cooked. The fluid exuded by the meat when it thaws is largely water, but it also contains some protein, B vitamins, and some mineral matter. The protein content of the drip from frozen meat has been found to be somewhat less than 10% regardless of kind—beef, lamb, pork, or veal.

Convenience and amount of time available to the homemaker may determine whether frozen meat is thawed slowly in the refrigerator, at room temperature, in water, or rapidly in a warm oven. All four methods were used in studies showing that, on the basis of weight, the loss from thawing is least, 0.55%, for steaks from rounds of beef thawed in a warm oven at 163° F. (73° C.), a little more, 0.65%, for steaks thawed at refrigerator temperature, and 1.42% for steaks thawed at room temperature. Steaks thawed in running tap water increase in weight by 1.44%. Palatability and vitamin content were determined after the steaks were cooked by braising. Thawing in water was least desirable. Thawing the frozen steaks at room and refrigerator temperatures appeared to allow slightly better retention of the B vitamins than thawing in a warm oven.

Analyses have been made for several B vitamins in the drip from beef rib steaks from carcasses that had been aged 5 days prior to freezing and storage at 0° F. (−18° C.), then thawed 14 to 15 hr. at room temperature. When the thawed steaks and drip were analyzed, significant proportions of the vitamins were found in the drip and would be lost if this portion were discarded. The drip contained 12.2% of the total thiamin, 10.3% of the riboflavin, 14.5% of the niacin, 9.4% of the pyridoxine, 33.3% of the pantothenic acid, and 8.1% of the folic acid.

Fish and Shellfish.—Preservation of fish by freezing has been practiced in this country for more than 100 years. Innumerable changes have occurred in methods of freezing and handling frozen fish and shellfish, especially since World War II. Attention has been focused mainly on the problems of retaining high quality in the fish and shellfish products, with relatively little direct study of the effects of freezing and frozen storage on the nutrients. Although conducted for a different purpose, the studies are helpful in using the few scattered data currently available on the retention of nutrients in frozen fishery products. This information should be helpful in the future for planning studies and applying data on nutrients in frozen fishery products.

Fish and shellfish are good sources of protein of high biological value; the content of protein for most kinds averages about 18 to 20%. A wider range occurs in the content of fat among the different species. Lean fish, such as cod and haddock, usually have less than 1% fat and fatty fish, such as mackerel and salmon, often have up to 10 to 14% or more fat. The content of fat for some species has been found to vary during the year and with location of the catch. These variations in the content of fat are probably dependent on feeding and activity patterns of the fish. The fat of fish is more highly unsaturated than is the fat on meats and protection from rancidity and the development of off-flavors is of major concern in preserving fish.

Temperatures of 0° F. (−18° C.) or below have been recommended for frozen storage of lean fish and below −20° F. (−29° C.) for fatty fish. Desiccation and loss of weight are problems and therefore relative humidities of 90% or more have been recommended for storage of seafoods. Prevention of oxidation is crucial in conserving values of many nutrients. Probably the conditions needed to maintain high quality in flavor, texture, aroma, and color during freezing and frozen storage of fish and shellfish are conducive to retaining high proportions of the important nutrients.

Changes described as denaturation have been found in protein during storage of frozen fish. This alteration in the protein may be important to quality and storage life of the fishery products, but to what extent, if any, it affects the nutritive properties of the protein is not clear.

Overall, the effects of processing, including freezing, on the nutritive value of fish indicates that the protein in fish held in frozen storage maintains its high biological value. Studies indicate 100% digestibility and 80% protein utilization in test rats receiving different forms of cod, namely frozen fillets and raw and cooked fresh cod meat. Work conducted cooperatively at the Universities of California and Wisconsin revealed no significant differences, except for methionine, in the percentages of amino acids from samples of fresh sardines and comparable samples of sardines that had been frozen in dry ice and held 44 days in frozen storage.

Fish and seafood furnish small but appreciable amounts of thiamin. They are good sources of riboflavin and of several other vitamins. Scattered studies indicate possible loss of some thiamin during storage of frozen fishery products and little, if any, loss of riboflavin, niacin, and pantothenic acid.

Frozen shrimp were found to have a small, steady drop in thiamin which, after 90 days of frozen storage, amounted to a loss of about a fourth of the content in the freshly frozen shrimp. Smaller losses or no loss were observed for riboflavin, niacin, and pantothenic acid.

Tests have been done on fresh, frozen, frozen stored, and smoked cod and haddock for content of thiamin, riboflavin, niacin, pantothenic acid, and vitamin B-12. Ranges in content for each vitamin in the fresh fish and in the frozen or smoked fish before and after storage were wide and overlapping. There was no indication that vitamin losses were being incurred in freezing or storage of frozen cod or frozen haddock.

When frozen fish thaws, it loses some fluid—the drip. Under some circumstances this may be considerable, exceeding 10% of the total weight of the fish. The drip contains some of the protein, and it is assumed that when this fluid is exuded from the tissues of the fish, some of the minerals and vitamins would also be lost, especially the water-soluble B vitamins. However, fat-soluble constituents may also be lost in the drip. In discussing changes in nutritive value of fish through handling and processing procedures, it is to be noted that approximately 10% of the loss of vitamin A that occurred on cooking fish was due to physical loss in the exudate.

Poultry.—Poultry is frozen in many forms: whole, halves, serving-size pieces, either raw or cooked, and if cooked, with or without gravy. Commercial freezing of poultry is an extensive industry; in addition, many families freeze poultry for later use.

Maximum storage times recommended by the USDA for home-frozen poultry that has been carefully wrapped to exclude as much air as possible, frozen quickly and held at 0° F. (−18° C.) are 1 month for sliced cooked poultry meat and sandwiches, 6 months for uncooked duck and goose, and 12 months for uncooked chicken and turkey.

It is important that the requirements for processing and packaging poultry be met to provide suitable stability under commercial conditions of time and temperature. Studies of moisture loss, peroxide development, off-odors in raw meat, and off-flavors in cooked meat emphasize the need for maintaining storage temperature at least as low as 0° F. (−18° C.) and for packaging to exclude air and reduce exposure to oxygen as much as possible. Until further specific information on the effect of freezing on the nutritive value of poultry meat becomes available, it would probably be satisfactory to assume that the nutrients in poultry are retained to about the same extent as those in red meats.

Dairy Products

The group of dairy products includes one of the oldest and most popular frozen foods. Ice cream is consumed on the average of about 18 lb., the equivalent of about 4 gal. per person annually in this country. In addition, a great deal of ice milk is used, as well as other frozen dairy confections. The freezing of these particular foods is not, of course, for preservation in the same sense that most foods are frozen; nevertheless, the frozen dairy foods materially extend the consumer market for milk and its products and make an important contribution to the American diet.

The nutrients in the ingredients used for the manufacture of ice cream and the other frozen dairy desserts are probably well retained. Data for the composition of the frozen dairy products in Agriculture Handbook No. 8 "Composition of Foods . . . raw, processed, prepared" were calculated on this assumption from average, year-round contents of nutrients in the ingredients prior to freezing, as losses of nutrients in manufacture and storage reported for the principal nutrients investigated have been small.

A study of the vitamin A value of butter was conducted cooperatively in the 1940's by several states and reported by the USDA. The effect of storage on the content of vitamin A and carotene in butter produced in different regions and held under various practical conditions was determined. Storage temperatures ranged from a low of −10° F. (−23° C.) to a high of 45° F. (7° C.). Length of storage varied from 15 to 30 days at the highest temperature used to 12 months at a temperature of 0° F. (−18° C.). Some samples of butter were held 8 months at about −6° F. (−21° C.). Practically no or very little change in either carotene or vitamin A content occurs during storage. No significant loss in the vitamin A potency of butter stored as long as 8 months at 14° F. (−10° C.) is found.

Frozen milk and frozen concentrated milk are being marketed to a limited extent, and research is continuing on the development of frozen milks satisfactory for beverage use. At present no information is available as to what extent, if any, the nutritive values of milk would be affected, but it is believed that neither the content nor the utilization of the more important nutrients would be reduced in these frozen milks.

CONTRIBUTION OF NUTRIENTS MADE BY FROZEN FOODS

Frozen foods are making a steadily increasing contribution to the nutritive value of the American diet. Data are available for calculating the nutritive value of frozen fruits and vegetables consumed by the American people. Fortunately, similar data are becoming available for frozen foods in other food groups—meat, poultry, eggs, dairy products, and grain products.

The amounts of nutrients furnished by frozen fruits and vegetables have been calculated from statistics on the civilian supplies of the different frozen foods and from data on the content of nutrients present in the fruits and vegetables. Proportions of the total amounts of each nutrient in the food supply furnished by frozen fruits and vegetables were shown in Table 14.1. In recent years frozen fruits and frozen fruit products have supplied from ¼ to ⅓ of the entire amount of ascorbic acid contributed by all forms of fruit. Frozen vegetables and frozen vegetable products have been providing a small but increasing proportion of the ascorbic acid contributed by all forms of vegetables. If the use of frozen forms of foods in other food groups increases as various predictions indicate, frozen foods can be expected to account for significant percentages of other major nutrients.

ADDITIONAL READING

ASHRAE. 1974. Refrigeration Applications. American Society of Heating, Refrigeration and Air-Conditioning Engineers, New York.
HARRIS, R.S., and KARMAS, E. 1975. Nutritional Evaluation of Food Processing, 2nd Edition. AVI Publishing Co., Westport, Conn.
JOHNSON, A. H., and PETERSON, M.S.1974. Encyclopedia of Food Technology. AVI Publishing Co., Westport, Conn.
KRAMER, A. 1973. Storage retention of nutrients. Food Technol. 28, No. 1, 50–60.
LACHANCE, P. A., RANADIVE, A. S., and MATAS, J. 1973. Effects of reheating convenience foods. Food Technol. 27, No. 1, 36–38.
PETERSON, M. S., and JOHNSON, A. H. 1977. Encyclopedia of Food Sicence. AVI Publishing Co., Westport, Conn.
SMITH, C. A., JR., and SMITH, J. D. 1975. Quality assurance system meets FDA regulations. Food Technol. 29, No. 11, 64–68.

TRESSLER, D. K., VAN ARSDEL, W. B., and COPLEY, M. J. 1968. The Freezing Preservation of Foods, 4th Edition. Vol. 2. Factors Affecting Quality in Frozen Foods. Vol. 3. Commercial Freezing Operations—Fresh Foods. Vol. 4. Freezing of Precooked and Prepared Foods. AVI Publishing Co., Westport, Conn.

15

Quality Compliance and Assurance

*Leonard S. Fenn, Amihud Kramer
and Bernard A. Twigg*

T he commercial producer of frozen foods needs to know not only the mechanics of getting his produce ready for market, but also all the trading requirements such as standards, specifications, regulations, and codes.

Quality standards and specifications, among other things, provide a common language for use in purchase and sale negotiations and other business transactions. Based on commercial quality levels, they serve as useful guides in packing the frozen product.

Regulations and codes primarily are rules governing the packing and marketing of frozen foods. Fundamentally developed in the interest of the consumer, they also encourage uniformity in packing and handling practices.

The field of "rules" under which the frozen food industry operates is both vast and complex. Basically, the so-called rules can be classed as either mandatory or voluntary. Both of these general categories deal with frozen food requirements and frequently with inspection activities performed by a federal, state, or municipal agency. Largely because both the mandatory and the voluntary categories are quite generally associated with inspectors, the difference between the two types can easily be confused.

The term mandatory applies to all standards, specifications and codes with which frozen food placed in the channels of commerce must comply. Requirements for compliance and the laws authorizing the development of the operating regulations spell out the scope of enforcement.

The term voluntary applies as long as compliance with the standard, specification, or code remains optional. Quality requirements in a grade standard or specification incorporated into a contract or marketing order convert it into an instrument in which compliance may be required. Nevertheless, these usages are subject to negotiation and are not enforced by regulation.

MANDATORY REQUIREMENTS

In carrying out statutory delegations of authority that require the compliance of frozen food producers and handlers, federal, state, and local government agencies are primarily concerned with the administration of three different types of mandatory requirements for frozen food in interstate commerce.

Food and Drug Administration—Definitions and Standards

Three kinds of standard may be established by the Food and Drug Administration: (1) a reasonable definition and standard of identity, (2) a reasonable minimum of quality, and (3) a minimum standard of fill of container.

A standard of identity defines the product, establishes its common or usual name, and limits the optional ingredients which may be used. This kind of standard is expected to establish what a buyer could reasonably expect to receive, based on a descriptive common or usual name.

A standard of quality establishes minimum quality requirements below which the product is considered inferior and subject to special labeling which will indicate low grade.

A standard of fill for frozen food establishes minimum requirements and procedure for ascertaining whether the package is filled as full as practicable with product.

The FDA usually develops a definition and standard on the basis of a proposal from an industry source pointing out a need for a regulation to promote honesty and fair dealing in the interest of consumers. However, the FDA may elect to propose food standards on its own initiative for the same purpose.

To develop standards, proposed regulations are published in the Federal Register, with a definite period for comments and objections. In the absence of reasonable objections the standard only needs to be republished to become effective. More often, objections are made which require development of definite issues and a scheduled hearing. Voluminous testimony may be offered and long delays may be experienced before a legal standard can be promulgated.

Standards of Identity.—Definitions and standards of identity have been promulgated by the FDA for a number of frozen products.

Frozen foods for which Food and Drug Standards have not been promulgated are covered under general regulations as "nonstandardized foods." The general labeling provisions must be met for the nonstandardized frozen foods shipped in interstate commerce. The National Association of Frozen Food Packers has summarized the FDA Standards in its Labeling Manual to provide guidance in good labeling practices. Legislation passed by the 89th Congress known as the Fair Packaging and Labeling Act became effective on July 1, 1967. This legislation is popularly known as the Truth-in-Packaging Law. Regulations developed by the FDA, required by the Act, initiated a basic overhaul of essentially all frozen food labels.

The label must identify the product and separately and accurately place the net quantity of the package contents on the principal display panel. The name and address of the manufacturer, packer, or distributor must be shown. If the label says anything about servings, it must state how much is in each serving. For many packages the label must show a dual net content statement in ounces and pounds and ounces. The type size of the net content declaration will have a relationship to the size of the package.

The packaging requirements of the Fair Packaging and Labeling Law delegates to the Federal Commerce Department responsibility for protecting the consumer's ability to make value comparisons in purchasing packaged food. This requirement would be accomplished through the development of voluntary standards when it is determined that undue proliferation of the weights, measures, or quantities of packages of comparable commodities impairs the ability of consumers to make value comparisons. The National Bureau of Standards has been given the responsibility of administering this voluntary program.

Agricultural Marketing Service—Meat and Poultry Inspection for Wholesomeness

Meat and poultry inspection, required by law, is provided by the Meat and Poultry Programs of the USDA. An important part of the mandatory requirements is that the meat or poultry is from healthy animals and is fit for food. Inspection and supervision during the processing of frozen meat and poultry products is required to insure that proper operating procedures and sanitary measures are carried out, thus assuring the production of a clean, wholesome frozen food. The plant and its processing facilities must also be approved for inspection during all operations.

Labels for frozen meat and poultry products must be approved before they are eligible for use. The same legal requirements which are applicable to labels under the FDA regulations apply also under the USDA regulations. Labels are required to show at least (1) name of the product,

(2) ingredient statement, (3) quantity of contents, (4) inspection legend, and (5) processor's or distributor's name and address.

Informative labeling of the product for compliance with its formulation is controlled through established guidelines and standards. In the FDA, these are known as "standards of identity" while in the Federal meat and poultry inspection program they are more commonly referred to as "standards of composition." Standards of composition usually specify the minimum amount and the kind of the more expensive ingredient(s), such as the meat or poultry or by-products, that must be included in the frozen food. The maximum amount of the cheaper ingredients or nonmeat and nonpoultry components that may be used in a product formula may also be set by specification. A standard of composition establishes the descriptive common or usual name for the food and so is very important for label approval. Label approval includes scrutiny of pictorial and promotional terms to assure their freedom from false or deceptive implications.

The component requirements for packing frozen meat products are contained in "Regulations Governing the Meat Inspection of the U.S. Dept. of Agr." This publication is identified as Service and Regulatory Announcement *C&MS-SRA-188*. Similar requirements for frozen poultry products are contained in "Regulations Governing the Inspection of Poultry and Poultry Products."

Processing plants which do not engage in interstate commerce are nevertheless required in some states to have inspection, and to identify their product(s) as complying with requirements similar to the federal mandatory program. On some of the programs, interagency agreements may be involved, designed to coordinate procedures for the consumers' protection to assure wholesomeness of these products.

The component requirements for frozen meat and poultry products have been established to a large extent by a more informal procedure than standards promulgated by the FDA; usually the requirements have been discussed in a preliminary way with packers of any formulated products involved. Efforts are made to eliminate problem areas before the formal proposed "rules" are published with an invitation for general comment by interested parties. In most cases, requirements have been developed without having to resort to formal hearing procedures. However, in the case of frozen chicken soup, a public hearing was held and a finding was made supporting the originally proposed requirements. The opponents of the proposed regulation then appealed to the Courts. The case was settled in favor of the USDA.

A listing of some frozen products on which component specifications have been adopted follows:

Meat Products

Lima Beans with Ham	Beef with Gravy
Chow Mein Vegetables with Meat	Gravy with Beef
Beef with Barbecue Sauce	Beef Pie
Pork with Barbecue Sauce	Pork Pie

Poultry Products

Poultry Dinners	Chop Suey with (Kind)	(Kind) Stew
Poultry Rolls	(Kind) Chop Suey	(Kind) Fricassee of
(Kind) Burgers	(Kind) Chow Mein	Wings
(Kind) Ravioli	with Noodles	(Kind) Noodles or
(Kind) Soup	(Kind) Tamales	Dumplings
(Kind) Fricassee	Noodles or Dumplings	Creamed (Kind)
Minced (Kind)	with (Kind)	(Kind) Cacciatore
Barbecue	(Kind) À-La-King	Sliced (Kind) with
		Gravy

State Regulations—AFDOUS Frozen Food Handling Code

AFDOUS is the abbreviation for the Association of Food and Drug Officials of the United States. It is the largest and most influential organization of food, drug, and public health officials from all levels of government within the United States. Public Health officials in the early 1950's became concerned with what they considered a potential health hazard of frozen foods. In 1956, AFDOUS passed a resolution calling for the development of a frozen food handling code. The code was to consider protection of frozen foods from adverse temperatures at all levels of distribution, in the consumers' interest, and also to consider the bacteriological quality of frozen prepared foods.

Studies of the stability of many kinds of frozen food by the Albany, Calif., laboratory of the USDA had pointed to 0° F. (−18° C.) as a good average commercial handling temperature. The studies showed that rates of quality change would be nearly undiscernible at this temperature for most products, although for certain kinds a still lower temperature would be better.

The National Association of Frozen Food Packers, cooperating with AFDOUS officials, studied retail handling practices at 3,063 retail stores located in 35 states. Both product temperatures and microbiological conditions were included in the extensive studies. Plant operations, over-the-road transportation, and warehousing of frozen food were studied by subgroups of AFDOUS in considering handling practices.

The AFDOUS handling code for frozen food was adopted at the Association's 65th Annual Conference; it provided for temperature limitations at all stages of distribution. The code did not include bacteriological provisions, but further studies in this area were recommended and are still under consideration. It did include a provision calling for laboratory testing for wholesomeness of frozen foods temporarily experiencing higher temperatures in the course of distribution.

The need for voluntary frozen food regulations for frozen fruits and vegetables was foreseen early in the development of the industry. Some acceptable documents for the use of buyers and sellers of frozen barreled strawberries, to facilitate orderly marketing, were in use in the 1920's. In the mid-30's packers in the Northwest concerned themselves with specifications for frozen peas. The serious obstacles encountered in selling frozen food to the Armed Forces during World War II brought about the first united efforts to develop quality standards and government inspection services.

The United States Department of Agriculture—Inspection and Standardization Programs

Standards are often described as a yardstick that can be used to measure the quality of a product. Thus they constitute the needed common measuring device upon which buyers and sellers can base their contracts. The USDA conducted its first program of participation in frozen food standards and inspection on the authority of an Agricultural Appropriation Act. Since the enactment of the Agricultural Marketing Act of 1946, standardization, classing, grading, and inspection programs have been carried out under the authority of the Act. The Agricultural Marketing Act of 1946 specifically authorizes the Secretary of Agriculture to "develop and improve standards of quality, condition, quantity, grade, and packaging," and "to inspect, certify, and identify the class, quality, quantity, and condition" of agricultural products when shipped or received in interstate commerce and to assess and collect such fees "as will be reasonable and as nearly as may be to cover the cost of the service rendered."

In developing and carrying out these programs, procedures must necessarily vary in accordance with the characteristics and manner of utilization of the frozen food; for example, procedures for frozen fruits and vegetables may vary from those for poultry or egg products.

Grade standards are developed in the USDA by the respective commodity agencies, and program procedures are carried out under published "Regulations" governing each product group. Both the grade standards and procedure regulations are considered "Rules" and therefore must be developed under the "Administrative Procedures Act" permitting public review and comments.

The grade standards, among other things, provide a common language for use in sales agreements, purchase specifications, purchase and sale negotiations, and other business transactions. They are used to some extent as a basis for designating a quality classification on packaged frozen foods (Fig. 15.1).

The standards would, of course, be of little value as acceptable documents between buyers and sellers without recourse to a grading service by a neutral party. One of the largest uses of standards, therefore, is for the classification of the grade by inspection services.

Inspection services are available on an "in-plant" basis or may be performed, for each type, on a statistical sampling plan and acceptance quality level. In-plant inspection procedure, frequently referred to as "Continuous Inspection" or "Contract Inspection," requires that the inspector observe plant sanitation and wholesomeness of raw material, as well as classification of the product quality (Fig. 15.2 and 15.3). Although the standards and services are classed as voluntary, compliance with contract requirements may establish certain processing and labeling restrictions not required by plants using inspection only on specified lots of frozen products.

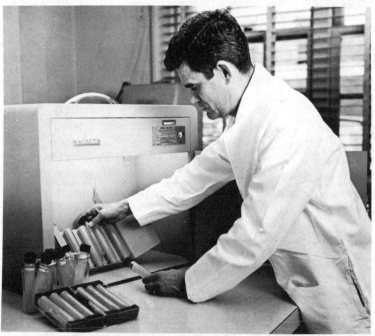

FIG. 15.1. SAMPLES OF FROZEN CONCENTRATED ORANGE JUICE JUDGED BY USDA COLOR STANDARDS UNDER CONTROLLED LIGHT

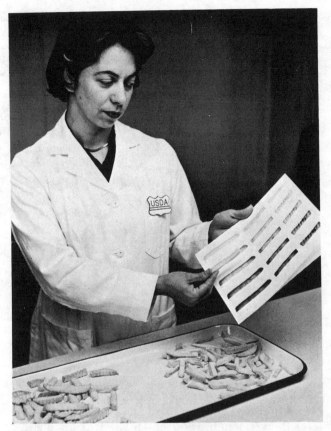

FIG. 15.2. COMPARING FROZEN FRENCH-FRIED POTATOES WITH USDA
COLOR STANDARDS

Grade standards have been developed in the USDA applicable to frozen fruits, vegetables and juices, and certain poultry products. Frozen
eggs and some other frozen products are inspected for wholesomeness
and sometimes the quality characteristics of the product are described
although no standards have been developed.

The voluntary U.S. grade standards promulgated on frozen fruits,
vegetables, and juices are under continuing review by the industry and
government and are frequently revised to keep them current with good
commercial production and marketing practices. Because of the frequent
changes made in grade standards, a revised list is published annually;
copies are available from Information Service, U.S. Dept of Agr.,
Washington, D.C.

FIG. 15.3. USDA STANDARDS SPECIALIST EXPLAINS FROZEN RED TART PITTED CHERRY QUAL-
ITY FACTORS TO A CONSUMER PANEL

A summary of the USDA Consumer and Marketing Services voluntary
and mandatory programs is contained in publication PA-661 entitled
"This is USDA's Consumer & Marketing Service."

United States Department of Interior—Standardization and Inspection Program

On March 22, 1958, a determination by the Bureau of the Budget was
reached under authority of the Fish and Wildlife Act of 1956 to transfer
all grade standard and inspection functions pertaining to fish and seafood
products from the USDA to the Bureau of Commercial Fisheries, U.S.
Dept. of the Interior.

The voluntary services are closely patterned after those in the Consumer & Marketing Service of the USDA. The inspection services are
financed from fees collected; the limited demand for this inspection on
lots of fish and fish products at various levels of distribution has contributed to limitation of the facilities available to perform this type of
inspection.

The Bureau of Commercial Fisheries, however, provides "in-plant" or
"continuous" inspection to plants for a large volume of products on which
quality standards have been promulgated (Fig. 15.4).

FIG. 15.4. U.S. DEPT. OF INTERIOR INSPECTION OF FISHERY PRODUCTS
IN SHRIMP PROCESSING PLANT

The U.S. Dept. of Interior inspector frequently checks each processing area
at his assigned plant. Here we observe the inspector visiting the preparation area where shrimp are peeled and deveined prior to the breading
operation.

The grade standards for fishery products were developed with the aid
of the fishery industry and the respective associations serving the individual groups such as shrimp packers. Quality grade designations are
permitted on the labels of product under the Bureau of Commercial
Fisheries' continuous inspection program. A number of grade standards
have been developed and promulgated on fish and fishery products. As in
the USDA, the inspection procedure is carried out under published
"Regulations" developed under the "Administrative Procedures Act" and
reviewed by interested parties.

Guideline Codes for Voluntary Compliance

The American people are eating better frozen food today than ever
before. The continued advances in providing wholesome high-quality
frozen food to the public can be attributed to a large extent to voluntary
self-imposed codes and requirements which usually exceed mandatory
regulatory requirements or minimum levels established under voluntary
procedures.

In-plant control is accomplished through a competent check system,
starting with the raw material at the plant's receiving dock, and continuing through each step in the preparation and processing to the finished
frozen product (Fig. 15.5). Many of the quality control procedures are

Courtesy of Seabrook Farms Co.

FIG. 15.5. QUALITY CONTROL LABORATORY IN A FROZEN FOODS PLANT

In the Quality Control Laboratory at Seabrook Farms Company, tests are run to check quality from the time the raw produce enters the plant through each processing and freezing step.

based on statistical assurance of compliance with the plant's standards. Proper attention to pertinent details in the preparation and processing of the raw material can mean the difference between a clean, wholesome product of the high quality desired, and a product of lower quality.

Frozen food industry trade groups have joined in an organized effort to promote continuing improvement of operating practices in the handling and distribution of frozen food. Recommended voluntary operating rules for the handling of frozen packaged foods, through all the steps of preparation and distribution, equal or exceed in strictness the standards established in the AFDOUS Code. The USDA, through its Federal Extension Service and the publications and service activities of the Western Utilization Research and Development Division, contributes substantially to maintenance of high standards for good practice in the handling of frozen foods.

The sound principle of relying heavily on voluntary compliance, not simply policing and enforcing rigid requirements, is recognized and employed by the FDA through its Bureau of Education and Voluntary Compliance. This activity provides an avenue of direct communication through which it can advise the industry on regulatory procedures, assist in organizing workshops, and develop educational material for industry's staff and employees so they may better protect the consumers' frozen food supply.

Buyers' Specifications

The current distribution of packaged frozen foods to consumers through both packer labels and private labels has developed an expanding group of "buyers' specifications." These specifications usually prescribe compliance with all of the requirements of the Federal Food, Drug, and Cosmetic Act and establish quality requirements, in most cases, that follow guidelines set by the grade standards of the USDA or U.S. Dept. of Interior. Additional requirements may be imposed, for example, on fill, labeling, and packaging.

Purchase Specifications of Government and State Agencies for Frozen Foods

The procurement of frozen food by government and state agencies is generally based on an applicable federal specification. However, the respective agencies, if unusual requirements arise, may develop their own specifications. For example, specifications have been developed by the Armed Forces for multicomponent frozen products and various items on which a Federal Specification is not available. These specifications are considered to be temporary. The Veterans' Administration has issued its own specifications for certain frozen food items to meet hospital feeding needs. The USDA also issues its own specifications for school lunch purchases; likewise, state procurement officers develop buying specifications. Both, however, follow the quality guidelines of the U.S. Standards for Grades. Compliance with the various state and federal government specifications in the purchase of frozen food is usually determined by the agency inspection. Uniformity in the quality requirements is quite consistent, inasmuch as voluntary grade standards establish the basic guidelines.

In addition to specifications of quality, military procurement and USDA purchases for the school lunch program also specify sanitary conditions which must be met by any plant supplying the product.

The "Index of Federal Specifications and Standards," published by the General Services Administration, Federal Supply Service, is for sale by the Superintendent of Documents, U.S. Government Printing Office, Washington, D.C. This index book, in addition, supplies information on where to obtain military and special specifications which are on the Defense Department's procurement list.

International Standards for Frozen Food

Efforts to establish standards for both products and packages so as to facilitate economic handling of goods between countries and in world trade have been going on for a number of years.

The first meeting to consider standards for frozen food was in Sept., 1964, when a group of experts representing the Economic Commission of Europe met in Geneva, Switzerland, and formulated plans for standardization of "Deep-Frozen Foods." Subgroups were set up at that meeting to develop proposals for definitions, packaging, and labeling requirements, a framework of standards, and a few pilot standards for individual products.

The second meeting, which was held Sept. 6–10, 1965, was broadened to an international basis and became the first Joint ECE/Codex meeting on the standardization of "Quick-Frozen" foods. A third meeting was held in Geneva in 1966; subsequent meetings are held annually to develop International Standards for frozen foods.

A "General Standard for Quick-Frozen Foods" has been developed for member governments. This document, similar in many respects to the AFDOUS Code, covers: (A) Definition of Quick-Frozen Foods, (B) Raw Material, (C) Treatments, (D) Hygiene Requirements, (E) Storage, Transportation and Distribution, (F) Refreezing, (G) Packaging, (H) Labeling, (I) Sampling and Methods of Analysis, (J) Food Additives, (K) Pesticides and Other Residues, and (L) Sphere of Application.

Individual product standards will be developed by the Joint ECE/Codex Alimentarius Group for Quick-Frozen Foods and other groups on a commodity as specifically assigned by the Codex Alimentarius Commission. The Commission under the direction of the Food and Agriculture Organization (FAO) and World Health Organization (WHO) is open for membership by all Member Nations interested in international standards. Food groupings have been made by the Commission as follows: (A) Fruits and Vegetables, (B) Fish and Fishery Products, (C) Meat and Meat Products, (D) Poultry and Poultry Products, (E) Fruit Juices, and (F) Ice Cream.

The government delegates of the various countries, which make up a group of experts to develop international trade standards on frozen foods, are usually accompanied by one or more trade advisors. The government delegates are generally representatives of the agency or agencies having the greatest interest in the international standards. The delegates, who are government officials, in general are permitted to select their trade advisors. This practice is an important safeguard to keep such standards in line with good commercial operating practices and avoid creation of trade barriers.

The general intention for individual product standards is to establish labeling requirements very similar to a U.S. FDA Standard of Identity and set up general wholesomeness and quality guidelines.

QUALITY ASSURANCE

Statistical quality control has been called the greatest advance in manufacturing during the last quarter century. Whatever the truth of this opinion, the fact is that Statistical Quality Control ("S.Q.C.") is a major contribution to manufacturing efficiency. It effects substantial savings by preventing waste, eliminating rework and reducing the amount of necessary inspection. It gives assurance of a high, uniform quality of products leaving the plant. By providing a common measure of product quality, it facilitates understanding between producer and consumer, and helps to insure acceptance of products. Statistical Quality Control is becoming recognized both in government and in private industrial plants as the hallmark of efficient management, and it has become standard operating procedure in inspection programs of the armed forces.

Actually, much of the early work in S.Q.C. was done in the food field—precisely because natural variability of the materials was so great, and factors affecting quality of products so many, that better means were required to distinguish cause and effect and thus speed remedial measures through changes in material, equipment, or production procedures.

Essentially S.Q.C. consists of optimal methods of sampling, reporting of test data particularly by control chart methods, and a decision-making function whereby significance of the reported data may be determined on a basis of statistical assurance.

Sampling

Sampling inspection is a well established part of the quality control operation in many frozen food processing plants. All too frequently, results of such inspection are taken at face value. It is assumed that the test value is identical with the true average value of the lot tested. Or, even more erroneously, the unit tested is assumed to be identical with each and every unit in the entire lot. These erroneous assumptions are not shared by many workers familiar with the inherent variability existing in biological material; they realize that inspection results become more representative of the lot as the frequency and size of sampling are increased. At times, such workers strive to increase inspection work to the maximum allowed by management. This approach may be wasteful, since an optimum inspection procedure should provide an estimate of lot quality at the minimum acceptable level of accuracy at the least cost. Such an optimum procedure may be selected only by the use of certain statistical tools. It is the purpose of this presentation to indicate the factors which influence the choice of specific sampling plans for specific purposes, and to outline some rapid statistical procedures by which these sampling plans may be constructed.

The factors which influence the sampling procedure to be selected may be listed as follows: (1) purpose for which the inspection is made; (2) nature of the material to be tested; (3) nature of the testing methods; and (4) nature of the lots being sampled.

Purpose of Inspection.—Not all inspection is done for the same purpose; hence, sampling plans must be selected which are capable of achieving the specific intended purpose.

Accept or Reject.—This situation is encountered where the freezer is offered a lot of raw material, and he may decide whether he will or will not buy the lot. In such cases, an *attributes* sampling plan is frequently suitable.

Evaluate Average Quality.—In many instances, there is no opportunity to reject lots, so that the quality level is determined primarily for the purpose of establishing payment on the basis of a predetermined sliding scale. Here actual measurements must be made so that a *variables* plan may be required.

Determine Uniformity.—Particularly where discrete units are packed rather than blends, the variability existing in a given lot is as important as average quality. Here again, a *variables* plan is indicated.

Nature of Material.—The size of the sample which is necessary to be representative of the lot is influenced by the following characteristics of the material.

Homogeneity.—Where an essentially homogeneous material is to be sampled, such as a true solution, one small sample is sufficient. As variability among units of the lot increases, there is need for increasing the number of units comprising the sample in order to have the sample representative of the lot.

Unit Size.—With liquid or semiliquid products, and with products consisting of small particle sizes, there is the need for establishing some discrete sample unit size, such as the volume of a container, or the contents of a probe. Where the unit size is larger, as for example an ear of corn, that unit may serve directly as a sample unit. Where the units are very large, such as a side of beef, it may be necessary to determine exactly how, where, and how much of the unit is to be removed for the sample.

History of Material.—Whatever is known of the history of the lot should be considered in setting up a sampling procedure. Sampling may be reduced if it is known that the lot originates from a reliable source. If the lot is part of a larger batch which had already been sampled repeatedly, sampling may be greatly reduced or even eliminated entirely for some lots. On the other hand, if the lot originates from a questionable source, sampling may be tightened.

Cost of Material.—Where the raw material in the sample itself constitutes a substantial part of the sampling cost, it may be necessary to restrict the amount used for sampling purposes. For this reason, for example, crab meat samples may be smaller than potato samples.

Nature of Test Procedures.—The nature of the tests to be determined will also influence the sampling procedure.

Importance of the Test.—All tests may be classified as being of critical, major, or minor importance. Inspection for critical factors, for example, factors that may have an effect on the health of the consumer should be sampled and tested more rigorously than major factors which do not constitute health hazards, but are of economic importance. Factors of minor importance may be sampled less thoroughly.

Destructive or Nondestructive.—Particularly where the raw material is costly, sampling may be reduced if the test procedure is destructive of the material tested. Thus, for example, peas tested for texture with a tenderometer may no longer be used, while asparagus examined for white butts may be returned for processing.

Time and Equipment Consumption.—A test that is time consuming and ties up elaborate and costly equipment and personnel is more costly to perform than some rapid, simple test. This must also be considered in setting up the frequency and size of sampling.

Nature of Lot.—Characteristics of the lot such as size, packaging, and loading help determine sampling procedures.

Size.—Accuracy of sampling is related almost entirely to the number of samples tested (n) and the acceptance number (c) rather than to size of the lot (N). The fact that lot size (N) is not important in establishing an acceptance sample plan indicates the fallacy of a straight percentage sampling plan. However, since more of an investment is involved with larger lots, it is desirable to increase sampling accuracy, justifying a larger sample number with a larger lot size.

Sublots.—Raw material frequently arrives at the processing plant not in bulk, but packaged. In such cases, there is the problem of how many units out of how many sublots are to be drawn.

Loading.—The manner in which the lot is stacked or loaded may present a problem in the random selection of sample units. The principle of randomness is basic to proper sampling. Every unit must have equal opportunity for being drawn. Where it is impossible or impractical to draw sample units from any location but the top or the end of a load, special facilities or arrangements must be made to overcome such difficulties. The factors which influence the selection of a sampling plan are summarized in Table 15.1.

The sampling procedure to be selected for a specific situation will depend on the conditions discussed previously. All sampling procedures, however, may be classified conveniently into four general categories based on whether an attributes or variables plan is required and whether lots to be tested are in bulk or packaged into sublots.

The four general categories into which sampling procedures may be classified are thus: (1) attributes, bulk-lots; (2) attributes, sublots; (3) variables, bulk-lots; and (4) variables, sublots.

Attributes, Bulk-Lots.—Since attributes (accept or reject) procedures are relatively easy to apply, it is well to consider their use, provided, of course, that the influencing factors are of such a nature that the requisites for their use are satisfied in whole or at least in large measure.

Some typical situations where attributes plans are applicable are in the sampling of supplies other than the raw food materials, such as cans, glass containers, cartons, labels, or in sampling for visual defects, such as insect or rodent damage, presence of foreign matter, diseased or discolored spots, etc. A number of attributes sampling plans are available.

U.S. Department of Agriculture.—A sample plan is provided by the USDA for the sampling inspection of certain processed foods (Table 15.2). The AQL for these sampling plans were established at 6% deviant, meaning any lots containing 6% or less deviant product, all of their various sampling plans will accept those lots 95% of the time or more, excepting the plan for 3 sample units and an acceptance number of 0. The buyer's risk of accepting lots that should be rejected may be substantially greater than that, and increase with decreasing size of the lot, and increasing size of the individual container. These sampling plans therefore tolerate a substantial buyer's risk because the sampling is destructive and testing procedures may be extensive, conditions which ordinarily militate against the use of attributes procedures. Double sampling plans are also available.

Example 1. Attributes Sampling Plan by Use of Dept. of Agriculture Tables.—An inspector is asked to certify a lot of frozen peas consisting of 400 cases each containing 48 10-oz. cartons, or a total lot size of 19,200 cartons. Referring to Table 15.3 he finds that the lot falls in size group 1, since the container is smaller than one pound, and in the lot size bracketed by 12,001–24,000. He therefore selects at random 13 cartons and certifies the lot provided not more than 2 of the cartons are out of grade.

Tables similar to the one shown as Table 15.3 have been prepared for frozen fruits, vegetables, and other foods. Copies of these plans are also available from the Agricultural Marketing Service, Washington, D.C.

TABLE 15.1

SUMMARY OF FACTORS INFLUENCING SELECTION OF SAMPLING PLANS

Factors			When to Use:			
	Attributes	Variables	Increased Sampling Frequency	Reduced Sampling Frequency	Increased Sample Numbers	Reduced Sample Numbers
Purpose of inspection	Accept or reject	Evaluate	Measure uniformity	Measure average quality	Increase precision	Reduce cost of sample
Nature of the material	Inexpensive	Costly	Variable	Homogeneous	Variable small units, unknown history, inexpensive	Homogeneous large units, known history, costly
Test procedures	Non-destructive, rapid	Destructive, time consuming	Less precise	More precise	Critical, rapid	Minor, time-consuming
Lot characteristics			Sublots	Bulk	Large	Small

TABLE 15.2

SINGLE SAMPLING PLANS FOR NORMAL INSPECTION (MASTER TABLE) BY ATTRIBUTES FROM MIL-STD-105D

Acceptable Quality Levels (normal inspection)

Sample code letter	Sample size	0.010		0.015		0.025		0.040		0.065		0.10		0.15		0.25		0.40		0.65		1.0		1.5		2.5		4.0		6.5		10		15		25		40		65		100		150		250		400		650		1000	
		Ac	Re	Ac	Re	Ac	Re	Ac	Re	Ac	Re	Ac	Re	Ac	Re	Ac	Re	Ac	Re	Ac	Re	Ac	Re	Ac	Re	Ac	Re	Ac	Re	Ac	Re	Ac	Re	Ac	Re	Ac	Re	Ac	Re	Ac	Re	Ac	Re	Ac	Re	Ac	Re	Ac	Re	Ac	Re	Ac	Re
A	2	↓	↓	↓	↓	↓	↓	↓	↓	↓	↓	↓	↓	↓	↓	↓	↓	↓	↓	↓	↓	↓	↓	↓	↓	↓	↓	↓	↓	↓	↓	↓	↓	0	1	1	2	2	3	3	4	5	6	7	8	10	11	14	15	21	22	30	31
B	3	↓	↓	↓	↓	↓	↓	↓	↓	↓	↓	↓	↓	↓	↓	↓	↓	↓	↓	↓	↓	↓	↓	↓	↓	↓	↓	↓	↓	↓	↓	0	1	1	2	2	3	3	4	5	6	7	8	10	11	14	15	21	22	30	31	44	45
C	5	↓	↓	↓	↓	↓	↓	↓	↓	↓	↓	↓	↓	↓	↓	↓	↓	↓	↓	↓	↓	↓	↓	↓	↓	↓	↓	↓	↓	0	1	1	2	2	3	3	4	5	6	7	8	10	11	14	15	21	22	30	31	44	45	↑	↑
D	8	↓	↓	↓	↓	↓	↓	↓	↓	↓	↓	↓	↓	↓	↓	↓	↓	↓	↓	↓	↓	↓	↓	↓	↓	↓	↓	0	1	1	2	2	3	3	4	5	6	7	8	10	11	14	15	21	22	30	31	44	45	↑	↑	↑	↑
E	13	↓	↓	↓	↓	↓	↓	↓	↓	↓	↓	↓	↓	↓	↓	↓	↓	↓	↓	↓	↓	↓	↓	↓	↓	0	1	1	2	2	3	3	4	5	6	7	8	10	11	14	15	21	22	30	31	44	45	↑	↑	↑	↑	↑	↑
F	20	↓	↓	↓	↓	↓	↓	↓	↓	↓	↓	↓	↓	↓	↓	↓	↓	↓	↓	↓	↓	↓	↓	0	1	1	2	2	3	3	4	5	6	7	8	10	11	14	15	21	22	30	31	44	45	↑	↑	↑	↑	↑	↑	↑	↑
G	32	↓	↓	↓	↓	↓	↓	↓	↓	↓	↓	↓	↓	↓	↓	↓	↓	↓	↓	↓	↓	0	1	1	2	2	3	3	4	5	6	7	8	10	11	14	15	21	22	30	31	44	45	↑	↑	↑	↑	↑	↑	↑	↑	↑	↑
H	50	↓	↓	↓	↓	↓	↓	↓	↓	↓	↓	↓	↓	↓	↓	↓	↓	↓	↓	0	1	1	2	2	3	3	4	5	6	7	8	10	11	14	15	21	22	30	31	44	45	↑	↑	↑	↑	↑	↑	↑	↑	↑	↑	↑	↑
J	80	↓	↓	↓	↓	↓	↓	↓	↓	↓	↓	↓	↓	↓	↓	↓	↓	0	1	1	2	2	3	3	4	5	6	7	8	10	11	14	15	21	22	30	31	44	45	↑	↑	↑	↑	↑	↑	↑	↑	↑	↑	↑	↑	↑	↑
K	125	↓	↓	↓	↓	↓	↓	↓	↓	↓	↓	↓	↓	↓	↓	0	1	1	2	2	3	3	4	5	6	7	8	10	11	14	15	21	22	30	31	44	45	↑	↑	↑	↑	↑	↑	↑	↑	↑	↑	↑	↑	↑	↑	↑	↑
L	200	↓	↓	↓	↓	↓	↓	↓	↓	↓	↓	↓	↓	0	1	1	2	2	3	3	4	5	6	7	8	10	11	14	15	21	22	30	31	44	45	↑	↑	↑	↑	↑	↑	↑	↑	↑	↑	↑	↑	↑	↑	↑	↑	↑	↑
M	315	↓	↓	↓	↓	↓	↓	↓	↓	↓	↓	0	1	1	2	2	3	3	4	5	6	7	8	10	11	14	15	21	22	30	31	44	45	↑	↑	↑	↑	↑	↑	↑	↑	↑	↑	↑	↑	↑	↑	↑	↑	↑	↑	↑	↑
N	500	↓	↓	↓	↓	↓	↓	↓	↓	0	1	1	2	2	3	3	4	5	6	7	8	10	11	14	15	21	22	30	31	44	45	↑	↑	↑	↑	↑	↑	↑	↑	↑	↑	↑	↑	↑	↑	↑	↑	↑	↑	↑	↑	↑	↑
P	800	↓	↓	↓	↓	↓	↓	0	1	1	2	2	3	3	4	5	6	7	8	10	11	14	15	21	22	30	31	44	45	↑	↑	↑	↑	↑	↑	↑	↑	↑	↑	↑	↑	↑	↑	↑	↑	↑	↑	↑	↑	↑	↑	↑	↑
Q	1250	↓	↓	↓	↓	0	1	1	2	2	3	3	4	5	6	7	8	10	11	14	15	21	22	30	31	44	45	↑	↑	↑	↑	↑	↑	↑	↑	↑	↑	↑	↑	↑	↑	↑	↑	↑	↑	↑	↑	↑	↑	↑	↑	↑	↑
R	2000	↓	↓	0	1	1	2	2	3	3	4	5	6	7	8	10	11	14	15	21	22	30	31	44	45	↑	↑	↑	↑	↑	↑	↑	↑	↑	↑	↑	↑	↑	↑	↑	↑	↑	↑	↑	↑	↑	↑	↑	↑	↑	↑	↑	↑

Reproduced from Table 11-A, Mil-Std-105D

⇩ = Use first sampling plan below arrow. If sample size equals, or exceeds, lot or batch size, do 100 percent inspection.

⇧ = Use first sampling plan above arrow.

Ac = Acceptance number.

Re = Rejection number.

TABLE 15.3

U.S. DEPARTMENT OF AGRICULTURE, SAMPLING PLANS FOR INSPECTION OF FROZEN OR SIMILARLY PROCESSED FRUITS, VEGETABLES, FISHERY PRODUCTS, AND PRODUCTS THEREOF CONTAINING UNITS OF SUCH SIZE AND CHARACTER AS TO BE READILY SEPARABLE

Container size group	Lot Size (Number of Containers)								
GROUP 1 Any type of container of 1 pound or less net weight.	2,400 or less	2,401–12,000	12,001–24,000	24,001–48,000	48,001–72,000	72,001–108,000	108,001–168,000	168,001–240,000	Over 240,000
GROUP 2 Any type of container over 1 pound but not over 4 pounds net weight.	1,800 or less	1,801–8,400	8,401–18,000	18,001–36,000	36,001–60,000	60,001–96,000	96,001–132,000	132,001–168,000	Over 168,000
GROUP 3 Any type of container over 4 pounds but not over 10 pounds net weight.	900 or less	901–3,600	3,601–10,800	10,801–18,000	18,001–36,000	36,001–60,000	60,001–84,000	84,001–120,000	Over 120,000
GROUP 4 Any type of container over 10 pounds but not over 100 pounds net weight.	200 or less	201–800	801–1,600	1,601–2,400	2,401–3,600	3,601–8,000	8,001–16,000	16,001–28,000	Over 28,000
GROUP 5 Any type of container over 100 pounds net weight.	25 or less	26–80	81–200	201–400	401–800	801–1,200	1,201–2,000	2,001–3,200	Over 3,200
	Single sampling plans								
Sample size (number of sample units)[1]	3	6	13	21	29	38	48	60	72
Acceptance number	0	1	2	3	4	5	6	7	8

[1] The sample units for the various container size groups are as follows: Groups 1, 2, and 3—1 container and its entire contents. Groups 4 and 5—approximately 3 pounds of product. When determined by the inspector that a 3-lb sample unit is inadequate, a larger sample unit or 1 or more containers and their entire contents may be substituted for 1 or more sample units of 3 lbs.

Derivation of a Specific Attributes Sampling Plan.—The two sets of sampling plans discussed above are very useful, and frequently will provide the quality control operator with a ready-made sampling schedule that will fit his particular requirements. The procedure by which the quality control operator can devise his own plan, however, is not at all difficult.

Attributes sampling plans are based on one of three frequency distributions, namely the hypergeometric, the binomial, or the Poisson. With the exception of very small lot sizes or very large proportions of defectives, the Poisson approximation of the binomial distribution is entirely satisfactory as a basis for devising attributes sampling plans. The theory of these distributions may be found in many sources.

The reader is directed to USDA's publications dealing with attributes standards for the latest attributes grade standards and inspection techniques.

Sequential Sampling.—The above discussion on attributes sampling plans has been limited to single plans, meaning that the entire sample is collected and examined, and the decision to accept, reject, or rework is made on the basis of the single sampling. Each of the two above mentioned systems also have multiple, or sequential, plans. Essentially, the sequential procedure makes it possible to reduce sampling numbers substantially if the lot is decidedly good or decidedly poor, and to concentrate additional effort on the borderline samples.

Military Standards—105D.—Another commonly used plan is described in Military Standard-105D. Upon determining the lot size, the acceptable quality level (AQL) and precision level required, sample numbers, and the limits for accepting or rejecting the lot for single, double, and multiple sampling plans may be found by reference to tables. The risk taken by the vendor and the buyer may be determined by reference to operation characteristic (OC) curves accompanying the tables. Usually the vendor's risk of not having his product accepted is less than 1 out of 20 if in fact his product is acceptable, while the buyer's risk of accepting a product that does not quite meet specifications is considerably greater and will vary, depending on the importance of the defects inspected, the size of the lot, and the reliability of the vendor. In these plans, the buyer's risk is reduced primarily by increasing sample numbers, thereby making the operating characteristic (OC) curve steeper.

Example 2. Sampling Plan by the Use of Military Standard—105D Tables.—A load of potatoes contains approximately 50,000 individual tubers. The processor agrees to accept the load provided there is not more than one percent diseased units. The single, normal inspection level is used. Entering Table I of Mil. Std. 105D (Table 15.4) find letter N under

TABLE 15.4

SAMPLE-SIZE CODE LETTERS FOR SAMPLING BY ATTRIBUTES FORM MIL-STD.-105 D[1]

Lot or Batch Size	Special Inspection Levels				General Inspection Levels		
	S-1	S-2	S-3	S-4	I	II	III
2 to 8	A	A	A	A	A	A	B
9 to 15	A	A	A	A	A	B	C
16 to 25	A	A	B	B	B	C	D
26 to 50	A	B	B	C	C	D	E
51 to 90	B	B	C	C	C	E	F
91 to 150	B	B	C	D	D	F	G
151 to 280	B	C	D	E	E	G	H
281 to 500	B	C	D	E	F	H	J
501 to 1,200	C	C	E	F	G	J	K
1,201 to 3,200	C	D	E	G	H	K	L
3,201 to 10,000	C	D	F	G	J	L	M
10,001 to 35,000	C	D	F	H	K	M	N
35,001 to 150,000	D	E	G	J	L	N	P
150,001 to 500,000	D	E	G	J	M	P	Q
500,001 and over	D	E	H	K	N	Q	R

[1] Special inspection levels may be used where relatively small sample sizes are necessary and large sampling risks can and must be tolerated. Unless otherwise specified, general inspection level II should be used; general inspection level I may be used when less discrimination is needed and III when greater discrimination is needed. In each case the higher the number the more discriminatory the plan.

column II for the general inspection levels for lot size of 35,001 to 150,000. Entering Table 11-A of Mil. Std. 105D (Table 15.2), we find the sample size of 500 and at an acceptable quality level (AQL) 1.0, 10 for acceptance, and 11 for rejection. In application, draw at random 500 tubers from the load, accept the load if 10 or less are rotted, reject the load if 11 or more of the 500 tubers are rotted.

A complete set of Military Standards-105D is available from the Government Printing Office, Washington, D.C.

Example 3. Sequential Sampling Plan Using Military Standard—105D Tables.—To illustrate again with the potato example (Example 2) and using the Military Standards—105D, the single sampling procedure called for taking the single sample of 500 tubers and accepting if 10 or less defects are found, and rejecting if 11 or more are found. For the sequential sampling plan, we take instead only 125 tubers as our first subsample (Table 15.5). If we find no defects among these 125 units, we may accept at this point. Similarly, if we find five or more defects among these 125 tubers we may reject, and the inspection is concluded with only 125 instead of 500 tubers examined. If, however, 1, 2, 3, or 4 defects are found, we must look at another 125 tubers. If in the total of 250 tubers, we find three or less defects, we accept. If we find eight or more, we reject. However, if we should find 4, 5, 6, or 7 defects among the 250 tubers, we must examine an additional 125 tubers. We continue this process until we

TABLE 15.5

SAMPLING PLANS FOR SAMPLE-SIZE CODE LETTER: N

Acceptable Quality Levels (normal inspection) — each cell shows **Ac / Re**

Type of sampling plan	Cumulative sample size	Less than 0.025	0.025	0.040	0.065	0.10	0.15	0.25	0.40	0.65	1.0	1.5	2.5	Higher than 2.5
Single	500	△	0 / 1	Use Letter M	Use Letter Q	1 / 2	2 / 3	3 / 4	5 / 6	7 / 8	10 / 11	14 / 15	21 / 22	△
Double	315	△	•	Use Letter M	Use Letter Q	0 / 2	0 / 3	1 / 4	2 / 5	3 / 7	5 / 9	7 / 11	11 / 16	△
Double	630			Use Letter M	Use Letter Q	1 / 2	3 / 4	4 / 5	6 / 7	8 / 9	12 / 13	18 / 19	26 / 27	
Multiple	125	△	•			# / 2	# / 2	# / 3	# / 4	0 / 4	0 / 5	1 / 7	2 / 9	△
Multiple	250					# / 2	0 / 3	0 / 3	1 / 5	1 / 6	3 / 8	4 / 10	7 / 14	
Multiple	375					0 / 2	0 / 3	1 / 4	2 / 6	3 / 8	6 / 10	8 / 13	13 / 19	
Multiple	500					0 / 3	1 / 4	2 / 5	3 / 7	5 / 10	8 / 13	12 / 17	19 / 25	
Multiple	625					1 / 3	2 / 4	3 / 6	5 / 8	7 / 11	11 / 15	17 / 20	25 / 29	
Multiple	750					1 / 3	3 / 5	4 / 6	7 / 9	10 / 12	14 / 17	21 / 23	31 / 33	
Multiple	875					2 / 3	4 / 5	6 / 7	9 / 10	13 / 14	18 / 19	25 / 26	37 / 38	

(Note in instruction region: Use Letter M, Use Letter Q, Use Letter P)

Acceptable Quality Levels (tightened inspection):

	Less than 0.040	0.040	0.065	0.10	0.15	0.25	0.40	0.65	1.0	1.5	2.5	Higher than 2.5

△ = Use next preceding sample size code letter for which acceptance and rejection numbers are available.

▽ = Use next subsequent sample size code letter for which acceptance and rejection numbers are available.

Ac = Acceptance number

Re = Rejection number

• = Use single sampling plan above (or alternatively use letter R).

= Acceptance not permitted at this sample size.

reach a cumulative sample size of 875, when we must reach a final decision. If, among the tubers examined which have by now reached a total of 875, we find 18 or less defective tubers, we accept. If we should find 19 or more defects, we reject. This procedure is illustrated in Table 15.5 (Table X-N-2 of Military Standards—105D). The operating characteristic curve for this multiple plan is practically identical to the one drawn for the single sampling plan. A double sampling plan is also illustrated in Table 15.5.

Sequential Sampling Derivation.—In order to take advantage of the possibility of inspecting fewer samples when lots are definitely good, or definitely poor, a sequential sampling plan can be constructed.

Attributes, Sublots.—All sampling procedures are based on the principle of randomness, implying that every unit in a lot has equal probability of being selected as part of the sample. We have noted that randomness can be achieved easily under continuous sampling conditions, but may be difficult to retain entirely when the sample is to be drawn before a lot is unloaded. The problem is further aggravated when the lot does not arrive in bulk, but is contained in sublots.

Example 4. Deriving a Two-Stage Attributes Sampling Plan by the Use of Military Standard—105D Tables.—We first consider an attributes plan under these circumstances, such as would be the case if the lot of 50,000 potatoes, discussed as Example 2, does not arrive in bulk, but in 100 crates, with about 500 potatoes in each crate. We are thus faced with the problem of how many tubers out of how many crates to remove for our sample.

The problem may be solved by developing two sampling plans, one for larger lot number (N_L) of 50,000 tubers and the other for the smaller lot number (N_3) of 100 crates. Since we are ordinarily satisfied with less precision for smaller lot sizes, we should arrive at a smaller sample number (n_s) for the smaller lot size. Then dividing the smaller sample number into the larger, we would obtain the number of tubers to be sampled from each crate, and the smaller sample number would be the number of crates to be sampled. Symbolically, $n_L / n_s = m$.

Using Military Standards—105D for our example, we note that the larger sample number is $n_L = 500$. Since the number of crates is 100, we refer to Table 15.2 and find code letter F under general inspection level II for lot sizes of 91–150. Entering Table 15.3 we find that the smaller sample number indicated is $n_s = 20$. Since the number of units to be removed from the subsample m is the larger sample number divided by the smaller or:

$$m = n_L / n_s = 500 / 20 = 25$$

We therefore arrive at a sampling plan in which 25 tubers are drawn from each of 20 crates.

If there is good reason to suspect substantial variability in the percentage of rot among crates, a screening procedure may be resorted to by which each crate (consisting of 500 tubers) would be sampled separately. In this case, code letter H would be used, so that 50 roots from each crate would be examined, and the crate accepted if only one or none of the tubers showed symptoms of rot (Tables 15.2 and 15.3).

Variables, Bulk-Lots.—As a rule variables plans require substantially less sampling than attributes plans in order to arrive at the same degree of precision. Thus, if the test data are of a variables nature in any event, that is, if we have measurements on some scale of values, and not just a statement of the unit being acceptable or not, then a variables procedure is likely to be more economical than an attributes plan.

Variables plans may be constructed and used in a manner similar to attributes plans, in determining acceptance, or rejection, or reworking of lots. In other situations it is extremely wasteful, in fact practically impossible to use attributes plans, and only variables plans will do. This is the case with situations where the buyer is committed to accepting the product, but he may pay on a sliding quality scale. Thus, the problem is not so much to find the minimal sampling plan which will provide the vendor with the assurance that the lot will be accepted if it in fact meets certain specifications. Nor is the problem to provide the consumer with the assurance that the lot will be rejected if in fact it fails to meet certain specifications. When buying on a quality scale basis both vendor and buyer are interested in finding a minimal sampling plan which will provide them with an evaluation as precise as possible of the real quality of the lot so that payment will be made equitably.

Some typical situations where variables sampling plans would apply are in the sampling of raw materials upon acceptance inspection, inspection during the process as well as the finished product, particularly for quantitative measurements such as chemical composition, weight or volume determinations, rheological properties, etc.

Prepared variables sampling plans similar to the prepared attributes plans discussed above are available. Note the similarity between Table 15.6 which provides the code letter for selecting the appropriate variables plan, and Table 15.2 which provides the code letter for the attributes plan. If the standard deviation of the lot is not known, then the proper code letter from Table 15.6 determines sample size as found in Table 15.7. A multiplier (k) for the standard deviation of the sample is then found in Table 15.7 for the proper acceptable quality level (AQL). If the variability in terms of the standard deviation of the lot is known, then

TABLE 15.6

SAMPLE-SIZE LETTER, BY INSPECTION LEVEL AND SIZE OF INSPECTION LOT FOR VARIABLES SAMPLING PLAN

Size of Inspection Lot	Sample-Size Letter for Inspection Level		
	I	II	III
Under 25	B	B	D
25– 50	B	C	D
50– 100	C	D	F
100– 200	C	E	F
200– 300	D	F	G
300– 500	E	G	H
500– 800	G	H	I
800– 1,300	H	I	K
1,300– 3,200	H	J	K
3,200– 8,000	I	K	M
8,000– 22,000	J	L	N
22,000–110,000	K	M	N
110,000–550,000	L	N	O
550,000 and over	N	O	O
Approximate relative number of items inspected	1.0	1.5	2.0

Notes:
(1) Inspection level II will be used for most products under normal inspection.
(2) For reduced inspection, use a level one level lower (if such is available) than that used for normal inspection.
(3) For tightened inspection, use the same inspection level and sample-size letter as for normal inspection.

Table 15.8 may be used instead of Table 15.7 at a substantial saving in sample numbers. Bowker and Goode (1952) give the operating characteristic curves for these plans, as well as for double sampling plans. Thus all that needs to be done is to determine the standard deviation within the sample.

Example 5. Comparison of Efficiency of Variables and Attributes Sampling Plan.—The following example will serve to illustrate the use of prepared variables plans and to show their advantages over attributes plans. A baker wishes to purchase 1,000 cartons of frozen egg whites. He wants the assurance that the fat content of the egg whites will not ordinarily (less than 5% of the time) exceed 1.0%. The test procedure calls for a fat extraction analysis which involves a labor cost of 1 man-hour at $3.00. The cost of removing a sample for analysis including the destruction of the sample material is $0.50. Thus the total cost of performing one sample analysis is $3.50. If an attributes procedure is used, and if we are to refer to Military Standards—105D (Table 15.2) we find the code letter J for the normal inspection level, and from Table 15.3 we find that this sampling plan requires 80 samples to be taken and analyzed at a cost of 80 times $3.50, or $280.00. We accept if we find no more than 2 samples with more

TABLE 15.7

SUMMARY OF SINGLE-SAMPLING VARIABLES PLANS, CLASSIFIED BY ACCEPTABLE QUALITY LEVEL AND SAMPLE-SIZE LETTER

Values of k for Acceptance Criteria of the Form $\bar{x} + ks \leq U$ or $\bar{x} - ks \geq L$

Sample-Size Letter	Single-Sample Size	Acceptable Quality Level Class, in Per Cent Defective														
		.024–.035	.035–.06	.06–.12	.12–.17	.17–.22	.22–.32	.32–.65	.65–1.2	1.2–2.2	2.2–3.2	3.2–4.4	4.4–5.3	5.3–6.4	6.4–8.5	8.5–11.0
B*	7	↓	↓	↓	↓	↓	↓	1.636	1.449	1.242	1.107	1.053	0.969	0.820	0.696	0.595
C	10	↓	↓	↓	↓	↓	↓	1.757	1.562	1.400	1.287	1.186	0.994	0.971	0.789	0.687
D	13	↓	↓	↓	↓	↓	1.957	1.764	1.583	1.472	1.371	1.189	1.132	0.978	0.926	0.772
E	16	↓	↓	↓	↓	2.116	2.018	1.822	1.694	1.437	1.378	1.217	1.180	1.059	0.906	0.815
F	20	↓	↓	↓	2.246	2.180	2.080	1.880	1.749	1.504	1.388	1.351	1.218	1.090	0.997	0.884
G	25	↓	↓	2.395	2.306	2.239	2.137	1.933	1.756	1.569	1.504	1.385	1.261	1.161	1.019	0.924
H	35	↓	2.653	2.480	2.389	2.319	2.234	2.031	1.811	1.672	1.552	1.390	1.284	1.196	1.068	1.004
I	50	↓	2.737	2.559	2.466	2.395	2.288	2.075	1.875	1.677	1.554	1.424	1.329	1.271	1.138	1.065
J	60	2.908	2.775	2.596	2.502	2.430	2.322	2.107	1.905	1.723	1.567	1.440	1.345	1.272	1.148	1.088
K	70	2.940	2.806	2.625	2.530	2.457	2.349	2.132	1.958	1.764	1.616	1.481	1.383	1.284	1.173	1.106
L	85	2.977	2.842	2.659	2.563	2.490	2.380	2.171	1.986	1.810	1.640	1.515	1.415	1.298	↑	↑
M	100	3.006	2.870	2.709	2.589	2.528	2.424	2.237	2.040	1.841	1.674	1.557	↑	↑	↑	↑
N	125	3.043	2.905	2.743	2.621	2.575	2.455	2.267	2.104	1.856	↑	↑	↑	↑	↑	↑
O	200	3.110	2.970	2.832	2.737	2.649	2.523	2.352	2.155	↑	↑	↑	↑	↑	↑	↑

Notes:
* Variables plans for sample-size letter A are not included because their use results in little saving over the corresponding attribute plans.
↓ Use the first sampling plans below arrow. If sample size is larger than inspection-lot size, use 100 per cent inspection or form larger inspection lots.
↑ Use the first sampling plan above arrow.
For tightened inspection use the same sample-size letter but choose a value for k from an acceptable-quality level class two classes lower than that used for normal inspection (if such a class and value of k are available).

TABLE 15.8

SUMMARY OF SINGLE-SAMPLING VARIABLES PLANS FOR KNOWN SIGMA, CLASSIFIED BY ACCEPTABLE QUALITY LEVEL AND SAMPLE-SIZE LETTER

Values of k' for Acceptance Criteria of the Form $\bar{x} + k'\sigma \leq U$ or $x - k'\sigma \geq L$

Sample-Size Letter	Single-Sample Size	.024–.035	.035–.06	.06–.12	.12–.17	.17–.22	.22–.32	.32–.65	.65–1.20	1.20–2.20	2.20–3.20	3.20–4.40	4.40–5.30	5.30–6.40	6.40–8.50	8.50–11.00
							Acceptable Quality Level Class, in Per Cent Defective									
B	5	↓	↓	↓	↓	↓	↓	1.748	1.522	1.278	1.117	1.027	1.015	0.786	0.669	0.546
C	6	↓	↓	↓	↓	↓	↓	1.812	1.586	1.343	1.272	1.181	0.973	0.936	0.851	0.701
D	7	↓	↓	↓	↓	↓	2.105	1.862	1.635	1.432	1.392	1.177	1.084	0.933	0.854	0.751
E	9	↓	↓	↓	↓	2.300	2.178	1.935	1.709	1.466	1.363	1.226	1.158	1.006	0.928	0.824
F	11	↓	↓	↓	2.433	2.352	2.231	1.988	1.761	1.518	1.400	1.356	1.210	1.093	0.980	0.876
G	13	↓	↓	2.579	2.473	2.392	2.270	2.028	1.801	1.558	1.539	1.396	1.250	1.160	1.020	0.916
H	16	↓	↓	2.624	2.518	2.437	2.315	2.073	1.846	1.664	1.532	1.351	1.253	1.178	1.065	0.961
I	20	↓	2.828	2.668	2.561	2.480	2.359	2.116	1.889	1.646	1.528	1.407	1.328	1.249	1.154	1.004
J	24	3.054	2.903	2.700	2.593	2.512	2.391	2.148	1.921	1.718	1.560	1.451	1.329	1.281	1.186	1.036
K	28	3.079	2.928	2.725	2.618	2.537	2.416	2.173	1.980	1.764	1.616	1.501	1.395	1.306	1.211	1.061
L	32	3.099	2.948	2.745	2.638	2.557	2.436	2.193	1.966	1.806	1.636	1.521	1.415	1.290	↑	↑
M	36	3.115	2.965	2.762	2.655	2.574	2.452	2.238	2.052	1.834	1.653	1.525	↑	↑	↑	↑
N	40	3.130	2.979	2.776	2.669	2.588	2.466	2.252	2.106	1.837	↑	↑	↑	↑	↑	↑
O	45	3.144	2.994	2.845	2.723	2.649	2.514	2.331	2.120	↑	↑	↑	↑	↑	↑	↑

Notes:
↓ Use the first sampling plan below arrow.   If sample size is larger than inspection-lot size, use 100 per cent inspection or form larger inspection lots.
↑ Use the first sampling plan above arrow.

than 1.0% fat, and reject if we have 3 or more samples with a fat content of more than 1.0%.

Thus the attributes plan is shown to be very simple, with very few statistical calculations to perform; however, the sampling cost may be not only alarming, but forbidding. Note that we have made no use whatsoever of these expensive individual fat analyses except to note whether they were above one per cent.

We now turn to Table 15.6 for a comparable variables plan, and find code letter I, which according to Table 15.7 indicates the necessity for only 50 samples at a total cost of $175.00, and the k value of 1.875 for an AQL of 1.0%. This time however, we must go to the trouble of obtaining mean, and standard deviation values for the 50 samples, as follows: The sum (Σ) of the 50 fat analyses is 36.42. Since the number of items (n) entering this sum is 50, then the mean ($\bar{x}$) = 36.42/50 = 0.728.

We also note that the highest fat value among these 50 analyses is 1.06, and the lowest is 0.41. Hence the range (R) is $1.06 - 0.41 = 0.65$. It is possible to estimate the standard deviation from the range by dividing by the proper d_2 value obtainable from Table 15.9. Thus the standard deviation of this set of 50 samples is:

$$s = R/d_2 = 0.65/4.498 = 0.144$$

TABLE 15.9

FACTORS OF d_2 FOR ESTIMATING STANDARD DEVIATION

DIVIDE THE AVERAGE RANGE (R) BY d_2 FOR AN ESTIMATE OF STANDARD DEVIATION (s), $s = R/d_2$

Number of Observations in R Group n	Factor for Estimating s from $\bar{R}$ $d_2 = \bar{R}/s$	Number of Observations in R Group n	Factor for Estimating s from $\bar{R}$ $d_2 = \bar{R}/s$
	. . .	21	3.778
2	1.128	22	3.819
3	1.693	23	3.858
4	2.059	24	3.895
5	2.326	25	3.931
6	2.534	30	4.086
7	2.704	35	4.213
8	2.847	40	4.322
9	2.970	45	4.415
10	3.078	50	4.498
11	3.173	55	4.572
12	3.258	60	4.639
13	3.336	65	4.699
14	3.407	70	4.755
15	3.472	75	4.806
16	3.532	80	4.854
17	3.588	85	4.898
18	3.640	90	4.939
19	3.689	95	4.975
20	3.735	100	5.015

We now have all the information needed to decide whether to accept or reject the lot. We will accept if the mean ($\bar{x}$) plus the standard deviation (s) multiplied times k (from Table 15.7) does not exceed ($\geqslant$) the upper control limit of 1.0% fat (U). Since in our example the mean value is 0.728, k is 1.875, and the standard deviation is 0.144, then:

$$\bar{x} + ks = 0.728 + (1.875)(0.144) = 0.998$$

Since 0.998 is less than the specified value of 1.0%, the baker may accept the lot.

A still further savings particularly in the analytical cost can be made if there is definite knowledge of the variability that can be expected among units within the lot that is to be purchased. Thus if it is definitely known that the standard deviation in fat content among cartons of frozen egg whites is typically 0.15, then by reference to Table 15.8 for lots of known sigma (σ, or standard deviation) we find that we need to sample only 20 containers for a sample size indicated by letter I, and our k value is now 1.889. Since we already know the standard deviation ($\sigma = 0.15$), we do not need to analyze each carton separately, but we may combine the material from the 20 samples and perform only one fat analysis. For the sake of some additional precision of the analytical procedure itself, we may prefer to do the fat analysis in duplicate. The cost of such a procedure, therefore, is the collection of 20 samples at $0.50 each, plus 2 analyses at $3.00 each, for a total cost of $16.00. The acceptance value is of course still calculated as $\bar{x} + ks$.

Derivation of Specific Variables Sampling Plans.—Although collections of sampling plans are useful, and make possible the development of a sampling plan rather quickly, the additional work and calculations involved in developing a variables plan designed for a specific use is nevertheless worth the effort, particularly if Fig. 15.6 is to be used to replace some calculations. In this figure the term k is maintained at the value of 2, and percent error (vertical scale) is equivalent to 2 standard deviations. Thus $X \pm 2s$ would determine product acceptance. If, for example, we find that $ks = 2.0$, but wish to reduce error to 0.4 so that we can accept at a value of $\bar{X} + 0.4$, we find the diagonal line on Fig. 15.6 on the vertical axis, and follow it diagonally until it intersects with the desired precision, in this case the horizontal 0.4 line. From this intercept we drop vertically to the horizontal scale and read the sample number, which is 25.

Variables, Sublots.—With variables plans, the elegant solution to nested (Sublot) sampling problems is by means of the analysis of variance particularly with a multi-stage situation. Where the nesting consists of only two stages a simpler graphic solution is possible.

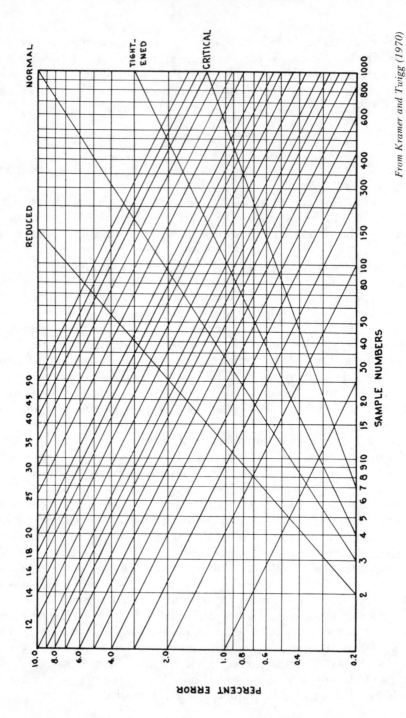

From Kramer and Twigg (1970)

FIG. 15.6. GRAPHIC METHOD FOR CALCULATING SAMPLE NUMBER TO REDUCE PERCENT ERROR TO DESIRED PRECISION

Example 6. Variables Sublot Plan with Two Stages.—We turn now to an example in which cantaloupes used to prepare frozen melon balls are procured not in bulk, but in crates of 12 melons each. The melons are to be tested for ripeness by removing a small plug of flesh and determining sugar content by means of a hand refractometer. We are faced with deciding how many cantaloupes out of how many crates will constitute the most economic sample. It may be assumed that sampling error will be reduced as the number of sublots from which a fixed number of cantaloupes is drawn is increased. On the other hand, cost will increase as the number of crates to be handled rises. Thus the problem is to determine at what point does the increased cost of handling additional sublots (crates in this case) begin to outweigh the reduction in sampling error.

We may begin by selecting *a priori* a series of sampling plans as shown in columns 1 and 2 of Table 15.10 where the number of crates from which 24 cantaloupes are removed varies from 2 to 24. We use each of these 7 plans in duplicate on as many lots as practicable, and calculate the standard deviation between duplicates (standard error) for each of the 7 plans separately (Table 15.10 column 3). Typically, the curve of closest fit between number of sublots (crates) sampled and the standard error follows a logarithmic line. We now estimate sampling cost in terms of minutes required to obtain each type of sample and to perform the analysis (Table 15.10 column 4). In this case, it was estimated that 15 min. are required to plug the 24 cantaloupes and obtain the sucrose equivalent value by the use of a hand refractometer, and 5 min. to remove, open, and return each crate. In order to reduce all the sampling plans to a common error basis, we determine how many times we would need to repeat each

TABLE 15.10

RELATIVE EFFICIENCY OF DIFFERENT SAMPLING PLANS FOR ESTIMATING SWEETNESS OF LOTS OF CANTALOUPES PACKED IN CRATES

1	2	3	4	5	6
No. of Crates	No. of Units from Each Crate	Standard Error (e)	Cost as Minutes of Labor	Sample Frequencies for 0.5% Error (n) [1]	Efficiency Index [2]
2	12	1.15	25	5.3	133
3	8	1.00	30	4.0	120
4	6	0.90	35	3.2	112
6	4	0.75	45	2.3	103
8	3	0.67	55	1.8	99
12	2	0.60	75	1.4	105
24	1	0.53	135	1.1	145

[1] Calculated as $\left(\frac{e}{0.5}\right)^2$.

[2] Calculated by multiplying columns 4 times 5.

sampling procedure in order to arrive at a given error value. In this instance we select the error value of 0.5%, and obtain the number of samplings required to reduce the error to 0.5% (Table 15.10 column 5) by the use of the following equation:

$$n = (e / 0.5)^2$$

where e is the value shown in column 3. All that now remains to be done is to multiply the values in column 5 by the values in column 4, to obtain the relative efficiency index shown in column 6 of Table 15.10. Since we wish to select the least expensive plan, we find the lowest value in column 6, and conclude that 3 cantaloupes removed from 8 crates constitutes the most efficient sampling procedure.

Continuous Control—Control Charts

Serious attention should be given by Quality Control to methods of recording and reporting. The most precise and accurate techniques for measuring quality may be used, and the most effective sampling plans employed, and yet control of quality will not follow unless results are posted in such a manner that action, when required, is clearly indicated. All too frequently control data are tabulated on sheets containing many columns of closely written figures, with little or no indication of their importance, so that a response in the form of an appraisal of the situation, or corrective action, is lost in the mass of the data.

Control records should begin with the specifications for the raw materials and samples and continue with records of production performance and end product quality. Acceptance inspection records are then in essence records of conformance, or lack of conformance to these specifications.

A major contribution towards improving the effectiveness of control records has been the development of the Shewhart control chart, which provides the opportunity not only of summarizing the pertinent information succinctly, but also of dramatically drawing attention to the need for action.

Indication for action provided by the control chart is based on statistical probability, usually better than 0.99, so that there is good assurance that action when called for is really needed, at the same time preventing changes when in fact they are not needed. Perhaps of greater importance than this statistical assurance is the psychological bonus of providing workers with definite information as to what they are expected to do, and immediate and continuous recognition of a job well done.

As with sampling plans, control charts may also be classified as attributes or variable types. Variables charts are reserved primarily for the important factors of quality which can be reported separately, while attributes charts are used when different types of quality attributes can be grouped together, and reported jointly. Again as with the sampling plans, variables charts are based on the normal distribution, while attributes charts are based on the binomial distribution, of which the Poisson distribution is a close approximation.

The control chart may be thought of as a frequency distribution curve but placed on its side, that is, with the quality measurements on the vertical scale, and the frequency extended on a time series (Fig. 15.7).

The Variables Control Chart.—The Shewhart control chart for variables, or the $\bar{X}$, R chart, is undoubtedly the most generally used statistical tool in any quality control program. Since it is a variables criterion, it is naturally reserved for use with inspection data that are obtained in the form of actual measurements on some numerical scale. Since construction and maintenance of such charts involves a recognizable amount of time and effort, they should not be used indiscriminately, but only where it can be definitely shown that their use improves the overall operation. Since one control chart can be used for only one quality attribute, the attributes for which the charts are used should be selected with care.

The most common, and usually the first application of variables charts in food manufacture is in weight control. It is noteworthy that control charts have been found to be extremely effective both when filling is done by machine or by hand. In fact a worker educational program based on control charts is potentially of greater worth than the control of mechanical fillers. Even where 100% weighing is done, the control chart can be very effective. Obviously, the value in such instances is largely psychological in that an out of control indicator is an alarm to the weighers that they are not watching the scales. Dramatic savings in material and improvements in uniformity can frequently be demonstrated in a very short time. Hence, this specific application has been used to "sell" a quality program. At times such selling is overdone, so that before the dust settles, control charts blossom out everywhere, whether they are needed or not.

On the other hand, $\bar{X}$, R charts should not be limited to weight control, since their fields of application are very wide. Some other area in which these charts may, and are being used in the frozen industry are: volume of containers, sizes or dimensions, color and other appearance properties; consistency, firmness, and other rheological properties; moisture, fat, protein, and other chemical or nutritional properties; mold, bacterial, and other microanalytical counts or measurements; yields, batch or continuous mixing operations, etc.

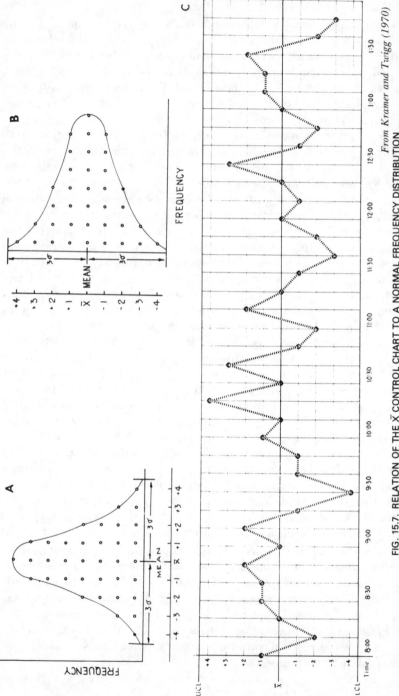

From Kramer and Twigg (1970)

FIG. 15.7. RELATION OF THE $\bar{X}$ CONTROL CHART TO A NORMAL FREQUENCY DISTRIBUTION

(A) Frequency distribution with frequency scale vertical; (B) Frequency distribution with frequency scale horizontal; (C) Control chart; frequency distribution extended into a time series.

Although such control charts are usually associated with the production line, they may also be maintained for charting quality of incoming materials. Thus, for example, a control chart can be maintained for all the raw material entering the plant, and a second for the same material after sorting. In addition to the use of such a pair of charts in maintaining control of quality, they may also serve as a basis for incentive pay to the workers on the sorting belt, and as basic data for cost accounting. Other personnel, suppliers, or sales performances by individuals or groups can also be charted in this manner. In fact any type of operation for which numerical evaluations are available can be charted, providing such charting does more than pay for itself.

Preliminary Considerations.—Certain decisions must be made, essentially on the how, what, and where of the problem before constructing the control chart.

What Is to Be Measured?—At times this is perfectly obvious, as in the case of weight control of fluid materials. At other times it is more difficult to determine just what to measure, as in the case of frozen peach halves which may contain from 5 to 9 halves. If, for example, the fifth half just fulfills the required weight, then the pack is filled economically. However, if the fifth peach just misses the required fill, the can must be overfilled by 20% to meet the minimum fill requirements. Obviously charting the fill alone will not provide the needed control. What is needed here is size control so that a given number of peach halves of a certain size will just meet the required minimum fill.

How Is It to Be Measured?—The method or procedure of obtaining the measurement should be objective, precise, and accurate. The procedure should be objective, in that it should be instrumental, rather than a matter of personal opinion. It should be accurate in that it actually measures the quality attribute it is purported to measure and it should be sufficiently precise to take advantage of the process capability. For example, if a filler can be adjusted to discharge quantities within a tolerance of ± one gram, it would be folly for the inspector to use a balance capable of weighing only to the nearest ounce.

Where Is It to Be Measured?—Ordinarily the best location for maintaining the control station and performing the necessary tests is right at the point of operation, rather than in the laboratory. If fill of container is to be checked, it is advantageous to do the weighing and the charting right at the filler. It may be difficult to perform some test procedures right on the production line, as for example some chemical determinations which must be done in a laboratory. The control charts, nevertheless, should be posted at the point where those responsible for the immediate operation may see a continuous record of their performance and take the necessary action immediately.

When Is It to Be Measured?—The frequency of measurement may be determined in a manner similar to the determination of sample size, particularly if the purpose of the measurement is to establish the grade, or quality level of lots for shipment. For example if the sampling plan calls for the inspection of 500 units in lots of 50,000, approximately every 100th unit could be inspected. However, if the main purpose of the inspection is to detect an out of control situation as soon as possible so that corrective action can be taken, then the major criterion is some knowledge of how rapidly, or when an out of control situation is likely to develop, and the cost of such an out of control situation, for each unit of time, as compared to the cost of inspection. For example if it is generally known that loads of raw material delivered to the plant are essentially the same in quality within each load, but may differ substantially from load to load, it would be logical to check the product thoroughly as soon as possible after a new load of material is used. If loads of berries enter a plant and are utilized in consecutive order, it would be advisable to time the mold determination to each new lot, in order to determine whether enough berries are sorted out, or too many.

Ordinarily a smoothly running operation requires that operations be performed at regular time intervals. This is desirable for quality control inspections too, but the time intervals should be varied to a sufficient extent to avoid anticipation of the moment of inspection by the operators.

How Many Observations at One Time?—The ordinary $\bar{X}$, R chart requires that a number of observations be made, otherwise there is no opportunity to obtain an average $(\bar{X})$, or a range (R). In some rare instances individual observations may be charted directly. This might be the case where the observation involves a long and expensive procedure. Ordinarily, 4 to as many as 10 observations are taken at one time interval, with 5 being the most usual, perhaps because the mean can be so easily determined from the sum of the 5 observations (multiply by 2 and divide by 10). At times, a logical number of observations becomes obvious, as in the case of a number of molds, in the line. The number of observations at one time can consist of one per mold, thus providing an opportunity for the range to cover all molds at every observation period. A theoretically correct sample size could be determined by following a procedure such as the one described for nested sampling (see Example 6).

Constructing the Variables Control Chart.—After first considering the questions of how, when, and where it is to be used, the control chart may be constructed and adjusted in accordance with the following three steps.

Determine Process Capability.—This is a preliminary study to determine just how well the uniformity of the particular quality characteristic may be maintained under the current operating conditions.

Adjustments to Meet Specifications at Minimum Costs.—Assuming the variability in quality is a characteristic of the process, should changes be made in the mean quality level to meet specifications or to reduce costs?

Adjustments to Improve Performance.—Can the control chart be used to indicate means of reducing variability in the quality level thereby resulting in even greater product uniformity and savings?

The reader is directed to *Quality Control in the Food Industry*, Vol. 1 and 2, by Kramer and Twigg for a full and complete development of the subject of quality assurance.

ADDITIONAL READING

AM. FROZEN FOOD INST. 1971A. Good commercial guidelines of sanitation for the frozen vegetable industry. Tech. Serv. Bull. *71*. American Frozen Food Institute, Washington, D.C.

AM. FROZEN FOOD INST. 1971B. Good commercial guidelines of sanitation for frozen soft filled bakery products. Tech. Serv. Bull. *74*. American Frozen Food Institute, Washington, D.C.

AM. FROZEN FOOD INST. 1971C. Food commercial guidelines of sanitation for the potato product industry. Tech. Serv. Bull. *75*. American Frozen Food Institute, Washington, D.C.

AM. FROZEN FOOD INST. 1973. Good commercial guidelines of sanitation for frozen prepared fish and shellfish products. Tech. Serv. Bull. *80*. American Frozen Food Institute, Washington, D.C.

ASHRAE. 1974. Refrigeration Applications. American Society of Heating, Refrigeration and Air-Conditioning Engineers, New York.

BAUMAN, H.E. 1974. The HACCP concept and microbiological hazard categories. Food Technol. *28*, No. 9, 30–34, 70.

BYRAN, F. L. 1974. Microbiological food hazards today—based on epidemiological information. Food Technol. *28*, No. 9, 52–66, 84.

BYRNE, C. H. 1976. Temperature indicators—state of the art. Food Technol. *30*, No. 6, 66–68.

CORLETT, D. A., JR., 1973. Freeze processing: Prepared foods, seafood, onion and potato products. Presented to FDA Training Course on Hazard Analysis in a Critical Control Point System for Inspection of Food Processors. Chicago, July and August.

DeFIGUEIREDO, M. P., and SPLITTSTOESSER, D. F. 1976. Food Microbiology: Public Health and Spoilage Aspects. AVI Publishing Co., Westport, Conn.

GRIFFIN, R. C., and SACHAROW, S. 1972. Principles of Package Development. AVI Publishing Co., Westport, Conn.

GUTHRIE, R. K. 1972. Food Sanitation. AVI Publishing Co., Westport, Conn.

HARRIS, R. S., and KARMAS, E. 1975. Nutritional Evaluation of Food Processing, 2nd Edition. AVI Publishing Co., Westport, Conn.

JOHNSON, A. H., and PETERSON, M. S. 1974. Encyclopedia of Food Technology. AVI Publishing Co., Westport, Conn.

KAUFFMAN, F. L. 1974. How FDA uses HACCP. Food Technol. *28*, No. 9, 51, 84.

KRAMER, A., and FARQUHAR, J. W. 1976. Testing of time–temperature indicating and defrost devices. Food Technol. *30*, No. 2, 50–53, 56.

KRAMER, A., and TWIGG, B. A. 1970, 1973. Quality Control for the Food Industry, 3rd Edition. Vol. 1. Fundamentals. Vol. 2. Applications. AVI Publishing Co., Westport, Conn.

MOUNTNEY, G. J. 1976. Poultry Products Technology, 2nd Edition. AVI Publishing Co., Westport, Conn.
NATL. ACAD. SCI. 1969. Classification of food products according to risk. An evaluation of the Salmonella problem. NAS-NRC Publ. *1683*. Natl. Acad. Sci.—Natl. Res. Council, Washington, D.C.
PETERSON, A. C., and GUNNERSON, R. E. 1975. Microbiological critical control points in frozen foods. Food Technol. *28*, No. 9, 37–44.
PETERSON, M. A., and JOHNSON, A. H. 1977. Encyclopedia of Food Science. AVI Publishing Co., Westport, Conn.
SMITH, C. A., JR., and SMITH, J. D. 1975. Quality assurance system meets FDA regulations. Food Technol. *29*, No. 11, 64–68.
STADELMAN, W. J., and COTTERILL, O. J. 1973. Egg Science and Technology. AVI Publishing Co., Westport, Conn.
TRESSLER, D. K., VAN ARSDEL, W. B., and COPLEY, M. J. 1968. The Freezing Preservation of Foods, 4th Edition. Vol. 2. Factors Affecting Quality in Frozen Foods. Vol. 3. Commercial Freezing Operations—Fresh Foods. Vol. 4. Freezing of Precooked and Prepared Foods. AVI Publishing Co., Westport, Conn.
USDA. 1967. Market Diseases of Fruits and Vegetables. Agriculture Handbook *66*. U.S. Govt. Printing Office, Washington, D.C.
WEISER, H. H., MOUNTNEY, G.J., and GOULD, W. A. 1971. Practical Food Microbiology and Technology, 2nd Edition. AVI Publishing Co., Westport, Conn.

16

Warehousing and Retail Cabinets

Willis R. Woolrich

P roper freezing alone will not ensure the successful distribution of frozen foods—successful storage is at least as important as freezing itself. Unless frozen foods are warehoused under proper conditions of storage humidity and temperature, all of the careful production and quality control methods that a packer may have used could easily be lost before the product is delivered to the final consumer.

WAREHOUSING

To avoid endangering the quality and/or nutritive value of frozen foods, consideration must be given to numerous related factors that are involved in the proper warehousing of frozen foods. These are warehouse design and operation temperature, humidity, odor control, and storing and handling methods.

Design and Operation

Classification of Storage Building Construction.—Freezer storage buildings are classified as either (a) curtain or envelope wall, or (b) insulated warehouse design.

The curtain or envelope wall cold storage building is a completely insulated structure within independent walls of the exterior enclosing structure. The outer walls are constructed independently of the interior insulated envelope package. A continuous heavy vapor seal is placed between the outer wall and the insulating package. Any vapor that does penetrate the vapor seal will travel on inward by the temperature differential of the vapor pressure and from thence will be deposited as frost, or remain within the insulation as ice crystals.

The insulated type of warehouse design differs from the curtain wall type in that all of the floors and the ceilings become an integral part of the supporting side walls. This makes it necessary to supplement the insulation at all outside junctures and floors and ceilings in their relation to the side walls to prevent excessive heat leakage. In cold storage buildings of this type built with steel reinforcing under direction of novice designers, unexpected spots of frost might be detected on the outer wall at each floor level if the supplementary insulation is not adequately applied. These spots show excessive heat leakage from the interior rooms.

The single floor freezer and cold storage plants lend themselves to curtain wall construction. Where real estate prices will permit, such a design is preferable to multistory cold storage houses of either the curtain or insulated types (Fig. 16.1).

Whether the cold storage house is a single floor or a multifloor structure, care should be taken not to extend the supporting concrete pillars through the freezer room floor and to the foundation piers without interrupting the concrete column structure with a block of insulating wood of sufficient strength to carry the load.

There are many freezer storage houses with continuous columns extending through the floors where heavy frosting can be seen below the floor slab.

Curtain Wall Multistory Warehouses.—Curtain wall single floor freezer warehouses are preferred by most managers. When the real estate is not available for the entire planned freezer warehouse complex, then multistory curtain wall construction is acceptable provided adequate exterior space is allowed for the incoming and outgoing truck and rail docks, and the essential commercial parking for employees, customers and salesmen is provided.

As in single floor curtain wall construction, the outer walls are carried up independent of the rest of the building and are truly curtain walls. The roof, interior columns, floors and a portion of the wall columns make up an independent structure from the exterior wall enclosure. In this design the insulation is a continuous envelope between the outer shell and the interior insulated structure (Figs. 16.2, 16.3, 16.4).

A large portion of those freezer storage structures of today are remodeled cold storage buildings that were built prior to 1950. Externally, they follow the pattern of the earlier cold storage warehouses. The room heights usually do not exceed 20 ft. for any floor. The freight elevator, still found on the interior of many buildings, is the earlier design practice for cold storage rooms above 32° F. (0° C.). Usually the first floor is given over to "in and out" rooms of the local processors and merchants but serves also as the receiving floor for the entire warehouse.

With this type of freezer warehouse great care must be taken of frosting of the outside walls and also of the detrimental temperature variations near the elevator shaft.

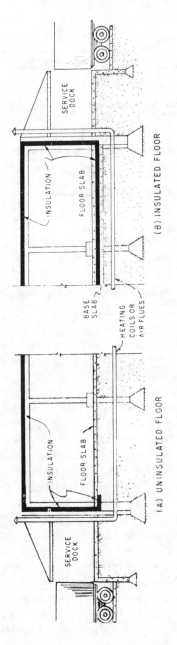

INSULATION

FLOOR SLAB

(B) INSULATED FLOOR

BASE SLAB

HEATING COILS OR AIR FLUES

SERVICE DOCK

INSULATION

FLOOR SLAB

(A) UNINSULATED FLOOR

SERVICE DOCK

FIG. 16.1. SINGLE FLOOR CURTAIN WALL

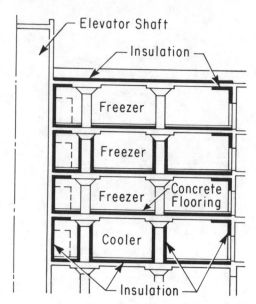

FIG. 16.2. ELEVATOR SHAFT WITH VESTIBULE CONTROLS FROM MULTI-
STORY FREEZER WAREHOUSING CURTAIN WALL

In remodeling such a warehouse for freezer use, if available land space will permit, the elevator shaft and the stairways should be relocated outside the insulated exterior walls and adequate vestibule space provided to prevent temperature variations in the elevator and staircase zone within the warehouse rooms.

Compound Compressors and Booster Units in Warehouse Freezer Practice.—Some confusion exists in refrigeration terminology concerning the semantics of freezers. In freezer warehouse practice, terms such as (1) carrying, (2) sharp, (3) quick, and (4) instant freezers are not clearly defined. However, from a thermodynamic consideration, the engine room unit essential for efficient freezing at or below 0° F. (−18° C.) should be either a compound or a booster refrigerant compressor, irrespective of the name attached to the freezer. For flexibility, the booster-type low temperature compressor is recommended.

Warehouse Freezer Arrangements.—The nature of the commercial loading of a freezer warehouse may determine the auxiliary equipment to be installed. Freezing of fresh fruits and vegetables may be by either still air or by blast in tunnel units. Still air freezes produce slower, but may be found advantageous for unwrapped packaged foods that are very sensitive to dehydration. For still air commercial freezing, the product is spread on trays and racks to allow free circulation during freezing. After freezing, the product is usually repacked in larger marketable packages to

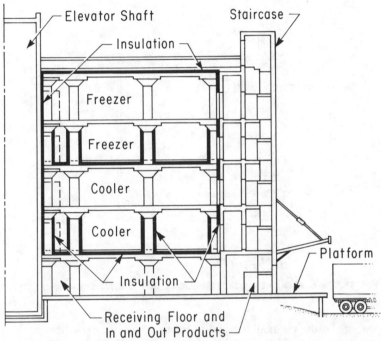

FIG. 16.3. ELEVATOR SHAFT AND STAIRCASE WITH VESTIBULE CONTROLS FOR MULTISTORY FREEZER WAREHOUSING

save storage space. Storage house managers discourage the use of much of their available freezer space for still air freezing since the area required per pound frozen is too great for the revenue usually charged.

Blast air freezers provide large freezing capacity in a small area. The higher rate of freezing reduces the time of crystallization of the liquid portion of the produce and this effects a reduction in dehydration.

Blast freezers are available in many designs from refrigeration equipment manufacturers, some built-in self-contained units, others constructed into a room space to meet the manufacturer's specification. Some blast freezers are designed as tunnels, others as pressurized frigid air arranged for passage through palletized produce carriers in the allotted fan distribution freezing section of the warehouse.

Factors Affecting the Selection of Compressor—Size and Type for Freezer Warehousing.—After the total load or number of B.t.u. to be removed in a 24-hr. period has been computed, the design engineer next selects the compressors to handle the load requirements. For a small cold storage installation, one might assume that a single large-capacity compressor would provide a more economical installation than several smaller compressors. Generally, this is not the case. Such an installation would

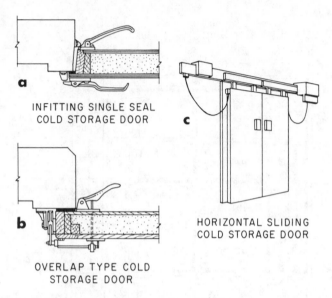

FIG. 16.4. TYPE a, b, AND c REFRIGERATOR DOORS. INFITTING, OVERLAP AND HORIZONTAL
SLIDING TYPES FOR FREEZER STORAGE

be practical only when all the rooms in the plant are to be held at the same temperature, when the heat load to be removed from these rooms is maintained at a nearly constant value, and when the periods of shutdown are the same for all the rooms in the plant. Naturally, these conditions are difficult to realize in present-day cold storage plants. The chilling and freezer rooms might be idle for several days or weeks. The single compressor would be operating at partial capacity to carry the rest of the plant. The partial capacity of a compressor is less efficient than the rated capacity. Likewise, the suction pressure from a blast freezer room or a brine cooler is less than the suction pressure from an ample storage room. These suction pressures must be all the same for a single compressor with one suction line. This means the higher of the suction pressures must be throttled through a pressure reducing valve to correspond with the lowest suction pressure. This method is wasteful of energy. An alternative is to use smaller capacity booster compressors to increase the low suction pressures to equal the highest suction pressure. Another alternative is to boost the lower suction pressures to an intermediate pressure and to throttle the higher pressures to the intermediate pressures.

The principal argument against a single compressor for an installation such as a cold storage plant is the lack of a stand-by compressor in event of failure of the one large compressor.

By using several compressors the plant operation can be made more flexible and the over all efficiency improved. In selecting the combination of sizes and types the purchaser should keep in mind the matter of repair parts. When all the compressors are of the same size and built by the same manufacturer, the spare parts to be kept on hand are reduced to a minimum.

It is recommended practice to install one stand-by compressor of a size equal to the largest compressor in the plant. This stand-by will then assume any load due to the shutdown of any one compressor in the plant.

From the standpoint of maintenance and long life, a slow-speed compressor is preferable to a high-speed compressor. Also, a compressor in operation 12 to 16 hr. per day will last longer than one operating 23 to 24 hr. per day. In other words, large slow-speed compressors will last longer operating intermittently and give less maintenance trouble than small high-speed compressors of the same capacity operating continuously. To obtain intermittent operation of the several compressors in the plant they should have a capacity greater than that required of them. For example, if 100 tons of refrigeration are required and a 100 ton capacity compressor is installed, it must work approximately 24 hr. a day to remove the heat load. If a 120-ton compressor is installed with an automatic control device, it will operate on and off approximately 85% of the time.

The labor cost factor of any freezer warehouse should be anticipated in the design in relation to the weekly income. Local high cost labor may indicate the necessity of mechanizing many of the operations. Automation usually requires a higher degree of skilled labor to man the equipment. Again local high labor cost might indicate the need of building a larger warehouse to reduce the proportionate overhead per $1,000.00 of product handled.

The plant layout should anticipate a minimum of overhead management and maintenance. The working manager is most essential in maintaining a low overhead cost of the main office. Especially in the beginning years, the smaller cold and freezer warehouses cannot support excess armchair administrators.

The Federal Construction Council in their Technical Report No. 38 prepared by the National Research Council on Refrigerated Storage Installations gave a very authoritative report with recommendations. These recommendations cover 74 typewritten pages.

As a sampling, six statements of the manuscript are selected:

(1) That vapor barriers are an absolute necessity for refrigerated spaces regardless of whether the insulation is permeable or impermeable.

(2) The single most significant factor determining the success or failure of a refrigerated storage facility is the quality of the vapor barrier and the care with which it is installed.

(3) Only one type of insulation, cellular glass (except on its surface) has been found free of moisture. All other insulating materials were found to contain some moisture.

(4) Refrigerator doors are a universal source of complaint. Such problems as freezing, sagging, rotting of wood members and physical abuse have been repeatedly mentioned. Few are satisfied with air operated doors, almost all prefer electrical operation. Vestibule doors are generally used for freezer spaces where traffic is frequent.

(5) The four possible approaches to solving the refrigerated warehouse floor problem seem to be the following: (a) design to provide ventilated crawl spaces under freezer floors; (b) when the slab is on grade, design to provide some continuous heating method beneath the slab; (c)

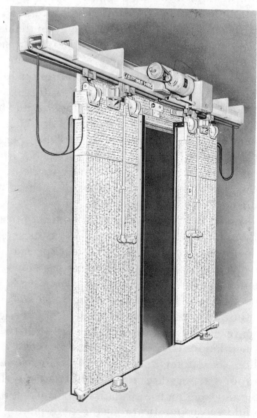

Courtesy of Butcher Boy Refrigerator Door Co., Harvard, Ill.

FIG. 16.5. BI-PARTING ELECTRIC TORQUE-O-MATIC HORIZONTAL SLID-ING FREEZER DOOR

Courtesy of The King Co., Owatonna, Minn.

FIG. 16.6. A REFRIGERATED AIR CURTAIN

Courtesy of Clark Door Co., Inc., Cranford, N.J.

FIG. 16.7. CLARK-PREST-O-MATIC ALUMA COLD DOOR

Courtesy of Jamison Cold Storage Door Co., Hagerstown, Md.

FIG. 16.8. AUTOMATIC ELECTROGLIDE MARK II JAMISON REFRIGER-
ATED WAREHOUSE DOORS IN ACTION

replacement of frost-susceptible soils with porous, low-capillarity fills; or (d) additional insulation.

(6) Where pallet operation is used, there is a noticeable trend to larger but lesser pallets. This trend has had a significant effect on the width and height of refrigerator doors. Some firms are using 12 and 16 ft. platforms.

Resume on Freezer Warehouse Design.—*Floors*.—Freezer floors classed as slab-on-fill or slab-on-ground should be provided with a source of controllable heating for the entire floor; this heat source may be warm air installed in tile ducts or by electric heating cable. Freezer floor slabs on piers should be well ventilated (Fig. 16.1 to 16.4).

Ceilings.—Ceilings of cold rooms should be suspended beneath the roof deck. The space between roof deck and the dropped insulated ceiling should be well ventilated to prevent condensation. A positive vapor seal and barrier to protect the insulation from moisture is essential (Fig. 16.1 to 16.4).

Doors.—Infitting doors are acceptable above 32° F. (0° C.). These should have double seal gaskets.

Overlapping doors should be used below 20° F. (−7° C.). If infitting doors are used from 32° F. (0° C.) to 20° F. (−7° C.), controlled electric heating cables should be installed in the door frames.

All freezer rooms should have vestibules if possible. This applies to elevator entrees as well as doors. (Fig. 16.4 to 16.8).

Roofs.—Cold storage roofs should be positively leak proof, preferably white to reflect solor heat, with ample roof pitch to drain readily (Fig. 16.1 to 16.4).

Walls.—Penetration by metallic or other highly conductive materials through insulation should not be permitted unless completely insulated.

Wall insulation should be installed as per the recommendation of the manufacturer. The design engineer should specify the vapor seal requirements.

Steel reinforced columns penetrating to other floors should be insulated, then protected with one-eighth sheet galvanized or stainless steel for a height of 6 to 8 ft. (Fig. 16.1 to 16.4).

Storage Site Drainage and Grade.—Footings and floors should be above the highest flood or water table on record.

All building run-off should be carried away from all walls, floors and footings. All floor slabs-on-fill should be one foot above grade to protect against flash flooding.

All grading should be well compacted to assure no settlement. On bentonitic and similar volcanic clays, the footings should penetrate to a solid foundation or the entire structure should be erected on a strongly reinforced slab-on-fill to "ride the waves" of alternate ground expansion and contraction as affected by excessive moisture or drouth.

The heating and air conditioning of the business and administrative offices of the cold and freezer storage warehouse should not be neglected in the total plant design. The heating unit should have sufficient capacity to furnish the necessary heat for the several offices, control and machine rooms, but all essential floor heating should be provided by the same central unit.

In tropical regions, such heating can be readily provided by the central electrical supply system for both the necessary floor and space heating. For northern zones, it may be more economical to use gas or oil space heaters for the offices and receiving rooms and a small gas-fired furnace with distributor fans to supply the necessary floor heating.

Sanitation

One seldom thinks of the term "sanitation" in connection with cold storage rooms maintained at 0° F. (−18° C.) or below as, with reasonable care, it is inconceivable that a low temperature room could be anything but sanitary.

Sharp freezing rooms are very likely to have odors that come from the unfrozen product when it was originally placed in the freezer. Some products continue to give off odors even at 0° F. (−18° C.). These odors should be removed or modified and equipment is available to handle this problem. Absorbers may be placed in the room to purify and deodorize by circulating the room air through activated carbon filters. Another type of air purifier is the germicidal ultraviolet light lamp.

Methods of Storing and Handling Products

From a theoretical standpoint, frozen foods should be stacked in solid piles in such a way as to reduce to a minimum air circulation around the products. Such a system is practical only when the storage has a cold floor and the products are not piled against warm walls, and when all of the foods moved into the storage are as cold or colder than the storage itself. Under such conditions, desiccation and oxidation will be less than when considerable air circulation is permitted.

In actual practice, frozen foods are often unloaded from trucks or refrigerated cars in which the products have warmed above the storage temperature. In this case, it is important that the food temperature be quickly reduced to that of the storage.

Frozen foods are warehoused in a number of different types of containers. Some meat is packed and stored in wooden boxes; likewise some poultry. Most frozen foods, including packaged frozen meats, are packed and stored in fiberboard containers. Large tin cans are commonly used for storing frozen cream and frozen eggs.

Whatever types of storage containers are used, they must be so placed in storage as to permit air circulation. Containers of frozen foods should never be stacked directly on the floor of the storage room. The cases may be piled on fork-type pallets or on floor racks, so that the first layer of containers is at least two, and preferably four inches above the floor. Permanent floor racks may be made by placing 2 × 4s on edge and nailing 1 × 4s or 1 × 6s at right angles and on top of the 2 × 4s. In locating the floor racks, it is necessary to have the 2 × 4s at right angles to the length of the ceiling coils in order not to interfere with the natural air circulation within the cold storage room.

Strips of wood (dunnage) should be used about every 2 or 3 ft. to separate layers of containers of frozen foods. Dunnage strips should be not less than ¾ in. thick and not more than 4 in. wide. In this way horizontal air circulation is assured. When containers are piled on pallets, the pallets themselves serve as dunnage strips. Care must be taken when using pallets, or floor racks and dunnage, to avoid having the containers overlap the edges. If this precaution is not taken, part of the bottom layer of cases will crush the ends. In addition to damage to the containers, the crushed part may hinder air circulation.

One of the problems encountered in warehousing frozen foods is the poor stacking strength of some containers, usually referred to as shipping cartons. Stronger containers are often needed especially in view of the increased use of mechanical, palletized handling. Hoover cautions that a square container should never be used as it does not stack well and recommends rectangular forms of shipping cases.

Vertical circulation is provided by not stacking the containers close to the wall or pipe coils. Six inches or more clearance should be provided on all four sides of the room. An occasional aisle is advisable, and should not be considered as so much wasted storage space. Containers should not be piled higher than 12 to 18 in. below the ceiling or six inches below the bottom of the ceiling coils. Frozen foods should not be stacked within five feet of any non-refrigerated spaces, such as openings to stairs or elevator wells.

When defrosting coils by scraping, a tarpaulin must be placed over the stacks of frozen products and directly under the coils to prevent frost and ice from falling on the cases.

In the handling of frozen foods today most warehouses use pallets and forklift trucks. The USDA Agricultural Marketing Service study of materials handling states:

About 75% of the labor employed in the public refrigerated warehouse industry is used to handle merchandise into, within, and out of the storage rooms.

The study indicated that the forklift truck can reduce handling costs in multistory warehouses considerably over the platform hand-truck type of operation where unit loads can be placed into storage without sacrificing too much storage space.

In the single-story warehouse, forklift trucks are a necessity. The distance between the loading platform and storage point is beyond the limits where a platform handtruck may be profitably employed. Also, manual labor cannot be used to advantage because of great stacking heights in the storage room.

For special handling of some vegetables and small fruits, plate freezers may become an adjunct of the total freezer warehouse plant. These units are of special usefulness in the freezer warehouses located in the product producng areas but are not generally recommended for storage-in-transit building designs.

Instrumentation of Refrigerated Warehouses

Instruments Essential in Warehouse Operation and Maintenance.— The five indicating instruments or facilities that the management of a freezer warehouse must have available for product protection are for temperatures, pressures, humidities, energy flow measurement, and odors. These are basic instruments for good management of any cold or freezer storage warehouse.

Clean Air Maintenance for Refrigerated Warehouses

Filtering and Air Purification.—Cold and freezer storage plants are under increasing pressure to improve the ambient air purity in the storage vaults and chilling rooms and freezers.

Air filtering has developed with ascending requirements of the percentage of foreign gases and toxic particles which it is necessary to remove from the ambient and enclosed air within the structures that are being upgraded in environmental cleanliness.

The ratings are usually specified by contaminants allowed per one million parts or per cubic foot or per cubic meter as sampled by the U.S. Dept. of Public Health. Maximum concentrations of gases are indicated as parts by volume per million parts of air. Toxic dusts, fumes, and mists are reported in milligrams per cubic meter of air sampled, and mineral dusts are indicated as millions of particles of dust per cubic foot of air sampled.

There are many types of air filters on the current market. The characteristics that distinguish the types of air cleaners are efficiency and dust-holding capacity.

Cleaners for food storage should be complemented by air treatment to sterilize the harmful bacteria and the contaminating mold growth usually associated with high humidity storage conditions.

In efficiency of dust and vapor removal, filters are ranked from 70 to 85% for central air conditioners to 85 to 99% for electrostatic and electronic precipitators.

Significant applications of ambient air treatment in cold and freezer storage plants are (1) electrostatic and electronic cleaning; (2) activated charcoal adsorption; and (3) ozone generation and violet ray lamps.

Electrostatic and Electronic Air Cleaners.—The electrostatic precipitation principle is used in the modern ionizing type of electronic air cleaners. These units are very efficient in protecting stored produce and meat against bacteria, radioactive dusts, smoke, fumes and odorous gases.

The electrostatic air cleaner has been used effectively for some years in collecting pollens and other minute particles from the ambient air. More recent changes of design involving electronic ionizing features have made these units much more applicable to refrigerated warehouse air purification.

Activated Charcoal Adsorption of Odors.—The phenomenon of the physical condensation of a gas on charcoal surfaces is classified as adsorption in contrast to absorption. Highly selective charcoal for gas masks and similar air filtering applications is usually prepared from coconut shells and peach kernels. With the increase in demand for activated charcoal, many new sources including coal and wood charcoal have been developed.

Activated charcoal is commercially sold in both cylindrical and flat plate cans. Its capacity to adsorb gaseous odors is so great that reactivation is not usually economical to the individual user. Most suppliers will offer instructions of high temperature treatment to revive the charcoal sufaces to their original adsorption capacity, but the individual usually finds it more economical to purchase a new charge in a replacement canister.

To activate the charcoal, it is exposed to heating in a neutral atmosphere, then the carbon particles are exposed to a high temperature oxidizing process to remove unwanted substances within the base carbon material. This leaves the carbon cellular surface greatly increased within its exposed interstices.

Ozone Generation and Ultraviolet Lamps.—Ozone is used by some warehousemen to prevent mold formations caused by required high humidities for storage of some produce and meats. Ozone generators that can be controlled to maintain a minimum concentration of 0.5 to 0.6 p.p.m. of ozone have been found effective. Ozone generators are used in the larger warehouses and ultraviolet lamps in the smaller installations. If ultraviolet lamps are used, they should be designed to produce ozone from the oxygen of the air as well as exert a bactericidal action.

Summary on Controls.—It is very important to remember that automatic controls can maintain satisfactory conditions only if the equipment they control has sufficient capacity. Normally, heating and cooling equipment is of slightly excess capacity for the design load and conditions. Therefore, without control of the usual variations due to weather and occupancy, the end results would at times be too high in winter or too low in summer. A manual on-and-off switch can be used instead of an automatic one, and the occupant could manually "cycle" the system to maintain comfort conditions. The only advantage of manual controls is the reduced installation and maintenance costs, but in most instances the disadvantages far outweigh this one advantage. Actually, most systems are combinations of manual and automatic controls.

Control of Active Corrosion in Cold Storage Warehouses

Some corrosion inhibitors can be built into the systems of a cold or freezer warehouse. By definition corrosion is the destruction of a metal or metallic alloy by chemical and / or electrochemical reaction associated with its industrial environment. There is galvanic corrosion created by the contact of two dissimilar metallic materials in association with an electrolyte or by two similar metals in electrolyte of varying degrees of concentration. This is sometimes called dissimilar metal corrosion.

Closely related to galvanic corrosion is the generated flowing current type often called "stray current corrosion." The corrosion by stray currents is usually associated with an electric circuit in which the deteriorating metal is serving as an auxiliary or leakage circuit; at the junction of the intended conductor and the auxiliary pathway there is a leakage of current, the electric current taking with it some of the anode.

The high cost of metallic corrosion can be reduced in some instances by adding inhibitors to the offending electrolyte. In purchasing condensers, cooling coils, fans, cooling towers and other cold storage warehouse equipment, care should be exerted to require all units and assemblies exposed to moisture, oxygen and active solutes to be manufactured and assembled with a full recognition of the galvanic or electrochemical series of metals that are put together in the exposed equipment (Table 16.1).

Stress Corrosion.—Corrosion is usually expedited when a metal is under stress. Stress corrosion is often evident where an internal stress is present in the metal from cold working, heat treatment or local welding heating effects.

Management Notes on Refrigerated Warehouse Operation

On Insulation Maintenance.—The single most significant factor determining the success or failure of a refrigerated storage facility is the quality of the vapor barrier and the care with which it is installed.

A good vapor barrier, properly installed, will permit the use of a wide variety of insulations, and provide many years of economical low-temperature storage.

Loose-fill insulations in some cases have been found to settle in both

TABLE 16.1

CORROSION REACTIONS BETWEEN WATER AND METALS (ELECTROCHEMICAL ORDER)

Sodium Potassium Calcium Barium Strontium Magnesium Aluminum Zinc Chromium Iron Cadmium Cobalt Nickel	The order given herewith indicates the activity between water (with included oxygen) and the metals from the alkali group at the top to the noble group at the bottom. The first five as a group are not used commercially in the handling of water or aqueous solutions. Any two of these metals with an electrolyte will produce electrochemical flow and corrosion. The wasting away of the metal in the above order to the one beneath is the characteristic of this electrochemical series.
Tin Lead Hydrogen Atimony Bismuth Copper Silver Gold Platinum	In plating, it is common procedure to establish the coating above in this order to the one below—thus zinc adheres electro-chemically to iron or nickel or copper or gold. In general the plating should go on the more noble position metal for protection. Reverse this process and an electro-chemical wasting away of the metal coated may be observed.

wells and ceiling areas; blown in loose fill insulations are especially noted for settling.

In almost every case the moisture content of ceiling samples is significantly greater than that of wall samples.

Where interior finishes are painted with enamel or metallic paints, serious operational difficulties have been encountered; most of the painted surfaces are flaking or blistering.

A number of floor heavage problems have been observed under freezers. The most satisfactory method of correction has been removal of the floor and installation of heating pipe-coils beneath the floor.

Defrost systems vary in type from manual removal of frost to a continuous defrost. Manual frost removal has been unsatisfactory, taking up to two weeks to complete. Hot gas, electric and continuous defrost systems, automatically controlled, are widely used with good results.

On Frost Action Under Floors.—The results have shown that frost formation occurs in certain types of soils, generally of a fine-grained texture, which are said to be frost-susceptible. A necessary factor for frost heavage is a supply of water. Freezing of the soil alone will not produce excessive heaving. Extensive heaving will occur only when conditions are such as to cause water to be drawn through the unfrozen portions of a soil bed to a surface within the soil at which freezing takes place.

On Wood Within the Vapor Barrier Envelope.—Wood is subject to decay, insect attack, and fire. The work of many researchers has supported the theory that fungus-caused decay usually will not occur if the moisture content of the wood is below 20%.

On Vapor Pressure Differentials and Moisture Infiltration.—The range of differential pressure between inside and outside vapor pressures is insufficiently recognized by many in the refrigerated structure field. These pressures are primarily due to the difference in inside-outside temperatures. There is a continuous drive of vapor from warm outer air to the cold inner air.

Aside from the basic considerations of temperature, humidity, and methods of storing and handling, other precautions should be observed in the storage of frozen foods. Every care should be taken to prevent any delay whatsoever in moving frozen foods from freezers to storage rooms and from storage rooms to refrigerator cars and trucks. The length of time that frozen foods are not in refrigerated space at 0° F. ($-18°$ C.) should be held to an absolute minimum.

Some Operational Data for Refrigerated Warehouse Managers

Refrigerated freezer warehouse costs are stated per cubic foot with all essential facilities. These costs will vary with the proportions of the warehouse reserved for (1) freezer and above 32° F. (0° C.) storage, (2) extent of installed facilities provided, and (3) overall dimensions of the complete warehouse.

Net operating income will be largely dependent upon (a) maintaining a high average product occupancy, (b) building investment costs, (c) direct labor commitment, (d) indirect and management labor overhead, (e) effective maintenance schedule, (f) monthly utilities cost, and (g) alertness in cost analysis.

Changing warehouse loadings and increasing of monthly business will require periodic changes in dock and loading facilities. Shipping platforms for rail cars should be approximately 54 in. above the rail and for trucks 52 in. above the paved driveway. Adjustable ramp facilities are recommended to serve special freezer car unloadings.

For effective, alert servicing of incoming and outgoing refrigerated product, the manager and his personnel should be physically located in adjacent offices and rooms for continuous observation.

The Operation of a Cold and Freezer Warehouse for Profit

The United States National Association of Refrigerated Warehouses, Washington, D.C., is made up of executives and administrators of commercial cold and freezer storage plants. Their Washington office has been a beneficial agency in education, operation, cost accounting, and management of the cooperating cold and freezer warehouses of the nation.

Definitions.—Some definitions presented by the National Association of Refrigerated Warehouses follow:

Gross Refrigerated Space.—That area from wall to wall, ceiling to floor, under refrigeration.

Gross Usable Space.—That portion of gross refrigerated space which is actually available for use as storage or revenue producing space—wholly available as storage space.

The gross refrigerated space must make allowances for probably 14% for aisles, 2% for posts and columns and 24% for refrigeration coils, fire protection sprinklers and handling space above the pile. The probable usable space of a modern refrigerated warehouse will be about 60% of the gross refrigerated space.

Net Occupied Space.—That portion of the gross usable warehouse that is actually used.

The gross usable space and 100% occupancy are synonomous. The stored product under refrigeration in relation to the gross usable space will represent that percentage of occupancy.

Package Density.—The weight in pounds per cubic foot of a package to be stored is the weight in pounds per cubic foot of the stored product.

Likewise, the terms storage period, storage rates and insurance, handling charges, delivery requirements, liability and liens are clearly specified by the Association.

Extra Service Charges.—Merchandise received or delivered before or after business hours will be provided by special arrangements.

Typical Charges for Special Services.—Charges per man-hour; a minimum charge for special clerical work per hr.; dunnage or material used in loading out cars—cost, plus 15%; extra book inventories are furnished per man-hour—no charge for one book inventory per month.

Warehouseman and Customer.—The warehouseman is expected to know his gross usable space, then determine what income per cubic foot he must consider essential based on a 60% occupancy to make an annual profit. This annual income per cubic foot to meet these conditions may vary as between two warehouses in the same city per year per cubic foot.

Each member cold storage warehouse company operates under its own terms and conditions but usually in full accord with the recommendations of the National Association of Refrigerated Warehouses.

The management of a property of 500,000 cu. ft. of "gross usable space" accepts foodstuffs property only under the following terms:

> All goods for storage shall be delivered at the warehouse properly marked and packed. The storer shall furnish at or prior to such delivery, a manifest showing marks, brands or size to be kept and account for separately and in the class or storage desired—otherwise the goods may be stored in bulk or assorted lots, in cooler or freezer at the discretion of this company and will be charged for accordingly.

FROZEN FOOD RETAIL CABINETS

Over the third of a century of advances since the first commercial low temperature frozen food retail cabinet was introduced to the trade, competition has kept this expanding industry alert to every new sales feature of merit. The initial requirements of a frozen food cabinet for retail sales were (1) a temperature of 0° F. ($-18°$ C.) with a minimum temperature variation, (2) storage space to be conveniently accessible and to offer sufficient space for an ample supply of the various kinds of product, and (3) the initial and operating costs must be as low as possible.

The use of heavy-duty meat cases for this purpose was tried and found to be unsuccessful due to difficulties in maintaining suitable low temperatures for the food. Early designs of display cases to meet the temperature requirements met with limited success, one of the major problems being the proper installation of the glass to avoid condensation and fogging. In many cabinets, exposure to the lights caused a change in color of some foods and warmed some cartons undesirably. Those few models which were acceptable from a refrigerating and mechanical standpoint had the disadvantage of high cost.

Although the ice cream cabinets on the market could maintain the proper temperatures, certain of their construction features and their general appearance did not make them completely adaptable for use in retail selling of frozen foods. After spending a considerable amount of time in research and testing, a completely new low-temperature cabinet designed exclusively for the retail merchandising of frozen foods was first placed in commercial service in the fall of 1934.

Over the period of more than three decades, there has been a revolutionary change in the essential equipment for the retailing of foods—both staple and refrigerated. The competitive supermarket with its offering of self-service in selection of foods has become accepted by the English-speaking world. The introduction of the effective retailing of frozen foods has been particularly responsible in affecting these revolutionary changes in techniques and display methods of food merchandizing.

Display Cabinets

The earlier type of retail cabinets functioned as display cases. The refrigerated display case, which provided a combination of limited storage space and the display feature, was available in two general types. One was the rolling glass top type and the other was an open-top, glass front case. The insulation usually was four inches of Fiberglas or equivalent. The refrigerating effect was obtained by the fin-type or plate-type coils, or a combination of both. Although the fin-type coils offered the advantage of a large amount of cooling surface in a relatively small space, the problem of frequent defrosting arose. The design and construction problem that had to be solved was in the arrangement of the plates of glass to avoid fogging and condensation which impaired transparency. Continued improvement in the design and construction resulted in modern, automatic defrosting display cases which are well suited for the display and storage of frozen foods and are now in common use in all supermarkets and all of the important retail grocery stores in the United States.

Automatic Defrosting Cabinets.—These cabinets defrost themselves during the night hours when stores are closed, or at other times depending on operating conditions. Some are so constructed that the water which accumulates during the defrosting is evaporated in the compressor compartment by the heat from the condensing unit; others dispose of the water by connection to drains.

Some of these cabinets have glass fronts while others have solid fronts, but all have some design of an open top. Many of these have covers for covering at night when the cabinet is not in use other than to serve as refrigerated storage space. The product in some cabinets is kept at a proper temperature by fans built in the cabinet blowing cold air through the evaporator coil, thence out over the product (Fig. 16.9). Frost collects on the evaporator coil, and by the use of a timer the frost is removed at regular intervals. This is done by running hot gas through the evaporator and drain pan, or applying heat by means of electric heaters. When heat is applied, the frost turns to water and runs off into a pan in the compressor compartment where either it is evaporated, or it flows out through a drain.

A common type of construction uses an all-steel welded angle iron frame and a steel inner tank to which is soldered the copper tubing carrying the refrigerant. The insulation, consisting of four inches of high grade Fiberglas, or equivalent insulant, is installed on all sides of the inner tank and vapor-sealed to prevent moisture infiltration. The outer panels and the metal top are put in place and secured by different methods to seal completely the insulation against infiltration of moisture.

Various manufacturers have their own special methods of cabinet

Courtesy of Tyler Refrigeration Div., Clark Equipment Co.

FIG. 16.9. AIR-SCREEN SALES CASE LOCATED IN GUIDONE'S FOOD PLACE, INDIANAPOLIS, IND.

construction, and in a number of instances, special types of insulation and inner tank construction.

The essential refrigerant tubing is soldered to the inner tank and is formed in a series of horizontal hairpin loops to afford a suitable temperature distribution. In addition, refrigerated divider plates are welded at suitable intervals across the tank in a vertical position, which serve the double purpose of stiffening the tank and affording a better distribution of the temperature to the products in the cabinet. The refrigeration unit is usually a hermetically sealed condensing unit installed in the lower left front part of the cabinet. The unit may be mounted on a track so that it can be easily pulled out from the front of the cabinet when servicing or repair is necessary.

With the rapid increase in the use of frozen foods in North America, new designs of cabinet and display cases come onto the market annually. There has been a move away from standardization as new designs appear.

Distributors and chain store corporations especially have made rapid progress in frozen food display equipment that meets their specifications and special needs. The result has been many excellent new offerings by both manufacturers and users of the display units.

Segregation of Unfrozen Retail Refrigerated Foods

Above 30° F. ($-1°$ C.) restaurant, dairy, meat, beverage and egg short-term commercial display cases and cabinets require judicious separation of foodstuffs to avoid odor contamination and product desiccation. Permissible short-time storage temperatures and humidities for each category can be somewhat higher or lower than for long-period cold storage but the differentials seldom will exceed 5° F. (3° C.).

The segregation by temperature and humidity categories most commonly encountered (many items of which require complete isolation) are: (1) fish, onions, cabbage and cataloupe, 35° to 40° F. (2° to 4° C.); (2) meats, fresh fruits, poultry and eggs and less odorant vegetables, 32° to 40° F. (0° to 4° C.); (3) wrapped heavy meat carcasses, 30° to 36° F. ($-1.1°$ to 2° C.); (4) bottled beverages, unfrozen canned goods and nuts in the shell, 35° to 45° F. (2° to 7° C.).

Many of the principal items of categories (1) and (2) will require segregation by odor transmittal. Number one class should be maintained at 95% R.H., number two at 90% R.H., number three at 85%, and number four can be dry storage.

Much more care is essential in retail management in maintaining judicious odor segregation of vegetables and fruits held above 30° F. ($-1°$ C.). Customers who object to foreign odors contaminating commercial cabinet stored foods are usually differentiating, by taste, foodstuffs that have been carelessly exposed to less inviting vegetables, fish or fruit odors.

This segregation will require a complexity of refrigerators, cabinets and display cases, including (1) reach-in refrigerators, (2) walk-in-coolers, (3) beverage coolers, (4) ready-to-eat and bakery foods refrigerators, (5) frozen food display cases (Fig. 16.10), (6) frozen food and ice cream closed cabinets and (7) auxiliary cold freezer storage houses.

Single Duty Units

Full vision display cases are designed to employ all glass or heavy clear plastic sliding panels for easy customer observation and are found most useful in food markets for both frozen and delicatessen food marketing.

Fish marketing display units are generally single-duty cases. The odor, temperature and humidity control feature is most essential for fish management. These may be ice or electric refrigerated, or in some designs ice and electric refrigerated with essential water drains. Experimental studies of the Scottish Aberdeen Fishery Research Station revealed that most freshly caught fish can be kept for seven days on ice, but beyond this time they should be frozen for maximum utilization.

Poultry display cases are designed to be held at 30° F. (−1° C.) to avoid visceral taints. The single-duty poultry display case must employ refrigeration that will prevent visceral taints and maintain a high humidity to prevent shrinkage.

Trends in Cabinet Design

The widespread expansion of the supermarket and self-service stores has forced a number of drastic changes in various phases of the retail store

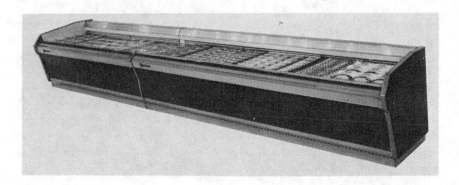

Courtesy of Friedrich Refrigerators, Inc.

FIG. 16.10. FROZEN FOODS DISPLAY CASE

field, not only in methods of store management but also in the store fixtures used for the sale of the product. As is well known, the fundamental principle of the supermarket seems to be an actual display of the product, usually in considerable quantity, in such a manner and location as to induce customers to serve themselves. The use of the standard type of retail storage cabinets in such supermarkets has proved to be of limited value insofar as volume of sales is concerned. This is due in part to lack of proper facilities to actually display the product and also to a natural reluctance of customers to open the cabinet and select the desired purchase. The regular type display cabinet for frozen foods has, of course, display value, but the self-service features are not as great as desired by supermarkets.

Many of the corporate food sales groups have their own designs of refrigerated cabinets and have these frozen food display cases built to their own specification, each with unique features of steel construction, insulation sealing, and automatic defrosting. Baked-on porcelain enamel resistant to discoloration and scratches is commonly specified on 18- or 16-gage steel sheet.

The cabinets in most general use today break down into three main classes. Each class is designed to fit into a particular type of store or to perform a certain job in a given location.

Courtesy of Hussmann Refrigeration, Inc.

FIG. 16.11. FULLY-INDEXED FROZEN FOOD DISPLAY CABINET

Rolling glass lid cabinets have double thick "Thermopane" easy rolling glass lids. These lids not only roll very easily so that it is a simple matter to get to the product in the cabinet, but during rush hours in the store when there is heavy traffic, the lids can be completely removed for periods of time to make an entirely open top cabinet.

Open-top glass cold-plate cabinets have quadruple glass "Thermopane" fronts to display the product better in the cabinet and, at the same time, provide adequate insulation to keep the product at proper temperatures. Refrigerated walls provide the refrigeration. This cabinet has an open top with a concealed night cover which can be used when the store is closed.

Automatic defrost display cabinets have "Thermopane" fronts and

Courtesy of Federal Refrigerator Manufacturing Co.

FIG. 16.12. MEAT MERCHANDISING ZERO SHELF REFRIGERATOR FOR
COMMERCIAL USE

open tops, but no divider plates. There is one large interior display area which can be set up in any way desired to store and display the product. These cabinets automatically dissipate all frost and condensation accumulations every 24 hr. or less as necessary.

R 12, 22, or 502, with the necessary regulation instrumentation of the liquid refrigerant into the evaporator, is installed with thermostatic control for starting and stopping the compressor motor and gives a close range of temperature variation within the cabinet. The noise of the motor, compressor and fan must necessarily be maintained at a very low level.

To carry out the display motif of the frozen food within the cabinet by electric lighting requires careful designing. A typical display cabinet for frozen foods will consume 500 to 1,000 kw. hr. per month for lighting, refrigeration and ventilation. This heat energy of 3.413 B.t.u. per kw. hr. must be dissipated within and by the cabinet surfaces by refrigeration, in addition to lowering and maintaining the temperature of the frozen food to a safe cold atmosphere for the foodstuffs on display, usually 0° F. (−18° C.) or below.

Display cabinets must be designed to (a) display the frozen product to the customers and (b) the articles on display must be easily accessible both to a child and a mature purchaser (Fig. 16.12). The American supermarket and the competing corner store must be offered useful display cabinets for frozen foods that will meet their needs adequately. This

Table 16.2

Temperature Ranges for Refrigerated Foods in Open and Closed Cabinets

Closed Cabinets	Temperature Range	
	°F.	°C.
Walk-in cooler—above freezing	35–42	2–6
Walk-in cooler—below freezing	−20–0	−29-(−18)
Reach-in refrigerator	35–45	2–7
Meat display cases	35–42	2–6
Vegetable display cases	35–48	2–9
Grocers dairy cabinet	35–42	2–6
Bakers dough retarders	40–50	4–10
Florist storage cooler	38–45	3–7
Frozen food cabinet	−10–0	−23-(−18)
Candy case	55–70	13–21
Ice cream cabinet	−10–0	−23-(−18)
Beverage cooler	35–42	2–6
Open display cases and cabinets		
Meat display	28–45	−2–7
Vegetable display	35–55	2–13
Dairy display	35–45	2–7
Florist display	40–50	4–10
Frozen food cases	−10–0	−23-(−18)
Candy display	55–70	13–21
Ice cream cabinet	−20–0	−29-(−18)

challenges the designer to bring forth models of practical usefulness. This requires thoughtful ingenuity to meet all of the economic limitations of the job.

Commercial food sales establishments such as modern grocery stores, meat shops, restaurants and dairy vendors need short-period, highly attractive retail display units. Usually the food on display is not kept by the merchant for more than three days and much is sold within 24 hr. The atmospheric conditions within such open display cases can be designated by temperature and humidity as an optimum retail display cabinet environment, as distinguished from the type of coolers or freezers that are commercially used by grocers, *et al.,* as closed units or rooms for standby refrigerated frozen food storage.

Table 16.2 gives the temperature ranges for both open and closed retail store cabinets. The open cabinets have been designed for refrigerated display over the temperature ranges specified. Usually, the lower temperatures are realized during the night and the higher temperatures prevail over the active afternoon business hours.

Many larger commercial food sales establishments maintain, adjacent to their stores and shops, an auxiliary cooler or freezer storage that complements their storage rooms that they have leased in the area from regional commercial cooler and freezer storage houses. Some of the larger retail merchants have turned to prefabricated portable cold, freezer, and blast freezer storage plants that are available for leasing from national corporations who specialize in such equipment. Leased cold storage equipment finds its place especially in supplying freezer space for seasonal peak deliveries of agricultural and orchard food products. Such freezer storage houses permit the buyer to take advantage of large bulk purchases of currently frozen perishable food stocks.

Modern Cabinet Construction.—Food display cabinets are made commercially with both wood or metal framing. More and more welded metal framing is being adopted by manufacturers to obviate the warping of the sections. A large portion of the display units are made up as endless cabinets to permit continuous extension construction. Endless cabinets that may eventually be extended require metal construction and the complete design permits the insertion or removal of sections to adapt to changing store plans. Metal display cabinets are more adaptable to tight gasket sealing and bolting. Stainless steel and aluminum are preferable to eliminate corrosion of the wetted surfaces. Frozen food open-type cabinets for 0° F. (−18° C.) service will require condensing units of approximately 1 to 1½ kw per 8 ft. of length in northern climates and 1½ to 2 kw per 8 ft. of length in semi-tropical zones.

The evaporators used vary with different manufacturers and services. An examination of current installations shows plate, finned tube, bare tube and special patented units such as combined evaporator and shelves.

Defrosting of Evaporators.—The defrosting procedures of display case evaporators is most important. Current designs include (1) hot-refrigerant defrosting, (2) electric heater defrosters, (3) tap or hot water defrosting, (4) condensing unit turned off and on to a scheduled time plan and (5) special patented heat source defrosting.

Hot Refrigerant Defrosting.—This defrosting method utilizes the heat of the compression of the refrigerant directly to the evaporator. This method may use the cold of the frosted evaporator to help condense the gaseous refrigerant or it may take the condensed refrigerant and sub-cool it by applying the warm liquid to the frosted surfaces.

A common procedure is to permit the hot refrigerant gas to by-pass the condenser through a solenoid control valve thence to the evaporator. This method must be well balanced in design since there is only a limited amount of heat available to complete the defrosting job.

Many variations have been employed under manufacturers' patents to utilize both the heat of the liquid of the condensed refrigerant and the hot gas coming from the compressor.

Electric Heater Defrosting.—This method is probably the most popular of the defrosting methods for display cases although it requires a longer period to accomplish the defrosting than some other methods. The electric heating method can be designed for assured reliability with greater ease and effective application than most other systems.

Water Defrosting.—Water defrosting by spray or flow is effective in melting the ice or frost in a short period of time, due to the high rate of heat transfer of wetted surfaces. If the system includes a fan for air cooling this must be shut off during the period of water delivery.

Water defrosting can be clock or manually regulated. If clock controlled, the timing schedule should include sufficient time for the water film and drain pans to dry sufficiently before the refrigerant cycle comes back into operation.

Defrosting by Scheduled Turning Off and On of the Compressor and Condenser.—This method of defrosting was one of the first to be used in cabinet systems. The condensing unit is turned off for a scheduled period of time sufficient for the evaporator to reach an above-freezing temperature and provide ample time for the condensate to drain off. This is the slowest method of defrosting, since the melting of the ice and frost is dependent upon the heat of the circulating air and equipment heat gain. For 0° F. (−18° C.) display cases the Off and On condensing unit procedure is not recommended. It is a useful method for temperatures just below freezing.

Special Patented Defrosting Methods in Display Cases.—Ingenious designs, usually involving parallel or dual complementary defrosting coils or plates with a secondary refrigerant circulating system, are installed in some sophisticated frozen food display cabinets. Such secondary refrigerant systems can be made most effective but usually they do add to the first and operating costs of the installation.

Doors of Closed Freezer Cabinets and Portable Cold Storage Houses.—Doors of freezer storage cabinets and portable freezer storage houses must be given special attention not required in cooler storage units. The likelihood of heavy condensation drip and freezing brings about such undesirable features as sticking door seals and overhead drip contamination of foods handled through the door openings.

Three types of refrigerator doors are manufactured in North America: (1) infitting, (2) overlap, and (3) sliding. Each type is available with various thicknesses of insulation and convenience hardware. Most state and province laws require that all cold storage doors must be readily opened from the inside as well as outside to prevent entrapment of the user.

For cooler storage doors, the overlap type was a favorite. With the great increase in freezer storage installations many manufacturers' changes were necessary on sliding door designs to reduce the sticking door hazard and many improvements were necessary to provide an insert door that would open readily without undue pressure. Each manufacturer of insert doors developed a product to reduce the sticking inconvenience and the current market can provide a non-sticking product.

The sliding type of freezer storage door is a later design of the three mentioned above. It requires less floor space as compared to the swing-out types and has less tendency to sag since there is little cantilever pressure on the hinges or guides.

The freezer door normally provides clear openings from 4 to 6 ft. The larger cold storage units that must provide tractor ingress and egress require the wider openings. Many of these door types are electrically operated from pull chain operated switches placed conveniently at the entrance for the tractor operator to reach without dismounting.

Evaporators for Commercial Cabinet and Display Refrigerators.— The principal types of evaporators used for cabinets and display cases are (1) bare-tube, (2) finned tube and (3) plate.

There are special problems in cabinet manufacture that make the bare tube evaporator a preferred design. The tube material is usually aluminum or copper and serves as a metal shelf support for single-product, single-low-temperature frozen foods or for above 32° F. (0° C.) procedure for water bath refrigeration. There are other special cases where the bare coil may be formed to fit the evaporator conformation such as ice cube and bucket-type freezers and other special designs.

The finned-tube evaporator has the most general application in both freezer and chiller designs. The fins are usually aluminum, bonded to either copper or aluminum. The latter are less likely to deteriorate from electrolytic action. Finned spacing may vary with the application or even on the same installed tubes to compensate for the anticipated heat transfer rate change within the tube. Fins may be found installed with spacings from 2 to 6 in.

Plate-type evaporators are built into evaporators, the plate being bonded to the refrigerant coil piping and serving as an evaporator surface of aluminum, copper or stainless steel. Patented methods of bonding are featured by some manufacturers.

Condensers and Condenser Cooling.—The early designs of water cooled cabinet condensers so common in the water-sufficient cities have given way to air-cooling, spray tower cooling or to evaporative condensers which use a minimum of water. Many cities in the water deficient regions prohibit the discharge of condenser water into municipal sewage mains, in order to conserve water for more essential usage.

Air cooled condensers in zones north and south of the 30th parallel should be designed to modulate the cooling air in controlled, warmed rooms above freezing temperature. Likewise, water cooled condensers must be protected against freeze-ups of condensing water in spray towers and evaporative condensers when outdoor temperatures drop below 32° F. (0° C.).

Insulation of Display Cabinets for Frozen Foods

The insulation of frozen food cabinets is usually a compromise with space limitations. At a temperature of 0° F. (−18° C.) an insulant of 5 to 6 in. would be thermally justified but most refrigeration designers must be satisfied with 2 to 3 in. of moisture-sealed insulation with an installed k factor of 0.30 or less. The compromise in thickness generally results in excessive sweating on the exposed surfaces. To correct this sweating electrical heaters or hot refrigerant coils are located to evaporate and dry out the excess moisture.

While frozen food display cabinets could be designed and manufactured to more nearly meet a more ideal insulation thickness standard that would improve the refrigeration efficiency, the space limitations in modern stores, especially in installation passages such as doors and halls, would prohibit any such unit enlargement.

Insulants of special merit in display cabinet construction are glass fiber, mineral wool, styrofoam, corkboard and special plastic foams of recent origin. Most important is the installation of a positive vapor seal about the insulation that does not permit any moisture to get into the insulant. The warranty and long time efficient service guarantee of reliable manufacturers and distributors of their vapor seal effectiveness is fully as important as the *k* factor of the insulation used.

ADDITIONAL READING

ASHRAE. 1972. Handbook of Refrigeration Fundamentals. American Society of Heating, Refrigerating and Air-Conditioning Engineers, New York.

ASHRAE. 1974. Refrigeration Applications. American Society of Heating, Refrigeration and Air-Conditioning Engineers, New York.

ASHRAE. 1975. Refrigeration Equipment. American Society of Heating, Refrigeration and Air-Conditioning Engineers, New York.

ASHRAE. 1976. Refrigeration Systems. American Society of Heating, Refrigeration and Air-Conditioning Engineers, New York.

BYRAN, F. L. 1974. Microbiological food hazards today—based on epidemiological information. Food Technol. *28*, No. 9, 52–66, 84.

BYRNE, C. H. 1976. Temperature indicators—state of the art. Food Technol. *30*, No. 6, 66–68.

JOHNSON, A. H., and PETERSON, M. S. 1974. Encyclopedia of Food Technology. AVI Publishing Co., Westport, Conn.

KRAMER, A., and FARQUHAR, J. W. 1976. Testing of time–temperature indicating and defrost devices. Food Technol. *30*, No. 2, 50–53, 56.

LACHANCE, P. A., RANADIVE, A. S., and MATAS, J. 1973. Effects of reheating convenience foods. Food Technol. *27*, No. 1, 36–38.

PETERSON, M. S., and JOHNSON, A H. 1977. Encyclopedia of Food Science. AVI Publishing Co., Westport, Conn.

RAPHAEL, H. J., and OLSSON, D. L. 1976. Package Production Management, 2nd Edition. AVI Publishing Co., Westport, Conn.

SACHAROW, S. 1976. Handbook of Package Materials. AVI Publishing Co., Westport, Conn.

SACHAROW, S., and GRIFFIN, R. C. 1970. Food Packaging. AVI Publishing Co., Westport, Conn.

TRESSLER, D. K., VAN ARSDEL, W. B., and COPLEY, M. J. 1968. The Freezing Preservation of Foods, 4th Edition. Vol. 1. Refrigeration and Refrigeration Equipment. Vol. 2. Factors Affecting Quality in Frozen Foods. Vol. 3. Commercial Freezing Operations—Fresh Foods. Vol. 4. Freezing of Precooked and Prepared Foods. AVI Publishing Co., Westport, Conn.

WILLIAMS, E. W. 1976A. Frozen Foods in America. Quick Frozen Foods *17*, No. 5, 16–40.

WILLIAMS, E. W. 1976B. European frozen food growth. Quick Frozen Foods *17*, No. 5, 73–105.

WOOLRICH, W. R. 1965. Handbook of Refrigerating Engineering, 4th Edition. Vol. 1. Fundamentals. Vol. 2. Applications. AVI Publishing Co., Westport, Conn.

WOOLRICH, W. R., and HALLOWELL, E. R. 1970. Cold and Freezer Storage Manual. AVI Publishing Co., Westport, Conn.

Appendix

SI UNITS AND CONVERSION FACTORS[1]

Since 1960, most of the countries of the world have made formal commitments to convert to the International System of Units (*Systeme International d'Unités*), abbreviated as SI.

This has been written to assist in the presentation of quantities in SI metric. The existence of other metric units which do not agree with recently adopted SI units is a source of confusion to many. As some have been accustomed to using the son-SI and/or British Imperial units, factors have been included for converting both to SI.

THE INTERNATIONAL SYSTEM OF UNITS (SI)

The SI system consists of seven base units, two supplementary units, and a number of derived units. The term, unit of measurement, SI symbol, and formula are given in Table 1.

Base units are defined as follows:

1. *Length:* The *metre* is the length equal to 1 650 763.73 wave lengths in vacuum of the radiation corresponding to the transition between the levels $2p_{10}$ and $5d_5$ of the krypton-86 atom.

2. *Mass:* The *kilogram* is the unit of mass; it is equal to the mass of the international prototype of the kilogram.

3. *Time:* The *second* is the duration of 9 192 631 770 periods of the radiation corresponding to the transition between the two hyperfine levels of the ground state of the cesium-133 atom.

4. *Amount of Substance.* The *mole* is the amount of substance of a system that contains as many elementary entities as the number of atoms in 0.012 kilogram of carbon-12.

Supplementary units are as follows:

1. *Plane Angle:* The *radian* is the unit of measure of a plane angle with its vertex at the center of a circle and subtended by an arc equal in length to the radius.

2. *Solid Angle:* The *steradian* is the unit of measure of a solid angle with its vertex at the center of a sphere and enclosing an area of the spherical surface equal to that of a square with sides equal in length to the radius.

---

[1]The following pages are adapted from ASHRAE's *1976 Systems Handbook* with permission from the American Society of Heating, Refrigerating and Air-Conditioning Engineers, Inc., New York, N.Y.

TABLE 1
SI UNITS

Term	Unit	Symbol	Formula
Base Units			
length	metre	m	
mass	kilogram	kg	
time	second	s	
electric current	ampere	A	
thermodynamic temperature	kelvin	K	
amount of substance	mole	mol	
luminous intensity	candela	cd	
Supplementary Units			
plane angle	radian	rad	
solid angle	steradian	sr	
Derived Units with Special Names			
electric capacitance	farad	F	C/V
quantity of electricity	coulomb	C	$A \cdot s$
electric potential difference	volt	V	W/A
electric resistance	ohm	Ω	V/A
electrical conductance	siemens	S	A/V
energy	joule	J	$N \cdot m$
force	newton	N	$kg \cdot m/s^2$
frequency (cycles per second)	hertz	Hz	cycle/s
illuminance	lux	lx	lm/m^2
inductance	henry	H	Wb/A
luminous flux	lumen	lm	$cd \cdot sr$
magnetic flux	weber	Wb	$V \cdot s$
magnetic flux density	tesla	T	WB/m^2
power	watt	W	J/s
pressure	pascal	Pa	N/m^2
stress	pascal	Pa	N/m^2
Derived Units without Special Names			
acceleration—			
angular	radian per second squared		rad/s^2
linear	metre per second squared		m/s^2
area	square metre		m^2
density	kilogram per cubic metre		kg/m^3
luminance	candela per square metre		cd/m^2
magnetic field strength	ampere per metre		A/m
moment of a force	newton-metre		$N \cdot m$
permeability	henry per metre		H/m
permittivity	farad per metre		F/m
specific heat capacity	joule per kilogram-kelvin		$J/kg \cdot K)$
thermal capacity (entropy)	joule per kelvin		J/K
thermal conductivity	watt per metre-kelvin		$W/m \cdot K$

Derived units are as follows:

1. *Force:* The *newton* is that force which, when applied to a body having a mass of one kilogram, gives it an acceleration of one metre per second per second.

2. *Energy:* The *joule* is the work done when the point of application of a force of one newton is displaced a distance of one metre in the direction of the force.

3. *Power:* The *watt* is the power which gives rise to the production of energy at the rate of one joule per second.

BASIC RULES

In using the SI system, the understanding of application of prefixes, basic rules of expression, and methods of conversion and rounding will be helpful.

Prefixes

Prefixes for multiple and submultiple units are listed in Table 2. Only one multiple or submultiple prefix is applied at one time to a given unit and is printed immediately preceding the unit symbol.

However, to maintain the coherence of the system, multiples or submultiples of SI units should not be used in calculations.

When expressing a quantity by a numerical value and a unit, prefixes should be chosen so that the numerical value lies between 0.1 and 1000, except where certain multiples or submultiples have been agreed to for particular use.

TABLE 2

MULTIPLE AND SUBMULTIPLE UNITS

Multiplication Factors		Prefix	SI Symbol
1 000 000 000 000	$= 10^{12}$	tera	T
1 000 000 000	$= 10^{9}$	giga	G
1 000 000	$= 10^{6}$	mega	M
1 000	$= 10^{3}$	kilo	k
100	$= 10^{2}$	hecto*	h
10	$= 10^{1}$	deka*	da
0.1	$= 10^{-1}$	deci*	d
0.01	$= 10^{-2}$	centi*	c
0.001	$= 10^{-3}$	milli	m
0.000 001	$= 10^{-6}$	micro	μ
0.000 000 001	$= 10^{-9}$	nano	n
0.000 000 000 001	$= 10^{-12}$	pico	p
0.000 000 000 000 001	$= 10^{-15}$	femto	f
0.000 000 000 000 000 001	$= 10^{-18}$	atto	a

*To be avoided if possible.

Multiple and submultiple prefixes representing steps of 1000 are recommended. Show force in mN, N, kN, and length in mm, m, km, etc. Use of centimetres should be avoided unless a strong reason exists.

Prefixes should not be used in the denominator of compound units, except for the kilogram (kg). Since the kilogram is a base unit of SI, this is not a violation.

With SI units of higher order such as m^2 and m^3, the prefix is also raised to the same order; for example mm^3 is 10^{-9} m^3 not 10^{-3} m^3. In such cases, the use of cm^2, cm^3, dm^2, dm^3, and similar nonpreferred prefixes is permissible. (N·m/kg) or joule per kilogram (J/kg), but it must not be converted to metre since pound-force and pound-mass are not equivalent units.

When non-SI units are used, a distinction should be made between (1) force and (2) mass, for example, lbf to denote force in gravimetric engineering units and lb for mass.

Common use has been made of the metric ton, also called *tonne* (exactly 1 Mg) in previously metric countries. This use is strongly discouraged, and such large masses should be measured in megagrams.

CONVERSION FACTORS

Table 3 can be used to convert either English (U.S. customary) units or non-SI metric units to SI. Conversion factors are written as a number greater.

Conversion and Rounding

Multiply the specified quantity by the conversion factor exactly as given in Table 3 and then round to the appropriate number of significant digits. For example, to convert 11.4 ft to metres: $11.4 \times 0.3048 = 3.474\ 72$, which rounds to 3.47 m. Do not round either the conversion factor or the quantity before performing the multiplication, as accuracy would be reduced.

The product will usually imply an accuracy not intended by the original value. Proper conversion technique includes rounding this converted quantity to the proper number of significant digits commensurate with its intended precision.

Table 4 is convenient for temperature interconversion for values between −40 and +298. For values below or above this range, use conversion factors given in Table 3.

TABLE 3

CONVERSION FACTORS TO SI UNITS

To convert from an English (customary U.S.) unit to its SI equivalent (listed in the center heading of that section) multiply by the SI factor to the left of the center line. To convert from a non-SI metric unit to SI, use the factor to the right of the line. Example: 1 ft/sec² × 3.048 000 E−01 = 0.304 8 m/s²; 5 gal × 1.000 E−02 = 0.05 m/s².

English Unit	Symbol	Multiplier		Multiplier	Symbol	Non-SI Metric Unit
		Acceleration (Linear), metre per second² (m/s²)				
foot/second²	ft/sec²	3.048 000*E−01				
inch/second²	in./sec²	2.540 000*E−02				
free fall, standard		9.806 650*E+00		1.000*E−02	gal	galileo
		Area, metre²(m²)				
acre (U.S. survey)[b]		4.046 873 E+03		1.000 000*E+02	a	Acre
circular mil		5.067 075 E−10		1.000 000*E−28	b	barn
foot²	ft²	9.290 304*E−02		1.000 000*E+04	ha	hectare
inch²	in.²	6.451 600*E−04				
mile² (International)		2.589 988 E+06				
mile² (U.S. survey)[b]		2.589 998 E+06				
yard²	yd²	8.361 274 E−01				
		Bending Moment or Torque, newton-metre (N·m)				
pound force-inch	lb$_f$-in.	1.129 848 E−01		1.000 000*E−07	dyn-cm	dyne-centimetre
pound force-foot	lb$_f$-ft	1.355 818 E+00		9.806 650*E+00	kg$_f$-m	kilogram force-metre
ounce force-inch	oz$_f$-in.	7.061 552 E−03				
		(Bending Moment or Torque) per Length, newton-metre per metre (N·m/m)				
pound force-foot/inch	lb$_f$-ft/in.	5.337 866 E+01				
pound force-inch/inch	lb$_f$-in./in.	4.448 222 E+00				
		Capacity (See Volume)				
		Concentration (See Mass per Volume)				
		Density (See Mass per Volume)				
		Electricity and Magnetism (See ASTM E 380)				

Energy (Includes Heat & Work), joule (J)

English Unit	Symbol	Multiplier	Non-SI Metric Unit	Symbol	Multiplier
British thermal unit (International Steam Table)*	Btu (IST)	1.055 056 E+03	calorie (Int. Steam Table)	cal (IST)	4.186 800*E+00
British thermal unit (mean)	Btu (mean)	1.055 87 E+03	calorie (mean)	cal (mean)	4.190 02 E+00
British thermal unit (thermochemical)		1.054 350 E+03	calorie (thermochemical)		4.184 000*E+00
			calorie (15°C)	cal_{15}	4.185 80 E+00
			calorie (20°C)	cal_{20}	4.181 90 E+00
			electron volt		1.602 19 E−19
kilowatt-hour		3.600 000*E+06	erg	erg	1.000 000*E−07
watt-hour		3.600 000*E+03	kilowatt-hour		3.600 000*E+06
watt-second		1.000 000*E+00	watt-hour		3.600 000*E+03
			watt-second		1.000 000*E+00

Energy per Area-Time, watt per metre² (W/m²)

English Unit	Symbol	Multiplier	Non-SI Metric Unit	Symbol	Multiplier
Btu (thermochemical)/foot²-second		1.134 893 E+04	calorie (thermochemical)/centimetre²-minute		6.973 333 E+02
Btu (thermochemical)/foot²-minute		1.891 489 E+02	erg/centimetre²-second		1.000 000*E−03
Btu (thermochemical)/foot²-hour		3.152 481 E+00	watt/centimetre²		1.000 000*E+04
Btu (thermochemical)/inch²-second		1.634 246 E+06			

Flow (See Mass per Time or Volume per Time)

Force, newton (N)

English Unit	Symbol	Multiplier	Non-SI Metric Unit	Symbol	Multiplier
pound-force	lb_f	4.448 222 E+00	dyne	dyn	1.000 000*E−05
poundal		1.382 550 E−01	kilogram-force	kg_f	9.806 650*E+00
kip (thousand pound-force)		4.448 222 E+03	kilopond		9.806 650*E+00

Force per Area (See Pressure)

English Unit	Symbol	Multiplier		Non-SI Metric Unit	Symbol	Multiplier

Force per Length, newton per metre (N/m)

English Unit	Symbol	Multiplier
pound-force/inch	lb_f/in.	1.751 268 E+02
pound-force/foot	lb_f/ft	1.459 390 E+01

Heat (See Energy)

Table 3. . . . Conversion Factors to SI Units (Concluded)

English Unit	Symbol	Multiplier	Symbol	Multiplier	Non-SI Metric Unit
				Volume (Includes Capacity), metre³ (m³)	
acre-foot (U.S. survey)[b]		1.233 489 E+03		1.000 000*E−03	litre (new)[c]
barrel (oil, 42 gal)		1.589 873 E−01		1.000 028 E−03	litre (old)[c]
board foot		2.359 737 E−03		1.000 000*E+00	stere
bushel (U.S.)		3.523 907 E−02			
foot³		2.831 685 E−02			
gallon (Canadian Liquid)		4.546 090 E−03			
gallon (U.K.)		4.546 092 E−03			
gallon (U.S. dry)		4.404 884 E−03			
gallon (U.S. liquid)		3.785 412 E−03			
gill (U.K.)		1.420 654 E−04			
gill (U.S.)		1.182 941 E−04			
inch³		1.638 706 E−05			
ounce (U.K. fluid)		2.841 307 E−05			
ounce (U.S. fluid)		2.957 353 E−05			
peck (U.S.)		8.809 768 E−03			
pint (U.S. dry)		5.506 105 E−04			
pint (U.S. liquid)		4.731 765 E−04			
quart (U.S. dry)		1.101 221 E−03			
quart (U.S. liquid)		9.463 529 E−04			
ton (register)		2.831 685 E+00			
yard-		7.645 549 E−01			
				Volume per Time (Includes Flow), metre³ per second (m³/s)	
cubic foot/minute	ft³/min	4.719 474 E−04			
cubic foot/second	ft³/s	2.831 685 E−02			
cubic inch/minute	in.³/min	2.731 177 E−07			
cubic yard/minute	yd³/min	1.274 258 E−02			
gallon (U.S. liquid)/day	gal/d	4.381 264 E−08			
gallon (U.S. liquid)/minute	gal/min	6.309 020 E−05			
				Work (See Energy)	

[a] The British thermal unit (International Steam Table) value of 1.055 056 E+03 was adopted in 1956. Some of the older International Tables use the value 1.055 04 E+03. The exact conversion is 1.055 055 852 62*E+03.

[b] Since 1893 the U.S. basis of length measurement has been derived from metric standards. *In 1959 a small refinement was made in the definition of the yard* to resolve discrepancies both in this country and abroad, which changed its length from 3600/3937 m to 0.9144 m exactly. This resulted in the new value being shorter by two parts in a million. At the same time it was decided that any data in feet derived from and published as a result of geodetic surveys within the U.S. would remain with the old standard (one foot equals 1200/3937 m) until further decision. This foot is named the U.S. Survey Foot.

As a result all U.S. land measurements in U.S. customary units will relate to the metre by the old standard. All the conversion factors in these tables for units referenced to this footnote are based on the U.S. Survey Foot, rather than the international foot.

[c] In 1964 the General Conference on Weights and Measures adopted the name litre as a special name for the cubic decimetre. Prior to this decision the litre differed slightly (previous value, 1.000028 dm³) and in expression of precision volume measurement this fact must be kept in mind.

TABLE 4

TEMPERATURE CONVERSION

The numbers in boldface type in the center column refer to the temperature, either in degree Celsius or Fahrenheit, which is to be converted to the other scale. If converting Fahrenheit to degree Celsius, the equivalent temperature will be found in the left column. If converting degree Celsius to Fahrenheit, the equivalent temperature will be found in the column on the right.

| Temperature | | | Temperature | | | Temperature | | | Temperature | | |
Celsius	°C or F	Fahr	Celsius	°C or F	Fahr	Celsius	°C or F	Fahr	Celsius	°C or F	Fahr
-40.0	-40	-40.0	-31.7	-25	-13.0	-23.3	-10	+14.0	-15.0	+5	+41.0
-39.4	-39	-38.2	-31.1	-24	-11.2	-22.8	-9	+15.8	-14.4	+6	+42.8
-38.9	-38	-36.4	-30.6	-23	-9.4	-22.2	-8	+17.6	-13.9	+7	+44.6
-38.3	-37	-34.6	-30.0	-22	-7.6	-21.7	-7	+19.4	-13.3	+8	+46.4
-37.8	-36	-32.8	-29.4	-21	-5.8	-21.1	-6	+21.2	-12.8	+9	+48.2
-37.2	-35	-31.0	-28.9	-20	-4.0	-20.6	-5	+23.0	-12.2	+10	+50.0
-36.7	-34	-29.2	-28.3	-19	-2.2	-20.0	-4	+24.8	-11.7	+11	+51.8
-36.1	-33	-27.4	-27.8	-18	-0.4	-19.4	-3	+26.6	-11.1	+12	+53.6
-35.6	-32	-25.6	-27.2	-17	+1.4	-18.9	-2	+28.4	-10.6	+13	+55.4
-35.0	-31	-23.8	-26.7	-16	+3.2	-18.3	-1	+30.2	-10.0	+14	+57.2
-34.4	-30	-22.0	-26.1	-15	+5.0	-17.8	0	+32.0	-9.4	+15	+59.0
-33.9	-29	-20.2	-25.6	-14	+6.8	-17.2	+1	+33.8	-8.9	+16	+60.8
-33.3	-28	-18.4	-25.0	-13	+8.6	-16.7	+2	+35.6	-8.3	+17	+62.6
-32.8	-27	-16.6	-24.4	-12	+10.4	-16.1	+3	+37.4	-7.8	+18	+64.4
-32.2	-26	-14.8	-23.9	-11	+12.2	-15.6	+4	+39.2	-7.2	+19	+66.2

TABLE 4

TEMPERATURE CONVERSION (CONCLUDED)

The numbers in boldface type in the center column refer to the temperature, either in degree Celsius or Fahrenheit, which is to be converted to the other scale. If converting Fahrenheit to degree Celsius, the equivalent temperature will be found in the left column. If converting degree Celsius to Fahrenheit, the equivalent temperature will be found in the column on the right.

Temperature			Temperature			Temperature			Temperature		
Celsius	°C or F	Fahr	Celsius	°C or F	Fahr	Celsius	°C or F	Fahr	Celsius	°C or F	Fahr
−6.7	+20	+68.0	+26.7	+80	+176.0	+60.0	+140	+284.0	+93.3	+200	+392.0
−6.1	+21	+69.8	+27.2	+81	+177.8	+60.6	+141	+285.8	+93.9	+201	+393.8
−5.5	+22	+71.6	+27.8	+82	+179.6	+61.1	+142	+287.6	+94.4	+202	+395.6
−5.0	+23	+73.4	+28.3	+83	+181.4	+61.7	+143	+289.4	+95.0	+203	+397.4
−4.4	+24	+75.2	+28.9	+84	+183.2	+62.2	+144	+291.2	+95.6	+204	+399.2
−3.9	+25	+77.0	+29.4	+85	+185.0	+62.8	+145	+293.0	+96.1	+205	+401.0
−3.3	+26	+78.8	+30.0	+86	+186.8	+63.3	+146	+294.8	+96.7	+206	+402.8
−2.8	+27	+80.6	+30.6	+87	+188.6	+63.9	+147	+296.6	+97.2	+207	+404.6
−2.2	+28	+82.4	+31.1	+88	+190.4	+64.4	+148	+298.4	+97.8	+208	+406.4
−1.7	+29	+84.2	+31.7	+89	+192.2	+65.0	+149	+300.2	+98.3	+209	+408.2
−1.1	+30	+86.0	+32.2	+90	+194.0	+65.6	+150	+302.0	+98.9	+210	+410.0
−0.6	+31	+87.8	+32.8	+91	+195.8	+66.1	+151	+303.8	+99.4	+211	+411.8
.0	+32	+89.6	+33.3	+92	+197.6	+66.7	+152	+305.6	+100.0	+212	+413.6
+0.6	+33	+91.4	+33.9	+93	+199.4	+67.2	+153	+307.4	+100.6	+213	+415.4
+1.1	+34	+93.2	+34.4	+94	+201.2	+67.8	+154	+309.2	+101.1	+214	+417.2
+1.7	+35	+95.0	+35.0	+95	+203.0	+68.3	+155	+311.0	+101.7	+215	+419.0
+2.2	+36	+96.8	+35.6	+96	+204.8	+68.9	+156	+312.8	+102.2	+216	+420.8
+2.8	+37	+98.6	+36.1	+97	+206.6	+69.4	+157	+314.6	+102.8	+217	+422.6
+3.3	+38	+100.4	+36.7	+98	+208.4	+70.0	+158	+316.4	+103.3	+218	+424.4
+3.9	+39	+102.2	+37.2	+99	+210.2	+70.6	+159	+318.2	+103.9	+219	+426.2
+4.4	+40	+104.0	+37.8	+100	+212.0	+71.1	+160	+320.0	+104.4	+220	+428.0
+5.0	+41	+105.8	+38.3	+101	+213.8	+71.7	+161	+321.8	+105.6	+222	+431.6
+5.5	+42	+107.6	+38.9	+102	+215.6	+72.2	+162	+323.6	+106.7	+224	+435.2
+6.1	+43	+109.4	+39.4	+103	+217.4	+72.8	+163	+325.4	+107.8	+226	+438.8
+6.7	+44	+111.2	+40.0	+104	+219.2	+73.3	+164	+327.2	+108.9	+228	+442.4

+7.2	**+45**	+113.0	**+105**	+40.6	+221.0	**+165**	+73.9	+329.0	**+230**	+110.0	+446.0
+7.8	**+46**	+114.8	**+106**	+41.1	+222.8	**+166**	+74.4	+330.8	**+232**	+111.1	+449.6
+8.3	**+47**	+116.6	**+107**	+41.7	+224.6	**+167**	+75.0	+332.6	**+234**	+112.2	+453.2
+8.9	**+48**	+118.4	**+108**	+42.2	+226.4	**+168**	+75.6	+334.4	**+236**	+113.3	+456.8
+9.4	**+49**	+120.2	**+109**	+42.8	+228.2	**+169**	+76.1	+336.2	**+238**	+114.4	+460.4
+10.0	**+50**	+122.0	**+110**	+43.3	+230.0	**+170**	+76.7	+338.0	**+240**	+115.6	+464.0
+10.6	**+51**	+123.8	**+111**	+43.9	+231.8	**+171**	+77.2	+339.8	**+242**	+116.7	+467.6
+11.1	**+52**	+125.6	**+112**	+44.4	+233.6	**+172**	+77.8	+341.6	**+244**	+117.8	+471.2
+11.7	**+53**	+127.4	**+113**	+45.0	+235.4	**+173**	+78.3	+343.4	**+246**	+118.9	+474.8
+12.2	**+54**	+129.2	**+114**	+45.6	+237.2	**+174**	+78.9	+345.2	**+248**	+120.0	+478.4
+12.8	**+55**	+131.0	**+115**	+46.1	+239.0	**+175**	+79.4	+347.0	**+250**	+121.1	+482.0
+13.3	**+56**	+132.8	**+116**	+46.7	+240.8	**+176**	+80.0	+348.8	**+252**	+122.4	+485.6
+13.9	**+57**	+134.6	**+117**	+47.2	+242.6	**+177**	+80.6	+350.6	**+254**	+123.3	+489.2
+14.4	**+58**	+136.4	**+118**	+47.8	+244.4	**+178**	+81.1	+352.4	**+256**	+124.4	+492.8
+15.0	**+59**	+138.2	**+119**	+48.3	+246.2	**+179**	+81.7	+354.2	**+258**	+125.5	+496.4
+15.6	**+60**	+140.0	**+120**	+48.9	+248.0	**+180**	+82.2	+356.0	**+260**	+126.7	+500.0
+16.1	**+61**	+141.8	**+121**	+49.4	+249.8	**+181**	+82.8	+357.8	**+262**	+127.8	+503.6
+16.7	**+62**	+143.6	**+122**	+50.0	+251.6	**+182**	+83.3	+359.6	**+264**	+128.9	+507.2
+17.2	**+63**	+145.4	**+123**	+50.6	+253.4	**+183**	+83.9	+361.4	**+266**	+130.0	+510.8
+17.8	**+64**	+147.2	**+124**	+51.1	+255.2	**+184**	+84.4	+363.2	**+268**	+131.3	+514.4
+18.3	**+65**	+149.0	**+125**	+51.7	+257.0	**+185**	+85.0	+365.0	**+270**	+132.2	+518.0
+18.9	**+66**	+150.8	**+126**	+52.2	+258.8	**+186**	+85.6	+366.8	**+272**	+133.3	+521.6
+19.4	**+67**	+152.6	**+127**	+52.8	+260.6	**+187**	+86.1	+368.6	**+274**	+134.4	+525.2
+20.0	**+68**	+154.4	**+128**	+53.3	+262.4	**+188**	+86.7	+370.4	**+276**	+135.6	+528.8
+20.6	**+69**	+156.2	**+129**	+53.9	+264.2	**+189**	+87.2	+372.2	**+278**	+136.7	+532.4
+21.1	**+70**	+158.0	**+130**	+54.4	+266.0	**+190**	+87.8	+374.0	**+280**	+137.8	+536.0
+21.7	**+71**	+159.8	**+131**	+55.0	+267.8	**+191**	+88.3	+375.8	**+282**	+138.9	+539.6
+22.2	**+72**	+161.6	**+132**	+55.6	+269.6	**+192**	+88.9	+377.6	**+284**	+140.0	+543.2
+22.8	**+73**	+163.4	**+133**	+56.1	+271.4	**+193**	+89.4	+379.4	**+286**	+141.1	+546.8
+23.3	**+74**	+165.2	**+134**	+56.7	+273.2	**+194**	+90.0	+381.2	**+288**	+142.2	+550.4
+23.9	**+75**	+167.0	**+135**	+57.2	+275.0	**+195**	+90.6	+383.0	**+290**	+143.3	+554.0
+24.4	**+76**	+168.8	**+136**	+57.8	+276.8	**+196**	+91.1	+384.8	**+292**	+144.4	+557.6
+25.0	**+77**	+170.6	**+137**	+58.3	+278.6	**+197**	+91.7	+386.6	**+294**	+145.6	+561.2
+25.6	**+78**	+172.4	**+138**	+58.9	+280.4	**+198**	+92.2	+388.4	**+296**	+146.7	+564.8
+26.1	**+79**	+174.2	**+139**	+59.4	+282.2	**+199**	+92.8	+390.2	**+298**	+147.8	+568.4

Index

Other AVI Books